HANDBOOK OF PHENOMENOLOGY

For Jim
  With warmest wishes
    Kay Toombs
    February 2007

Philosophy and Medicine

VOLUME 68

Founding Co-Editor
Stuart F. Spicker

Editor

H. Tristram Engelhardt, Jr., *Department of Philosophy, Rice University, and Baylor College of Medicine, Houston, Texas*

Associate Editor

Kevin Wm. Wildes, S.J., *Department of Philosophy and Kennedy Institute of Ethics, Georgetown University, Washington, D.C.*

Editorial Board

George J. Agich, *Department of Bioethics, The Cleveland Clinic Foundation, Cleveland, Ohio*
Nicholas Capaldi, *Department of Philosophy, University of Tulsa, Tulsa, Oklahoma*
Edmund Erde, *University of Medicine and Dentistry of New Jersey, Stratford, New Jersey*
Eric T. Juengst, *Center for Biomedical Ethics, Case Western Reserve University, Cleveland, Ohio*
Christopher Tollefsen, *Department of Philosophy, University of South Carolina, Columbia, South Carolina*
Becky White, *Department of Philosophy, California State University, Chico, California*

The titles published in this series are listed at the end of this volume.

HANDBOOK OF PHENOMENOLOGY AND MEDICINE

Edited by

S. KAY TOOMBS

*Philosophy Department, Baylor University,
Waco, Texas, U.S.A.*

KLUWER ACADEMIC PUBLISHERS
DORDRECHT / BOSTON / LONDON

A C.I.P. Catalogue record for this book is available from the Library of Congress.

ISBN 1-4020-0151-7 (HB)
ISBN 1-4020-0200-9 (PB)

Published by Kluwer Academic Publishers,
P.O. Box 17, 3300 AA Dordrecht, The Netherlands.

Sold and distributed in North, Central and South America
by Kluwer Academic Publishers,
101 Philip Drive, Norwell, MA 02061, U.S.A.

In all other countries, sold and distributed
by Kluwer Academic Publishers, Distribution Center,
P.O. Box 322, 3300 AH Dordrecht, The Netherlands.

Printed on acid-free paper

All Rights Reserved
© 2001 Kluwer Academic Publishers
No part of this publication may be reproduced or utilized in any form or by any means, electronic, mechanical, including photocopying, recording or by any information storage and retrieval system, without written permission from the copyright owner.

Printed in the Netherlands.

TABLE OF CONTENTS

ACKNOWLEDGEMENTS   vii

S. KAY TOOMBS / Introduction: Phenomenology and Medicine   1

SECTION ONE / PHENOMENOLOGY AND MEDICINE

JOHN B. BROUGH / Temporality and Illness: A Phenomenological Perspective   29
PATRICK A. HEELAN / The Lifeworld and Scientific Interpretation   47
FRANCES CHAPUT WAKSLER / Medicine and the Phenomenological Method   67
FREDRIK SVENAEUS / The Phenomenology of Health and Illness   87
PER SUNDSTRÖM / Disease: The Phenomenological and Conceptual Center of Practical-Clinical Medicine   109
RICHARD M. ZANER / Thinking About Medicine   127

SECTION TWO / THE BODY

SHAUN GALLAGHER / Dimensions of Embodiment: Body Image and Body Schema in Medical Contexts   147
MAUREEN CONNOLLY / Female Embodiment and Clinical Practice   177
GLEN A. MAZIS / Emotion and Embodiment Within the Medical World   197
BRUCE WILSHIRE / The Body, Music, and Healing   215

SECTION THREE / LIVED EXPERIENCE

ARTHUR W. FRANK / Experiencing Illness Through Storytelling   229
S. KAY TOOMBS / Reflections on Bodily Change: The Lived Experience of Disability   247

TABLE OF CONTENTS

IRENA MADJAR / The Lived Experience of Pain in the Context
  of Clinical Practice   263
JO ANN WALTON / The Lived Experience of Mental Illness   279

SECTION FOUR / CLINICAL PRACTICE

CARL EDVARD RUDEBECK / Grasping the Existential Anatomy:
  The Role of Bodily Empathy in Clinical Communication   297
PAUL KOMESAROFF / The Many Faces of the Clinic: A Levinasian
  View   317
IAN R. MCWHINNEY / Focusing on Lived Experience: The
  Evolution of Clinical Method in Western Medicine   331
PATRICIA BENNER / The Phenomenon of Care   351
ERIC J. CASSELL / The Phenomenon of Suffering and its
  Relationship to Pain   371

SECTION FIVE / MEDICAL ETHICS

MARK J. BLITON / Imagining a Fetus: Insights From Talking
  With Pregnant Women About Their Decisions To Undergo
  Open-Uterine Fetal Surgery   393
CATRIONA MACKENZIE / On Bodily Autonomy   417
MICHAEL C. BRANNIGAN / Medical Feeding: Applying Husserl
  and Merleau-Ponty   441

SECTION SIX / RESEARCH

MAX VAN MANEN / Professional Practice and 'Doing
  Phenomenology'   457
CHRISTINA PAPADIMITRIOU / From Dis-ability to Difference:
  Conceptual and Methodological Issues in the Study of
  Physical Disability   475

NOTES ON CONTRIBUTORS   493

INDEX   495

ACKNOWLEDGEMENTS

I should like to express my deep appreciation to the many people who have assisted in the preparation of this volume. In particular, I would like to thank the Herbert H. and Joy C. Reynolds Endowment Fund for University Excellence, the Baylor University Philosophy Department, the College of Arts and Sciences and the Office of the Provost and Vice President for Academic Affairs of Baylor University for providing funding for a summer sabbatical that allowed me to work on the project.

I am also deeply grateful to the authors in this volume who have contributed their time, expertise, and enthusiasm for phenomenology and medicine – and who have patiently gone through the process of review and revision in such a timely and gracious fashion. Not only have I learned a great deal from their work but I have enjoyed our "conversations" via email and am very grateful for the opportunity to develop new and stimulating avenues for international discussions about ongoing work.

H. Tristram Engelhardt, Jr. and Ruiping Fan first persuaded me to consider undertaking this project. In retrospect, I am very glad that I accepted their kind invitation – although "in the midst of things" I must admit I was a little less sure about my wisdom in doing so.

The preparation of the manuscript would not have been possible without the assistance of Rodney Culpepper who coaxed the computer to perform all the necessary formatting tasks that it absolutely refused to do for me.

I am also grateful to Anne Ultee, Publishing Editor, Humanities Division, Kluwer. She has been both an understanding editor and a delightful correspondent.

Lastly, I want to thank my beloved husband, Dee. Throughout the preparation of this volume he has held me up (both literally and figuratively). My work – and my life – would simply not be possible without him.

S. Kay Toombs
Baylor University
Waco, Texas
U.S.A.

S. KAY TOOMBS

INTRODUCTION: PHENOMENOLOGY AND MEDICINE

I. INTRODUCTION

As the fields of philosophy of medicine and bioethics have developed in the United States, the philosophical perspective of phenomenology has been largely ignored. Yet, the central conviction that informs this volume is that phenomenology provides extraordinary insights into many of the issues that are directly addressed within the world of medicine. Such issues include: the nature of medicine itself; the distinction between immediate experience and scientific conceptualization; the nature of the body – and the relationship between body, consciousness, world and self; the structure of emotions; the meaning of health, illness and disease; the problem of intersubjectivity – particularly with respect to achieving successful communication with another; the complexity of decision-making in the clinical context; the possibility of empathic understanding; the theory and method of clinical practice; and the essential characteristics of the therapeutic relationship – i.e. the relationship between the sick person and the one who professes to help.

A. A Phenomenological Approach

Although the interests and conceptions of phenomenology among particular phenomenologists differ in important ways[1], there are some commonalities that allow one to speak of "a phenomenological approach." The most important is derived from Edmund Husserl's injunction: "to the things themselves." That is, the phenomenologist is committed to setting aside his or her taken-for-granted presuppositions about the nature of objects or "reality" in an effort to begin with what is given in immediate experience, the phenomena as encountered, precisely as they are encountered. One of the primary aims of an explicitly phenomenological approach is to let what is given appear as pure phenomenon (the thing-as-meant) and to work to describe the invariant features of such phenomena.

A phenomenological approach thus involves a type of radical disengagement, a "distancing" from our immediate ongoing experience of everyday life in order to make explicit the nature of such experience and the essential intentional structures that determine the meaning of such experience. As such, phenomenology is an essentially reflective enterprise.

S.K. Toombs (ed.), Handbook of Phenomenology and Medicine, 1–26.
© 2001 *Kluwer Academic Publishers. Printed in the Netherlands.*

The common sense world itself (and our experiencing of it) becomes the focus of reflection. Our attention shifts from that of engagement in the world to a focused concern for the sense and strata of the very engagement itself (Zaner, 1970, p. 51). The task is to elucidate and render explicit the taken-for-granted assumptions of everyday life and, particularly, to bring to the fore one's consciousness of the world. In rendering explicit the intentional structures of consciousness, phenomenological reflection discloses the meaning of experience (Toombs, 1992, pp. xi-xvi).

In order to describe phenomena as they are encountered, the phenomenologist attempts to effect a systematic neutrality. That is, in the phenomenological attitude "one places in abeyance one's taken-for-granted presuppositions about the nature of 'reality,' one's commitments to certain habitual ways of interpreting the world" (Toombs, 1992, p. xii). In particular, one sets aside any theoretical commitments derived from the natural sciences in order to describe what gives itself directly to consciousness. As Merleau-Ponty (1962, pp. vii-xxi) notes, this radical reflection does not deny the existence of the physical, social and cultural world. Rather, it reveals the "prejudices" and taken-for-granted presuppositions that are not explicitly recognized in our spontaneous, unreflective experience. Thus, the phenomenologist adopts a particular kind of attitude ("stance") towards inquiry and the objects of inquiry.

B. *The Domain of Medicine*

In turning attention to the field of medicine as the focus of inquiry, a phenomenological approach is particularly helpful in the following ways: (1) phenomenology provides an explication of the fundamental and important distinction between the immediate pre-theoretical experiencing of the world of everyday life and the theoretical, scientific account of such experience. This distinction is particularly important in furthering our understanding of the relation between medical science and clinical practice, as well as in recognizing the differences between, say, the immediate experience of illness vs. the conceptualization of illness as a disease state, and the body-as-experienced vs. the body as the object of scientific inquiry; (2) phenomenology provides a method for engaging in the radical reflection on experience – the phenomenological "reduction" or epochē. As noted above, the phenomenological "reduction" not only attempts to "bracket" (set aside) hitherto unquestioned assumptions, but simultaneously to clarify (and render explicit) the taken-for-granted presuppositions that "color" our experience. In the process phenomenology reveals the extent to which theoretical, social, cultural, and professional "habits of mind" influence our understandings and interpretations of experienced phenomena – in many instances circumscribing our "knowledge" about the phenomena themselves (Waksler, 2001; van Manen, 2001); (3) with its emphasis on firsthand or direct description phenomenology can provide a rigorous, detailed account of the manner in which we experience, and interpret, the world of everyday life in the context of medicine and medical practice. To give one example: in turning to "the things themselves," the phenomenologist focuses on the phenomenon of illness *as it is immediately encountered or experienced* (illness-as-lived). Such an analysis discloses certain invariant (or typical) features of the experience apart from the varieties of its concrete instantiations. Understanding the phenomenon of illness in all its richness and complexity is of enormous practical value in the clinical context; (4) certain central themes of investigation in phenomenology are of particular relevance in thinking about issues related to medicine. Such themes include: the body, the nature of human being and existence – *Being-in-the-World*; temporality; the constitution of meaning; the theory of intentionality; the structures of the Lifeworld – *Lebenswelt*; the analysis of intersubjectivity, the turn to lived experience.

The essays in Section One illustrate the relevance of phenomenology in the domain of medicine. John Brough explores Husserl's *phenomenology of time* and temporal awareness, noting that "phenomenology seeks to disclose the fundamental temporal framework in which we experience the contents of our lives, whatever they may be." Brough explores some of the ways in which temporality pervades our experience in everyday life and then turns his attention to the phenomenon of time in the context of illness. He demonstrates that illness "provides a unique window on to the many levels of time that inform our lives," and that it also reveals "resources in our temporal being" that enable us to cope with the various disruptions that are a part of illness.

Patrick Heelan draws upon the phenomenological analysis of the *Lifeworld* (as found in the works of Husserl, Heidegger, Gadamer, and Merleau-Ponty) in order to provide a critique of the hypothetical-deductive account of modern science (the 'received view') and to develop new themes about the way theory is related to praxis and how medical science is related to clinical practice. In particular, he aims to address the role of modern science in an important sector of the life-world, the world of medical practice. In noting the different roles of theory and practice, Heelan explores the "Janus-like" face of scientific theories and entities and the role of metaphor in scientific discourse and medical science. He shows how distortions in communication occur when the Janus-like face of a scientific fact goes unnoticed leading –

among other things – to confusion in the public debate about "such contentious practices as abortion, cloning, disease prevention."

In discussing *phenomenological method*, Frances Waksler explicates two central features of the phenomenological approach that are "of particular significance for medical practitioners": the primacy of the subject's perspective and the suspension of belief (*the phenomenological epochē*). Waksler shows that the phenomenological method of inquiry provides a rigorous way to examine the activities of medical practice and theorizing, as well as illuminating alternative ways of thinking and acting within the domain of clinical medicine.

Fredrik Svenaeus focuses on the phenomena of *health and illness* noting, "The ultimate test of a phenomenology of illness must be the extent to which it does justice to the different forms of illness afflicting the individuals encountered each day by doctors, nurses and others working in the clinic." Drawing upon Heidegger's analysis of *Dasein,* Svenaeus argues that health can be understood as "homelike *Being-in-the-World.*" Illness can, thus, be understood as a kind of "unhomelikeness" with a basic structure that can be identified. Recognizing this basic structure allows us to comprehend the particular ways in which individuals experience themselves as ill, as well as providing a vocabulary by means of which we can talk about different illnesses.

Per Sundström turns his attention to the phenomenon of *disease*. He examines the relationship between disease and illness, noting that in the clinical context most of the time the two go together, "disease/illness presenting itself as a mixed subjective/objective phenomenon, which is disturbing, disabling, or even calamitous to the individual patient, *and ...* recognizable by the medically trained professional as a deviation from normal functioning, or shape, of the body/organism." Sundström explores the pluridimensional meanings/conceptions of disease in this clinical sense, an exploration that is particularly important since "disease" is, in fact, the "phenomenological and conceptual center of practical-clinical medicine."

Richard Zaner reflects on the astonishing developments in the newly emerging molecular medicine and asks (1) what do these developments mean with respect to *the nature of medicine itself* – in particular, what are the implications of the shift from restorative medicine to a "new paradigm" that "is centrally concerned to conquer, replace and transcend nature," thereby altering the human condition itself? and (2) how are we to think about the *self* in light of the Genome Project? How is the *self* known and experienced? Is there a self at all? How exactly are we to understand the mystery of *human being*?

II. THE BODY AS THEME OF PHENOMENOLOGICAL INVESTIGATION

The phenomenological exploration of the body is of particular relevance for medicine. As Gallagher (2001) notes, "Perhaps the most general and most obvious fact about medicine is that it concerns the body. If one eliminates the body one eliminates the subject and object of medical science and practice." Yet, although the body is indeed the central concern of medicine, inquiries into the nature of the body in Western medical thought and practice are typically focused on the body as a type of mechanism – the body-as-machine. The body is considered to be a particular type of material object, one that can be observed, scientifically analyzed, and understood exclusively in terms of its anatomical and physiological characteristics.

Phenomenologists such as Edmund Husserl (1989), Maurice Merleau-Ponty (1962), and Jean-Paul Sartre (1956) – and, following them, Richard Zaner (1981, 1964); Drew Leder (1990); H.Tristram Engelhardt (1977, 1973); Erwin Straus (1963, 1966); Stuart Spicker (1970); Hans Jonas (1966); Shaun Gallagher (1986); and S. Kay Toombs (1992a), to name a few – have focused instead on the phenomenon of embodiment, i.e. the body as a living, embodying organism. In setting aside the mechanistic construal of the body-as-machine, in order to explore the nature of the living, experiencing body, phenomenologists have distinguished the *lived body* from the objective physiological body. Behnke (1997, p. 66) notes that this distinction is often expressed in the German words "*Leib*" and "*Körper*."

Leib is usually translated "lived body," sometimes "animate organism," "living body," or simply "Body," and is related to the phenomenological use of the French "*corps propre*." *Körper* is usually translated "physical body," sometimes "material body," "object body," or simply "body."
... Fundamental to the Body-body distinction is that the turn to the Body involves the turn to experiential evidence in contrast to the body as investigated by such natural sciences as anatomy and physiology.

A. The Lived Body

Although it is not possible in the context of this introduction to give a full account of the phenomenological account of embodiment, certain key features of the lived body can be identified:

➤ The lived body is not an object akin to other physical, animate objects in the world but, rather, the medium through which, and by means of which, I apprehend the world and interact with it (Husserl, 1982, p. 97, 1989, p. 159; Merleau-Ponty, 1962; Sartre, 1956, pp. 429-30, 436).

➤ In all its worldly involvements the lived body exhibits a bodily intentionality that reveals a dynamic relation between body/world – that is, the lived body is an embodied consciousness that simultaneously engages, and is engaged in, the surrounding world (Merleau-Ponty, 1962, p. 79; Gallagher, 2001; Mazis, 2001; Brannigan, 2001; Rawlinson, 1986, pp. 42-43). My embodying organism is always experienced as "in the midst of environing things, in this or that situation of action, positioned and positioning relative to some task at hand" (Zaner, 1981, p. 97).

➤ In this continuous body/world interaction, the lived body synthesizes the various senses and movements (Merleau-Ponty, 1962, 1968) into a unity of experience.

The experiencing body...is not a self-enclosed object, but an open, incomplete entity. This openness is evident in the arrangement of the senses: I have these multiple ways of encountering and exploring the world – listening with my ears, touching with my skin, seeing with my eyes, tasting with my tongue, smelling with my nose – and all of these various powers or pathways continually open outward from the perceiving body, like different pathways diverging from a forest. Yet my experience of the world is not fragmented; I do not commonly experience the visible appearance of the world in any way separable from its audible aspect, or from the myriad textures that open themselves to my touch....Thus, my divergent senses meet up with each other in the surrounding world, converging and commingling in the things I perceive. We may think of the sensing body as a kind of open circuit that completes itself only in things, and in the world...it is primarily through my engagement with what is not me that I effect the integration of my senses, and thereby experience my own unity and coherence (Abrams ,1996, p. 125).

Given this ongoing body/world interaction, illness is not simply a problem with the isolated physiological object-body but, rather, a problem with the whole embodying organism/environment (Gallagher, 1986, 2001; Rudebeck, 2001; Wilshire, 2001; Brannigan, 2001; Mazis, 2001, Toombs, 2001).

➤ The body/world relation reflects the fact that the lived body represents not just one's bodily being but one's contextual Being-in-the-world.

➤ As the means by which one interacts with the world, the lived body makes possible the existential projects that are expressive of one's personhood (Husserl, 1989, pp. 195-96; Toombs, 1999). Consequently, the disruption of bodily capacities has a significance that far exceeds that of simple mechanical dysfunction (Toombs, 1995a, 1994).

➤ The lived body is both intentional and orientational locus – that is, the lived body is the spatial and temporal center around which the rest of the world is grouped. My body has the central mode of givenness of "Here" whereas all other things (including my fellow human beings) are given as located "There" in relation to my body (Husserl, 1982, pp. 116-117, 1989, pp. 61, 165-66; Schutz 1973, 1962a, 1962b). Since surrounding space is intimately related to the positioning of the body (e.g. the designations of "near," "far," "high," "low," depend upon my bodily "Here") changes in bodily bearing (e.g. the loss of upright posture) and the disruption of bodily capacities inevitably disrupt the experience of surrounding space (Toombs, 2001; Straus, 1966; Toombs, 1992, p. 65).

➤ The body's "Here-ness," its "proximity," also means that I cannot physically distance myself from my body – although I may separate myself from certain parts of it under certain circumstances (Engelhardt, 1973, p. 41; Toombs, 1992a, 1992b). Thus, there is a symbiotic relationship between body and self. This symbiotic relationship with body can represent a source of threat in the event that the body malfunctions – since whatever happens to my body also happens to *me* (Plügge, 1970).

➤ The term "embodiment" refers both to the unity of body/consciousness and the fact that, since I am "embodied," my "own" body is always "with" me. For example, in reaching for the glass, I am in some sense aware that it is "my" arm that moves (Husserl, 1989, p. 61; Merleau-Ponty, 1962, pp. 70, 90; Sartre, 1956, pp. 425-27) and that it is "my" body that effects the action. However, this relation with "my" "own" body is not a relation of "ownership" or "possession" – in the same way that I own, or possess, an automobile or a house (Gallagher, 1986; Toombs, 1999; Mackenzie, 2001).

➤ One's experiential awareness of the body is limited in a variety of ways. Not only does the body's orientational locus make it impossible for me to apprehend all aspects of my body directly (e.g. I cannot walk around my body to view the back, nor can I directly view the interior of my body), but the body is "uncanny" in that it includes "events, processes and structures over which I have no control and of which I have no awareness" (Zaner, 1981, pp. 52-53; Leder, 1990; Gallagher, 1986).

➤ In the experience of "uncanniness" the lived body is apprehended as in some sense "other than me" (Straus, 1970, p. 139). This sense of bodily alienation is particularly profound in the experience of bodily breakdown and malfunction – when the body is apprehended as a neurophysiological organism (a "hidden presence" that is necessarily beyond one's control) (Sartre, 1956; Sacks, 1984; Cole, 1991; Murphy, 1987).

➤ As an embodied being I exhibit a certain corporeal style (bodily identity) – way of walking, gesturing, and so forth – that is unique. This bodily comportment not only identifies the lived body as "mine" (Merleau-Ponty, 1992, p. 150) but also reflects the body as a social and cultural entity (Husserl, 1989, p. 282; Behnke, 1989).

B. Lived Body and Medicine

Perhaps the most obvious fact about the phenomenological concept of lived body is that it bears no resemblance to the notion of the "body" in medicine. As Zaner (1984, pp. 154-170), Foucault (1975, p. 111), and Leder (1992a, p. 17) point out, Western scientific medicine is " based, first and foremost, not upon the lived body, but upon the dead, or inanimate, body" – a mechanistic and reductionistic understanding that has extensively shaped medical practice and theory (Leder, 1992b).[2] Under the medically trained "gaze" of the healthcare professional, the body assumes the status of a scientific object, i.e. it is construed as an anatomical, neurophysiological organism and, more particularly, as a mass of cells, tissues, organs, and so forth, according to the categories of natural science (Toombs, 1992a, pp. 76-81). As a scientific object, a particular body is simply an exemplar of *the* human body (or of a particular class of human bodies) and, as such, it may be viewed independently from the person whose body it is. Thus, the notion of body in medicine (often characterized as a "Cartesian" paradigm of body) effectively separates body/mind, body/world, and body/person (Zaner, 1981; Leder, 1984a; Young, 1997; Sheets-Johnstone, 1992; Toombs, 1992a).

Critics point out that, although the mechanistic/materialistic conception of body has "given rise to therapeutic triumphs" in modern medicine, it has also resulted in limitations and distortions in medical practice (Leder, 1992a). For example, if health care practitioners focus their attention almost exclusively on the body/object as a malfunctioning anatomico-physiological entity and ignore (or de-emphasize) the patient's *lived experience* of bodily disturbance, then important factors of illness (such as interpretations, emotions, desires, worldly involvements, cultural background, and so forth) are ignored – despite the fact that such factors play a crucial role in the course of disease and in therapeutic effectiveness (Cassell, 2001, 1966; Engel, 1997a, 1977b, 1987, 1976; Kleinman, 1988; Toombs, 1992a; Leder, 1992a, 1984a; Mazis, 2001).

Furthermore, in limiting the clinical "gaze" to the biological body and, in effect, abstracting the sick person from the body, this model *disembodies* (Young, 1997; Connolly, 2001) and *dehumanizes* those seeking medical care (Baron, 1981; Frank, 1991, 1995; Stetten, 1981; Donnelly, 1986; Sacks, 1984; Toombs, 1992a).[3] As Leder (1992a, p. 24) notes, "Patients often complain that they have been dealt with by their health-care providers or institutions in a dehumanized fashion: as if they were but a disease entity, or a piece of meat to be prodded, punctured, and otherwise ignored."

The process of *disembodiment* is not only dehumanizing for the patient, but it also directly affects the quality of the clinical encounter by constraining the healthcare professional's ability:
➢ to communicate effectively with patients (Komesaroff, 2001; Baron, 1985, 1981; Cassell, 1985a, 1985b; Mischler, 1984; Robillard, 1994; Rawlinson, 1982);
➢ to comprehend physical symptoms (Rudebeck, 2001, 1992; Lebacqz, 1995; McWhinney, 1986);
➢ to respond to the needs of the chronically and terminally ill (Jennings, *et al*, 1988; Rabin, 1982; Toombs, 1995b, 1995c) and those whose disease is not easily correlated with demonstrated pathoanatomical or pathophysiological findings (Engelhardt, 1982, p. 50; McWhinney, 1983; Baron, 1985; Donnelly, 1986);
➢ to recognize and alleviate suffering (Cassell, 2001, 1991a, 1991b, 1982; Carmody, 1995; Kleinman, 1988; Diamond, 1996, Toombs, 1996b);
➢ to grasp, and respond to, the experiential characteristics of illness (Frank, 2001; Komesaroff, 2001; Mazis, 2001; Svenaeus, 2001; Walton, 2001; Kestenbaum, 1982a, 1982b; Baron, 1981; Madjar and Walton, 1999; Sacks, 1983; Van Manen, 1994);
➢ to comprehend the lived experience of disability (Papadimitriou, 2001; Toombs, 2001; Robillard, 1999; Frank, 2000; Sacks, 1984);
➢ to understand the phenomenon of pain (Cassell, 2001; Madjar, 2001; Behnke, 1989, p. 9; Scarry, 1985; Leder, 1984b; Schrag, 1982; Buytendijk, 1962);
➢ to attend fully to the particular problems of the unique individual who is seeking help – a task that both characterizes the traditional role of healer (McWhinney, 2001; Toulmin, 1976, pp. 46-47; Toombs, 1996a; Cassell, 1979, 1966) and that also morally grounds the profession of clinical medicine (Komesaroff, 2001; Pellegrino, 1983, 1979a; Pellegrino and Thomasma, 1981; Zaner, 1997; Schwartz and Wiggins, 1985, 1988; Rudebeck, 1992, 2001; Benner, 2001, 1994; Bishop and Scudder, 1990; Toombs, 1992, pp. 89-119). As Pellegrino (1982, p. 160) states,

It is the needs arising from the experience of illness in *this* person that provides the source of professional morality of those who profess to heal. Their moral obligations are rooted in the phenomenology of illness; since the experience of illness gives rise to expectations in the one who is ill. These expectations become promises each time the professional presents himself or herself to the person who is ill and offers to help. The promise of help shapes the nature of every healing act and defines the requirements for successful healing – even when cure is not possible. To heal is "to make whole again" and that entails confronting and ameliorating the ways illness wounds the *humanity* of the one who is ill (emphasis mine).

Furthermore, in the context of clinical medicine, ethical discussion and decision-making must necessarily address the existential predicament of the particular suffering individual who seeks help – i.e. the person by whom, and for whom, a right and good decision must be made (McWhinney, 2001; Komesaroff, 2001, 1995b; Siegler and Singer, 1988, p. 759; Pellegrino, 1979b; Welie, 1998). Thus, *clinical ethics* must (among other things) explicitly attend to the unique life situation and lived body of the individual (Bliton, 2001; Zaner, 1996, 1993, 1988, Komesaroff, 1995b).

Moreover, since the human body-as-lived (and not merely the biological body) is necessarily the focus of ethical deliberations in the context of medicine, it is vital that discussions in the field of *bioethics* take into account the lived experience of embodiment (Leder, 1995; Cahill and Farley, 1995; Komesaroff, 1995a). For instance, an understanding of embodiment provides relevant and important insights into many issues that are the subject of debate in contemporary bioethics. I will mention just a few examples: *abortion* (Mackenzie, 1995); *issues related to pregnancy/procreation* (Kahn, 1996; Shanner, 1996; Duden, 1993; Young, 1990; Depraz, 2001); *organ transplantation* (Varela, 2001; Nancy, 2000; Leder, 1999; Toombs, 1999); *reproductive technologies and practices* (Diprose, 1995); *the concept of autonomy* (Mackenzie, 2001, Mackenzie and Stoljar, 2000; Agich, 1995; Benner, et al., 1994;); *issues of gender in clinical medicine* (Connolly, 2001; Young, 1992; Rothfield, 1995; Bordo, 1992; Miller, 1978; Sherwin, 1991; Holmes and Purdy, 1992; Lebacz, 1995); *withholding/withdrawing artificial nutrition/hydration* (Brannigan, 2001a, 2001b); *the self/body relationship* (Brannigan, 2001; Charmaz, 1991; Gadow, 1982; Gallagher, 2001; Komesaroff, 2001; Madjar, 2001; Mazis, 2001; Rudebeck, 2001; Svenaeus, 2001; Toombs, 2001, 1994) – including *the notion of "ownership" with relation to one's body and body parts* (Mackenzie, 2001; Ricoeur, 1992; Gallagher, 1986; Toombs, 1999); *concepts of health* (Svenaeus, 2000, 2001; Gadamer, 1996) *and disease* (Sundström, 2001); and *the nature of the doctor/patient relationship* (Cassell, 1997; Komesaroff, 2001; Pellegrino, 1983; Toombs, 1992a; Young, 1997; Zaner, 1988).

III. PHENOMENOLOGICAL REFLECTIONS

A. Embodiment in the Medical Context

In reflecting on embodiment in the medical context, Shaun Gallagher focuses on, and extends, several of the concepts that informed Merleau-Ponty's analysis of the lived body. In particular, Gallagher makes a clear distinction between the concepts of body image and body schema noting that these concepts continue to play a role in medical research, explanation, and treatment, as well as in philosophical analyses of questions pertaining to cognition and personal identity. In the medical context, for example, the distinction between body image and body schema can be usefully employed in the systematic explanation of certain pathological conditions (including aplasic phantoms, unilateral neglect, deafferentiation, and schizophrenia), as well as providing the basis for developing therapies that address disorders involving pathological distortions of body image. Gallagher also explores the manner in which these aspects of embodiment contribute to the sense of self and personal identity, as well as to the senses of agency and ownership.

Maureen Connolly turns her attention to the issue of gendered embodiment. In particular, she notes the incompatibility between biomedical and lived body approaches to the body – noting specific ways in which the lived body is effectively "eliminated" in medicine. After providing a brief overview of feminist and phenomenological work on female embodiment, Connolly discusses how female embodiment is constructed in the context of clinical practice – a construction that prohibits the healthcare practitioner from understanding (and responding to) the lived experience of illness and pain, that prevents effective communication, and that often results in differential (and sometimes dismissive) treatment for women. Connolly ends by suggesting concrete ways in which a "phenomenological sensibility" can inform praxis and provide an alternative way of approaching (and inhabiting) the clinical encounter.

Glen Mazis considers the significance of the "emotional body" in the context of serious illness and injury. Following Merleau-Ponty, he notes the constant interweaving of lived body and world – an embodied engagement that is, at its heart, an emotional engagement. Mazis asks (1) in illness what are some of the emotional dimensions of a world that has at its center the "body uncanny" (Zaner, 1981), particularly since the only cultural response to one's predicament is the biomechanical colonization of the body, and (2) how can attention to the emotional dimension transform the practice of medicine (i.e. shift the emphasis from that of purely medical intervention to "healing" – making whole – an approach that specifically takes into account the emotional dimension of bodily being.)

Bruce Wilshire also explores the significance of the interweaving of body/world/community in the process of healing. Noting that we are already sensuously aware of our surroundings from the time we are in the womb – particularly through the capacity to hear the sounds emanating from within our own bodies, our mother's body, and from the environing world – Wilshire describes how Native American medicine calls upon this primal embodied connectedness to the world in order to arouse the "immunological and regenerative capacities of the body that bring about healing." In focusing on the phenomenon of auditory perception, Wilshire considers Rudolf Steiner's phenomenological analysis of the engaged body as hearer and music maker in order to grasp the role that music can play in the healing process. Wilshire notes that, as is the case with indigenous ways of regarding sickness, Steiner considers "disease as dis-ease, lack of balance or flow." It seems that music is, thus, particularly well suited to restoring the sense of embodied equilibrium (a point that is illustrated concretely in Mazis' essay on the "emotional body" when he recounts Oliver Sacks' specific experience of regaining physical function through the power of music to "move" him.)

B. Lived Experience

In turning to lived experience phenomenology provides concrete insights into what it is like, what it means, to experience illness and disability. An important avenue of such disclosure is the first-person narrative of illness – a form of story-telling that has proliferated in recent years (Hawkins, 1993). Arthur Frank turns the "phenomenological gaze" to the illness narrative itself, in order to consider not so much what is disclosed about illness as lived experience but rather to analyze "how such narrative disclosures originate in an attitude of consciousness ... *telling illness*, comprising both speaking and hearing as reciprocal aspects of shared telling." In considering several illness narratives in light of Alfred Schutz's notion of "finite provinces of meaning," Frank demonstrates that "telling-illness" is more than simply a consciousness of being ill. Rather it is an awareness of "living illness as a story in which one is both narrator and narrated, organizing consciousness and consciousness that is organized." Frank concludes by considering how the phenomenological analysis of "telling illness" raises important questions such as "How should one respond to the illness storyteller?" "What is the moral relationship between story-teller and listener?" "What does the moral occasion of illness require of each one of us?"

Kay Toombs reflects on her own experience of neurological disease – multiple sclerosis – in order to illuminate the lived experience of bodily change (acquired disability). In particular, she considers the meaning of loss of mobility. In noting the dynamic relation between lived body and world, she describes, for example, the multiplicity of ways in which the experience of space and time change as one moves from the world of the "upright" to the world of the wheelchair user. Toombs also discusses the significance of loss of upright posture in terms of one's relationship with others, one's sense of self, the exercise (and perception) of autonomy, and the interrelationship between the emotions and bodily being. She suggests that understanding the lived experience of disability has important practical implications in the clinical context, as well as in the wider society.

Irena Madjar considers what it is like to experience pain – particularly in the somewhat paradoxical context of clinical practice where the one whose aim is to relieve pain and suffering is also the one who is sometimes required to inflict pain. She notes that there are some essential qualities to the phenomenon of lived pain – its embodied painfulness, its wounding nature, its socially mediated expression (acceptable ways of being with others when one is in pain) – as well as some commonly held meanings with respect to the pain experience (for instance, in searching for the meaning of pain, many sufferers assume that pain is an inevitable consequence of illness, or that it is some kind of punishment or sign of failure). In focusing on the meaning of pain, Madjar draws attention to the inadequacy of biomedical descriptions and the need for healthcare professionals to move beyond standardized terms and numerical values in order to comprehend that lived pain is much more than simply "a pure sensation." "The key," she says, "is our attentiveness to the lived experience of the person in pain, and our willingness, individually and as members of health care teams, to work as much *with* as *on* our patients."

Although noting that the term "mental illness" is, in itself, problematic, Jo Ann Walton considers some of the elements that are central to the lived experience of those who are diagnosed with "mental illness" and who experience themselves as persons "with a mental illness." She discusses several phenomenological studies that have examined the experience of schizophrenia, postpartum depression, and anorexia nervosa. Such studies bring to light important features of the illness experience. For instance, in describing the ways in which the participants in her study of people with schizophrenia looked "both backward on their illness and forward toward the future," Walton finds a parallel with Heidegger's analysis of the temporal dimension of hope in everyday "normal" existence. Furthermore, so called "mental illnesses" are revealed to be particular ways of being-in-the-world (and not simply a disturbance of mental – as opposed to physical – functioning).

C. *Clinical Practice*

The essays on clinical practice attest to the practical import of rigorously examining "the things themselves" in the context of the clinical encounter. Four of the authors are physicians, another has extensive experience in the field of nursing and patient care. Carl Edvard Rudebeck explores the role of bodily empathy in clinical communication. In distinguishing between the symptom and the symptom presentation, Rudebeck clarifies the meaning structure of the body in lived experience – a meaning structure that he identifies as "the existential anatomy." Noting that the intentionality of bodily experience can be characterized along a continuum between outward intentionality (away from the body towards one's involvements in the world) and reflexive or inward intentionality (toward the body-as-nature), Rudebeck provides a descriptive account of the complex intentionalities of both symptoms, symptom presentations, and the communicated experience of the symptom. He notes that the first step in clinical diagnosis is grasping the patient's experience. "The richer, and more differentiated the doctor's perception of the symptom presentation ... the easier it is for the doctor to reach a valid judgment." In addition to providing an insightful and illuminating account of the symptom presentation, Rudebeck's analysis shows how the ground of lived experience provides the basis for a type of empathy that he calls "bodily empathy." Since doctor and patient share their "existential anatomies" in important ways, the doctor's own bodily experience can provide insights into presented symptoms. Rudebeck distinguishes several dimensions of bodily empathy and suggests ways in which these can be developed.

In his exquisite account of the "many faces of the clinic," Paul Komesaroff focuses on the irreducible "intricacy and complexity" of the encounter between patient and carer. In reflecting on his relationship with a patient, Rebecca (whom he has known for about ten years), Komesaroff describes the heterogeneity that is typical of clinical discourse – the patient's story is "fluid, unpredictable and fecund" moving "seamlessly between different kinds of language." It cannot be reduced to the language of scientific medicine, nor does it come out as a single, coherent account. This variability of linguistic usage in the clinical dialogue is "not a purely technical matter: it is literally the means by which the patient reveals her 'self'" – a "self" that is not a unified subject independent from social life or interpersonal experience, but rather one that is formed and generated in the relation between "self" and other (Levinas). The face-to-face confrontation with another thus has a particular ethical significance in that it is the "starting point for meaning, value and truth." In the clinical encounter, the patient herself is formed and presented in the clinical dialogue. Komesaroff considers what it means to be a partner in the clinical relationship – where one is both called upon to witness, confront, and be responsive (and responsible) to the unique other who presents herself. He concludes, "the clinical encounter is inherently a moral relationship, but ... the conditions of morality here are very different from those presupposed in classical ethical theory."

In noting that modern medicine and its institutions are "strictly divided along dualistic mind/body lines" even though "the evidence accumulates that all illness is a function of the human organism as a whole, at all levels, from the molecular to the cognitive and affective, and that the organism's environment, both social and physical, is an important influence on healing," Ian McWhinney traces the evolution of clinical method in Western medicine from its roots in Greek antiquity to the present. He proposes that the "modern clinical method" – with its dominance on abstraction and the devaluation of experience – be transcended by "the patient-centered clinical method," a method that enjoins the clinician to focus explicitly on the patient's lived experience. After describing the patient-centered method in detail, McWhinney concludes, "The new method should not only restore the Hippocratic ideal of friendship between doctor and patient, but also make possible a medicine which can see illness as an expression of a person with a moral nature, an inner life, a unique life story: a medicine that can heal by a therapy of the word and a therapy of the body."

Patricia Benner focuses on the phenomenon of care both in terms of its ontological structure in the human lifeworld of concerns and possibilities (i.e. what is it about our Being-in-the-world that is the ground for care or concern?), as well as in terms of particular caring practices that are related to specific lifeworlds and contexts (such as clinical medicine). She illustrates how an understanding of the ontological dimension of care enables us to reflect on (and develop) caregiving practices that are optimally responsive in terms of meeting the needs of the one who seeks care.

In reflecting on the phenomenon of suffering and its relationship to pain, Eric Cassell states, "The path, stretching from the physiological response evoked by a physical stimulus to the disintegration of the person experienced as suffering, is important to understand not only for itself but also as an exemplar of mind-body interactions." He demonstrates that, while there *is* a distinction between pain and suffering, it is a mistake to assume that the experience of pain itself is a purely physical sensation detached from the meanings, predictions, interpretations and behaviors associated with it (all of which are a function of the person experiencing the pain). Thus, pain is "subjective in the sense of experienced by the subject – the person ... [and] ... in the sense of expressive of the subject – the person." Pain progresses to suffering when the injury to the patient broadens to the extent that the integrity of the person is threatened. Cassell notes that, like pain, suffering is personal and individual. It can come from "any physical, social, or emotional process that leads to a loss of integrity of the person." Unless suffering is directly addressed, it may continue even if the pain or other physical or social processes are controlled. In explicating the complexity of pain and suffering, Cassell shows that "the involvement of persons in their own illness and manifestations of disease can be traced from the earliest symptoms to the ultimate personal injury, suffering. It follows that there can be no true appreciation of sickness and its treatment without an understanding of persons."

D. Medical Ethics

Mark Bliton considers the relevance of phenomenological method in the context of his engagement in *clinical ethics* consultation. He notes the complexity of decision-making in the clinical setting where "not only are there apparent differences among personal beliefs, professional opinions, and responsibilities, whose meanings may not be reducible to some common theme, but the meanings of physiological factors, let alone moral ones ... are embedded in complex webs of cultural and social relationships." Thus, responsible ethical inquiry must "always be guided by careful attention to the moral relationships and understanding of the individuals actually making the decisions and living with their choices." Bliton illustrates by relating his experience of talking with more than 150 pregnant women (and partners) about their decisions to undergo open-uterine fetal surgery. From these conversations he identifies a number of themes that pervade such moral deliberations. These include: the manner in which the experience of pregnancy is influenced by prenatal ultrasound imaging (and, particularly, the ways in which ultrasound shapes the perceived moral status of the fetus); the impact of uncertainty on the process of moral deliberation; the ways in which the meanings of disability influence the decision "to do" something; and the role of religious language and faith in decision-making. Bliton's findings raise important issues not only for the particular context of open-uterine fetal surgery but in terms of ethical deliberations about the status of the fetus, considerations about what it means to say that an individual "freely" chooses a certain course of action, and in terms of defining the nature of clinical ethics.

Catriona Mackenzie critiques prevailing conceptions of autonomy that are widely held in mainstream *bioethics*. In particular, she questions "maximal choice" conceptions of bodily autonomy which include "the view that an agent's bodily autonomy is always enhanced by maximizing the range of bodily options available to her; and the view that any expansion of a person's bodily capacities, or of her instrumental control over her bodily processes ... enhances her bodily autonomy." Mackenzie suggests that the phenomenological notion of embodiment offers an alternative view of bodily autonomy that explicitly recognizes the "complex relationships among the biological, social and individual dimensions of embodiment," as well as the intimate connection between body and selfhood. Following Ricoeur, Mackenzie identifies two different senses of "belonging" with respect to the body ("selfhood" and "ownership") – a distinction that "helps explain why the notion of bodily ownership misunderstands the sense in which a person's body belongs to her." She proposes a concept of bodily autonomy that takes into account whether or not a person has an "integrated bodily perspective" (a notion that explicitly recognizes that the person is an *embodied* agent whose body is constitutive of selfhood), and whether or not the person has the capacity for normative critical reflection.

Michael Brannigan suggests that the work of Husserl and Merleau-Ponty can provide insight into a variety of issues in *medical ethics* – including the issue of how best to decide whether or not to provide nutrition and hydration via nasogastric feeding to particular individuals. In bringing a phenomenological perspective to bear on the case of Claire Conroy, Brannigan critiques the prevailing distinction between "ordinary" and "extraordinary" treatment and considers how Merleau-Ponty's ontology of the body might have specifically informed the decisions made on Ms. Conroy's behalf. In particular, he is concerned to show that what is "morally obligatory" with respect to treatment decisions "rests upon a phenomenological understanding, one that considers treatment in terms of the entire context" of the particular individual who will be the recipient of such medical intervention.

E. Research

Max van Manen considers what it means to engage in phenomenological research in the domain of public and professional practice. Noting that phenomenological inquiry involves the taking up of a particular attitude or style of thinking (i.e. "an attitude of reflective attentiveness to what it is that makes life intelligible and meaningful to us"), van Manen both outlines "how some of the methodological features of the phenomenological reduction are engaged by methods inventively borrowed from the broader field of the social and human sciences, from the humanities, and from other practical contexts," and demonstrates how phenomenological research can deal with practical topics in the domain of medicine. Since understanding the meaning of illness is of crucial significance in the care of patients, van Manen illustrates how one might engage in a research study focused on a certain aspect of Alzheimer's dementia. How, for example, can one understand the profound forgetfulness or memory loss in Alzheimer's patients? Beginning with experiential accounts of the ordinary experience of forgetting someone's name and including lived experience descriptions from Alzheimer's patients, van Manen briefly describes the reflective process of phenomenological analysis and writing. He points out that, even in this brief account, "some aspects of the meaning of name forgetting readily suggest themselves" – for instance, the way in which memory is inextricably tied to selfhood. Thus, the tragedy of forgetting in Alzheimers is that the disease "slowly seems to rob a person of his or her selfhood." It is vital for practitioners to gain such knowledge of the experiential (or pathic) dimensions of Alzheimer's disease, since clinical decisions always relate to the suffering individual in the context of his or her particular life situation (as is the case with all illnesses). Furthermore, "implicitly or explicitly professional practices are shaped on the basis of how one conceives the inner life and the effects of Alzheimers on the nature of selfhood of the patients suffering from it." Phenomenological research is, therefore, an important resource for those engaged in the practice of clinical medicine since professional practice needs to be "not only pragmatic and effective, but also philosophic and reflective so that one can act and interact with competence and tact, and with human understanding."

Christina Papadimitriou discusses her phenomenological research practices as a fieldworker studying the phenomenon of physical disability. In particular, she is concerned to identify theoretical and conceptual biases in the study of disability that both limit our understanding of the lived experience of disability, and that cause us to "see" persons with disabilities as radically Other than we are. Papadimitriou describes her use of two phenomenological practices (bracketing and empathic listening) that have been particularly helpful in assisting her both to understand, and communicate with, her disabled informants. She ends by suggesting that, rather than seeing the "normal" and "pathological" as separate, we can understand them as "varieties situated along a continuum of modes of being-in-the-world." Disabled embodiment may thus be conceived as a form of human diversity – "difference" as opposed to "dis-ability" – a conception that has important social and political implications.

IV. CONCLUSION

The essays in this volume demonstrate that the philosophical perspective of phenomenology is an extraordinarily fruitful approach for examining theoretical and practical issues in the domain of medicine, clinical practice, and medical ethics. Clinical medicine is necessarily at the intersection of science and humanity. It is a *human* science that has, at its center, the suffering *embodied* human being who seeks help and the professional who responds (is responsive) to the call for assistance. As a discipline, medical ethics is also primarily concerned with issues that relate to (or have implications for) the care of the sick person. In eschewing abstraction in favor of a commitment to focus on "the things themselves," and in emphasizing the importance of clarifying the meaning and structure of the lifeworld and existence, phenomenology is thus particularly well suited for engaging in philosophical reflections that pertain to medicine.

Moreover, in providing the means to examine phenomena in all their richness and complexity, the philosophical perspective of phenomenology can powerfully illuminate and clarify theoretical and practical issues in a variety of disciplines. Some of the authors who have contributed to this volume are academic philosophers, some are engaged in other academic disciplines, and several are practicing healthcare professionals. However, all have found phenomenology particularly insightful in their work. Their essays (and experience) attest to the fact that phenomenology is an invaluable practical tool for those working in the field of clinical medicine, for those engaged in clinical practice, and for all who are interested in the philosophy of medicine.[4]

Baylor University
Waco, Texas
U.S.A.

NOTES

[1] See, for example, Spiegelberg, 1985; Embree, Behnke, Carr, *et al*, 1997; Toombs, 1992 (notes to pp. xi-xvi). For an excellent introduction to phenomenology, see Zaner, 1970.
[2] Engelhardt (1982, pp. 41-57) notes that, largely as a result of developments in the science of pathoanatomy in the 19th century, the primary focus of medicine went inside the body and disease thus became identified with pathoanatomical lesions or pathophysiological disturbances. The live body thus became explicable in terms of the dead body (Zaner, 1988, p. 134).
[3] The process of "dehumanization" that occurs in the clinical encounter has been noted in numerous illness narratives (See, for example, Hawkins, 1993; Frank, 1998, 1995, 1991; Murphy 1987; Rabin, 1985; Price, 1994; Duff, 1993).
[4] Obviously the essays in this volume are not intended to provide a comprehensive assessment of all the work that has been carried out utilizing a phenomenological approach to explore issues related to the domain of medicine. Rather the intent of the volume is to give examples that concretely demonstrate the phenomenological perspective to be a viable and relevant alternative to more traditional philosophical approaches to the philosophy of medicine. It should perhaps be noted that phenomenology has had an important influence in the field of psychiatry (see for example the entry on psychiatry in the *Encyclopedia of Phenomenology*, Schwartz and Wiggins, 1997, pp. 563-568). However, given the necessary constrictions with regard to the length of this

volume, a decision was made to include an essay on the lived experience of mental illness – rather than to survey the field of phenomenology and psychiatry as a whole (a topic that warrants a complete volume).

BIBLIOGRAPHY

Abrams, D.: 1996, *The Spell of the Sensuous*, quoted in E. Thompson, 'Human consciousness: From intersubjectivity to interbeing. A report to the Fetzer Institute, 1999', unpublished manuscript.

Agich, G.: 1995, 'Chronic illness and freedom', in S.K. Toombs, D. Barnard and R.A. Carson (eds.), *Chronic Illness: From Experience to Policy*, Indiana University Press, Indiana, pp. 129-53.

Baron, R.J.: 1985, 'An introduction to medical phenomenology: I can't hear you while I'm listening', *Annals of Internal Medicine* **103**, 606-11.

Baron, R.J.: 1981, 'Bridging clinical distance: An empathic rediscovery of the known', *The Journal of Medicine and Philosophy* **6**, 5-23.

Behnke, E.A.: 1997, 'Body', in L. Embree, E.A. Behnke, D. Carr, et al. (eds.), *Encyclopedia of Phenomenology*, Kluwer Academic Publishers, Dordrecht, The Netherlands, pp. 66-71.

Behnke, E.A.: 1989, 'Edmund Husserl's contribution to phenomenology of the body in Ideas II', *SPPB Newletter* **2**, 5-18.

Benner, P.: 2001, 'The phenomenon of care', in this volume, pp. 345-63.

Benner, P. (ed.): 1994, *Interpretive Phenomenology: Embodiment, Caring and Ethics*. Sage, Thousand Oaks, California.

Benner, P., Janson-Bjerklie, S., Ferketich, S., et al.: 1994, 'Moral dimensions of living with a chronic illness, autonomy, responsibility and the limits of control', in P. Benner (ed.), *Interpretive Phenomenology: Embodiment, Caring and Ethics*. Sage, Thousand Oaks, California, pp. 225-54.

Bishop, A.H. and Scudder, J.R.: 1996, *Nursing Ethics: Therapeutic Caring Presence*, Jones and Bartlett, London.

Bishop, A.H. and Scudder, J.R.: 1990, *The Practical, Moral, and Personal Sense of Nursing: A Phenomenological Philosophy of Practice*, SUNY, New York.

Bliton, M.: 'Imagining a Fetus: Insights from talking with pregnant women about their decisions to undergo open-uterine fetal surgery', in this volume, pp. 396-408.

Branningan, M.C.: 2001a, 'Medical Feeding: Applying Husserl and Merleau-Ponty', in this volume, pp. 432-445.

Brannigan, M.C.: 2001b, 'Re-Assessing the ordinary/extraordinary distinction in withholding/withdrawing nutrition and hydration', in M.C. Brannigan and J.A. Boss (eds.), *Healthcare Ethics in a Diverse Society*, Mayfield Publishing Company, Mountain View, California, pp. 516-22.

Bordo, S.: 1992, 'Eating disorders: The feminist challenge to the concept of pathology', in D. Leder (ed.), *The Body in Medical Thought and Practice*, Kluwer Academic Publishers, Dordrecht, The Netherlands, pp. 197-213.

Brough, J.: 2001, 'Temporality and illness', in this volume, pp. 29-46.

Buytendijk, F.J.J.: 1962, *Pain: Its Modes and Functions*, University of Chicago Press, Chicago.

Cahill L. and Farley, M. (eds.): 1995, *Embodiment, Morality and Medicine*, Kluwer Academic Publishers, Dordrecht, The Netherlands.

Cassell, E.J.: 2001, 'The phenomenon of suffering and its relationship to pain', in this volume, pp. 364-83.

Cassell, E.J.: 1997, *Doctoring: The Nature of Primary Care Medicine*, Oxford University Press, New York.
Cassell, E.J.: 1991a, *The Nature of Suffering*, Oxford University Press, New York.
Cassell, E.J.: 1991b, 'Recognizing suffering', *Hastings Center Report* 21, 24-31.
Cassell, E.J.: 1985a, *The Theory of Doctor-Patient Communication*, Vol.1, *Talking With Patients*, MIT Press, Cambridge, Massachusetts.
Cassell, E.J.: 1985b, *The Theory of Doctor-Patient Communication*, Vol. 2, *Talking With Patients*, MIT Press, Cambridge, Massachusetts.
Cassell, E.J.: 1982, 'The nature of suffering and the goals of medicine', *The New England Journal of Medicine*, 306, 639-45.
Cassell, E.J.: 1979, 'The subjective in clinical judgment', in H.T. Engelhardt, Jr., *et al*, (eds.), *Clinical Judgment: A Critical Appraisal*, D. Reidel, Dordrecht, The Netherlands.
Cassell, E.J.: 1966, *The Healer's Art*, J.B. Lippincott, New York.
Carmody, J.T. and Carmody, D.L.: 1995, 'The body suffering: Illness and disability', in L.Sowle Cahill and M.A. Farley (eds.), *Embodiment, Morality, and Medicine*, Kluwer Academic Publishers, The Netherlands, pp. 185-97.
Charmaz, K.: 1991, *Good Days, Bad Days: The Self in Chronic Illness and Time*, Rutgers University Press, New Jersey.
Cole, J.: 1991, *Pride and a Daily Marathon*, Gerald Duckworth and Co., London.
Connolly, M.: 2001, 'Female embodiment and clinical practice', in this volume, pp. 176-95.
Depraz, N.: 2001, 'The lucid body in the light of the lived experience of pregnancy', The Sorbonne, Paris, unpublished manuscript.
Diamond, S.: 1996, 'Case story: Lifelong effects of chronic atopic eczema' *Making the Rounds in Health Faith and Ethics* 2, 1-4.
Diprose, R.: 1995, 'The body biomedical ethics forgets', in P.A. Komesaroff (ed.), *Troubled Bodies: Critical Perspectives on Postmodernism, Medical Ethics, and the Body*, Duke University Press, Durham, pp. 202-21.
Donnelly, W.: 1986, 'Medical language as symptom: Doctor talk in teaching hospitals', *Perspectives in Biology and Medicine* 30, 81-94.
Duden, B.: 1993, *Disembodying Women: Perspectives on Pregnancy and the Unborn*, Harvard University Press, Cambridge, Massachusetts.
Duff, K.: 1993, *The Alchemy of Illness*, Random House, New York
Embree, L., Behnke, E.A., Carr, D., *et al.* (eds): 1997, *Encyclopedia of Phenomenology*, Kluwer Academic Publishers, Dordrecht, The Netherlands.
Engel, G.L.: 1987, 'Physician-scientists and scientific physicians: Resolving the humanism-science dichotomy', *The American Journal of Medicine* 82, 107-11.
Engel, G.L.: 1977a, 'The care of the patient: Art or science?' *The Johns Hopkins Medical Journal* 140, 222-32.
Engel, G.L.: 1977b, 'The need for a new medical model: A challenge for biomedicine', *Science* 196, 129-36.
Engel, G.L.: 1976, 'Too little science: The paradox of modern medicine's crisis', *The Pharos*, 127-31.
Engelhardt, H.T., Jr.: 1982, 'Illnesses, diseases, and sicknesses', in V. Kestenbaum (ed.), *The Humanity of the Ill*, University of Tennessee Press, Knoxville, pp. 142-56.
Engelhardt, H.T., Jr.: 1977, 'Husserl and the mind-body relation', in D. Ihde, *et al.* (eds), *Interdisciplinary Phenomenology*, Martinus Nijhoff, The Hague, pp. 51-70.
Engelhardt, H.T., Jr.: 1973, *Mind-Body: A Categorial Relation*, Martinus Nijhoff, The Hague, The Netherlands.
Frank, A.W.: 2001, 'Experiencing illness through storytelling', in this volume, pp. 227-43.
Frank, A.W.: 1998, 'Just listening: Narrative and deep illness', *Families, Systems & Health* 16, 197-212.

Frank, A.W.: 1995, *The Wounded Storyteller: Body, Illness and Ethics*, University of Chicago Press, Chicago.
Frank, A.W.: 1991, *At the Will of the Body*, Houghton Mifflin, Boston.
Frank, G.: 2000, *Venus on Wheels: Two Decades of Dialogue on Disability, Biography, and Being Female in America*, University of California Press, California.
Foucault, M.: 1975, *The Birth of the Clinic: An Archaeology of Medical Perception*, A.M. Sheridan Smith (trans), Vintage Books, New York.
Gadamer, H.J.: 1996, *The Enigma of Health*, J. Gaiger and N. Walker (trans.), Stanford University Press, Stanford, California.
Gadow, S.: 1982, 'Body and self: A dialectic', in V. Kestenbaum (ed.), *The Humanity of the Ill, Phenomenological Perspectives*, The University of Tennessee Press, Knoxville, pp. 86-100.
Gallagher, S.: 2001, 'Dimensions of embodiment: Body image and body schema in medical contexts', in this volume, pp. 147-75.
Gallagher, S.: 1986, 'Lived body and environment', *Research in Phenomenology* 16: 139-70.
Hawkins, A.H.: 1993, *Reconstructing Illness: Studies in Pathography*, Purdue University Press, West Lafayette, Indiana.
Heelan, P.: 2001, 'The Lifeworld and scientific conceptualization', in this volume, pp. 47-66.
Heidegger, M.: 1926/1962, *Being and Time*, J. Macquarrie and E. Robinson (trans.), Harper and Row, New York.
Holmes, H.B. and Purdy, L.M.: 1992, *Feminist Perspectives in Medical Ethics*, Indiana University Press, Indiana.
Husserl, E.: 1989, *Ideas Pertaining to a Pure Phenomenology and to a Phenomenological Philosophy: Second Book: Studies in the Phenomenology of Constitution*, R. Rojcewicz and A. Schuwer (trans.), Kluwer Academic Publishers, Dordrecht, The Netherlands.
Husserl, E.: 1982, *Cartesian Meditations: An Introduction to Phenomenology*, D. Cairns (trans.), 7th Impression, Martinus Nijhoff, The Hague, The Netherlands.
Jennings, B., et al.: 1988, 'Ethical challenges of chronic illness', *The Hastings Center Report*, pp. 1-16.
Jonas, H.: 1966, *The Phenomenon of Life: Toward a Philosophical Biology*, Dell Publishing Co., New York.
Kahn, R.F.: 1996, *Bearing Meaning: The Language of Birth*, University of Illinois Press, Chicago.
Kestenbaum, V. (ed.): 1982a, *The Humanity of the Ill, Phenomenological Perspectives*, The University of Tennessee Press, Knoxville.
Kestenbaum, V.: 1982b, 'The experience of illness', in V. Kestenbaum (ed.), *The Humanity of the Ill, Phenomenological Perspectives*, The University of Tennessee Press, Knoxville.
Kleinman, A.: 1988, *The Illness Narratives: Suffering, Healing and the Human Condition*, Basic Books, New York.
Komesaroff, P.A.: 2001, 'The many faces of the clinic: A Levinasian view', in this volume, pp. 311-24.
Komesaroff, P.A. (ed.): 1995a, *Troubled Bodies: Critical Perspectives on Postmodernism, Medical Ethics, and the Body*, Duke University Press, Durham.
Komesaroff, P.A.: 1995b, 'From bioethics to microethics: Ethical debate and clinical medicine', in P.A. Komesaroff (ed.), *Troubled Bodies: Critical Perspectives on Postmodernism, Medical Ethics, and the Body*, Duke University Press, Durham, pp. 62-86.
Lebacqz, K.: 1995, 'The 'Fridge': Health care and the disembodiment of women', in L.Sowle Cahill and M.A. Farley (eds.), *Embodiment, Morality, and Medicine*, Kluwer Academic Publishers, The Netherlands, pp. 155-67.

Leder, D.: 1999, 'Whose body? What body? The metaphysics of organ transplantation', in M. Cherry (ed.), *Persons and Their Bodies: Rights, Responsibilities, Relationships*, Kluwer Academic Publishers, Dordrecht, The Netherlands, pp. 233-64.
Leder, D.: 1995, 'Health and disease: The experience of health and illness', in W. T. Reich (ed.), Vol. 2. *Encyclopedia of Bioethics*, Simon & Schuster Macmillan, New York.
Leder, D.: 1992a, 'A tale of two bodies: The Cartesian corpse and the lived body', in *The Body in Medical Thought and Practice*, D. Leder (ed.), Kluwer Academic Publishers, Dordrecht, The Netherlands, pp. 17-35.
Leder, D.: 1992b, 'Introduction', in D. Leder (ed.), *The Body in Medical Thought and Practice*, Kluwer Academic Publishers, Dordrecht, Holland.
Leder, D.: 1984a, 'Medicine and the paradigms of embodiment', *The Journal of Medicine and Philosophy* 9, 29-43.
Leder, D.: 1984b, 'Toward a phenomenology of pain', *Review of Existential Psychology and Psychiatry* 19, 255-66.
Leder, D.: 1990, *The Absent Body*, University of Chicago Press, Chicago.
Mackenzie, C.: 2001, 'On bodily autonomy', in this volume, pp. 409-31.
Mackenzie C. and Stoljar, N.: 2000, 'Autonomy refigured', in C. Mackenzie and N. Stoljar (eds.), *Relational Autonomy: Feminist Perspectives on Autonomy, Agency and the Social Self*, Oxford University Press, New York, pp. 3-31.
Mackenzie, C.: 1995, 'Abortion and embodiment', in P.A. Komesaroff (ed.), *Troubled Bodies: Critical Perspectives on Postmodernism, Medical Ethics, and the Body*, Duke University Press, Durham, pp. 38-61.
Madjar, I.: 2001, 'The lived experience of pan in the context of clinical practice', in this volume, pp. 259-73.
Madjar, I. and Walton, J.A. (eds.): 1999, *Nursing and the Experience of Illness: Phenomenology in Practice*, Routledge, London.
Marcel, G.: 1956, *Being and Having: An Existentialist Diary*, K. Farrer (trans.), Harper and Row, New York.
Mazis, G.: 2001, 'Emotion and Embodiment within the Medical World' in this volume, pp. 196-213.
McWhinney, I.R.: 2001, 'Focusing on lived experience: The evolution of clinical method in Western Medicine', in this volume, pp. 325-44.
McWhinney, I.R.: 1983, 'Changing models: The impact of Kuhn's theory on medicine', *Family Practice* 1, 3-8.
Merleau-Ponty, M.: 1968, *The Visible and Invisible*, A. Lingis (trans.), Northwestern University Press, Evanston, Ill.
Merleau-Ponty, M.: 1962, Phenomenology of Perception, C. Smith (trans.), Routledge and Kegan Paul, London.
Miller, J.: 1978, *The Body in Question*, Random House, New York.
Murphy, R.F.: 1987, *The Body Silent*, Henry Holt, New York.
Nancy, J-L.: 2000, *L'intrus*, Galilée, Paris, France.
Papadimitriou, C.: 2001, 'From dis-ability to difference: Conceptual and methodological issues in the study of physical disability', in this volume, pp. 466-83.
Pellegrino, E.: 1983, 'The healing relationship: The architectonics of clinical medicine', in E. Shelp (ed.), *The Clinical Encounter: The Moral Fabric of the Physician-Patient Relationship*, D. Reidel, Boston, pp. 153-72.
Pellegrino, E.: 1982, 'Being ill and being healed: Some reflections on the grounding of medical morality', in V. Kestenbaum (ed.), *The Humanity of the Ill, Phenomenological Perspectives*, The University of Tennessee Press, Knoxville.
Pellegrino, E., and Thomasma, D.: 1981, *A Philosophical Basis of Medical Practice: Toward a Philosophy and Ethic of the Healing Professions*, Oxford University Press, New York.

Pellegrino, E.: 1979a, 'Toward a reconstruction of medical morality: The primacy of the act of profession and the fact of illness', *The Journal of Medicine and Philosophy*, **4**, 32-56.
Pellegrino, E.: 1979b, 'The anatomy of clinical judgments: Some notes on right reason and right action', in H.T. Engelhardt, Jr., *et al.* (eds.), *Clinical Judgment: A Critical Appraisal*, D. Reidel Publishing, Dordrecht, The Netherlands.
Plügge, H.: 1970, 'Man and his body', in S.F. Spicker (ed.), *The Philosophy of the Body: Rejections of Cartesian Dualism*, Quadrangle Books, Chicago, pp. 293-311.
Price, R.: 1994, *A Whole New Life: An Illness and a Healing*, Atheneum, New York.
Rabin, D. and Rabin P. (eds.): 1985, *To Provide Safe Passage: The Humanistic Aspects in Medicine*, Philosophical Library, New York, pp. 29-37
Rabin, D.: 1982, 'Occasional notes: Compounding the ordeal of ALS: Isolation from my fellow physicians', *The New England Journal of Medicine* **307**, pp. 506-509.
Rawlinson, M.C.: 1986, 'The sense of suffering', *The Journal of Medicine and Philosophy* **11**, 39-62.
Rawlinson, M.C.: 1982, 'Medicine's discourse and the practice of medicine', in V. Kestenbaum (ed.), *The Humanity of the Ill: Phenomenological Perspectives*, University of Tennessee Press, Knoxville, pp. 69-85.
Ricoeur, P.: 1992, *Oneself as Another*, K. Blamey (trans.), Chicago University Press, Chicago.
Robillard, A.B.: 1999, *Meaning of a Disability: The Lived Experience of Paralysis*, Temple University Press, Philadelphia.
Robillard, A.B.: 1994, 'Communication problems in the intensive care unit', *Qualitative Sociology* **17**, 383-95.
Rothfield, P.: 1995, 'Bodies and subjects: Medical ethics and feminism', in P.A. Komesaroff (ed.), *Troubled Bodies: Critical Perspectives on Postmodernism, Medical Ethics, and the Body*, Duke University Press, Durham, pp. 168-201.
Rousso, H.: 1984, 'Fostering healthy self-esteem', *The Exceptional Parent*, 9-14.
Rudebeck, C.E.: 2001, 'Grasping the existential anatomy: The role of bodily empathy in clinical communication', in this volume, pp. 291-310.
Rudebeck, C.E.: 1992, 'General practice and the dialogue of clinical practice', *Scandinavian Journal of Primary Health Care*, Supplement 1, 3-87.
Sacks, O.: 1985, 'The disembodied lady', in *The Man Who Mistook His Wife For a Hat and Other Clinical Tales*, Summit Books, New York, pp. 45-52.
Sacks, O.: 1984, *A Leg to Stand On*, Summit Books, New York.
Sacks, O.: 1983, *Awakenings*, E. P. Dutton, New York.
Sartre, J.P.: 1956, *Being and Nothingness: A Phenomenological Essay on Ontology*, H.E. Barnes (trans.), Pocket Books, New York.
Scarry, E.: 1985, *The Body in Pain: The Making and Unmaking of the World*, Oxford University Press, New York.
Schrag, C.: 1982, 'Being in pain', in V. Kestenbaum (ed.), *The Humanity of the Ill: Phenomenological Perspectives*, The University of Tennessee Press, Knoxville.
Schutz, A. and Luckmann, T.: 1973, *The Structures of the Life-World*, R.M. Zaner and H.T. Engelhardt, Jr. (trans.), Northwestern University Press, Evanston, Illinois.
Schutz, A.: 1962a, 'On multiple realities', in M. Natanson (ed.), *The Problem of Social Reality*, Vol. 1, *Alfred Schutz: Collected Papers*, Martinus-Nijhoff, The Hague, pp. 207-59.
Schutz, A.: 1962b, 'Symbol, reality and society', in M. Natanson (ed.), *The Problem of Social Reality*, Vol. 1, *Alfred Schutz: Collected Papers*, Martinus-Nijhoff, The Hague, pp. 287-356.
Schwartz, M.A. and Wiggins, O.: 1997, 'Psychiatry', in L. Embree, E.A. Behnke, D. Carr, *et al.* (eds.), *Encyclopedia of Phenomenology*, Kluwer Academic Publishers, Dordrecht, The Netherlands, pp. 562-568.

Schwartz, M.A. and Wiggins, O.: 1988, 'Scientific and humanistic medicine: A theory of clinical methods', in K.L. White (ed.), *The Task of Medicine, Dialogue at Wickenburg*, The Henry J. Kaiser Family Foundation, Menlo Park, California, pp. 131-71.

Schwartz, M.A. and Wiggins, O.: 1985, 'Science, humanism, and the nature of medical practice: A phenomenological view', *Perspectives in Biology and Medicine*, **28**, 331-61.

Shanner, L.: 1996, 'Bioethics through the back door: Phenomenology, narratives, and insights into infertility', in L.W. Sumner and J. Boyle (eds.). *Philosophical Perspectives on Bioethics*, University of Toronto Press, Toronto.

Sheets-Johnstone, M.: 1992, 'The materialization of the body: A history of Western medicine, a history in process', in M. Sheets-Johnstone (ed.), *Giving the Body Its Due*, SUNY Press, New York, pp. 132-158.

Sherwin, S.: 1991, *No Longer Patient: Feminist Ethics and Health Care*, Temple University Press, 1991.

Siegler, M. and Singer, P.A.: 1988, 'Clinical ethics consultation: Godsend or "God squad"?' *American Journal of Medicine* **85**, 759-60.

Spicker, S. (ed.): 1970, *The Philosophy of the Body: Rejections of Cartesian Dualism*, Quadrangle Books, Chicago.

Spiegelberg, H.: 1985, *Phenomenological Movement*, 3rd Revised Edition, Martinus Nijhoff, The Hague.

Spiegelberg, H.: 1972, *Phenomenology in Psychology and Psychiatry*, Northwestern University Press, Evanston, Illinois.

Stetten, D., Jr.: 1981, 'Coping with blindness', *The New England Journal of Medicine* **305**, 458-60.

Straus, E.: 1966, *Phenomenological Psychology*, Erling Eng (trans.), Basic Books, New York.

Straus, E.: 1963, *The Primary World of the Senses*, J. Needleman (trans.), Basic Books, New York.

Sundström, P.: 2001, 'Disease: The phenomenological and conceptual center of practical-clinical medicine', in this volume, pp. 109-26.

Svenaeus, F.: 2001, 'The phenomenology of health and illness' in this volume, pp. 87-108.

Svenaeus, F: 2000, *The Hermeneutics of Medicine and the Phenomenology of Health: Steps Towards a Philosophy of Medical Practice*, Kluwer Academic Publishers, Dordrecht, The Netherlands.

Toombs, S.K.: 2001, 'Reflections on bodily change: The lived experience of disability', in this volume, pp. 244-58.

Toombs, S.K.: 1999, 'What does it mean to be some*body*? Phenomenological reflections and ethical quandaries', in M. Cherry (ed.), *Persons and Their Bodies: Rights, Responsibilities, Relationships*, Kluwer Academic Publishers, Dordrecht, The Netherlands, pp. 73-94.

Toombs, S.K.: 1996a, 'The individual in clinical practice', *AllmanMedicin, Swedish Journal of General Practice* **17**, Supplement 1, pp. 19-22.

Toombs, S.K.: 1996b, 'Heeding the call of suffering', *Making the Rounds in Health Faith and Ethics* **2**, 5-6

Toombs, S.K.: 1995a, 'Sufficient unto the day: A life with multiple sclerosis', in S.K. Toombs, D. Barnard and R.A. Carson (eds.), *Chronic Illness: From Experience to Policy*, Indiana University Press, Indiana, pp. 3-24.

Toombs, S.K.: 1995b, 'Chronic illness and the goals of medicine." *Second Opinion* **21**, pp. 11-19.

Toombs, S.K.: 1995c, 'Healing and incurable illness', *Humane Medicine: A Journal of the Art and Science of Medicine*, pp. 98-103.

Toombs, S.K.: 1994, 'Disability and the self', in T.M. Brinthaupt and R.P. Lipka (eds.), *Changing the Self: Philosophies, Techniques, and Experiences*, SUNY Press, New York, pp. 337-57.

Toombs, S.K.: 1992a, *The Meaning of Illness: A Phenomenological Account of the Different Perspectives of Physician and Patient*, Kluwer Academic Publishers, Dordrecht, The Netherlands.
Toombs, S.K.: 1992b, 'The body in multiple sclerosis: A patient's perspective', in D. Leder (ed.), *The Body in Medical Thought and Practice*, Kluwer Academic Publishers, Dordrecht, The Netherlands, pp. 122-37.
Van den Berg, J.H.: 1972, *A Different Existence: Principles of Phenomenological Psychopathology*, Duquesne University Press, Pittsburgh.
Van Manen, M.: 2001, 'Professional practice and "doing phenomenology"', in this volume, pp. 448-65.
Van Manen, M.: 1994, 'Modalities of body experience in illness and health', *Qualitative Health Research: An International Interdisciplinary Journal*, Sage Periodicals Press, Vol. 8, 7-24.
Varela, F.J.: 2001, 'Intimate distances: Fragments for a phenomenology of organ transplantation,' *Journal of Consciousness Studies* **8**, 259-71.
Waksler, F.: 'Medicine and the phenomenological method', in this volume, pp. 67-86.
Walton, J.: 2001, 'The lived experience of mental illness', in this volume, pp. 274-88.
Walton, J.: 1999, 'On living with schizophrenia', in I. Madjar and J. Walton (eds.) *Nursing and the Experience of Illness: Phenomenology in Practice*, Routledge, New York.
Wilshire, B.: 2001, 'The body, music, and healing', in this volume, pp. 214-24.
Young, I.M.: 1990, *Throwing Like a Girl and Other Essays in Feminist Philosophy and Social Theory*, Indiana University Press, Bloomington.
Young, I.M.: 1992, 'Breasted experience: The look and the feeling', in D. Leder (ed.), *The Body in Medical Thought and Practice*, Kluwer Academic Publishers, Dordrecht, The Netherlands, pp. 215-30.
Young, K.: 1997, *Presence in the Flesh: The Body in Medicine*, Harvard University Press, Cambridge, Mass.
Zaner, R.M.: 2001, 'Thinking about medicine', in this volume, 127-44.
Zaner, R.: 1997, 'Medicine', in L. Embree, E.A. Behnke, D. Carr, *et al.* (eds.), *Encyclopedia of Phenomenology*, Kluwer Academic Publishers, Dordrecht, The Netherlands.
Zaner, R.M.: 1996, 'Listening or telling? Thoughts on responsibility in clinical ethics consultation', *Theoretical Medicine* **17**, 255-77.
Zaner, R.: 1993, *Troubled Voices: Stories of Ethics and Illness*, The Pilgrim Press, Cleveland, Ohio.
Zaner, R.: 1988, *Ethics and the Clinical Encounter*, Prentice Hall, Englewood Cliffs, New Jersey.
Zaner, R.: 1981, *The Context of Self: A Phenomenological Inquiry Using Medicine as a Clue*, Ohio University Press, Athens.
Zaner, R.: 1970, *The Way of Phenomenology: Criticism as a Philosophical Discipline*, Western Publishing, New York.
Zaner, R.: 1964, *The Problem of Embodiment: Some Contributions to a Phenomenology of the Body*, Martinus Nijhoff, The Hague.

SECTION ONE

PHENOMENOLOGY AND MEDICINE

JOHN B. BROUGH

TEMPORALITY AND ILLNESS: A PHENOMENOLOGICAL PERSPECTIVE

I. SOME ASPECTS OF TEMPORALITY

In ordinary living, we are absorbed in the routine of our days, focused on objects that momentarily hold our attention, on tasks that must be done, and on a host of other activities that distract, delight, and sometimes disturb us. In the midst of this busy engagement, we scarcely give a thought to time. True, calendars are ubiquitous in our lives – How many days are left before Christmas? When does spring break begin? – and we consult our watches in anticipation of an appointment with the doctor or glance at the clock to see whether it is time to take our medicine. This common attending to time, however, is largely oblivious to the myriad ways in which time permeates our experience. All of the tasks and objects and activities that form the stuff of our daily lives are soaked with time. Collectively, they form the realm of temporality; and our consciousness of them, whatever else it may be, is time-consciousness. We ourselves are not only beings in time but beings whose very fabric, mental and physical, is temporal. We compare our consciousness to a stream, and we know and feel that our bodies age. Both we, and the world we experience, are temporal through and through, though this remains, for the most part, veiled from us.

In his phenomenology of temporality, Edmund Husserl (1966, p. 276; 1991, p. 286) sought "to lift the veil a little from this world of time-consciousness, so rich in mystery, that up until now has been hidden from us." Husserl saw time and our temporal awareness as the most difficult of all phenomenological problems and perhaps the most important. He made no pretense of having solved the problem, but he is among the handful of philosophers who have cast real light on it. Husserlian phenomenology, understood broadly as reflection on experience with the aim of revealing its essential structures, does not fabricate new worlds as surrogates for our ordinary universe. Instead, it seeks to reveal what was there all along in our experience but that we had overlooked or glimpsed only obscurely. Phenomenology does this by turning toward the conscious subject and its acts of perceiving, remembering, judging, and so on, which make manifest the world and the things and events within it. The technical term Husserl uses to describe the nature of consciousness as disclosing the world, whether inner or outer, is "intentionality," which is not restricted to the ordinary sense of intention as involving some kind of purpose, but refers to consciousness in its character of always being the consciousness of something. To be conscious is to have, in the broadest possible sense of the term, an object, something of which we are aware. The phenomenology of temporality, then, explores the consciousness of time in the full panoply of its intentionality. But since temporality is present everywhere in our experience, as phenomenon and as the very foundation of our awareness, the phenomenology of time-consciousness is a special case of intentional analysis, for it not only exemplifies intentionality but makes it possible.

Phenomenology proceeds by making distinctions and revealing relationships. In the case of the phenomenology of temporality, it begins by drawing a distinction between, on the one hand, "the world time that I fix in relation to earth and sun" (Husserl, 1966, p. 124; 1991, p.128), and, on the other hand, the time and temporal objects that I experience when I perceive, remember, suffer pain, have a sense of my continuity and identity, and so on. The former is the objective time that I measure with clocks and calendars; the latter is the time of personal experience. This time is "not an objective time and not a time that can be determined objectively. This time cannot be measured; there is no clock and no other chronometer for it. Here one can only say: now, before, and further before, changing or not changing in the duration, etc." (Husserl, 1966, p. 339; 1991, p. 351). What phenomenology investigates is this "appearing time, appearing duration, as appearing" (Husserl, 1966, p. 5; 1991, p. 5), the time of my lived experience.

Two cardinal features mark appearing time: its 'spreadoutness' and its flowing character. Time is not something condensed into a single point; it stretches out. Thus we speak of an extent of time or of an expanse of time. Time is also a process of movement, of passage: a ceaseless flowing. We say that time flies, or that it creeps in its "petty pace from day to day." Even when we say that it stands still, we are describing an extraordinary phenomenon that presumes the background of time's passing: "All experiences flow away. Consciousness is a perpetual Heraclitean flux; what has just been given sinks into the abyss of the phenomenological past and then is gone forever . . ." (Husserl 1966, p. 349; 1991, p. 360). All that Husserl says about the appearing of temporal objects and about our consciousness of them depends on these two features.

The phenomenology of temporality focuses on the specifically temporal ways in which the object appears, regardless of what the object's content might be. In that sense, Husserlian phenomenology of time is formal. Its concern is with the temporal setting in which we experience life's rich and varied content, and with the basic structures of time-consciousness that make such experience possible. Within this formal setting, objects will appear to us in a variety of different ways, depending on the kind of objects they are:

from the front or side or back if they are three-dimensional things in space, as sharp or throbbing if they are pains, as loud or soft if they are sounds. Beyond such differences, however, they will have in common as temporal objects that they appear to us as enduring, as changing or not changing in the course of their duration, and as simultaneous or successive with respect to other objects, all through a consciousness in a constant state of flux. Above all, they will have in common that they display themselves in the specifically temporal modes of appearance, that is, as now, past, and future.

Like their cousins the spatial perspectives, now, past, and future are not things or parts of things. They are the temporal ways in which things appear, their "running off" modes (Husserl 1966, p. 364; 1991, p. 375). The front of a house is part of the house; one can paint it or decorate it for the holidays. The spatial perspective from which one views the front of the house in a particular perception – at an oblique angle, for example – is not part of the house; one cannot paint or decorate it. Similarly, the now in which the house appears to me is not in any sense a component of the house. It is simply my temporal perspective on the house. And unlike the spatial perspective, which may remain constant, the temporal perspective continually changes: a "new" now replaces the old now; what appeared as now comes to appear as past, or, in the case of the unchanging house, as enduring from the past into the now.

Now, past, and future are also not objective time-points, fixed forever according to clock and calendar. Indeed, they are precisely the changing ways in which such points – noon on July 4th, 2001, along with everything I experience at that moment, for example – appear to me. It is precisely because now, past, and future are not themselves things or even fixed points in objective time that they can serve as the ever-changing ways in which the whole order of temporal objectivity presents itself.

Among the temporal modes of appearance, the now enjoys a certain privilege. It is the absolute point of reference for our conscious life: only in relation to the living now do some things appear as past and others as future. While it is true that we can never stand still in time as we can in space, the now nonetheless provides a perpetual vantage point from which we can survey time's spread and time's passing. It is, in this sense, continually the same: it is in the now that we always live; it is there that we are "actual," even when our attention is directed toward past and future. Indeed, such direction would be impossible if we were not first of all centered in the now. But the now also remains, so to speak, in the thick of time's flow. What this means, however, is not that the now itself changes, but that what appears in it does. Husserl (1966, p. 318, 1991, p. 330) observes that now and past exclude one another, which means that, strictly speaking, it is a phase of the object or of its experience that becomes past, not the now itself. As a temporal mode of appearing, the now cannot become past, any more than the past can become now. Each "new now," then, is really a new phase of the temporal object that we are experiencing, and each "past now" is an elapsed phase of the object, a phase that once was but no longer is now.

The now as point of orientation in our conscious lives implies another aspect of the privilege of the now: its role as cornucopia of all that is new in our experience, the "generative point" (Husserl 1966, p. 25; 1991, p. 27) of conscious life, "constantly filled in some way." It is in the now that we experience new notes of a melody we are hearing, new aspects of a sculpture we are contemplating, a sudden pain, the news that we have a disease. Since the now is not a thing but a mode of appearance, it can play host to many experienced objects and events at once. Its hospitality is limited only by the boundaries of what we can experience. And since "the actually present now is one now . . . however many objectivities are separately constituted in it" (Husserl, 1966, p. 71; 1991, p. 74), its many contents will share a common now, making them simultaneous. Through its openness to new contents, the now also becomes the source of new temporal positions. "Each actually present now creates a new time-point because it creates . . . a new object-point" (Husserl, 1966, p. 66; 1991, p. 68). This means that the now is the original source of individuation as well. An object or event is individuated by appearing at a particular point in time, which first happens in the now. In that sense, the now is "a continuous moment of individuation" (Husserl, 1966, p. 66; 1991, p. 68). An object is the individual it is because it appears in the now and at the temporal location established in that now.

Once an object or event has appeared in the new objective time-point established in the now, it will occupy a fixed place in objective time, forever assigned to precisely that point. As such, it will stand in unalterable relations of before and after to other objects affixed to other time-points. The moment at which I learned of the death of a parent will always come after an early morning walk on the beach and before a series of phone calls making travel arrangements. Such relations of before and after represent the fixed and unchanging dimension of personal time. They are not, however, immune to time's flow. Fixed though they are, they nevertheless recede further and further into the past in relation to the actual now. Husserl (1966, p. 64; 1991, p. 67) writes, "Time is fixed, and yet time flows." The fixed time appears in the flowing time: it is "in the flow of time, in the continuous sinking down into the past [that the] nonflowing, absolutely fixed, identical, objective time becomes constituted" (Husserl, 1966, p. 64; 1991, p. 67). We live in both temporal dimensions, of course: in the relations of before and after that, for better or worse, make up the course of our lives, and in the ever changing temporal modes in which they present themselves. "Thus the matter is the same matter, the temporal position the same temporal position, only the mode of givenness has changed: it is givenness of the past" (Husserl, 1966, p.68; 1991, p. 70).

Although the now has a privileged position among the temporal modes of appearance, we never experience the now in isolation. Husserlian phenomenology resists the prejudice of the now in both of its aspects. On the one hand, in opposition to certain strains of poststructuralist thought that claim that we can never be aware of the present because the present has always already escaped into absence before we can experience it, Husserl insists that we can indeed be conscious of something as actually now. A pain that suddenly assaults me appears as now, as actually present. It is true that if it endures, it will appear as both past and now, but that is not to say that it does not originally announce its unwelcome presence in the actual now. It is simply not the case that I am first aware of something only when it has become past.

The other and more common side of the prejudice of the now holds that we can be aware only of what is now. After all, are not past and future, respectively, no longer and not yet? And would they not therefore fall outside the scope of consciousness? From the phenomenological perspective, this view is plainly contrary to experience. We are always aware of the now as centered within a horizon of past and future. Now, past, and future are inseparable in our experience, even if they are distinct as modes of appearance; to have one is necessarily to have the others. This means that the now is no more free-standing than any other temporal mode. Husserl (1966, p. 68; 1991, p. 70) writes, it "is a relative concept and refers to a 'past,' just as 'past' refers to a 'now'." If we flatter ourselves that we have reached an indivisible, unextended now-point, we have in fact abandoned experience for abstraction: "the now is precisely only an ideal limit, something abstract, which can be nothing by itself" (Husserl 1966, p. 40; 1991, p. 42). And if we think that everything we experience must be compressed into such a now, we have taken a position that is experientially absurd. "That all reality lies in the indivisible now-point, that in phenomenology everything ought to be reduced to this point – these are sheer fictions and lead to absurdities" (Husserl, 1966, p. 169; 1991, p.174). Indeed, we cannot imagine what it would be like to be aware exclusively of the punctual now, for the now is never separated in our experience from its companion modes of past and future. If the now – or better, that which is now – were by some strange twist given all by itself, it would not in fact be given as now, since it appears as now only in relation to past and future. The now-point "is not something *toto coelo* different from the not now but is continually mediated with it" (Husserl, 1966, p. 40; 1991, p. 42). We are indeed creatures of the now, but also inescapably creatures of past and future, thanks to the very fact that we live in the now. Hence we have either the whole temporal package or nothing. In that sense, to be aware at all is to be aware of more than the now; and to try to imagine that one is aware only of the now is, in effect, to try to imagine that one is aware of nothing whatsoever. Consciousness and time-consciousness are one.

What must the structure of time-consciousness be if it is to present temporal objects in the modes of present, past, and future, of now, no longer, and not yet? Given time's flowing character and its 'spreadoutness,' the consciousness of time must accomplish two things. First, each of its phases "must reach out beyond the now" and "must be overlapping" (Husserl 1966, p. 227; 1991, p. 234). If consciousness did not reach out beyond what immediately appears as now and "overlap" what is no longer and what is not yet now, there would be no awareness of extended temporal objects such as melodies and pains in their duration or succession, no awareness of their past and future phases. "Each perceptual phase has intentional reference to an extended section of the temporal object and not merely to a now-point" (Husserl, 1966, p. 232; 1991, p. 239). But reaching out beyond the now is not by itself sufficient to account for time-consciousness. The second thing that consciousness must accomplish if we are to be aware of temporal objects is modification. Reaching out beyond and overlapping lets us preserve phases that have elapsed and look forward to phases yet to come. However, if the elapsed notes of a piece of music to which I am listening were preserved in consciousness without modification, I would hear an instantaneous burst of sound, not a melodic succession. There would be no sense of the music as spread out and flowing in time, which is to say that one would not hear the music at all. What consciousness preserves when it reaches out beyond the now to what has elapsed, it preserves in the modification of the past; and when it reaches out beyond the now in the other direction, it is aware of phases yet to come in the mode of the future.

This awareness of temporal objects in their modes of now, past, and future comes about through a threefold intentionality that forms the structure of each phase of time-consciousness. The first element of this structure, which Husserl calls "primal impression," is the original consciousness of the now. Primal impression is the intentional moment of the phase of consciousness that presents a new phase of the object as now and a new time-point as now. It is "the absolutely original consciousness" in which, for example, an actual pain "stands before us 'in person,' as present itself, as now" (Husserl, 1966, p. 325; 1991, p. 338). Since it is the original consciousness of the now, "the primal impression is something absolutely unmodified, the primal source of all further consciousness and being" (Husserl, 1966, p. 68; 1991, p. 70). Just as the now has a privileged role among the temporal modes, the primal impression as original consciousness of the now has a privileged status as well. On the other hand, just as the now cannot stand alone but is always attended by the past and future, so the primal impression is a dependent, non-selfsufficient moment of the actual phase of intending consciousness, not the whole phase and not an independent act (Husserl, 1966, p. 326, n.1; 1991, p. 338, n.60). It always occurs in company with the two other intentional moments essential to our awareness of time: retention and protention.

What Husserl calls "retention," and occasionally "fresh" or "primary" memory, is the impressional consciousness of the just past phases of the temporal object. Retention has "the character of a perception of the 'just having been.'" (Husserl, 1966, p. 191, n.2; 1991, p.197, n.6). It is the "'direct' consciousness . . . of the-having-been-present (more precisely, of the just-having-been-present: the 'just' pointing to a certain intuitive temporal position)" (Husserl, 1966, p. 191; 1991, pp.197-198). Perception or impression is the act in which all origin lies, and so retention, as the impressional consciousness of the past, is the moment of consciousness in which we first become aware of the past. It is important to note that retention presents rather than re-presents what is past. "Secondary" or ordinary memory, in which, for example, I recall the removal of a skin cancer two years ago, presumes that the past has already been originally constituted in primary memory. Secondary memory then re-presents the past event in the sense that it enables me to live through an entire experience or a substantial part of it once again. It re-presents a now that is no longer actual and moments of the past that are no longer actually retained. It is as if I were perceiving it again, only in modified way: as "quasi-present" (Husserl, 1966, p. 290; 1991, p. 301), "as if seen through a veil" (Husserl, 1966, p. 48; 1991, p. 50). As such, ordinary memory does not perceive, or give, the remembered now, past, and future impressionally. "Only in primary memory do we see what is past, only in it does the past become constituted – and constituted presentatively, not re-presentatively" (Husserl, 1966, p. 41; 1991, p. 43). Of course, as a mode of modifying time-consciousness, primary memory does not preserve what is past as now; rather, it presents the past as past. It is the immediate consciousness of "just having been" (Husserl, 1966, p. 165; 1991, p.169) in which what has elapsed is "still present . . .only as just past" (Husserl, 1966, p. 212; 1991, p. 219), and in which I hold on to what is just past as it recedes into oblivion.

The final moment in the triple intentionality of original time-consciousness is "an intention directed toward what is to come" (Husserl 1966, p. 232; 1991, p. 240). Husserl calls this awareness or openness to the immediate future "protention." Both primal impression and retention are "filled" modes of intentionality. The pain of which I am aware as now is actually there and present. The pain retained as just past is equally definite; it has actually occurred and is now being retained as it elapses. Protention, by contrast, is empty. True, I will normally "expect" that what I am now experiencing will continue to unfold for me, that, for example, the pain I now feel will go on. But the future phases of the pain, of which I am conscious in protention, have in fact not yet arrived, indeed, might never arrive (the pain might suddenly cease). In that sense, "the only thing determined [about protention] is that something or other will come" (Husserl, 1966, p. 106; 1991, p. 111). It is, however, precisely in this unfailing consciousness of something more to come attending every moment of our conscious lives that our sense of the future is first constituted. It is this primitive awareness of the future that makes possible the full-blown act of expectation, the sibling of secondary memory, pointed toward the future rather than the past. In expectation, I explicitly anticipate a particular event, running through it as if I were perceiving it, filling out my empty intentions with imaginative constructions of what will occur.

Thanks to primal impression, retention, and protention, each phase of my flowing conscious life is aware of what is immediately now, past, and future. Together they constitute my first and primitive sense of the fundamental temporal modes. Retention and protention, however, do not furnish our only access to past and future. As we indicated above, ordinary or secondary memory and expectation are also ways of being conscious of past and future; indeed, they are unique and vital ways. Retention, for example, allows no freedom: it is the immediate consciousness of what is just past as it elapses and fades into emptiness. One might say that it does its work "automatically." Protention is similarly restricted: it is the immediate consciousness of what is just about to occur. Secondary memory and expectation do involve freedom, however, and this is true in at least two senses. Through them I can range freely across the broad landscape of my life, in memory returning to and reliving portions of my more distant past and in expectation anticipating, perhaps in great detail, what is to come. It is true that we would lack even a rudimentary sense of past and future without retention and protention. But it is equally true that if the only intentional arrows time-consciousness had in its quiver were retention and protention, our sense of the temporal – and hence of the world and of ourselves – would be narrow and poor, constricted to the actual now and to meager stretches of the immediate past and future. That this is not the case we owe to secondary memory and expectation. Together they open up the broad temporal sweep of our lives.

A final achievement of time-consciousness is the sense we have our own continuity. Time-consciousness is not only the awareness of enduring temporal objects but also self-consciousness, the consciousness of the continuity of the self in the face of time's relentless flow. Just as we experience "pain as a continuum" (Toombs, 1990, p. 232), we experience ourselves as a continuum. There is undeniably, of course, an episodic aspect to life: we are absorbed in a book and suddenly spill our coffee; we conclude an afternoon's work, suffer from a headache, have dinner, and then go to a concert. Still, our lives are not strings of episodes without connection or ground. The various experiences through which we live are all moments of an abiding and continuous sense of self, which is precisely what time-consciousness as self-consciousness makes possible. All of these fleeting experiences are implicitly experienced as belonging to the same consciousness, to me. This is true of my present experiences, but also of my remembered past experiences, of which I am aware as belonging to the same consciousness that is now doing the remembering. Particular experiences, this pain I am now living through, for example, will eventually cease; but as long as we experience anything at all, we will be aware of our continuing selves.

Phenomenology thus avoids a Humean atomism that would pulverize our lives into disconnected experiential bits, but equally a Cartesian substantialism that would freeze the dynamic flow of conscious life. "There is one, unique flow of consciousness," Husserl (1966, p. 80; 1991, p. 84) observes, "in which both the unity of [a] tone in immanent time and the unity of the flow of consciousness itself become constituted at once." The two go hand in hand: In experiencing temporal objects, we always also experience ourselves; we experience ourselves as continuing and identical beings even as we are aware of our lives as in constant flux. The threefold intentionality of impression, retention, and protention, together with memory and expectation, binds our lives together. These moments of time-consciousness are precisely the self's way of becoming aware of itself as continuing and abiding in the midst of the myriad changes that it undergoes in time. From the phenomenological perspective, each of us is always, and at once, one and many, abiding and changing.

II. REFLECTIONS ON TEMPORALITY AND ILLNESS

The preceding account will no doubt confirm in the reader's mind just how formal the phenomenology of temporality is. This should come as no surprise, since, as we have seen, phenomenology seeks to disclose the fundamental temporal framework in which we experience the contents of our lives, whatever they might be. Hence that everything we perceive appears to us in the temporal modes of now, past, and future would be part of temporality's formal framework. Two things, however, must be kept in mind when thinking about such universal features of temporality. First, however invariable and universal they may be, they are neither identical to nor dependent upon the clock time determined by the rotation of the earth or its movement around the sun. Clock time has its own formal structure and, thanks to that structure, can be measured with great precision by sophisticated instruments. It plays an indispensable role in a world dominated by technology. Precise measurement of temporal increments must be possible, for example, if aircraft are to maintain safe separation from another in the skies and if patients are not to be burned by excessively prolonged radiation. Still, the formal structures of clock time differ from the forms of experienced temporality.

This brings us to the second point. While the formal structures of lived time are necessarily and universally present throughout our experience, they are also flexible. What we call a "second" in clock time is rigid; it cannot be expanded or contracted. Indeed, if it could, records in swimming or running, for example, would be meaningless. The "now" of lived time, on the other hand, while certainly a form, is not a fixed standard by which we measure durations precisely. The now, as we have seen, is a form in the sense of being a manner or mode in which temporal objects appear. This means that while it makes sense to say that a swimmer broke the record in the 200-meter butterfly by an "astounding two seconds," it does not make sense to say that he or she broke it by an "astounding two nows." As such, the now is an elastic form, and so too are the experienced past and future. Depending on what content "fills" them, they can expand or contract.

These flexible temporal forms belong to all of our experience, whether ordinary or extraordinary. The extraordinary, however, bring home their flexibility most vividly. Nowhere is this more evident than in the experience of serious illness. "When life-threatening disease looms over us, everything changes" (Murray, 2000, p. 59).

Patients often report that when they first received the extraordinary news that they had a life-threatening or debilitating illness, their awareness of time altered radically. In recounting the moment when she learned that she might have cancer, Kate Scannell, a physician, describes an enormously complex experience. She reports that the wait on the phone for word from the pathologist seemed endless. "How long does 10:20 last?" she asked (Scannell, 2000, p. 55). No statement could capture the difference between subjective, lived time and clock time more starkly. From the point of view of the clock, 10:20 will last exactly 60 seconds and then give way to 10:21. In the face of an extraordinary situation, however, the lived now can suddenly dilate to the point at which it dominates one's experience, becoming an oppressive presence, forcing past and future to contract to almost nothing. The hospitable 'now' has welcomed into experience something new, as it always does, but what is new turns out to be devastating news. It pins one to a 'now' that will not give way. Scannell (2000, p. 55) recalls that she felt "melded to the moment" and that every fiber of her being "resisted forward time." Similarly, T. Jock Murray (2000, p. 58) writes of one of his patients suffering from multiple sclerosis that "life seemed to stop for her when she had her first attack and learned the diagnosis." Kay Toombs (1990, p. 237) observes that time seems to "'stand still'" in such circumstances, with past and present coalescing "into a stagnating present."

There are a number of things that might easily be overlooked with respect to this phenomenon of the "expanded" now. The most important of these is that the form of temporality, particularly the fundamental structure of now, past, and future in their mutual interdependence, is not abrogated when the now expands. Indeed, the now seems to crowd out past and future precisely because they retain such an intense presence in my field of awareness. "If this news is true," I say to myself, "my past is now past in an altogether new sense." What has come before in my life stands in stark contrast to the new reality that has suddenly overwhelmed my being in the now. I am at once disoriented and reoriented. I know that the past is my past, but it now seems cut adrift, no longer the phase of my life coherently preceding my present and forming the ground of my harmonious passage into the future. "If this is true," I think, "then the future for which I have lived and hoped and planned is suspended, perhaps even closed off forever." I find myself "stranded" (Scannell, 2000, p. 57) in the present because I know that the news I have received has irrevocably changed my life, and hence my past and future. That realization painfully and violently stretches the now, compresses the past, and shortens the future (Murray, 2000, p. 59), but it does so because time's spread remains intact for me. Experienced temporality can be so dramatically distorted precisely because the fundamental temporal forms – here the inseparability of now, past, and future – abide as the setting for my experience. Furthermore, if I experience time as having stopped or as standing still in such situations, this is possible only against the background of time as flowing – as flowing around me, so to speak, while I am frozen in place. Time's spread and time's flow remain the constant frame for my temporal experiences, no matter how odd and exceptional the shapes they assume.

The revelation that one has a serious illness, therefore, can strand one in the now precisely because of its significance for one's past and future. But this is also what makes possible the reverse side of entrapment in the now:

movement back and forth in time. Scannell (2000, p. 55) reports that the news of her diagnosis unmoored her from objective time and caused her to drift "chaotically through future, past and present." Objective time gives us no choice but to embrace its rigid course and to pass through its phases. It sweeps us along inexorably, and there is nothing we can do about it: We are born and move from childhood through youth to middle age, old age, and death. We can imagine it otherwise, but it is not and cannot be otherwise. And yet, thanks to memory and expectation, we can also move about in time; time ceases to be a straight line and becomes the landscape of our personal lives (Murray, 2000, p. 58). In real time, we are bound; in time-consciousness, we can be free. When one learns that one has a life-threatening illness, one abruptly confronts, personally and in one's own life, objective time's inevitable course, which usually remains in the background of our experience. But at the same time this triggers in the immediate present that chaotic movement between present, past, and future that Scannell describes. For in the landscape of our personal lives, the past is not, so to speak, just something left far behind us on the tracks, nor is the future simply something distant and remote lying up ahead. Past and future are with us now as the present horizons of our being. It is this that makes the news so disturbing. The freedom to move about the temporal landscape of our lives in memory and expectation is brought into play by the revelation and intensifies its impact. There is a collision between the two times in which we live.

We have seen that the awareness of events in our lives as before and after is one of the formal characteristics of temporal experience and a fundamental aspect of the sense a person has of his or her continuity. Most of the events that become arranged on the "time line" of one's life are routine. Some, however, mark major changes: marriage, the birth of a child, the death of a parent, retirement. After such events, things are not what they were before. This is particularly true in the case of a serious or life-threatening illness. Scannell (2000, p. 56) writes that when she learned that she might have cancer, "I felt my life organizing into a sharply divided 'before' and 'after,' with cancer cleaving the middle." The experience of before and after presumes the temporal continuity of our lives, but such cataclysmic news, portending not merely a change in life but intimating its closure, may signal the end of one self and the beginning of another, thus challenging the very sense of continuity that time-consciousness constitutes. When she heard her diagnosis, Scannell (2000, p. 56) noted that her "sense of self as continuous through time began to fracture," and she "desperately craved the continuity of [her] past and future self."

In the presence of a life-threatening illness patients thus realize that everything has changed and that they now have a new life with "a different view of the future" (Murray, 2000, p. 58). But just as the dilated now presumes past and future, the shattering of one's self in illness, as real and painful as it may be, still depends on the deep and abiding sense of the self's continuity established in time-consciousness. It is always against the background of this continuous sense of self that all of our experiences announce themselves in the hospitable now, even the most disturbing. No matter how disruptive the latter may be, they cannot finally destroy the sense of self. What a truly disruptive experience *can* do, as we have seen, is transform our way of experiencing what is now, past, and future. The content that has burst into the now may swell and overwhelm one, but the flow of one's inner time has not ceased, nor has one's awareness of time. In fact, one is overwhelmed precisely because of what the devastating news means for *one's own* future. It is openness to the future of the same continuing self that is now jeopardized. If that were not the case, the unwelcome revelation would mean nothing. As it is, it interrupts the ordinary temporal flow of one's conscious life precisely because it focuses one so intensely on what is now, doing so, as we have seen, against the background of one's past and future. Its disruptive force ultimately depends on an abiding sense of the self's continuity, no matter what happens in the "content" of one's life. Furthermore, it is thanks to this deeper and always present sense of continuity that one has the chance, over time, to restore a shattered life and to come to terms with one's situation. In one sense, a patient's life may change radically; in another sense, it is still *her* life. This deep, underlying continuity of the self serves as the ground for that temporary sense of fracturing that may overwhelm one in certain extraordinary circumstances. But once the initial shock of the revelation has subsided, it also enables the patient to achieve a less frenzied and more profound and enhanced grasp of the temporal continuity of life.

When patients become ill, they may begin to see time running backward as well as forward, with life spread out as on a landscape. They may think of where they have been and where they come from, of their families and relationships, of what came before and what comes after. They think of their grandparents as well as their grandchildren (Murray, 2000, p. 62).

The phenomena that emerge so vividly in serious illness – dilating nows and fractured selves, the conception of time as landscape rather than as beads on a string – are concrete ways in which the differences between subjective, private, inner time and objective, public, clock time manifest themselves. These differences, however, do not mean that public time and private time are altogether disconnected in our experience. In fact, even to speak of "the dissociation of chronological time and personal time" (Murray, 2000, p. 61) presumes that the two regularly play off against each other. As it is, Stephen Kern

(2000, p. 4) observes, "It is not possible for a human being to experience private time in complete isolation from public time." It is certainly true, as Toombs (1990, p. 238) writes, that a present illness "represents not so much an isolated instant along a time-line as it does a present-now which must be considered against the horizon of past and future," but that personal temporal horizon is often also involved with the horizon of clock and calendar time. A grandparent may hope to be well enough by a certain date to attend a grandchild's wedding in another city, a situation in which personal time with its memories and expectations intersects poignantly with the objective time of the calendar. Or if a pain seems to go on endlessly while "really" lasting no more than an hour, that "hour could not in fact seem slow if we did not in some way know that it was *an* hour (the 60-minute kind) that went slowly" (Kern, 2000, p. 4). The interplay of clock time and lived time are essential to such experiences.

Yet another sense in which objective time and subjective time relate to one another involves what might be called "embedded" time. This is the species of objective time that shows itself in mountain ranges, in buildings new and old, and, above all, in our bodies as we age or become infirm. It is time engraved in stone or flesh, and in our abilities to move, breathe, hear, see, speak, and present ourselves to others. This is the time that make-up artists and actors exploit in their crafts, which they can do because we constantly live with embedded time, experiencing it in ourselves and in everyone we meet. We encounter it in the aged faces of old acquaintances we have not seen in years. We confront it in a parent who still seems vigorous but, we suddenly realize, no longer has the stamina to tour three museums in a day. And I may find that my personal sense of time, which ordinarily I do not separate from the time embedded in my own body, collides with it. I may think that I can walk at the same speed and cover the same distance as swiftly as I did twenty years ago, but be brought up short by my slow pace and the surprisingly long time it takes to cover a familiar route. Or I may think that my hearing is as sharp as it ever was, but gradually realize that I am having difficulty hearing students in discussion groups. Or I may just assume that my aging face seems as youthful to others as it does to me, but notice that I am being treated by younger people with the respect and deference given to those who are seen as much older than they are. As these experiences reveal, embedded time is unique in that it interacts with both calendar time and subjective time. It shows how thoroughly our lives are saturated by time of every sort.

That inner time and objective, clock time play off against each other does not mean, of course, that the distinction between the two collapses. They remain, in Toombs' (1990, p. 203) observation, incommensurable. In inner time, one experiences temporal objects "as temporal wholes which span present, past, and future"; in objective time, one breaks the object down into a series of "isolated, discrete, now-points along a given time-line." The points in objective time can be measured by the clock; the experience in inner time cannot. Thus the now of the pain I live through may appear to me to be endless, while the procedure inflicting it may by the clock take only 45 seconds. Toombs (1990, p. 234-35) argues this incommensurability entails that the subjective dimension of time is "incommunicable." On the other hand, we possess a common language for clock time precisely because clock time is objective and public. A military commander can instruct his officers to synchronize their watches to a specific hour and minute and be confident that his meaning will be understood. A patient and a physician's receptionist both know what it means to set an appointment for 2:00 PM on Friday, March 30. On the other hand, it would not occur to the military commander to ask his men to synchronize their subjectively felt, temporally extended fears and anxieties as the time of battle approaches. And while the patient might well report being in pain to the receptionist, any attempt to make the lived pain transparent to the person on the other end of the telephone would be futile. For pain, unlike the objective time of clocks and calendars, is private and unsharable. "Physical pain 'resists' language and is inexpressible. There are no words to describe adequately one's experience of pain to another" (Toombs, 1990, p. 235).

Now it is certainly true that I alone can directly experience my pain and its subjective duration. No one else can have my pain or my emotions as I have them, that is, as first-person experiences. Inner time, the correlate of inner time-consciousness, does not seem to be communicable, and yet as J. B. Priestley (1964, quoted in Murray, 2000, p. 58) observes, "on a deeper level, perhaps we begin to share it again, in ways which we cannot fully understand."

Priestley may mean that we can have an intimation of the inner temporal experience of others because, at the deepest level of our being, we all share the same formal structures of time and time-consciousness and, at least in a general sense, the same content as well. Whatever we are conscious of, we are all conscious of it as occurring in the three fundamental temporal modes; and as far as content is concerned, we have all had experiences in which a pain or an emotion has flooded the now of our temporal field, threatening the banks of past and future. The grounds of a possible communication are there formally and materially, then. But how could this possibility be realized? Art affords some of the best examples, for art might be said to parallel and perhaps to express the inner, subjective world. I may not be able to share my private world of imagination with anyone else any more than I can share my pain, but my imagination – or at least the artist's imagination – is capable of creating a work of art, public and intersubjectively available, that embodies a glimpse into subjective life. Sometimes whole bodies of work seem to do this, as in the case Edvard Munch's paintings. Or "Chariots of Fire" achieves it cinematically in the case of competitive runners. Running affords interesting examples of temporal experience. Clock time plays an obvious role: One can measure a sprinter's time in the 100-meter dash, and everyone will know what the published time means. The runner, however, will have a different experience of time, a time that is private or subjective and in which pain and emotion find their place. Through the use of slow motion, "Chariots of Fire" manages to convey a sense of this personal time as dramatically slowed or arrested and the emotions surrounding it. In one scene, for example, we view a race from a distance. The runner who is the protagonist suddenly stumbles and falls; the camera moves in on the tumbling figure. The noise of the crowd, which presumably still fills the stadium, is suddenly suppressed, a signal that we have entered the experience of the runner. For what seems like an agonizingly long moment, expressed in slow motion, he struggles to get back on his feet. He succeeds, and with a look of determination begins to run again, now no longer in slow motion, but still with the sound of the crowd muted. In all of these respects, one could say that the images on the screen, which distort objective time and the objective situation, manage to render available to others something of the runner's pain and courage as he himself experienced it. In another episode in the film, a runner who desperately desired to win is defeated. His loss is captured in slow motion saturated by a grinding crash of sound, and then repeated over and over again – on the screen, of course, for all to see, but really in the runner's mind as he compulsively relives his defeat, unable to let it go. Objectively, the race was run once in a few seconds of real time, but it echoed again and again in far slower time in the athlete's inner time. This suggests that while each of us can experience such temporal variations in the first person within ourselves, we can also glimpse them publicly, as depicted in the visual images of painting and films.

The visual arts, however, are not the only ways to give some access to subjective temporal experience. Poetry may achieve the same manifestation. In her poem "Otherwise," for example, Jane Kenyon (1996, p. 214) describes in a series of lapidary images the temporal routine of her day: getting out of bed, eating breakfast ("cereal, sweet / milk, ripe flawless / peach"), her work in the morning, a nap with her mate at noon, and so on. After the lines devoted to each of these moments of the day save the last, she writes, "It might have been otherwise." However, after the description of the final event of her day ('I slept in a bed / in a room with paintings / on the walls, and /

planned another day / just like this day"), come the lines: "But one day, I know / It will be otherwise." This poem is about time on many levels: certainly on the level of life's routine, and specifically about what it *feels* like to experience that routine, but also on the level of the existential realization of the contingent character of what we do and what we count on, and finally of the realization of our mortality. "Otherwise" was written when Kenyon was in good health. A few years later, in 1994, she learned that she had a virulent form of leukemia. She died fifteen months later at the age of 47. One of the few poems she wrote during her illness was "The Sick Wife," in which she expresses her sense of lost freedom and vitality against the background of the seeming freedom of everything that she simultaneously experiences around her – children, retired couples, dry cleaning swinging on hangers – as she waits in the car for her husband in a shopping center parking lot: "How easily they moved / with such freedom, / even the old and relatively infirm." The last lines of the poem formulate and reveal her inner state: "The cars on either side of her / pulled away so briskly / that it made her sick at heart" (Kenyon, 1996, p. 221). Simultaneity is a temporal form we all share, but how we take what appears in it can differ greatly depending on whether we are well or ill. Simultaneity has the capacity to bring individuals together in a common experience, but the illness of the sick housewife serves to isolate her painfully from those with whom she shares the same time. She knows that they have the benefit of the ordinary and routine that she once had but that might have been otherwise then and that now is otherwise. They have the freedom that freedom from her illness would restore.

This essay has attempted to explore some of the ways in which temporality pervades our experience, and particularly our experience of illness. Indeed, the phenomenon of illness provides a unique window on to the many levels of time that inform our lives, from the superficial to the profound, from the experience of the dilating now in the presence of pain to our sense of threatened continuity when illness looms. It also suggests the resources in our temporal being that enable us to live with illness and come to view our lives, beyond the ephemeral time of our quotidian routine, as integral moments in the passage of generative time, a totality in which we see our lives whole and ourselves and others as members of one generation within many generations.[1] In that experience, personal temporality merges with the temporality shared by us all.

Georgetown University
Washington, D.C.,
U.S.A.

NOTES

[1] For a discussion of generative time, see K.Held.: 2000, 'Generative Experience of Time', in J.B. Brough and L.Embree (eds.), *The Many Faces of Time*, Kluwer Academic Publishers, Dordrecht, The Netherlands.

BIBLIOGRAPHY

Husserl, E.: 1996, 'Zur phänomenologie des inneren zeitbewusstseins (1893-1917)', R. Boehm (ed.), *Husserliana X*. Martinus Nijhoff, The Hague.
Husserl, E.: 1991, *On the Phenomenology of the Consciousness of Internal Time (1893-1917)*, J. Brough (trans.), Kluwer Academic Publishers, Dordrecht, The Netherlands.
Kenyon, J.: 1996, *Otherwise: New and Selected Poems*, Graywolf Press, St. Paul, Minnesota.
Kern S.: 'Time and Medicine', *Annals of Internal Medicine* **132**, 3-9.
Murray, T.J.: 2000, 'Personal Time: The patient's experience', *Annals of Internal Medicine* **132**, 59.
Priestley, J.B.: 1964, *Men and Time*, Aldus Books, London.
Scannell, K.: 'Leave of absence', *Annals of Internal Medicine* **132**, 55-57.
Toombs, S.K.: 1990, 'The temporality of illness: Four levels of experience', *Theoretical Medicine* **11**, 227-241.

PATRICK A. HEELAN

THE LIFEWORLD AND SCIENTIFIC INTERPRETATION

I. THE 'RECEIVED VIEW' OF MODERN SCIENCE

Though some would still say that it was under the spiritual aegis of Plato, current historical scholars prefer to hold that it was in fact under the aegis of a newly discovered mathematical Aristotle that something like the hypothetical-deductive account of modern science emerged in the 17th century and eventually became the 'received doctrine' or 'view' inherited by most professional scientific researchers today.[1] The hypothetical-deductive method focuses on the permanence of the categorial inventory of things, the objectivity of mathematical ideas, and the necessity of the laws of Nature. Its privileging of theory is a legacy of the theological notion common to two millennia of secular and religious thought that the order in Nature comes from God (the ordering Demiurge or Creator) whose ideas it expresses.

Many aspects, however, of Nature as we now experience it are left out of this account. Absent are historicity, contingency, contextuality, emergence, the role of religion and human culture, and most particularly, the embodiment of mind in the everyday life-world.[2] These aspects pervade those new channels of research that are centered on evolution, development, history, anthropology, technology, culture, and especially medicine. We find them also in retrospect in anomalies associated with the old classical sciences. These new features refer to dynamic processes in Nature that change not just the states of a given thing but the things themselves, through evolutionary emergent processes, or they change things contingently but irreversibly as in quantum theory, or non-linearly (following small destabilizing changes) as in the 'butterfly effect.' All of these effects are found in medical practice. In addition, reflective medical practice has to have some approach to the Mind/Body and Mind/Brain problem since the dualism underlying the 'received view' can be the source of both blindness in clinical practice and systematic failure of the imagination in clinical medical research.

Human subjects and research communities are not apart from Nature, they also belong to Nature, for (despite the approach of the 'received view') there is no gap between Nature and Culture. Nature is the human ecological niche in the Cosmos, and Culture is the location of all that we know about Nature, including science. Through new technologies in the cultural life-world, researchers direct, shape, produce, and give meanings to new scientific phenomena and change the meanings of old phenomena. By these means the cultural life-world is changed with significant consequences of progress or decline both for individuals and for communities. It is in science's relation to human culture that the traditions of phenomenology[3] and Heideggerian hermeneutics[4] can make a contribution – indeed, I would say, a correction – to the understanding of how modern science and its theories and technologies serve medical science and clinical practice.

Scientific medical research begins, is carried through, and eventually ends in the everyday life-world with the production of resources for medical practice, such as pharmacological products and imaging devices. It is there that its nature is to be researched, and there that its long term influence is to be weighed. Since the life-world is where all inquiry begins and ends, philosophy also begins and ends there, and it is in relation to the contemporary life-world that philosophy is itself a resource. In contrast, the 'received view' begins in the life-world and ends in a meaningful construction about the life-world that takes the form of an ideal representational model of Nature. The gap between the life-world and the scientific model of Nature is bridged by a postulate, let me call it the 'mirroring postulate,' one of the same kind that is commonly thought to link geometry and the life-world. Just as geometrical objects float, as it were, off the page or blackboard and take their place in the ideal realm of the Mind, so too do scientific models or theories. The 'received view' as a philosophy is, then, no more than a hypothetical-deductive theory like scientific theories and 'invented' likewise on the basis of a postulate, as Brisson and Meyerstein (1995) so cleverly show in their comparison between the Big Bang Theory and Plato's *Timaeus*. This, of course, does not lessen the value of a theory as the resource it has proved to be, but it limits the validity of philosophical claims often made for a theory.

I take a philosophical account to be universal and fundamental in its scope in the study of knowing, being, truth, goodness, etc. Clearly the 'received view' satisfies neither the universality nor the fundamentality criterion since the basic 'mirroring' postulate is not self-justifying. For the reasons given above, it is then important to be clear both why philosophy begins and why it ends in an understanding of the life-world. Phenomenology is a philosophical orientation that gives privilege to what is given *('die Sache selbst')* in the life-world (Heidegger, 1996, pp. 27-29 in the Niemeyer German edition). Moreover, since the life-world is the locus of the ups and downs of individual human lives as well as of core historical development, (a certain kind of) hermeneutics is necessary to close the gap between life and understanding (see, e.g., Husserl, 1970, 1999; Kockelmans, 1993; Heidegger, 1996, pp. 27-39), theory and practice (Heelan, 1997, 1998; Heelan and Schulkin, 1998), being and language (see, e.g., Halliday and Martin, 1983, pp. 3-21; Bakhtin, 1986, pp. 103-131; Heelan, 1991b, 1994), history and change (see, e.g., Heelan, 1991a, Husserl, 1970; Ricoeur, 1981, pp. 43-62 and passim.), and medical science and clinical practice (see, e.g., Duden, 1993; Leder,

1990; Miller, 1978; Sheets-Johnstone, 1990; Young 1997; Zaner, 1981).

The principal authors followed here are Husserl, Heidegger, Gadamer, Merleau-Ponty and the philosophical orientation they brought. Since the strength of this orientation is in the human sciences and few writers of this orientation have been close to the grass roots problems of the natural sciences, this essay will have to go beyond what is presently found in the basic phenomenological corpus, to develop new themes about the way theory is related to praxis and how medical science is related to clinical practice. Husserl (1970) and Heidegger (1996, 1997a) criticized the ethos of modern science for a certain cultural bias it tends to encourage. This bias is in the way it privileges explanatory theories ('ideas' or 'mental objects') over life-world processes ('phenomena'), with the consequence that the cultural hegemony of (what Aristotle called) 'calculative thinking' is promoted over the more foundational 'meditative thinking.' Calculative thinking is ordained toward the management and control of things and people through rigid frameworks of organized thought. Meditative thinking has as its basic concern cultural meaning and meaning-change, and their consequences in terms of cultural progress or decline (Heidegger, 1968, pp. 3-35; 1966, p. 46, and Husserl, 1970).

No minimizing is intended, however, of the great benefits that can and do flow from the 'calculative thinking' of scientific inquiry. I aim to address in particular the role of modern science in an important sector of the life-world, the world of medical practice, for the purpose of correcting some of the biases that inevitably follow from confusing the 'received view' of science with a more general, I mean, philosophical, viewpoint.

A. Lifeworld and Being

We come now to the difficult task of specifying what our philosophical terms mean. Just as 'theory' in the 'received doctrine' means 'an ideal representational and computational model that virtually replaces the sensible phenomenon' *for the purposes of science*, so the meanings of philosophical terms are derived *from the practical context of philosophical inquiry* (see, e.g., Husserl, 1970, pp. 121-147; Heelan, 1988). In phenomenology, the Lifeworld (I capitalize the term) is the *practical and pre-theoretical* 'field' ('space' or 'domain') of and for human understanding, including philosophical understanding. One may enter into the understanding of the Lifeworld in a number of ways: (1) by a special reflection on the everyday life-world, (2) by a critique of Mind/Body or Mind/Brain Dualism presupposed by the 'received view,' (3) by linguistics and a critique of language, and (4) by studying the essential embodiment of the human mind in the brain, senses, active physical bodily powers, and in technologies, Nature, and the lifeworld as a whole.

(1) The philosophical notion of the Lifeworld can be derived by a special kind of reflection from the everyday life-world by dropping all abstract theoretical or explanatory thinking in the sense of the 'received view.' Any affirmation that *abstract entities, theories, models, categorial list of contents by abstract kinds exist is excluded except as tools and resources for practical activity.* By a species of reflection of this kind one discovers one's Lifeworld. In it one finds oneself, as subject, (in Heidegger's words) as 'having-been-thrown' into this Lifeworld with others, 'located' at some place and time in human history not by one's choice, and conscious of having no more than a finite lifespan (Heidegger, 1996, pp. 63-114, 236-263; also Natanson, 1970, p. 103; Husserl, 1970; Ihde, 1983). Each subject inherits among other things a practical language, a culture, a community, a set of cares – perhaps, more than one of each – that give meaning, structure, and purpose to the Lifeworld one shares with one or more communities. Although the Lifeworld is not of the subject's own creation or choice, it nevertheless permeates the subject's life experience at conscious and unconscious levels.

The Lifeworld is an attempt to *show, to point out,* the historical river of individual human existence. What is thus revealed is the *ontological* dimension of human experience, prior to (because a grounding condition of) the *epistemological*. This ontological dimension is the human being-in-the-Lifeworld, or being-in-the-world, for short. In it, the individual human is *there-in-Being* – or *Da-sein* to use Heidegger's (1996, pp. 15-17) term.

(2) The notion of Lifeworld can also be reached through a critique of the Cartesian Mind/Body or cognitive science's Mind/Brain model that underlies the 'received view.' This takes the Mind to be the home of all clear and distinct logical thinking and of the personal Ego, it is distinct and separate from the Body and Brain where the imagination lies that is the home of sensory images, perception, and emotions. This kind of dualism was attacked by Merleau-Ponty (1971) and others who applied this critique to the field of medical practice, such as, to mention a few, Duden (1993), Leder (1990), Miller (1978), Sheets-Johnstone (1990), Young (1997), Zaner (1981). The human-mind-as-embodied pervades not just the brain and body but also the Lifeworld (see Varela, Thompson, and Rosch, 1993; and Heelan, 1983a/1988).

(3) The same conclusion has been reached through the functional linguistics, say, of Halliday and Martin (1993) and by Lakoff and Johnson (1999) who concluded that all discourse using abstract thoughts and scientific theories depends in metaphorical ways on the structures of the bodily senses and on converting action verbs denoting processes, into nouns denoting 'things,' e.g., 'X moves' into 'the movement of X.' They claim there are no Cartesian 'clear and distinct ideas,' and we do not have disembodied ways of thinking about disembodied ideas. All

such thinking uses virtual bodily activities as a kind of lexico-grammatical metaphor for expressing and structuring abstract thought and abstract thought by the reverse metaphor derives its meaning from human systematic actions in the Lifeworld.

(4) The notion of the Lifeworld can also be reached by studying the way the human mind is essentially embodied. Its embodiment is the living organism we call our body. But what is 'the body in question'? There are a variety of answers: the mind is embodied exclusively in the brain, possibly also in the muscular tissues, or (the answer I prefer) the embodiment extends also into the environment that sustains life and provides the meaningful physical space in which we live our conscious lives in society (see, e.g., Merleau-Ponty, 1962; and Heelan, 1983a/1988, 1997, 1998). This critique of dualism often takes the form of a critique of the 'medical body,' that is, of the body conceived as a material object constituted of separable physical parts that collaborate among themselves and with the outside environment in more or less independent ways to maintain the operational physical integrity of the body as a functional whole. In this view what is missing are the dynamic processes and physical exchanges that mutually transform the functional ontology of the bodily parts, as well as the engagement of the whole body in the Lifeworld through conscious and unconscious functional exchanges (see, e.g., Duden, 1993; Leder, 1990; Sheets-Johnstone. 1990; Young, 1997; Zaner, 1981).

B. Interpretation and Meaning

Human understanding functions by *interpretation* and the product of interpretation is *meaning*.[5] Meaning is nothing physical. It is not a text, a behavior, a neural network, a computation, not even a sign or a medium, nor any relationship among things, though all of these may be generated by and are productive of meaning. It is not a private 'domain' accessible only by some kind of introspection. But it is a public 'domain' where people share the products of human understanding first by common habits of action (in which diverse networks are recognized) and then by the use of language and language-like media. Meaning is the public domain in which people understand one another, argue with one another, give reasons, establish goals, set up norms, define kinds, etc. – more or less effectively according to the purpose, intelligence, language skills, and education of the parties involved. Meaning, moreover, is historical because language is constitutive of history; it is also deeply affected by human temporality and historical forgetfulness because the community/Lifeworld milieu in which it is transmitted has gains and losses over time. Meaning is local and social, because it is the product of active local interests and social communities and constitutive of their interests. It is then neither once-and-for-all fixed, nor ever in total flux. Finally, though subject to change under transmission, meaning is not on this account devoid of truth, but it is the place where *truth makes its appearance*.

The appropriate philosophical approach to the method or process of interpretation is the 'hermeneutical circle' (or 'hermeneutical spiral') (Heidegger, 1996, pp.150-153; Gadamer, 1995, pp. 265-341). Briefly, following Heidegger, any inquiry is initiated by the unexpected breakdown of a life-world task, such as, to identify what belongs and what does not belong to (what is believed to be) a named life-world phenomenon (of a certain kind) but is poorly understood, or to understand why a life-world phenomenon that is thought to be well-understood, fails to occur under regular circumstances. On reflection, these failures involve the ontological structure of the Lifeworld where alone phenomena are generated and have their place.[6] The hermeneutical circle has three phases. The first phase is the initial problem state (Heidegger's *'Vorhabe'*).

As an illustration, it is worth turning to Fleck's (1979, German orig. 1935), *Genesis and Development of a Scientific Fact*. Fleck was a Polish bacteriologist, serologist, historian of syphilis, and epistemologist of medicine from Lvov. In his book, he traces the change in the initial clinical picture of a group of venereal symptoms to the final 'invention' of the underlying phenomenon, a disease-entity, syphilis. The initial stage he gives as comprising 'two points of view developed side by side, together, often at odds with one another: (1) an ethical-mystical disease entity of "carnal scourge," and (2) an empirical-therapeutic disease entity' (Fleck, 1979, p. 5). The former is the archaic 'thought style' of astrology that related the planets to ethics, and ethics to good and bad blood; the latter is the group of blood related venereal symptoms for the treatment of which the therapeutic use of mercury ointment was helpful.

The second phase brought into play a variety of theory-based procedures (Heidegger's *'Vorsicht'*). These were the diagnostic devices available at that time. The archaic 'thought styles' entrenched in the 'thought collective' of the research community were – had to be – maintained during the search in order to keep the elements of the problem in focus. They would gradually be re-shaped and transcended so as to be relevant to what would later become by 'invention' a theory of syphilis. The theoretical part, however, was implemented with the aid of practical theory-designed diagnostic tests on the blood serum, such as were to become the Wassermann Test. Such tests usually involved some tentative 'theory' or 'model' of the underlying hidden dimensions of the task, and performing the tests ultimately became the function of a new set of expert practitioners, in this case, serologists. In line with this thinking, a causative pathological agent was discovered in the lymphatic ducts that was named, *Spirochaeta pallida*. The scientific model of syphilis was then complete, since by that time the other venereal symptoms, those of gonorrhea and chancre, had been re-classified as separate disease entities.

Fleck is quite insistent that the model is not to be interpreted literally, that it is not the disease, nor a simple 'mirror' of the disease, and that the model needs to be prudently used by experienced specialists in the light of the general condition of the patient. When interpreted in this way, hermeneutically, a new phenomenological fact (Heidegger's *'Vorgriff'*) emerges in the Lifeworld; it is a new disease entity, syphilis. The solution is (1) the sought-for recognition of what makes the named phenomenon in question to be what it is (*'die Sache selbst'*), syphilis, and (2) the understanding of a set of theoretical conditions that *ceteris paribus* had the power to produce (*'die Sache selbst'*) syphilis.

If in the pursuit of a solution the first round or 'circle' of inquiry is not successful, something nevertheless has been learnt, and a new phase or 'circle' of the inquiry begins from there with a new *Vorhabe*, a new *Vorsicht*, and a new *Vorgriff*, and so on for as many circles as are necessary to bring success, assuming, of course, success in this case is possible.

Note that the interpretative process begins in the Lifeworld and ends in the production of a phenomenon (*die Sache selbst*) in the Lifeworld. It uses thought but does not end just in thought. Moreover, it seeks not primarily control, but an ontological solution in the production of a phenomenon (*die Sache selbst*). The outcome of the inquiry then is not just an abstract account, nor a theory, nor a model, though all of these are usually involved, but the understanding of a phenomenon that is culturally defined as the product of a successful experimental praxis. Since the design of the praxis is usually based on a theory, we are brought back to the question: what is the role of 'theories' and 'abstract concepts' in determining an ontological solution?

II. TRUTH AND MEANING

In the hermeneutical perspective, meaning is the product of a human understanding that works through interpretation. Interpretation, however, functions through the construction of meaning by common action, theory, and language, with common action involving the human body which in science is allied also with technology. Theoretical meanings contribute an implicit abstract component (see below) and common action contributes a constitutive cultural or practical component.

Turning to the classical sense of meaning, it grasped the object's inner or essential intelligibility, abstracting from what was individual, variable, sensory-based, material, technological, and historical. This mental object was conceived as a universal representation of the kind of thing (the natural kind) it represented and was true if it 'conformed to' reality. The function of Mind was to 'mirror' the forms of Nature. Language and life did not enter into the shaping of what is so presented as meaning and truth. Powerful and historically significant as was the classical notion of meaning and truth, there came a time, however, when the function of language and culture eventually could not be ignored. Tarski, for instance, proposed to define truth as a property of statements. Thus, let 'p' ('Snow is white') be a statement, then: 'p' ('Snow is white') is true if and only if p (snow is white). From the hermeneutic perspective such an account turns out either to beg the question or to be vacuous. For consider, how is the meaning of the sentence 'p' ('Snow is white') arrived at? To be meaningful words need a context of use and a users' community, but there are an infinite variety of contexts of use and of users' communities for the sentence 'p' ('Snow is white'), giving different meanings, yet none is specified in Tarski's definition. Turning to the other half of the definition, how is it determined that p (snow is white)? By experience, of course! But either experience presupposes an ability to use language correctly which begs the question or it is indeterminate and so cannot function as a criterion. Tarski's logical definition, however, was proposed within a philosophical framework different from the one used in this paper and within that framework was unquestioned until recently.

Returning to the duality of theoretical and practical meaning discussed earlier, Heidegger's approach to truth was through his return to the Greek term, *aletheia* (literally 'uncovering'), for truth (Heidegger, 1996, pp. 213-4, 1977b). It signaled: (1) a change in the notion of truth from the classical model of full transparency of natural forms, towards one of only partial transparency, that is historical, local, practical, and contextual.[7] (2) It made truth a property of the act of revealing, or 'uncovering,' what is presented to the subject's, Da-sein's understanding. (3) It is never complete, and what is given in a partial way is given to the extent that Da-sein is 'open' to receive what is being offered, and is 'free' to accept it. This is the 'truth,' '*aletheia*,' that is at the root of logical truth (the truth of propositions) and the other forms of truth (e.g., the truth of names). None of these derivative forms of truth have validity without reference to the pre-conditions of authenticity, namely, 'openness' and 'freedom,' that govern speaking, writing, listening, reading, and related practices for the meaningful and truthful use of these or other media.[8]

We do not then take Truth about things in the Lifeworld to be a classical conformity between a universal abstract mental representation and that which is represented categorially, nor a property of statements about it, but a property of meanings with individual, local, variable, sensory-based, material, technological, and historical origins and uses. The meanings we entertain about things are disclosed by praxes that are implicitly theory-laden, and usually over time multiply theory-laden, impermanent and changing. They are local, because expert communities are exclusive and gather locally around elite practitioners. They are technological because, if they are scientific entities, they are revealed only with the aid of technologies. They are historical because it happens that, when the particular theory-ladenness of the praxes becomes explicit, new, better or, at least, different, praxes can often be engineered with a consequent transformation of cultural meaning. It can happen, say, that a new electronic gun functions as a hammer, or that hammers eventually disappear entirely in a world of plastics and high tech. Such a spiral of meaning change turns endlessly within the historicality of Being (see Heidegger, 1996, p. 9) as such examples show. Such a process of theory-driven cultural change brings new historical perspectives into play and, through forgetfulness, inevitably puts old ones out of play. The big mistake of modernity was to commit itself to the classical notion of truth that could only be retained by supposing that scientific theory could be separated logically or ontologically from temporality and culture. Fleck's (1979) study of the genesis and development of a theory and clinical praxis for syphilis illustrates the dependence of research on hermeneutics, temporality, and culture.

III. TRADITION AND MEANING

Meaning is articulated and transmitted only through the medium of texts, speech, actions, empirical procedures, and other public signs expressive of meaning.[9] While these serve as 'conduits' for meaning, it is not the structure of the 'conduit' that constitutes meaning.[10] Nor are meanings transported in conduits like coal or water. Meanings have to be re-created by interpretation at each reading of the source texts by the addressees, or if by other readers, according to their interest in the source texts, etc. Consequently, if two readers belong to different communities, say, a professional community and a lay community, it is entirely possible that the meanings derived by perfectly legitimate processes of interpretation will not agree, yet each can yield a truth, say, one literal and one metaphorical, or both metaphorical in different ways, lexical or grammatical (see Lakoff and Johnson, 1999; Halliday and Martin, 1993; Bazermann, 1988, pp. 291-297)

Although there are many ways in which words and texts are used, it is not the case, however, that anything goes. Scientists like Einstein and Heisenberg were well aware that rational hermeneutic inquiry acknowledges the existence of *traditions of interpretation* that give today's readers and inquirers a version of the meaning of past sources that is culturally privileged by the current goals of the linguistic and cultural environment of the community with special 'ownership' rights in the subject matter. Such traditions of interpretation from the short term perspective tend to possess a static quality that from the long term perspective shows up as an inauthenticity to be overcome for the sake of growth of knowledge (cf. Heidegger, 1996, p. 9; Heelan, 1975a). Within the sciences such traditions of interpretation approximate to what Fleck (1979, pp. 125-145), and Duden (1993) called 'thought styles,' T.S. Kuhn (1970) called 'paradigms,' and Crombie (1994) called 'styles of scientific thinking.'

IV. HERMENEUTICS OF THEORY AND PRAXIS

Returning to the role of theory in the 'received view': it is important to understand how in this view theory serves to 'explain' a phenomenon and how an 'explanation' is used to solve a scientific problem by 'computation' or 'calculation.' It does so, say, by predicting the occurrence or non-occurrence of a phenomenon. A theory then 'explains' by providing an ideal computational model that purports to represent the causes, conditions, and circumstances relevant to the occurrence (or non-occurrence) of the phenomenon. The model (as an objective ideal reconstruction) replaces the Lifeworld phenomenon for scientific purposes, and these purposes are achieved by using the model computationally to predict real Lifeworld outcomes. The fact that the beliefs surrounding the 'received view' work successfully for prediction and control in broad ranges of scientific phenomena, does not, however, imply that scientific theory works for the reasons set forth in the 'received view.' It could be successful for other reasons. However, since the reasons given by the 'received view' deeply distort our understanding of ourselves and our Lifeworld by misunderstanding the role of theory in shaping how we think about phenomena, it is well to probe what is really implied by the meaning of theory in science or ordinary life.

I follow Heidegger once again (1996, pp. 69-70). He begins with a worker engaged in a building project, using a hammer, and the hammer unexpectedly breaks. Let us suppose that a replacement can't be found and that he has to have one made. His problem: what are the specifications of a hammer (of the kind he needs to finish the job)? The answer to this question will be a theory (about hammers) that explains a hammer's ability to do the hammer's job. What is a hammer's job? It is the 'meaning' of a hammer. In this case it is a *cultural praxis-laden meaning* within the context, let us say, of the building trade. Note that without a specification of context, the question is relatively indeterminate. In the context of the 'received view,' however, the hammer is a physical entity specified by a theory that lays out the physical specifications under which *it can become the host of the cultural meaning of a hammer, but whether or not it is so assigned this function is a separate and contingent matter.* This second meaning is a *theory-laden meaning*. The two meanings are not independent for the

THE LIFEWORLD AND SCIENTIFIC INTERPRETATION   57

theory-laden meaning makes sense only if a local contingent existential condition is fulfilled, namely that the hammer-referent is *praxis-laden* in the conventional sense.

If this condition is fulfilled, the hammer is a public cultural reality, constituted by a socio-cultural meaning. But what if the existential condition is not fulfilled? 'It' would not be a hammer, it would not be theory-laden, and would have no more title to being listed in the hammer category among any categorial listing of the furniture of the Lifeworld than any old boot that could be used to drive in a nail. 'It' would become (in Heidegger's words) 'a mere resource' (*'Vorhanden'* or *'Bestand'*) for hammering or other indeterminate functions, or just 'nothing in particular' (1996, pp. 42-43; 1977a).

Despite the fact then that (hammer-) theory 'explains' (hammering-) praxis, the language of theory and the language of praxis belong to different though locally and contingently coordinated perspectives. Coordination does not imply, however, isomorphism between the two perspectives,[11] for someone working on a carpentry project could, perhaps, be served on this occasion by an old boot or something other than a hammer. Since theory and praxis are merely coordinated but not isomorphic, they can be taken as axes for a kind of cultural phase space within which there are zones of uncertainty between explanatory theory and Lifeworld praxis, suggesting a Heisenbergian indeterminacy principle in the theory-praxis phase space.

Reflecting on the fact that individual things in our experience are never without a possible human purpose, everything in our experience, including scientific entities, bears some resemblance to a hammer, or other tool or equipment. There are then (at least) two perspectives on everything in the Lifeworld: (1) a praxis-laden cultural perspective that is constitutive of the Lifeworld, and (2) a theory-laden perspective – possibly multiply theory-laden – that explains this cultural perspective but does not constitute it. *Being, as we said above, is not represented by theory, but by the Lifeworld.* And the Lifeworld is a humanly meaningful contingent process that is subject historically both to possible development and possible decline.

V. WHEN 'THEORETICAL ENTITIES' ARE ALSO CULTURAL
ENTITIES OF THE LIFEWORLD

The 'theoretical entities' that enter into scientific theories of the human body, e.g., carbon or sodium atoms, water molecules, genes, proteins and their properties, energy, momentum, wave length, spin, and much more, belong in the first place to the imaginative or fictive world of scientific models. Some in addition, however, belong to public fora of the Lifeworld. Which? The criterion of belonging to the Lifeworld is the possibility of realizing the theoretical entity as a Lifeworld phenomenon.[12] This is achieved in the first instance by standard processes of measurement, because when a variable is measured it shows itself as present under the aspect measured. It is then a phenomenon in the Lifeworld and one shaped by the practices associated with the standard measurement setup (see Heelan, 1989, 1983b). In addition to the public Lifeworld forum of basic laboratory research, there are other public fora in which the theoretical entity has a presence with the status of a cultural phenomenon. These feature, for example, technology, clinical medicine, pharmaceuticals, finance, politics, religion, art, media. All of these fora – like that of the basic laboratory – are local fora in which a scientific entity, always in some technological context, can play the role of a dedicated cultural resource (for the life of clinical medicine, pharmaceuticals, finance, politics, religion, art, media) and by this means can become part of the local furniture of the Lifeworld. However, there is a difference between measurements (a) that are invasive and destroy the local interfaces with the living organism (e.g., by fixating and dying a cell), and (b) those that do not (e.g., certain imaging techniques, such as MRI). It is clear that the products of the former are no longer living parts of the body and so are mere resources with multiple uses like any tool or equipment. Only the products of the latter are living phenomena since they retain their role as functioning parts of a living organism and, under the Lifeworld criteria stated above, may not actually have (or have the use of) all the phenomenological properties accessible only under invasive conditions (a). This suggests a Heisenbergian phenomenological indeterminacy principle between measurable profiles of type (a) and measurable profiles of type (b).

In all such local public fora, the scientific entity and its data are meaningfully bivalent and emulate the relationship between theory and praxis in the study of a hammer. If removed from actual or dedicated association with all such local fora the putative 'data' are no longer data at all since they no longer have the capacity to make manifest in the Lifeworld the functioning presence of anything specific.[13] Having no determinate Lifeworld meaning, they should be treated possibly as indeterminate resources – or, perhaps, just as noise.

By way of illustration from the classical sciences, consider that when new theory-based technologies are added to the Lifeworld, as happened in Italy in the *quattrocento* with the invention of perspectival projection and the *camera obscura*, terms of a new kind, scientific and theoretical, came to be introduced into everyday language with new practical measurement-based cultural meanings, particularly, in this case, for space and time. The techniques of mathematical perspective, for instance, revolutionized the Lifeworld of Italy and later of Europe, through art, architecture, urban planning, navigation, warfare, and much more. The geometry-filled productions of craft skills in optics, astronomy, map making, painting, music, weapons' design, and other arts and skills prepared the way for the elevation of artistic craft skills to scientific theory skills, where the product of the new art came to be regarded not as works of art, but as works of geometrical reason or science (Crombie, 1994, pp. 499-680). Among other things the works of geometrical reason changed the public urban space of Europe from a quilt of diverse local spaces and times into a single universal space and uniform cosmic time based on measurement with rigid rulers and mechanical clocks synchronized with the stars. For those who looked for a unified cosmology, the way was prepared for Galileo and the Copernican revolution (see Heelan, 1983a/1988, chap. 11). It was Galileo who helped convert works of thoughtful art – his deft experiments, such as, timing balls rolling down an inclined plane – into a world of science and reason, a move on which we now look back with more critical eyes.

It follows from what has been said that the furniture of the Lifeworld is not fixed with respect to categorial kinds. Along with 'natural' physical phenomena, such as 'trees,' it comprises invented 'cultural' phenomena. Among them are those created by institutions, such as the Wassermann Test created by medical research laboratories. Some are categorized resources but not functionally assigned ('*Bestand*' or '*Vorhände*'), such as, chemicals, reagents, and appliances in central storage available for a variety of uses (Heidegger, 1977a; 1996, pp. 42-43). Others are functionally assigned ('*Zuhände*'), such as those actually used in surgery (Heidegger, 1996, p. 69). Only the latter enjoy a meaning in the Lifeworld that is actually specified for definite tasks and, consequently, are part of the furniture of the Lifeworld.

VI. METAPHOR AND THE JANUS-LIKE FACE OF SCIENTIFIC ENTITIES

Fleck, in his history of the scientific theory of syphilis, recognized the fact that in establishing the meaning of the new scientific term, 'syphilis,' some special usage had to be negotiated between scientific terms normed by the 'thought collective' of the research community and everyday terms normed by the 'thought collective' of everyday life. To quote Duden (1993, pp.69).

... as a practising bacteriologist, [Fleck] knew that his eyes were caught, not only in the norm imposed by the collective of the laboratory, but equally by the thought style characteristic of his everyday family life. It is this double anchorage – in the laboratory and at the table – that makes the scientist a conduit through which scientific facts become confused with cultural interpretations. As a result, scientific facts ... have a Janus-like face.

What would result if the Janus-like face of a scientific fact were to go unnoticed or were to be flouted or ignored by convention in public fora of communication? This happens all too frequently, partly as a consequence of the widespread acceptance of the 'received view' and partly because of the limitations on public discourse caused by the difficult subject matter. The result is distortions in communication. Two systematic errors become possible. Each, in its most innocent form, leads to the more or less conscious use of a figure of speech, something like a metaphor.

(1) The post-scientific praxis-laden perspective of the laboratory Lifeworld zone is simply re-described metaphorically in terms of the theoretical scientific meanings (see Halliday and Martin, 1993; Bazermann, 1988, pp. 201-299). Under such conditions, theoretical descriptions replace practical descriptions. But if the metaphorical character of the predication is not recognized, it is easy to take the replacement to be exclusive and ontological, and to think that the Lifeworld conditioning of phenomena no longer exists, and that only the theoretical scientific world exists. For example, perceptual space is assumed to be modeled by Euclidean geometry, colors by electromagnetic wavelengths, sounds by pitch and loudness, and syphilis by a positive Wassermann Test, when all such predications are no more than metaphors apart from the collaboration of the human senses, language, and cultural environment. Perceptual space, color, sound, and syphilis exist only as the product of interpretation through which they in their Lifeworld involvement become intelligible as phenomena of human experience. Modern scientific medicine then has been often charged with a weakness for reducing patients to a bundle of anatomical parts and physiological processes, each having its scientific model at the level of chemistry, molecular biology, or physiology, and with little regard for the human life in which they are engaged and that uses such systems to cope, well or ill, satisfactorily or unsatisfactorily with the challenges of the patient's Lifeworld. All would agree that, ultimately, scientific medical models should not replace the Lifeworld of the patient and should be at the service of the patient's quality of life as lived in and tested by his or her Lifeworld.

(2) The theory-laden perspective of the scientific zone is simply re-described metaphorically in terms of pre-scientific ('naive' or 'folk') Lifeworld meanings. In other words, since the scientific terms are not well understood in many public fora, they are simply filled with the old familiar pre-scientific Lifeworld meanings that the 'received view' wants to replace, possibly with a more or less conscious sense that this involves a metaphorical construction. From this awareness comes a warning, generally heeded by historians of science and medicine, that needs, however, to be heeded also by ethicists, media pundits, and public policy makers who, confusing the context of science with that of Lifeworld ontology, so easily and offhandedly fill scientific terms with prescientific Lifeworld meanings in their public discourse. For example, in such discourse, scientific terms such as 'cells,'

'organs,' and 'bacteria' are treated as ('naive' or 'folk') 'things' like machine replaceable parts violating their natures as integral parts of a living organism, for unlike machine parts these organic terms are constituted by the continuous flow of chemical exchanges across their interfaces with surrounding tissues.

What follows from the non-recognition of these (at their innocent best) metaphorical transitions is, for instance, confusion in the public debate about such contentious practices as abortion, cloning, disease prevention, AI (artificial intelligence), and much more, where scientific model terms, such as, 'fetus,' 'genotype,' 'bacteria,' and 'neural networks,' are filled in public discourse with meanings taken from related practical everyday ('naive' or 'folk') contexts, making them falsely synonymous with the everyday uses of the related everyday terms. In this case, the everyday terms would be 'child,' 'adult,' 'cause of disease,' and 'intelligence,' etc., respectively. This usage may be good politics, but it is itself a form of cultural disease.

Despite the problems created by possible metaphorical usages due to the complementarity of explanatory scientific theory and preventive medical practices, scientific theories have been a very positive force in shaping the contemporary Lifeworld. There is no need to press this point. The bacterial theory of infection led to a host of new cultural practices dealing with food handling, personal hygiene, sewage and water systems, the urban environment, and the treatment of bacterial diseases. But these practices, of course, have to be carefully designed and prudently implemented. However, as scientific theories grow and change, a train of new and often contentious practical problems are emerging: for instance, genetic theory has led to noisy debates as to whether or under what conditions genetically modified (GM) foods should be admitted to the food chain. New cultural practices found to be effective also lead in their turn to new scientific theories, which in turn lead to better medical practices, which may lead to better scientific theories, and so on. Though often treated by public fora and sometimes even by the medical profession as stripping the mystery from Nature and as exposing what is constitutive of what 'really is,' scientific theory is in fact no more than a tool for, or a way of coping with some living function of the human body constituted as meaningful by a lifestyle in the patient's Lifeworld. Because of the zone of uncertainty between theory making and cultural practices, and another between pre-scientific and post-scientific Lifeworld terms, there is an inescapable tension in the public mind that can – and often does – result in changes, possibly also in confusion, concerning conditions for meaning-fulfillment and concerning policy norms. Noting such changes, one captures something about the historicity and contingency of hermeneutic truth.

A critical example from medicine illustrates how the multivalence of scientific descriptions can create new moral perplexities in the Lifeworld. Barbara Duden

(1993), historian of the woman's body in clinical medicine, questions the scientific term 'fetus' that belongs to contexts of scientific imaging and biology, and asks whether it is being abused in public fora of discussion when substituted for the term 'child' that is used in the Lifeworld context of pregnancy and maternity. Has the separateness of contexts between model-scientific, pre-scientific Lifeworld-processes, and post-scientific Lifeworld-processes been illegitimately suppressed in our medical culture, in the media, and in public policy discourse? The terms 'fetus' and 'child' are, of course, correlative (each in its own context reveals something about what the other term refers to) but they are not isomorphic and interchangeable. A 'fetus' is a term whose primary owner is the medical profession. A living fetus is recognized by sonographical and other imaging techniques apart from the mother's context in the everyday life-world. Even while inseparable from the living tissue of the mother, the fetus is generally described as a 'thing,' as if, like a pre-scientific machine part, it had an existence separate from the mother. Duden notes with some concern that ethical rules and legislation in Western countries concerning pregnancies are presently being written in terms of the 'fetus' where that term slurs the difference between the fetus as part of the scientific model, the fetus as an organic part of the post-scientific Lifeworld, and the child as an element in the mother's (usually) pre-scientific pregnant life. Duden is unhappy with this and asks: should the difference between the two cultural perspectives be recognized and an accommodation found that defers to the special cultural role of the mother in decision-making about the child?

VII. SUMMARY

In the assessment of scientific theory and practice, medical science and clinical practice, the critique of the 'received view' of science made via the phenomenological orientation of Husserl, Heidegger, and Merleau-Ponty towards the Lifeworld and Heidegger's hermeneutics (or interpretation) of experience has made it possible to assign different roles to theory and practice: technological design for the purposes of environmental control to theory, and ontological understanding for the purpose of human culture to practice. Scientific theories then have a 'Janus-like face,' one side looks in the direction of computational and technological control which is not constitutive of scientific knowledge but is merely a resource or tool for multiple practices, the other looks in the direction of human culture which is ultimately constitutive of ontological scientific knowledge.

This bivalence underscores the prevalence of metaphor in scientific discourse and, in particular, in medical science whether inside and outside the scientific community under conditions where modern culture and the 'received view' tend to mask the presence of metaphor in such discourse. It was shown, however, that under the broader analysis of phenomenology, metaphor is as fundamental for true scientific discourse as literality is for the 'received view' (see Bazermann, 1988; Fiumara, 1995; Heelan, 1998; Hesse and Arbib, 1986; and Lakoff and Johnson, 1999). Since the theoretical is mathematical and the practical is empirical, it makes no sense to predicate mathematical models literally of the Lifeworld; at best, the two must come together consciously in some unambiguous but metaphorical way guided by professional experts in the spirit of (what Aristotle called) *'phronesis'* (prudent action), aware that they are seeking no more than a consensus about a set of relevant soluble Lifeworld issues.

Georgetown University
Washington, D.C.
U.S.A.

ACKNOWLEDGEMENTS

This paper owes much to helpful discussions with B. Babich, H. Byrnes, E. Hurley, and J. Schulkin and many others.

NOTES

[1] The 'received view' is the name given to the analytic/empiricist mainline Anglo-American tradition in the philosophy of science that stems from the influence of Descartes, Kant, and particularly, the Marburg School of Neokantianism (cf. Friedman, 2000), whose principal exponents in America were R. Carnap, C. Hempel, H. Feigl, and the contributors to the *Minnesota Studies in the Philosophy of Science Series* and the *Boston Studies in the Philosophy of Science Series*. For an account of the 'received view', see the Introduction in Suppe (1977).
[2] The 'life-world' or 'everyday world' are terms used in a non-technical sense. In the technical phenomenological sense, I use the term 'Lifeworld'.
[3] The phenomenological tradition goes back to Edmund Husserl. Among those most important to mention within the context of this paper are, B. Babich, R. Crease, H. Dreyfus, B. Duden, T. Glazebrook, J. Grondin, A. Gurwitsch, D. Ihde, T. Kisiel, J. J. Kockelmann, D. Leder, M. Merleau-Ponty, M. Natanson, O. Pöggeler, J. Rouse, R. Safranski, A. Schutz, K. Young, and R. Zaner.
[4] Hermeneutical phenomenology stems chiefly from M. Heidegger. I have been guided also in a curious way, by the works of B. Lonergan (1971, 1994) in developing Heidegger beyond what he said in order to lay out a more detailed position than Heidegger himself gave.
[5] From the point of view of contemporary functional linguistics, see, for example, the work of Halliday and Martin (1993), Bakhtin (1986).
[6] I am assuming Husserl's analysis of a phenomenon as the object of a noetic-noematic intentionality-structure, where the structure is group-theoretic relative to the connected set of variations (e.g., perceptual profiles) that maintain the invariance of the interest the subject has in the phenomenon, see Husserl (1999), pp.163-185, and Heelan (1991b).
[7] See, for example, Heidegger (1996), pp. 34-39. Polanyi, probably influenced by Heidegger, seems to

say the same in different terms: the *explicit meaning* conceals a *tacit meaning*; see Polanyi (1964), pp. x-xi.

[8] The Heideggerian notion of authenticity, openness, and freedom connect with Bakhtin's interest in how the heteroglossia of meanings and the polyphony of voices in any community of speakers can be reconciled with 'universal reason.' He writes: 'But [in any dialogue] in addition to [the] addressee (the second party), the author of the utterance, with a greater or lesser awareness, presupposes a higher *superaddressee* (third), whose absolutely just responsive understanding is presumed.... In various ages and with various understandings of the world, this superaddressee and his ideally responsive understanding assume various ideological expression (God, absolute truth, the court of dispassionate human conscience, the people, the court of history, science, and so forth).' Bakhtin (1986), p. 126.

[9] See the work of Bakhtin (1986), Bazermann (1988), and Halliday and Martin (1993) who address these questions from the functional linguistic standpoint. Also relevant are the writings of other scholars, such as, M. Holquist, T. Givon, who share this point of view.

[10] However, the structure of the conduit gives rise to what scientists call 'information.' Following Shannon and Weaver's information theory, scientific 'information' is defined as the capacity of a signal channel to encode messages. It does not connote the interpretation of the signals nor for what purpose this message-encoding capacity is used.

[11] By *isomorphism* is meant a one-to-one translatability of any statement in one language into a unique statement in the other language. The two context-dependent languages refer to the same things but from different, often interacting and mutually interfering, perspectives. I have argued that these languages are related among themselves within a lattice structure which includes a least upper bound (lub) and a greatest lower bound (glb) as well as complements. This thesis is presented in Heelan (1983a/1988), chaps. 10 and 13, and in Heelan (1975b).

[12] For a fuller account of what constitutes a 'phenomenon' in Husserlean phenomenology, see Heelan (1983a/1988; 1991b), and Heidegger (1967).

[13] There is the alternative strategy of re-evaluating the interpretive context of the experiment to pursue another goal. For a detailed study of data, see Heelan (1989), also (1983a/1988).

BIBLIOGRAPHY

Babich, B.: 1994, *Nietzsche's Philosophy of Science*, SUNY Press, Albany, New York.

Bakhtin, M.M.: 1986, 'The Problems of the Text in Linguistics, Philology and the Human Sciences: An Experiment in Philosophical Analysis', in C. Emerson and Michael Holquist (eds.), *Speech Genres and Other Late Essays*, University of Texas Press, Austin, Texas, pp. 103-131.

Bazermann, C.: 1988, 'How language realizes the work of science', in *Shaping Written Knowledge: The Genre and Activity of the Experimental Article in Science*, University of Wisconsin Press, Madison, Wisconsin.

Brisson, L., and Meyerstein, F. W.: 1995, *Inventing the Universe: Plato's Timaeus, Three Big Bang, and the Problem of Scientific Knowledge*, SUNY Press, Albany, New York.

Crease, R.: 1995, *The Play of Nature*, University of Indiana Press, Bloomington, Indiana.

Crombie, A. C.: 1994, *Styles of Scientific Thinking in the European Tradition: A History of Argument and Explanation in the Mathematical and Biomedical Sciences and the Arts*, Vols. 1 – III, Duckworth, London.

Dilthey, W.: 1989, *Introduction to the Human Sciences*, R. Makkreel (trans.), Princeton University Press, Princeton, New Jersey.

Dreyfus, H.: 1991, *Being-in-the-World: A Commentary on Heidegger's Being and Time*, MIT Press, Cambridge, Massachusetts.

Duden, B.: 1993, *Disembodying Women: Perspectives on Pregnancy and the Unborn*, Harvard University Press, Cambridge, Massachusetts.

Fiumara, G.C.: 1995, *The Metaphoric Process: Connections between Language and Life*, Routledge, London and New York.
Fleck, L.: 1979, *Genesis and Development of a Scientific Fact*, University of Chicago Press, Chicago, Illinois.
Friedman, M.: 2000, *The Parting of the Ways*, Open Court, New York.
Gadamer, H-G.: 1995, *Truth and Method*, 2nd edition, J. Weinscheimer and D.Marshall (trans.), Continuum Press, New York.
Glazebrook, T.: 2000, *Heidegger's Philosophy of Science*, SUNY Press, Albany, New York.
Grondin, J.: 1994, *Introduction to Philosophical Hermeneutics*, Yale University Press, New Haven, Connecticut.
Gurwitsch, A.: 1966, *Studies in Phenomenology and Psychology*, Northwestern University Press, Evanston, Illinois.
Halliday, M.A.K. and Martin, J. A.: 1993, 'General Orientation', in *Writing Science: Literacy and Discursive Power*, Falmer Press, London/Washington, pp. 3-21.
Heelan, P.A. and Schulkin, J.: 1998, 'Hermeneutic Philosophy and Pragmatism: A Philosophy of Science', *Synthese* **115**, 269-302.
Heelan, P.A.: 1998, 'The Scope of Hermeneutics in Natural Science', *Studies in the History and Philosophy of Science*, **29**, 273-298.
Heelan, P.A.: 1997, 'Why a Hermeneutical Philosophy of Natural Sciences?' *Man and World*, **30**, 171-198.
Heelan, P.A.: 1994, 'Galileo, Luther, and the Hermeneutics of Natural Science', in T. Stapleton (ed.), *The Question of Hermeneutics: Festschrift for Joseph Kockelmans*, Kluwer, Dordrecht/Boston, pp. 363-375.
Heelan, P.A.: 1991a, 'Hermeneutical Phenomenology and the History of Science', in D. Dahlstrom (ed.), *Nature and Scientific Method: William A. Wallace Festschrift*, The Catholic University of America Press, Washington, D.C., pp. 23-36.
Heelan, P.A.: 1991b, 'Hermeneutic Phenomenology and the Philosophy of Science', in H. Silverman (ed.), *Gadamer and Hermeneutics: Science, Culture, and Literature*, Routledge, New York, pp. 213-228.
Heelan, P.A.: 1989, 'After Experiment: Research and Reality', *Amer.Philos. Qrtly.*, **26**, 297-308.
Heelan, P.A.: 1988, 'Husserl, Hilbert, and the Critique of Galilean Science', in R. Sokolowski (ed.), *Edmund Husserl and the Phenomenological Tradition*, The Catholic University Press, Washington, D.C., pp. 157-173.
Heelan, P.A.: 1983a/1988, *Space-Perception and the Philosophy of Science*, University of California Press, Berkeley and Los Angeles, California.
Heelan, P.A.: 1983b, 'Natural Science as a Hermeneutic of Instrumentation', *The Philosophy of Science*, **50**, 181-204.
Heelan, P. A.: 1975a, 'Heisenberg and Radical Theoretic Change', *Zeit. f. allgemeine Wissenschaftstheorie*, **6**, 113-138.
Heelan, P.A.: 1975b, 'Hermeneutics of Experimental Science in the Context of the Life World', in D. Ihde, and R. Zaner (eds), *Interdisciplinary Phenomenology*, Martinus Nijhoff, The Hague, pp. 7-50.
Heidegger, M.: 1996, *Being and Time*, Joan Stambaugh (trans.), SUNY Press, Albany, New York. References in the text are to page numbers in the Niemayer German edition found in the margins of this and other translations
Heidegger, M.: 1977a, *The Question Concerning Technology and Other Essays*, W. Lovitt (trans.), Harper Colophon, New York.
Heidegger, M.: 1977b, 'On the Essence of Truth', in D.F. Krell (ed.), *Martin Heidegger: Basic Writings*, Harper and Row, New York, pp. 117-139.
Heidegger, M.: 1968, *What is Called Thinking?* J.G. Gray (trans.), Harper and Row, New York.

Heidegger, M.: 1967, *What Is a Thing?* W.B. Barton, Jr., and V. Deutsch (trans.), Regnery, Chicago, Illinois.
Heidegger, M.: 1966, *Discourse on Thinking*, E.M. Anderson and E.H. Freund (trans.), Harper and Row, New York.
Hesse, M. and Arbib, M.: 1986, *The Construction of Reality*, Cambridge University Press, Cambridge.
Husserl, E.: 1999, *The Essential Husserl: Basic Writings in Transcendental Phenomenology*, D. Welton (ed.), Indiana University Press, Bloomington, Indiana.
Husserl, E.: 1970, *The Crisis of European Science and Transcendental Philosophy*, D. Carr (trans.), Northwestern University Press, Evanston, Illinois.
Ihde, D.: 1983, *Existential Technics*, SUNY Press, Albany, New York.
Kockelmans, J. J.: 1993, *Idea for a Hermeneutic of the Natural Sciences*, Kluwer, Dordrecht, The Netherlands.
Kuhn, T.S.: 1970, *The Structure of Scientific Revolutions*, University of Chicago Press, Chicago. Illinois.
Lakoff, G., and Johnson, M.: 1999, *Philosophy in the Flesh*, Basic Books, New York.
Leder, D.: 1990, *The Absent Body*, University of Chicago Press, Chicago.
Lonergan, B.: 1994, *Insight: A Study of Human Understanding*, Vol. 4, *Collected Works*, University of Toronto Press, Toronto.
Lonergan, B.: 1971, *Method in Theology*, University of Toronto Press, Toronto.
Merleau-Ponty, M.: 1962, *The Phenomenology of Perception*, Routledge and Kegan Paul, London.
Miller, J.: 1978, *The Body in Question*, Random House, New York.
Natanson, M.: 1970, *The Journeying Self: A Study in Phenomenology and Social Role*, Addison-Wesley, Reading, Massachusetts.
Pöggeler, O.: 1963, *Martin Heidegger's Path of Thinking*, trans. D. Magurshak and S. Barber, Humanities Press, New Jersey.
Polanyi, M.: 1964, *Personal Knowledge: Toward a Post-Critical Philosophy*, Harper and Row Torchbooks Edition, New York.
Ricoeur, P.: 1981, *Hermeneutics and the Human Sciences: Essays on Language, Action and Interpretation*, J. B. Thompson (ed. and trans.), Cambridge University Press, Cambridge.
Rouse, J.: 1987, *Knowledge and Power: Toward a Political Philosophy of Science*, Cornell University Press, Ithaca, New York.
Safranski, R.: 1998, *Martin Heidegger: Between Good and Evil*, Harvard University Press, Cambridge, Massachusetts.
Schutz, A.: 1973, *The Problem of Social Reality*, Vol. 1, *Collected Papers*, Martinus Nijhoff, The Hague.
Sheets-Johnstone, M.: 1990, *The Roots of Thinking*, Temple University Press, Philadelphia, Pennsylvania.
Suppe, F.: 1977, *The Structure of Scientific Theories*, 2nd ed. University of Illinois Press, Urbana, Illinois.
Varela, F.J., Thompson, E., and Rosch, E., 1993, *The Embodied Mind: Cognitive Science and Human Experience*, MIT Press, Cambridge, Massachusetts.
Young, K.: 1997, *Presence in the Flesh: The Body in Medicine*, Harvard University Press, Cambridge, Massachusetts.
Zaner, R.: 1981, *The Context of Self: A Phenomenological Inquiry Using Medicine As a Clue*, Ohio University Press, Athens, Ohio.
Zaner, R.: 1970, *The Way of Phenomenology*, Pegasus, New York.

FRANCES CHAPUT WAKSLER

MEDICINE AND THE PHENOMENOLOGICAL METHOD

> Allow the possibility of the facts,
> and all the actors comport themselves
> as persons would in their situation.
> Horace Walpole, *The Castle of Otranto*

I. INTRODUCTION

If he is to be one who thinks for himself...an autonomous philosopher with the will to liberate himself from all prejudices, he must have the insight that all the things he takes for granted *are* prejudices, that all prejudices are obscurities arising out of a sedimentation of tradition – not merely judgments whose truth is as yet undecided – and that this is true even of the great task and idea which is called "philosophy" (Husserl, 1970, p. 72, emphasis in original).

My goal in this paper is to suggest that phenomenology can serve the medical profession by providing a different way of looking at medical theory and practice, a way that is grounded in both philosophy and lived experience. Central to a phenomenological perspective is an exploration of the taken-for-granted assumptions (what Husserl refers to above as "prejudices") that undergird thought and action. Such an exploration can bring forward for examination features of medical theory and practice that otherwise may go unnoticed, unexamined, and unassessed. Husserl (Palmer translation, 1971, p. 77) states, "The term 'phenomenology' designates two things: a new kind of descriptive method which made a breakthrough in philosophy at the turn of the century, and an *a priori* science derived from it; a science which is intended to supply the basic instrument ... for a rigorously scientific philosophy and, in its consequent application, to make possible a methodical reform of all the sciences."[1] Phenomenology provides both (1) an intentionally minimal set of philosophical assumptions about the nature of humans and the world and (2) a set of methodological principles based on those assumptions that provide rules for engaging in phenomenological analysis.

I come to this project as a phenomenological sociologist. As such, and given phenomenology's focus on the perspective of the one who knows, I here work from phenomenology as *I* know it, the phenomenological method as I conceive of and use it, the ways that I have found it helpful in sociological analysis, and what I see as the particular contributions it can make to medical theory and practice.[2] I rely primarily on the work of

Edmund Husserl, the founder of phenomenology. Although many sociologists work from Alfred Schutz' version of Husserl's ideas, I prefer the purity of Husserl's own version of his thought.[3]

Two central features of the phenomenological approach are crucial to its understanding, radical in relation to other modes of investigation, and of particular significance for medical practitioners: the primacy of the subject's perspective and suspension of belief (the phenomenological epoché). For these two features I discuss the philosophical positions that phenomenology takes and the methodological principles that flow from such positions. I then suggest some implications of these features for investigating medical theory and practice.

The relation between phenomenology and medicine is neither arbitrary nor artificial but implicit in phenomenology's general project of providing what Tymieniecka (1962, p. xx) in *Phenomenology and Science* describes as "a methodological basis for *all* fields of inquiry." Phenomenology is relevant to medicine not only through phenomenological sociology but more directly, i.e., as phenomenological medicine, a direct link I note without pursuing here. What follows emerges from phenomenological sociology as I understand it.

II. THE SUBJECT'S PERSPECTIVE

By my living, by my experiencing, thinking, valuing, and acting, I can enter no world other than the one that gets its sense and acceptance or status...in and from me, myself (Husserl, 1960, p. 21).

Phenomenology takes as its starting point the examination of phenomena as they appear in a subject's experience. For Husserl the individual is the locus of that which can be taken as certain. The world is analytically accessible only as it is present in a subject's experience. It is this experience that provides the grounding for phenomenological analysis. Husserl (1970, pp. 97-8, emphasis in original) characterizes this approach as "transcendental," by which he intends

the motif of inquiring back into the ultimate source of all the formations of knowledge, the motif of the knower's reflecting upon himself and his knowing life in which all the scientific structures that are valid for him occur purposefully, are stored up as acquisitions, and have become and continue to become freely available. Working itself out radically, it is the motif of a universal philosophy which is grounded purely in this source and thus ultimately grounded. This source bears the title *I-myself*, with all of my actual and possible knowing life and, ultimately, my concrete life in general. The whole transcendental set of problems circles around the relation of *this*, my "I" – the "ego" – to what it is at first taken for granted to be – my soul – and, again, around the relation of this ego and my conscious life to the *world* of which I am conscious and whose true being I know through my own cognitive structures.

Phenomenology thus directs analysts to their own *experiences*, the structure of which is accessible to them in reflection. Husserl takes pains, however, to make clear that by focusing on one's own experience one is not denying the "world outside." What he is emphasizing is that that world outside is accessible *only* through the one who knows – but in that way the whole world *is* accessible.

I take Husserl's claim as a warrant to use one's own experience as the grounding of what one takes to be knowledge but not as an invitation to solipsism. In one's experience are others who have experiences. Just as I allow that I am the focal point of my experiences – experiences that include all the world as I perceive it – so do I allow that the same is true of others, including those who are the focus of my attention, analytic or practical. Nonetheless, their experiences are only available to me through me. The significance of the focus on the subject's perspective is that one is directed to a detailed consideration of *what indeed one knows* and *how it is* that one comes to know it.

Phenomenological exploration of one's taken-for-granted "knowledge" discloses a mixture of ideas – not only what one might view as verified truth but also unexamined tradition, guesses, suppositions, beliefs, and hopes – some of which, upon disclosure, one may choose to retain but others one may want to reject. To bring forward such ideas is to make one's taken-for-granted "knowledge" accessible for evaluation – for acceptance, rejection, re-formulation, and refinement.

> Let us direct our attention to the fact that in general the world or, rather, objects are not merely pregiven to us all in such a way that we simply have them as the substrates of their properties but that we become conscious of them (and of everything ontically meant) through subjective manners of appearance, or manners of givenness, without noticing it in particular; in fact we are for the most part not even aware of it at all. Let us now shape this into a new universal direction of interest; let us establish a consistent universal interest in the "how" of the manners of givenness...(Husserl, 1970, p. 144).

It is this "manner of givenness" of medical theory and practice whose examination can provide insight into one's own formulation of the world and the formulations of others.

III. SUSPENSION OF BELIEF: THE PHENOMENOLOGICAL EPOCHÉ[4]

> The whole world as placed within the nature-setting and presented in experience as real, taken completely "free from all theory," just as it is in reality experienced, and made clearly manifest in and through the linkings of our experiences, has now [within the phenomenological epoché] no validity for us, it must be set in brackets, untested indeed but also uncontested. Similarly all theories and sciences, positivistic or otherwise, which relate to this world, however good they may be, succumb to the same fate (Husserl, 1962, p. 100).

Husserl's technique for clarifying of presuppositions any field of investigation and proceeding with a presuppositionless analysis involves a process he terms "epoché." Epoché (bracketing) is divided by Husserl into two forms: *phenomenological bracketing*, in which causal explanations and all presuppositions are removed (neither believed nor denied but suspended), and *eidetic bracketing*, in which the particularity of an instance is suspended and the enduring features of the phenomenon are disclosed.[5] It is phenomenological bracketing that is of particular relevance here (eidetic bracketing being more relevant to more theoretically oriented projects).

Of the phenomenological epoché Husserl (1970, p. 76, emphasis in original) writes that it

> has in truth a hitherto unheard-of radicalism, for it encompasses expressly not only the validity of all previous *sciences*...but even the validity of the pre- and extrascientific *life-world*, i.e., the world of sense-experience constantly pregiven as taken for granted unquestioningly and all the life of thought which is nourished by it.

It is important to note that Husserl is not advising *disbelief* in the world and all its characteristics; he is recommending *suspension of both belief and disbelief* as a method for looking at the world in a different, a fresh way, one clarified of preconceptions. As a phenomenologist one must suspend belief in the existence or non-existence of everything except one's own certainty, as a subject, that one is indeed an experiencing subject. This method leads to a constant questioning of experience, a refusal to accept existing explanations and descriptions – whether common-sense or scientific – simply on the basis of their existence and authority, and a refusal to reduce experience to its mere material content. The world, nonetheless, remains accessible in all its fullness, available for analysis in the multiplicity of ways it appears in experience. Although suspending belief in reality, the analyst nonetheless has available for analysis all that which is *taken as reality*, including the processes whereby such reality is established.

In *The Social Construction of Reality* Berger and Luckmann (1967, p. 3, emphasis added) reflect this phenomenological position, writing,

"Sociological interest in questions of 'reality' and 'knowledge' is...initially justified by the fact of their *social relativity*." "Reality" and "knowledge" are seen as outcomes of acts, as decisions, rather than as objective states. Reality is "a quality appertaining to phenomena that *we recognize* as having a being independent of our own volition" (Berger and Luckmann, 1967, p. 1, emphasis added). Within the phenomenological epoché emphasis is not on what is "real" in any absolute sense but on what is "taken to be real" and how that "taking to be real" comes about. Of knowledge Berger and Luckmann (1967, p. 1) write, "It will be enough, for our purposes...to define 'knowledge' as the certainty that phenomena are real and that they possess specific characteristics." How one arrives at that certainty is a topic of phenomenological analysis. Within the window into lived experience, phenomenology gives invaluable information all that is taken to be knowledge is suspended. Indeed to assess the validity of knowledge as one investigates it is, in the words of Berger and Luckmann (1967, p. 13), "somewhat like trying to push a bus in which one is riding". Instead, the analytic focus of phenomenology is on "whatever passes for 'knowledge' in a society, regardless of the ultimate validity or invalidity (by whatever criteria) of such 'knowledge'" (Berger and Luckmann, 1967, p. 3). Within the phenomenological epoché the social constructions of "reality" and "knowledge" become available for analysis and understanding.

IV. MEDICAL THEORY AND PRACTICE AS EVERYDAY LIFE ACTIVITIES

There are...problems which arise from the naïveté through which objectivist science takes what it calls the objective world for the universe of all that is, without noticing that no objective science can do justice to the {very}[6] subjectivity which accomplishes science (Husserl, 1935, pp. 294-5).

Engaging in phenomenology and engaging in medical practice are significantly different kinds of endeavors, for the latter is activity grounded in the world of everyday life and, as such, is "subjectively accomplished." In *The Unity of Mistakes: A Phenomenological Interpretation of Medical Work* Paget (1988, p. 49) writes,

In clinical medicine, knowledge is embedded in a particular activity, the care and treatment of the sick. It is not a form of knowledge but a method of acting and thinking about illness. In use, knowledge takes characteristic shape in acts that are experiments with knowledge – trials, as it were.

In medical practice, activity requires not the suspension of belief and disbelief but detailed engagement with both as the grounds for action.

Although less obviously the case, medical theorizing is also located in the world of everyday life. In his argument that sociology can be viewed as a natural observational science, the enthnomethodologist[7] H. Sacks (1989, p. 213) notes:

> My consideration of these issues led me to make an extremely simple observation: The doing of natural science, indeed the doing of biological inquiries, was something which was reportable, first, and second, the reports of the activities of doing science did not take the form that the reports of the phenomena under investigation took.
>
> That is to say, sticking for the moment to the second point, biologists in reporting their researches to their colleagues could effectively do so without first casting their investigative activities into biological operations. They could report what they did in ways which were accessible to naturalistic observation, and reports so accessible were adequate reports.

The import of Sacks' observation is that while scientific findings are routinely presented as "objective," standing above the world of everyday life, the actual way such findings are achieved and reported are located in that everyday world. Those who do research on genetics, for example, do not describe their *own* research activities in genetic terms but in everyday life terms. They attribute their findings not to their genes but to their scientific but nonetheless mundane activities.

"Doing science" is guided both by scientific criteria and by practical everyday life concerns. Paget (1988) quotes a physician describing his research experiences.

> I would do a procedure five, sometimes five or six times before the darn thing worked: I would have made every conceivable error before I finally, you know, ran out of things that could possibly go wrong. I controlled most of the variables before the darn thing worked: technical things. I did many, many runs before they would work out.

To have a preconceived idea of what "working out" looks like brings everyday life criteria into scientific research. Phenomenology can recognize and direct attention to the taken-for-granted criteria by which such "working out" is formulated and determined and to the ways in which preconceived ideas influence outcomes.

By bringing forward for consideration the subject and the subject's actions, phenomenology recognizes that medical theorizing is activity engaged in by individuals and groups located in the world of everyday life. "Analytic language tends to break up the unity of a sentient figure, thinking and acting and acting and thinking" (Paget, 1988, p. 50). Detailed phenomenological examination of how theorizing is accomplished can display its grounding in the subjects who do the theorizing.

Medical theory and practice are located in specific cultures and in specific historical times whose ideas, values, and morality all guide what counts not only as scientific explanation but as what indeed is in need of scientific explanation. Any search for "solutions" presupposes "problems" but what are defined as problems change from place to place and time to time. (See, for example, Engelhardt, 1981a.) The search for solutions can obscure the evaluative process by which problems come to be defined as such. By recognizing and delineating the social context within which medical work is conceived and conducted, "the {very} subjectivity which accomplishes science" can be explored.

Phenomenology is not a method for acting in the world but for understanding that action. It provides a way to grasp what it is that people do in and through their action and to grasp the knowledge that serves them as the basis for those acts. It is indeed radical, for "[w]ith one blow he [the psychologist; here, the medical practitioner] must put out of effect the totality of his participation in the validities explicitly or implicitly effected by the persons who are his subjects; and this means all persons" (Husserl, 1970, p. 239). Kohák (1978, p. 86) emphasizes that this is possible, "We cannot think away experience and retain the reality of the world, but we can think away the reality of the world without changing the structure of experience."

V. THE SUSPENSION OF MEDICAL KNOWLEDGE

This greatest of all revolutions [phenomenology] must be characterized as the transformation of scientific objectivism – not only modern objectivism but also that of all the earlier philosophies of the millennia – into a transcendental subjectivism (Husserl, 1970, p. 68).

Recognizing the primacy of the subject's perspective enables that subject to suspend what is taken to be medical knowledge, both theoretical and practical – neither rejecting nor uncritically accepting it. This suspension makes available to medical theorists and practitioners a different vantage point from which to view a rich world of otherwise obscured experiences. As Carr (1970, p. xxx) notes, "The true introduction to phenomenology – and thus to philosophy, for Husserl – is to be found precisely by turning *away* from the opinions of philosophers [and, in the context here, of medical professionals] and by turning *toward* 'the things themselves'."

When not pre-formulated by medical theories, the experiences of both medical practitioners and their clients appear in new ways. Paget (1988, p. 21) writes of medical practice that her description of medical work is "at odds with many contemporary descriptions because they refer to medicine as an occupation or a profession rather than as the subject's experience of doing work – doctoring."

A recurrent theme of my description of medical work is that it is a process of discovery: medical work is discovered in action. *Discovering* is not like seeing or observing. Patients do not wear their illnesses as they wear apparel. One apprehends, one infers, one tests, one experiments, one tracks, one follows the course of events in order to disclose the nature of illness and affect it (Paget, 1988, p. 19, emphasis in original).

But Husserl (ante 1928, p. 310, emphasis added) writes, "Determinative thinking, judging, inferring, generalizing, particularizing, which are accomplished in the actual sphere of experience, take the appearing things, characteristics, regularities as the true ones; *but this truth is a relative and 'subjectively conditioned' truth.*"

If indeed medical work can be viewed as Paget describes it, then practitioners may find phenomenology a useful tool for exploring the details of that work, as it is locally enacted by practitioners, and for exploring the "subjectively conditioned truth" on which it is based. Phenomenology provides a method for clearing away presuppositions and for elucidating taken-for-granted assumptions, thereby disclosing the grounds of apprehension, inference, testing, etc. and of determinative thinking, judging, etc.

Looking at medical theory and practice from a phenomenological perspective can be difficult initially. Most challenging may be *acknowledging* the primacy of the subject's perspective and *suspending* belief and doubt, for both may run counter to one's typical ways of being in the world as well as to the tenets of medical science. However, once the legitimacy of these processes is recognized and undertaken, one can readily engage in the phenomenological gaze. As Husserl (1970, p. 237) writes, "[O]ne cannot transform certainty into doubt or negation, or pleasure into displeasure, love into hate, desire into abhorrence. But one can unhesitatingly abstain from any validity, that is, one can put its performance out of play for certain particular purposes."

It is this "putting out of play" that characterizes the phenomenological epoché. Given the potential difficulty in accepting the legitimacy of these two central features of phenomenology, in the remainder of this paper I want to suggest what "recognizing the primacy of the subject's perspective" and "suspending belief and doubt" might look like for medical practitioners. The data on which I draw comes from a variety of sources, chosen to give a sample of the range of ideas that become available for serious consideration when a phenomenological perspective is adopted. The central ideas that pervade these illustrations are the recognition that (1) different groups have different, perhaps mutually exclusive, perspectives and have what they take to be supporting evidence for their perspectives; (2) what is taken to be scientific theory and scientifically based practice is intertwined with personal preferences, practical considerations, values, social concerns, and political decisions; and (3) not only diagnostic categories but indeed the very notions of "health" and "illness" and "normality" and "abnormality" are social constructions (See, for example, Engelhardt, 1981b and Douard, 1995). By displaying what becomes available when the subject's perspective serves as starting point and when belief and doubt in the world are suspended, I hope to convey both the radical nature of the phenomenological method and its legitimacy as a method for examining medical theory and practice.

VI. ALTERNATIVE HEALING PRACTICES

Western medical practitioners are faced with a dilemma when considering alternative healing practices. If such alternative methods conflict with the tenets of Western medicine, and if those tenets are held to be true in an absolute sense, alternative healing practices may well require rejection even in the face of *experiential* evidence to the contrary. When alternative practices themselves reject the legitimacy of the Western conception of science, an impasse occurs, with each denying the validity and legitimacy of the other.

From a phenomenological perspective, Western medical formulations *are* formulations, a way of looking at the data but not the original data itself. Porkert (1975, p. 62) notes that conventional standards of scientific measurement are "oriented upon or derived from natural data *but they are neither imposed by nor do they directly express natural data.*" Alternative healing practices can be seen as derivations from natural data just as Western medicine is such a derivation. Phenomenological suspension allows for the understanding and appreciation – though not necessarily acceptance – of the experiences of those who use alternative healing methods. Husserl

(1962, p. 88) urges this respect for what he terms "the original right of all data."

Western medical science is *a* method but to take it as *the* method and the standard against which all other methods are assessed is to obscure the possibilities available in those other methods. The suspension of "knowledge" as knowledge makes possible the exploration of its character as what Husserl describes as "subjectively conditioned truth." What is taken to be "knowledge" can then be explored in terms of the taken-for-granted assumptions it embodies – some of which disclose themselves as less than scientific.

Are alternative healing practices best judged by the standards of Western medicine? According to Porkert (1975, p. 63), speaking of traditional Chinese medicine, "the attempt to judge Chinese medical theory by the utterly alien procedural standards of Western medicine is...[one] way in which its highly consistent data are vitiated and obscured." Using Western medicine as an absolute standard inevitably leads to the denigration of alternative healing practices (e.g., their success as more legitimately attributable merely to placebo effects, "spontaneous remissions," or the psychosomatic nature of the condition treated) and especially a denigration of alternative explanations. That a host of healing practices may not be analyzable or verifiable in Western medical terms does not, however, thereby render them false (nor, alternatively, true). Furthermore, the experiential evidence offered for such healing practices as chiropractic, acupuncture, homeopathy, herbalism, Tellington touch, Bach flower remedies, spiritual healing, shamanism – or indeed any alternative method – remains. Clearly there is a difference between scientific and experiential evidence, but simply because experiential evidence is not "scientific" in Western medical terms does not render it meaningless, especially to those who have the healing experiences.

Readers of the foregoing list of alternative healing practices might arrange the items from more to less credible but, *and this point is crucial*, they will do so in different ways. The *grounds* used for such an arrangement can be explored through the phenomenological epoché and may display themselves as not scientifically based. When the subject's "prejudices" are recognized, and when Western medical theory is suspended, alternative healing methods can be appreciated as methods, with their own theories, practices, and confirmatory evidence.

VII. NON-SCIENTIFIC ASPECTS OF SCIENCE

A consequence of the location of medical theory and practice within the world of everyday life is that ideas from both may be confounded. Within the phenomenological epoché, what is taken to be science may disclose itself as firmly located in the world of everyday life.

In practical life I have the world as a traditional world, no matter where the tradition comes from; it may even come from second-hand scientific acquisitions, even false ones, which I get from the newspaper or from school and which I may transform in one way or another in my own motivations or {through} those of my fellows who accidentally influence me (Husserl, ante 1930, p. 326).

Some of what is taken to be scientific knowledge may, within the reduced sphere of the epoché, display itself as grounded in personal preferences, practical considerations, values, social concerns, or political decisions rather than in scientific criteria. Consider the following two examples in terms of such motivations intertwined with medical considerations.

A. *Diagnostic and Statistical Manual of Mental Disorders*

The non-scientific, everyday life considerations that make their way into the creation of a nosology is illustrated in detail by Kirk and Kutchins in *The Selling of DSM: The Rhetoric of Science in Psychiatry* (1992). In offering a nosology, the creators of DSM impose a language on experience in a way that can come to be taken as definitive of that experience. Husserl (1936, p. 362, emphasis in original) was attentive to this distorting feature of language, noting that even in "{ordinary} human life, and first of all in every individual life from childhood up to maturity, the originally intuitive life which creates its originally self-evident structures through activities on the basis of sense-experience very quickly and in increasing measure falls victim to the *seduction of language.*" Awareness of this seductive feature of language is directly relevant to medical practice for, as Paget (1988, p. 30) notes, "the work of medical practice is done in and through discourse – both the diagnosis and the treatment of illness are communicated in interviews." Medical theory is also a linguistic activity, one that rests heavily on categorization ("naming"). Diagnostic categories, especially when they embody non-scientific aspects, have important consequences for medical discourse for categorizers, those categorized, and for any who use those categories.

Kirk and Kutchins make clear that the categories of DSM embody just such non-scientific aspects. The proliferation of categories in each succeeding version of DSM may be taken simply to indicate scientific refinement but there is reason to believe that other considerations are also involved, as implicit in Kirk and Kutchins' (1992, p.199) statement that "[d]iagnoses have rapidly expanded in number to cover broader areas of social behavior and to slice narrower disorders out of broader established ones. Newer categories of disorder contend for territory with established ones, as each revision of the diagnostic manual proclaims the latest nosological redistricting."

Kirk and Kutchins chronicle political jockeying involved in the creation of DSM, e.g., in excluding homosexuality as a disorder (for what appear to be social reasons, most notably politicking by those so labeled) and in including Post-Traumatic Stress Disorder (for what seem to be insurance considerations). Furthermore, "Diagnosis in mental health, now more than ever before, is a business act, as well as a clinical one" (Kirk and Kutchins, 1992, p. 233). The practical considerations of getting treatment reimbursed may direct medical practitioners to one rather than another diagnosis but once that diagnosis is made, these practical considerations may be obscured, the diagnosis reified.

Not only a particular nosology appearing in any one version of DSM but the very possibility of such a nosology can be questioned.

[M]ental health agencies are founded on a concept that many think is of dubious merit: that there are discrete "disorders" that are "mental." There is little agreement among mental health experts that disorders can be easily or clearly separated from nondisorders. Even more divisive among scholars is the notion that the mental can separated clearly from the physical, behavioral, social, or moral (Kirk and Kutchins, 1992, p. 226; see also the classic critique by Szasz, 1974).

From a phenomenological perspective, any nosology is an imposition on what Husserl (1962, p. 88) refers to as "the original right of all data". That diagnostic categories may be useful should not obscure their socially constructed nature and their potential to distort the ambiguous phenomena that they purport to describe.

Like most professionals, psychotherapists believe in what they do. They see people in pain and, as best they can, try to help them. It is uncertain work. Their clients' suffering is often diffuse, connected in so many intricate ways to their past lives, their personalities, and their current social relationships and social circumstances. Their problems seldom can be quickly understood or crisply described. In offering help, there are few certainties about what interventions will be most effective, who should administer them or for how long (Kirk and Kutchins, 1992, p. 196).

Although the uncertainties of psychotherapeutic work make a nosology such as DSM tempting as a potential source of certainty, its examination within the phenomenological epoché can disclose the costs, both intellectual and practical, of its employment.

Suspending diagnostic categories makes available the experiences of those being categorized. As Kohak (1978, p.85) notes, "What an experience is – for instance, the structure or 'logic' of paranoidal fear – is an experiential rather than an empirical question. We grasp it in experience seen from a subject's standpoint. We cannot deduce it, induce it, or construct it out of third-person observation of 'objective' movements."

Attention to the subject's standpoint discloses the subject's view of the matter, which may be quite different from the view attributed by another, even when (unlike the doctors in the following example) that attribution is made seriously.

One [schizophrenic] patient, who believed he had a rat in his throat and asked the doctors to look at it, was told *sardonically* by the doctors that the rat was too far down to see. When the patient recovered he recalled that "I would have been grateful if they had stated quite plainly that they did not believe that there was a rat in my throat" (from Torrey, 1988, p. 286, cited in Doubt, 1996, p. 4, emphasis added).

Suspending diagnostic categories also allows the reformulation of the very idea of "mental disorders," seeing them not necessarily as deficits and distortions but, alternatively, as possibilities. "[M]ental illness effaces, but it also emphasizes; on the one hand, it suppresses, but on the other, it accentuates; the essence of mental illness lies not only in the void that it hollows out, but also in the positive plenitude of the activities of replacement that fill that void" (Foucault, 1987, p. 17).

While DSM provides a nosology, in doing so it forestalls the creation of other kinds of nosologies and, as well, obscures the experiences of those categorized. When non-scientific concerns influence the construction of a nosology, the consequent distortions can be hidden beneath the patina of scientific authority. The phenomenological epoché allows a view beneath the patina. Subjecting scientific language to the phenomenological epoché can disclose its everyday life constituents and make available alternative formulations of the defined experience.

Although it may not be as readily apparent as with categories of "mental disorder," personal, practical, evaluative, social, and political considerations are also implicated in the construction of illness/disease/disability categories. Within a phenomenological perspective, "normality" and "health" are displayed as judgments made in specific contexts and in terms of particular sets of standards. Of the concepts "health" and "disease" Engelhardt (1981b, p.31) writes, "The concepts are ambiguous, operating both as explanatory and evaluatory notions. They describe states of affairs, factual conditions, while at the same time judging them to be good or bad. Health and disease are normative as well as descriptive." He makes the further point (Engelhardt, 1981a, p. 279) that although "vice and virtue are not equivalent to disease and health, they bear a direct relation to these concepts. Insofar as a vice is taken to be a deviation from an idea of human perfection, or 'well-being,' it can be translated into disease language."

B. Deafness and Other "Disabilities"

The designation of deafness as a "medical condition"[8] obscures the non-scientific elements underlying this classification. Medical practitioners and those who are deaf may have dramatically different perspectives on the medical status of deafness. The perspective adopted has important implications for the activities that follow.

[T]he more we view the child born deaf as tragically infirm, the more we see his plight as desperate, the more we are prepared to conduct surgery of unproven benefit and unassessed risk.[9] Our representation of deaf people determines the outcome of our ethical judgment (Lane, 1992, p. 238).

Yet for the deaf, a child born deaf may be seen as perfectly healthy. Is deafness then a medical condition or a feature of a culture? Phenomenology cannot resolve these differing perspectives; it can, however, recognize that they exist, that they are fundamentally different ways of viewing deafness, and that very different consequences, medical and non-medical, follow from these differing formulations.

Lane describes the perspective of those in the deaf community who argue that deafness is simply a way of being in the world, not a medical problem. In doing so he discloses the political and ethical dimensions that can lie concealed within a medical perspective.

There is now abundant scientific evidence that, as the deaf community has long contended, it constitutes a linguistic and cultural minority. I expect most Americans would agree that our society should not seek the scientific tools or use them, if available, to change a child biologically so he or she will belong to the majority rather than the minority – even if we believe that this biological engineering might reduce the burdens the child will bear as a member of a minority. Even if we could take children destined to be members of the African-American, or Hispanic-American, or Native American, or Deaf American communities and convert them with bio-power into white, Caucasian, hearing males – even if we could, we should not. We should likewise refuse cochlear implants for young deaf children even if the devices were perfect (Lane, 1992, pp. 236-37).

A phenomenological consideration of deafness raises the questions: What needs fixing? Who decides? And, how do they do so?

The issues raised here about deafness can apply to blindness, autism, and the variety of differences discussed by Oliver Sacks (1995) in *An Anthropologist From Mars*, as well as to mental retardation and such medical procedures as the separation of conjoined twins. (Notably, in the present context, the former designation, "Siamese" twins, has come to be viewed as *politically* problematic.) Both Sacks' (1995) discussion of Virgil in "To See and Not See" and Wilkie Collins' (1995/1872) description of the heroine in his novel, *Poor Miss Finch*, suggest that "curing" blindness may be viewed by those "cured" as no such thing but rather as a fundamental change in their way of being in the world whose destructive consequences can only be overcome by a return to blindness.

For those with what are medically described as "disabilities," the focus on cure (rather than, for example, on environmental change, assistance with daily living, or acceptance of different ways of being in the world) can appear a political and evaluative decision made by the non-disabled in general (those folk singer Fred Small refers to as the "temporarily able bodied") and by members of the medical profession in particular. The promise of a cure urges waiting rather than acting to bring "normality" to the on-going day-to-day process of living.

The key tension that exists in nearly all stories about disability is seen most clearly in the "cure narrative." It's a story of uneasy obliteration, for it always ends in the "loss" of the object upon which it depends: Either the disability is eradicated and surrenders its seductive interest, or it resists conforming to the narrative of its rehabilitation and the evidence of that refusal must itself be obliterated. This is what Paul Longmore means by the "cure or kill" paradigm: both get rid of disability (Mitchell, 1997, p. 21).

The search for "a cure," despite its initial appearance as unquestionably worthy, can be disclosed as taking for granted that all for whom the "cure" is designed both see the need for cure and are willing to wait for it. "Cure" makes attributions to the "to-be-cured" that they may reject. It directs attention to the future and away from the present, away from the construction of lives out of the resources that people do have available. Volkman (2000, p. 4), referring to disability in general and his own in particular, writes, "I have a trait, not a disease. No disease, no cure."

Seen through the phenomenological epoché, judgment of what is "wrong" and what is to be done can be disclosed as having a distinctly personal aspect, one grounded in the subjectivity of the one making the judgment. Subjective considerations may underlie the medical treatment plans offered or the decision to withhold treatment. Johnson (1997, p.10) in *Ragged Edge* – a publication of the disability rights movement – writes that in the euthanasia debate, the non-disabled "who support the right to die seem to have an almost panicky horror of living disabled...[T]he right to die is nearly always framed as a good for 'the terminally ill.' In right-to-die circles, severely disabled people are all said to be 'terminally ill' even if their disabilities are nonlife-threatening."

To speculate about what one might feel if one were disabled is a questionable basis on which to make attributions to those who are disabled. Upon becoming disabled, one may at first indeed want to die but whether such a wish is taken as "sensible" or simply a "stage" in the process of reconceptualizing oneself is of serious consequence for the medical activities that follow. Such considerations apply not only to those with disabilities but to those who may not be able to speak for themselves and are dependent upon the very consequential decisions of others, e.g., fetuses, children, and the elderly. When immediate medical decisions are called for, they may be influenced by whether or not the medical practitioner or other decision-makers would "want to live that way."

Phenomenology does not lead to the idea that "disability" is a meaningless designation. In the world of everyday life and in medical work it is indeed a very significant one. What phenomenology does do is direct attention to the experiences of those who label and of those who are so labeled. What O. Sacks (1995, pp. xviii-xix) says of disease applies equally to "disability," that its study,

for the physician, demands the study of identity, the inner worlds that patients, under the spur of illness, create. But the realities of patients, the ways in which they and their brains construct their own worlds, cannot be comprehended wholly from the observation of behavior, from the outside. In addition to the objective approach of the scientist, the naturalist, we must employ an intersubjective approach too, leaping, as Foucault writes, "into the interior of morbid consciousness, [trying] to see the pathological world with the eyes of the patient himself."

What appears to the physician as "morbid" or "pathological" may appear to the one so characterized as what one *is* and only possibly in need of fixing.

VIII. CONCLUSION

As a phenomenologist I can, of course, at any time go back into the natural attitude, back to the straightforward pursuit of my theoretical or other life-interests; I can, as before, be active as a father, a citizen, an official, as a "good European," etc., that is, as a human being in my human community, in my world. As before – and yet not quite as before. For I can never again achieve the old naïveté;

I can only understand it (Husserl, 1970, p. 210).

The ideas presented in this paper do not entail a rejection of medical theory and practice. Neither are they intended to suggest that, within the world of action, and specifically within the everyday world of medicine, all ideas are equal, nothing is known, no judgment is possible. The claim is that by looking at that world phenomenologically it is possible to examine the *grounds* of ideas and actions, both one's own and those of others. An awareness of those grounds makes it possible, within the world of everyday life and action, to accept, modify, or reject the ideas disclosed in the process of phenomenological examination. It is this *awareness* that phenomenology provides.

Husserl (1935, p. 295) criticizes works in which "the scientist does not become a subject of investigation." By taking themselves as subjects of investigation, medical theorists and practitioners can examine the presuppositions that undergird their perspectives.

[W]hen we become conscious of..."presuppositions" and accord these their own universal and theoretical interest, there opens up to us, to our growing astonishment, an infinity of ever new phenomena belonging to a new dimension, coming to light only through consistent penetration into the meaning- and validity-implications of what was thus taken for granted – an infinity, because continued penetration shows that every phenomenon attained through this unfolding of meaning, given at first in the life-world as obviously existing, itself contains meaning- and validity-implications whose exposition leads again to new phenomena, and so on (Husserl, 1970, p. 112).[10]

The phenomenological epoché makes available for analysis the taken-for-granted sphere of medical theory and practice. It also brings forward for recognition, consideration, and understanding the alternative mundane, everyday-life worlds inhabited by others. What are in both spheres taken to be "reality" and "knowledge" can be seen as social constructions and examined as such.

With a recognition that Western medical formulations are constructions and not direct expressions of natural data comes an appreciation of other (albeit at times perforce contradictory) theoretical formulations. To understand that at least some of what is taken to be scientifically based may disclose itself as a blend of personal preferences, tradition, practical considerations, values, social concerns, political decisions, and general everyday life "knowledge" is to grasp the full range of what is taken to be medical "knowledge."

The epigraph that begins this paper captures the spirit of the phenomenological method in its urging to "allow the possibility of the facts." In a similar vein Kohák (1978, p.87, emphasis added) writes, "The *possibility* of experience, rather than its contingent factual counterparts, thus constitutes the primordial subject of rigorous study."

Allowing the possibility that the world might be different from the ways it appears in everyday life and in medical theory and practice, examining that world free of "prejudices" and with a respect for the data itself, provides a different way of looking and different things to be seen. The understanding that follows can profoundly inform action. To suspend judgment, to direct attention to "modes of appearing," to acknowledge the multiplicity of perspectives (including one's own), and to recognize the socially constructed nature of "knowledge" and "reality" is to clarify one's perception of the world in which one lives, in which others live, and the worlds that we inhabit together.

Wheelock College
Boston
U.S.A.

NOTES

[1] I have chosen to work from Palmer's translation of Husserl's *Encyclopedia Britannica* article for the reasons offered by Spiegelberg (1971).

[2] My view of phenomenology draws heavily upon lectures given by Erazim Kohák and discussions with him. I have been particularly influenced by his interpretation of Husserl's thought. Nonetheless, the interpretations presented here are my own and ought not to be attributed to Kohák or others. Kohák, for example, criticizes Schutz' subjectivistic reading of Husserl; my own interpretation might be judged even more subjectivistic.

Some of the ideas about medicine presented herein have been explored in a course I have taught for a number of years, "Cross-Cultural Perspectives on Health and Illness."

[3] Without denying Schutz' contributions, I do see him as departing in significant ways from Husserl's phenomenological method. This departure is both explicit and intentional. Schutz (1967, p. 97) states, "As we proceed to our study of the social world, we abandon the strictly phenomenological method. We shall start out by simply accepting the existence of the social world as it is always accepted in the attitude of the natural standpoint, whether in everyday life or in sociological observation." I see Schutz as notably, and avowedly, straying from Husserl by *beginning* with the natural standpoint when Husserl so clearly brackets that standpoint. For a detailed critique of Schutz' interpretation of Husserl, see Costelloe, 1996.

[4] It is my application of the epoché, and particularly of phenomenological bracketing, to which some may object as going beyond Husserl, of being more radical than he was recommending. Spiegelberg (1965, p. 690), for example, minimizes the importance of the epoché and accuses Husserl of being unclear and inconsistent in his recommendations regarding its use. Nonetheless, in my own surveying of Husserl's works I have found repeated recommendations regarding the way in which it is to be applied and the importance of doing so (see, for Husserl's explicit statements (1964a, p. 22) and (1962, pp. 161-2) as well as the quotations presented above; see also Kohák, 1967). Even if it is true that Husserl is not completely clear about the method of the suspension of belief, there is little

question in my mind that he viewed the process as of vital importance.
⁵ These distinctions, including the way in which they are worded, are drawn from Kohák's lectures; see also Kohák, 1978, pp. 35-40.
⁶ In his translation of *Crisis*, Carr made some editorial additions for the sake of clarity. In his work he indicates these by brackets. Since I use brackets here for my own insertions, I have indicated his additions with { }.
⁷ The links between phenomenology and ethnomethodology are both complex and disputed. It is sufficient here to note that ethnomethodological studies of scientific practice (e.g., Lynch, 1988, Jordan and Lynch, 1992) provide empirical evidence for the claim I am making that scientific theorizing is located in the world of everyday life.
⁸ For the consequences for women when childbirth is viewed as a "medical condition," see Oakley (1980). For the implications of disease designations for the chronically ill, see Kleinman (1995).
⁹Here Lane is referring specifically to cochlear implants but his point applies more broadly to medical interventions for what clients do not see as medical problems.
¹⁰Here Husserl is speaking specifically of Kant; the sentence begins (on p. 111): "When we proceed, philosophizing with Kant, not by starting from his beginning and moving forward in his paths but by inquiring back into what was thus taken for granted (that of which Kantian thinking, like everyone's thinking, makes use as unquestioned and available),..." Rather than restricting the applicability of Husserl's claim to Kant, I take it as having relevance for phenomenology in general.

BIBLIOGRAPHY

Berger, P.L. and Luckmann, T.: 1967, *The Social Construction of Reality: A Treatise in the Sociology of Knowledge*, Doubleday, Garden City, New York.
Caplan, P.J.: 1995, *They Say You're Crazy: How the World's Most Powerful Psychiatrists Decide Who's Normal*, Addison-Wesley, Reading, Massachusetts.
Carr, D.: 1970, 'Translator's Introduction', in Edmund Husserl, *The Crisis of European Sciences and Transcendental Phenomenology: An Introduction to Phenomenological Philosophy*, Northwestern University, Evanston, Illinois, pp. xv-xliii.
Collins, W.: 1995/1872, *Poor Miss Finch*, Oxford University Press, New York.
Costelloe, T.M.: 1996, 'Between the Subject and Sociology: Alfred Schutz's Phenomenology of the Life-World', *Human Studies*, **19**, 247-266.
Douard, J.W.: 1995, 'Disability and the Persistence of the "Normal"', in S.K. Toombs et al (eds.), *Chronic Illness: From Experience to Policy*, Indiana University Press, Indiana, pp.154-175.
Doubt, K.: 1996, *Towards a Sociology of Schizophrenia*, University of Toronto Press, Toronto.
Engelhardt, H.T., Jr.: 1981a, 'The Disease of Masturbation: Values and the Concept of Disease', in A.L. Caplan et al. (eds.), *Concepts of Health and Disease: Interdisciplinary Perspectives*, Addison-Wesley, Reading, Massachusetts, pp.267-280.
Engelhardt, H.T., Jr.: 1981b, 'The Concepts of Health and Disease', in A.L. Caplan et al. (eds.), *Concepts of Health and Disease: Interdisciplinary Perspectives*, Addison-Wesley, Reading, Massachusetts, pp.267-280.
Foucault, M.: 1987, *Mental Illness and Psychology*, A. Sheridan (trans.), Foreword by Hubert Dreyfus, University of California, Berkeley.
Husserl, E.: 1970/1954, *The Crisis of European Sciences and Transcendental Phenomenology: An Introduction to Phenomenological Philosophy*, D, Carr (trans.), Northwestern University Press, Evanston, Illinois.
Husserl, E.: 1964a, *The Idea of Phenomenology*, W.P. Alston and G. Nakhnikian (trans.), Martinus Nijhoff, The Hague.

Husserl, E.: 1964b/1928, *The Phenomenology of Internal Time-Consciousness*, Indiana University Press, Bloomington.
Husserl, E.: 1962/1913, *Ideas: General Introduction to Pure Phenomenology*, Collier Books, New York.
Husserl, E.: 1960, *Cartesian Meditations: An Introduction to Phenomenology*, Martinus Nijhoff, The Hague.
Husserl, E.: 1936, 'The Origin of Geometry,' in Husserl, E., 1970.
Husserl, E.: 1935, 'The Vienna Lecture', in Husserl, E., 1970.
Husserl, E.: *ante* 1930, 'The Attitude of Natural Science and the Attitude of Humanistic Science, Naturalism, Dualism, and Psycholphysical Psychology', in Husserl, E., 1970.
Husserl, E.: *ante* 1928, 'Idealization and the Science of Reality – The Mathematization of Nature', in Husserl, E., 1970.
Johnson, M.: 1997, *Ragged Edge*, Jan./Feb., 9-12.
Kirk, S.A. and Kutchins, H.: 1992, *The Selling of DSM: The Rhetoric of Science in Psychiatry*, Aldine de Gruyter, New York.
Kleinman, A.: 1995, 'The Social Course of Chronic Illness: Delegitimation, Resistance, and Transformation in North American and Chinese Societies', in S.K. Toombs et al (eds.), *Chronic Illness: From Experience to Policy*, Indiana University Press, Indiana, pp.176-189.
Kohak, E.: 1978, *Idea and Experience*, University of Chicago Press, Chicago and London.
Lane, H.C.: 1992, *Mask of Benevolence: Disabling the Deaf Community*, Knopf, New York.
Mitchell, D.: 1997, 'The Frontier That Never Ends', *Ragged Edge*, Jan./Feb., 20-21.
Oakley, A.: 1980, *Women Confined: Towards a Sociology of Childbirth*, Schocken, New York.
Paget, M.A.: 1988, *The Unity of Mistakes: A Phenomenological Interpretation of Medical Work*, Temple University, Philadelphia, Pennsylvania.
Palmer, R.E.: 1971, '"Phenomenology": Edmund Husserl's Article for the Encyclopaedia Britannica (1927), New Complete Translation', *The Journal of the British Society for Phenomenology*, **XX**, 2, 77-90.
Porkert, M.: 1975, 'The Dilemma of Present-Day Interpretations of Chinese Medicine', A. Kleinman. et al. (eds.), *Medicine in Chinese Culture*, U.S. Dept. of Health, Education, and Welfare, Washington, D.C.
Sacks, H.: 1989, 'Lectures 1964-1965, *Human Studies* **12**, Issues 3-4.
Sacks, O.: 1990, *Seeing Voices: A Journey into the World of the Deaf*, Harper Collins, New York.
Sacks, O.: 1995, *An Anthropologist on Mars*, Vintage Books, New York.
Schutz, A.: 1967/1932, *The Phenomenology of the Social World*, G. Walsh and F. Lehnert (trans.), Northwestern University Press, Evanston, Illinois.
Spiegelberg, H.: 1965, *The Phenomenological Movement: A Historical Introduction*, 2nd Ed. Martinus Nijhoff, The Hague.
Szasz, T.S.: 1974, *The Myth of Mental Illness: Foundations of a Theory of Personal Conduct*, Harper & Row, New York.
Tymieniecka, A.T.: 1962, *Phenomenology and Science in Contemporary European Thought*, Noonday, New York.
Volkman, M.: 2000, "Cure, Revisited," (letter to editor), *Ragged Edge*, Sept./Oct., 4.
Waksler, F.C.: 1995, 'Introductory Essay: Intersubjectivity as a Practical Matter and a Problematic Achievement', *Human Studies*, **18**, 1, January, 1-7.
Walpole, H.: 1931/1764, 'The Castle of Otranto' in H. R. Steeves (ed.), *Three Eighteenth Century Romances*, Charles Scribner's Sons, New York.

FREDRIK SVENAEUS

THE PHENOMENOLOGY OF HEALTH AND ILLNESS[1]

'Being ill is above all alienation from the world' (Buytendijk 1974, p. 62).

I. INTRODUCTION

During the last three decades we have witnessed what might be called a revival of the philosophy of medicine and health. For different reasons the insight has become more widespread that the success of medical science by itself will never provide us with answers to all problems of clinical practice. Medical science itself does not provide us with a language in which we can carry out a philosophy of medicine and health; that is, a language in order to broach such questions as 'What is medicine?' and 'What is health?'[2] In this paper I will suggest that phenomenology provides us with such a language in its focus on lived experience – that is the feelings, thoughts and actions of the individual person living in the world. Such a theory of health will of course be normative, since it will fundamentally rely on the individual's interpretation and evaluation of his situation and not only on a biological investigation of his body.[3] The physiology of the body, however, certainly affects and sets limits to the different ways we are able to experience and interpret our being-in-the-world. To develop a phenomenological theory of health is, therefore, not intended as an attempt to replace biomedical research. In light of the successful history of modern medicine that would certainly be an absurd project. Phenomenology is meant to enrich our understanding of health in adding to the disease-level analysis a level of analysis that addresses the questions of how the physiological states are lived as meaningful in an environment.

The ultimate test of a phenomenology of illness must be the extent to which it does justice to the different forms of illness afflicting the individuals encountered each day by doctors, nurses and others working in the clinic. Thus, the way to a phenomenology of illness must indeed follow the well-known injunction of Edmund Husserl, the founder of phenomenology: 'back to the things themselves.' In this case this will mean: back to the ill persons themselves, back beyond theories of diseases to the experiences of the persons suffering from these diseases. Therefore, although this paper is not based on case studies, but rather on the philosophical explication of essences characteristic to illness through interpretations of the works of Martin Heidegger and other phenomenologists, it is nonetheless intended to inspire empirical challenge and confirmation.

II. THE PHENOMENA OF HEALTH AND ILLNESS

Phenomenology today is not one but several philosophical, sociological and psychological theories.[4] They all derive their origin, however, from the philosophy of Edmund Husserl. As mentioned above, the common root comes down to Husserl's claim, at the beginning of this century, that it is necessary to go back to 'the things themselves' – and the way they are presented to us – in order to find a solid ground for philosophy. The conceptualization of appearance, the *eidos* or meaning of experience, is to be found in the phenomena themselves by adopting a certain mode of attention. This project has been taken up and developed in many different ways, and some of them will be addressed in this paper.

How should we proceed when we want to carry out a phenomenological analysis of health? In many ways the phenomenon of illness seems to be far more concrete and easy to grasp than the phenomenon of health. When we are ill, life is often penetrated by feelings of meaninglessness, helplessness, pain, nausea, fear, dizziness, or disability. Health, in contrast, effaces itself in an enigmatic way. It seems to be the absence of every such feeling of illness, the state or process in which we find ourselves when everything is flowing smoothly, running the usual way without hindrance. Is it possible then only to characterize health in a negative way? Is it only the absence of illness? And should, consequently, the phenomenology of health focus only upon illness?

The distinction between disease as a state or process causing biological malfunction and illness as the lived experience of the person is an important one. It has formed a platform for thinking about medicine as something more than, and different from, applied biology in psychology, sociology and anthropology during the last three decades of the twentieth century.[5] Illness, not disease – that is, the lived experience of being ill – will also be my own point of departure when I try to formulate a phenomenological theory of health and illness in what follows.[6] In my attempt to formulate a phenomenological theory of health, however, illness is not meant as a psychological characterization of the life of the person, in contrast to the 'real' diseases of somatics. Phenomenology, as we will see, is not a psychological theory, and it is not some dualistic attempt to overcome materialism. To focus attention on lived experience as a structure of meaning means to bracket both materialism and dualism in order to get at the foundation – the lifeworld-based underpinnings – of the knowledge expressed in philosophical and scientific theories.

But is illness our only possibility for phenomenological analysis? Cannot health itself be explicated phenomenologically? Hans-Georg Gadamer (1993, pp. 143-44) writes in his book, *Über die Verborgenheit der Gesundheit*:

So what possibilities do we really have when it comes to the question of health? Without doubt it is part of our nature as living beings that the conscious awareness of health conceals itself. Despite its hidden character, health nonetheless manifests itself in a kind of feeling of well-being. It shows itself above all where such a feeling of well-being means that we are open to new things, ready to embark on new enterprises and, forgetful of ourselves, scarcely notice the demands and strains which are put upon us. This is what health is.

In spite of its character of withdrawal, Gadamer consequently sees a possibility of conceptualizing health. Unfortunately he does not provide us with this phenomenological theory of health in his book. What he does give are some ingenious, albeit unsystematic, hints concerning how we should look upon the phenomenon of health:

Health is not a condition that one introspectively feels in oneself. Rather it is a condition of being there (Da-Sein), of being in the world (In-der-Welt-Sein), of being together with other people (Mit-den-Menschen-Sein), of being taken in by an active and rewarding engagement with the things that matter in life. . . . It is the rhythm of life, a permanent process in which equilibrium re-establishes itself. This is something known to us all. Think of the processes of breathing, digesting and sleeping for example (Gadamer, 1993, pp. 144-45).

Here Gadamer is on the verge of a truly phenomenological account of health. As we will see, the terminology he is using is taken from the phenomenology of Martin Heidegger, which is also where I will begin my search for such a theory. First, however, some general and systematic problems of approaching health phenomenologically need to be addressed.

There are not many examples of phenomenologists who have attempted to work out a general theory of health.[7] In spite of this I think that the approach of phenomenology, which focuses upon the *meaning* of phenomena, opens up promising possibilities for the conceptualization of health. But the phenomenon of health does not quite seem to fit the structure of intentionality that we find in Husserl. To do phenomenology according to Husserl means to study our experience as *consciousness* of an object. To experience something, according to Husserl, is always to be conscious of it, to be directed towards it in an act.[8] The terms that Gadamer uses to characterize health in the last quotation – being there, being in the world, being with people, to be busy with projects in life – indicate that health is something we live *through* rather than *towards*. To illustrate let me give an example of a very common episode of illness: having a bad cold. We all know what it is like to be in this condition – dizzy, tired, shivering, listless, the throat is sore and the head feels as if it is stuffed with cotton. Our own body and the world around us offer severe resistance, it takes great effort not to simply stay in bed and sleep through the whole day.

Illness is thus obviously an obstruction to health and its transparency;

everything that goes on without us paying explicit attention to it when we are healthy – walking, thinking, talking – now offers resistance. The body, our thinking, the world, everything is now 'out of tune', colored by feelings of pain, weakness and helplessness. I will suggest in this paper that this way of being in the world in illness is best understood as a form of homelessness. But to understand this claim fully, we first need to learn more about phenomenological analysis and its vocabulary

We can see clearly from the example of having a bad cold that, in order to understand what changes when you fall ill, it is not enough to study what type of acts you live through that day in contrast to a healthy day; that is, the simple fact that you do not approach as many objects or subjects that day as on a healthy day is not the crucial issue, though it is important. What is crucial is the very character of the acts you *do* perform: they acquire a different texture (dizziness, pain, resistance), and as the subject of these acts, you experience yourself differently. The attunement and embodiment of the self in its being in the world is vital for understanding illness as I will try to explicate in more detail below. Some phenomena – and health and illness obviously belong to this domain – thus seem to demand a phenomenological approach which does not start with the scaffoldings of consciousness, intentionality and object, but rather investigates the meaning of human experience situated in the world as acting, attuned and embodied. Let us now turn to a version of phenomenology which I think provides such an approach – the philosophy of Martin Heidegger.

III. HEIDEGGER'S PHENOMENOLOGY

Heidegger, in his first main work *Sein und Zeit* (1927), widened the domains of phenomenology.[9] Husserl was mainly concerned with epistemology – the theory of knowledge, with an emphasis on the theory of science, whereas Heidegger rather focused on the everyday world of being and understanding. Accordingly, his phenomenology is what he calls a 'fundamental ontology' – investigating different modes of what it means to be, rather than what it means to know.

Instead of the subject of knowledge, Heidegger takes as his starting point what he calls 'Dasein', the 'being-there' of human existence. This being-there means that we are situated or 'thrown' (*geworfen*) into the world that we live in. We are always already *there* (*da*), involved in daily activities. But the term 'Da-sein' also signifies that we have a relation to our own existence in asking what it means to *be* there at all (rather than not to exist). *Dasein* is the only being that asks the fundamental ontological question of what it means to be (*die Frage nach dem Sein*). Therefore philosophy has to start with an analysis of the understanding human being – *Dasein*. This – as we shall see – is the only way to start, since every other being in the world attains its meaning through the understanding of *Dasein*. Human understanding as a 'being-there' in the world is accordingly the starting point for philosophy as a phenomenology of 'everydayness' – an investigation of the everyday forms of understanding carried out by human beings.

According to Heidegger, when we study our relationship to the world, we should not view the world as a collection of objects outside of consciousness, towards which we are directed by way of the latter. We should instead study the 'worldliness' of the world, the way we are *in* the world, giving it meaning through our actions; the world indeed being nothing other than a cultural, intersubjective *meaning-structure*, lived in by us and, ultimately, a mode of ourselves. Human understanding is, consequently, for Heidegger, always a being-there in the sense of being-in-the-world (*in-der-Welt-sein*). The hyphens indicate that *Dasein* and world are thought as a unity and not as subject and object. The world is not something external, but is constitutive for the being of *Dasein*.

The concept 'worldliness' (*Weltlichkeit*) in *Sein und Zeit* indicates that the structure of the world is built up by the understanding actions, thoughts and feelings of human beings situated in the world and not by any properties that belong to the world in itself as a collection of objects (things, molecules, atoms). Heidegger can, therefore, write that worldliness essentially is an *existential* – that is, something belonging to *Dasein*, to understanding human beings, and not to the world in itself. The meaning-structures of the world are made up of relations between that which Heidegger calls tools (*Zeuge*). That is, the meaning of phenomena, according to Heidegger, is not primarily dependent upon how things look, but upon how they are being used. This makes the connection between the structure of the world and *Dasein* more lucid. For how could the world itself as something independent of human beings lead us to an understanding of the function of any tool? A tool always refers to its user. We will only learn what a hammer is by using it, never by staring at it (Heidegger, 1986, p. 69).

The relations between the different tools are explicated as an 'in order to' (*um zu*) (Heidegger, 1986, p. 68). The tools in this way relate to each other; their meanings are determined by their places within the totality of relevance. One uses a hammer in order to nail the palings, in order to raise the walls, in order to build the house, in order to find shelter from the rain, etc.[10] One need not pay attention to these different levels of sub-goals at all times. Indeed, some of the sub-goals are never explicitly attended to, but are only revealed in a theoretical analysis of the activity.

The final meaning of every tool is the existence of human being – *Dasein*. The understanding of *Dasein* is the activity in relation to which every phenomenon (tool) takes on meaning. In this activity, as pointed out, we most often do not pay explicit attention to any of the tools. We are absorbed in the activity – as in the example of hammering above. Heidegger's (1986, p. 69) analysis of the worldliness of *Dasein* stays true to this phenomenological fact by focusing upon the practical relations between tools in the world, instead of upon the intentionality of consciousness, which he holds to be a special form of theoretical activity.

Heidegger conceptualizes the being-in-the-world, the 'worldliness' of human existence, by stressing several different aspects of this existence. Since these aspects belong to the only being that truly exists – *Dasein* – and not to things, they are called 'existentials' (*Existenzialien*). Human beings *exist*: that is, they have a relation to their own being, and they are open to the world as a possibility for themselves. This openness to the totality of tools is a pattern not only of action, but also of thinking, feeling and talking. These three modes of being must, however, not be conceived of as attributes of a subject – qualities of a thing – but as a meaning pattern that binds human being and the being of the world together. They must likewise articulate a being that is not merely contemplative but *acts* in the world, as the tool pattern makes obvious. The three main existentials which Heidegger chooses in order to make sense of our being-in-the-world are understanding (*Verstehen*), attunement (*Befindlichkeit*) and discourse (*Rede*). These three existentials are thought of as intertwined; they always work together inseparably as an attuned, articulated understanding situated in the world. The choice of understanding instead of cognition makes a relation to action possible. Understanding can include the active and incarnated sides of life; and, as we will see, so can the concepts of attunement and discourse.

IV. HEALTH AS HOMELIKE BEING-IN-THE-WORLD

Let us now return to health. Heidegger never wrote anything substantial about the subject.[11] There is no analysis of the structures of the healthy versus the ill existence in *Sein und Zeit*. Although a whole chapter in the second division of *Sein und Zeit* is devoted to an analysis of the meaning of death, Heidegger never links the ontological interpretation of death as the finitude of human existence to an analysis of the meaning of illness and health. He makes clear that the existential interpretation of death is prior to any biology or ontology of life, but he also mentions that the medical and biological inquiry into life and dying could be of importance in the analysis of the meaning-structure of human existence (Heidegger, 1986, p. 247). The relation between biology and phenomenology is not further discussed, however, nor is a phenomenology of health and illness developed.

It is tempting to interpret the famous opposition of authenticity and inauthenticity in *Sein und Zeit* as a difference between health and illness, at least when one considers mental health or the 'health' of a culture; and some interpreters have indeed done so.[12] But I do not think this is an attractive approach, at least not if one intends to preserve the normal, everyday meanings of the terms 'healthy' and 'ill'. Authenticity and inauthenticity in Heidegger's phenomenology are two different modes of understanding.

The authentic, philosophical understanding (which in *Sein und Zeit* is reached in the moment of anxiety) would obviously not always be identical with a healthy being-in-the-world, however. The phenomenologist could be temporarily ill, but still have an authentic understanding, or he could indeed reach it while (or even through) being ill. Also, the phenomenologist would always have to return to the inauthentic understanding of his daily life, since it provides the necessary background for authentic understanding, by being precisely that which is thereby explored as a pattern of meaning (as the existentials and as the totality of relevance of the world). It would, therefore, be absurd to regard every human being living in an inauthentic mood of understanding as ill. In *Sein und Zeit* Heidegger associates the term 'das Man' with a certain form of everyday existence lacking genuine selfhood and reflection; but though there is certainly some contempt in his description of modern public life, he did not mean to suggest that all human beings who never reach authentic understanding are ill.

Despite the shortcomings of the attempt to identify health with authentic existence, it will be instructive here to go back to the account Heidegger (1986, p. 188) gives of existential anxiety:

In anxiety one has an 'uncanny' feeling. (In der Angst ist einem 'unheimlich'.) Here the peculiar indefiniteness of that which Dasein finds itself involved in with anxiety initially finds expression: the nothing and nowhere. But uncanniness means at the same time not-being-at-home. In our first phenomenal indication of the fundamental constitution of Dasein and the clarification of the existential meaning of being-in in contradistinction to the categorical signification of 'insideness', being-in was defined as dwelling with .. being familiar with . . .

What authentic anxiety makes evident is essentially the same phenomenon that is brought to attention, not in healthy, but in *ill* forms of life – the *not* being at home in the world. Even in our everyday modes of being-in-the-world, Unhomelikeness – (*Unheimlichkeit*)[13] – which is taken to an extreme in the authentic mood, is a basic aspect of our existence; but there it is hidden by a dominating being *at home* in the world and is therefore covered up

(*Verborgen*) (Heidegger, 1986, p. 277).

The being-at-home of human being-there (*Da-sein*) – 'dwelling with, being familiar with' – is consequently, at the same time, a being not quite at home in this world. The familiarity of our lifeworld – the world of human actions, projects and communication – is always also pervaded by a homelessness: this is my world but it is also at the same time not entirely mine, I do not fully know it or control it. This is not a deficit, but a necessary phenomenon: I am delivered to the world (*geworfen*) with other people, and being together with them (*Mitdasein*) is a part of my own being (Heidegger, 1986, pp. 117 ff.). Thus, the world I live in is certainly first and foremost *my* world (and not the 'objective' world of atoms and molecules), but to this very 'mineness' also belongs otherness in the sense of the meaning of the world belonging to other people. The otherness of the world, however, is not only due to my sharing it with other people, but also to *nature* (as opposed to culture) as something resisting my understanding.[14] We will return to this alien nature of the world later, in discussing illness and the body.

Health is to be understood as a being at home that keeps the not being at home in the world from becoming apparent. In illness, the not being at home, which is a basic and necessary condition of human existence, related to our finitude and dependence upon others and otherness, is brought to attention and transformed into a pervasive homelessness. One of two *a priori* structures of existence – not being at home and being at home – wins out over the other: unhomelikeness takes control of our being-in-the-world. The basic alienness of my being-in-the-world, which in health is always in the process of receding into the background, breaks forth in illness to pervade existence. This unhomelikeness will be the central theme in our following interpretation of illness.

V. THE ATTUNEMENT OF ILLNESS

When we try to grasp the difference between healthy and ill ways of being-in-the world, the *attunement* of our being-in-the-world seems to be the phenomenon to focus upon. To be healthy or ill is, however, certainly not identical with just having a good or a bad feeling. Attunement is not the quality of a thing – of an isolated human subject – but rather, as stated before, a being delivered to the world, a finding oneself in the meaning-structure of the world as an understanding existence. Recall the quotation from Gadamer (1993, pp. 144-45) I gave above:

Health is not a condition that one introspectively feels in oneself. Rather it is a condition of being there (Da-Sein), of being in the world (In-der-Welt-Sein), of being together with other people (Mit-den-Menschen-Sein), of being taken in by an active and rewarding engagement with the things that matter in life. . . . It is the rhythm of life, a permanent process in which equilibrium re-establishes itself.

We now recognize Gadamer's terms as derived from the phenomenology of *Sein und Zeit*, and we are able to understand this healthy, rhythmic being-in-the-world as a form of attunement, which is certainly not just a state that one feels in oneself, but which nevertheless has a 'tune', a feeling component to it.

Citing another passage from Gadamer's (1993) book, I concluded that the phenomenon of illness seems to be easier to grasp than the phenomenon of health. When we are healthy everything 'flows', the mood in which we find ourselves does not make itself heard or seen. It could possibly be felt as a certain form of rhythm in our being-in-the-world connected to time and to the way we are incarnated – to our breathing and to the beating of our hearts (Gadamer 1993, p. 166). It should not, however, be confused with other *positive* moods like well-being or happiness. Such moods also color our understanding, but in a much more obvious and manifest way than health. Health is a non-apparent attunement, a rhythmic, balancing mood that supports our understanding in a homelike way without calling for our attention.

The homelikeness of health can obviously not be a balance between two (or several) entities. The 'balance of health' must phenomenologically refer to the way human being finds its place in the world as a meaning pattern – a being-*in*-the-world. Health is, thus, not a question of a passive state but rather of an active process – a *balancing*.[15] As an illuminating comparative example in exploring balance and its relation to health, we will therefore choose a human activity, namely the everyday activity of riding a bicycle.

The healthy mood – the healthy attuned understanding – is always in the process of receding, since it is a mood of homelikeness. In the same way as you do not think of your homelike being-in-the-world when you are healthy, you do not think of your physical balance when you are riding a bicycle. These moods are similar in this respect, though the first one is more complex, since it is not only an attunement of one activity (bicycling), but of one's entire being-in-the-world. Both are, however, unobtrusive *attunements* supporting our action and understanding. But are they really moods? Can health be *felt*? As developed above, health is obviously not only a feeling in the normal sense, but as a background attunement it does offer the possibility of a kind of direct attention.[16] You could seek this out for yourself phenomenologically by trying to neutralize every mood that is coloring your understanding and focus attention upon the way you are *in* the world as a taking place in a meaningful context. You will not manage to extirpate all attunement, but rather will reach a basic form of balanced, homelike attunement that *supports* your understanding in transcending to the world.[17] Heidegger (1986, p. 134) is indeed right when he states that human beings are always in *some* mood. The health-mood could be seen as a sort of borderline case of the transparency of attunement. It is there all the time in the alternation between different more intrusive moods, sometimes left alone as the pure background mood of homelike understanding.

If you fall off the bicycle or get ill, however, you will notice. When your balance is challenged – when you ride over a stone, for example – you will make efforts to regain your balance in order not to fall down, but eventually will not succeed and will fall over. The moods of illness, in contrast to healthy attunement, seem to manifest themselves in obtrusive ways, coloring our whole existence and understanding. This is not always an immediate experience – like falling off a bicycle – but sometimes a gradual process, during which (in the same way as in the example of riding a bicycle) one strives constantly to 'stay upright', to keep one's balance, but finally has to give in to illness.

VI. THE LIVED BODY AND THE BROKEN TOOL

Let us now return to Heidegger's analysis of the person's being-in-the-world as an attuned understanding played out in the meaning-patterns formed between different *tools*. How is it possible to get explicit, philosophical hold of these meaning-patterns if we are always absorbed in pre-intentional activity? I have already given a preliminary answer to this question by emphasizing the role of anxiety as the prerequisite for reaching authentic understanding. The moment of anxiety signifies a certain form of interruption of all activity in which the world loses its meaning as a totality of relevance for human existence and stands out as perfectly irrelevant – that is, as lacking sense. The focus of understanding thereby changes from the things and goals *in* the world to the *world itself* as a meaning-structure of *Dasein* (being-there-in-the-world) (Heidegger, 1986, p. 186). In authentic understanding the senselessness makes it possible to explicate the world-structure, not as a collection of things, but as a meaning-structure of the phenomenologist's being-in-the-world. In this attitude (the phenomenological attitude) the irrelevant senselessness can be developed into a positive theoretical understanding.

I do not think that anxiety is the only mood that offers the possibility of phenomenological analysis. In his later works Heidegger emphasizes other moods (Held 1993). One of the points with respect to Heidegger's focus upon anxiety in *Sein und Zeit* is, however, that the possibility of phenomenological analysis is not entirely up to oneself, but also depends upon the world in which one lives and into which one has been thrown. One's attunement, which is where every analysis starts, does not arise in one on account of one having chosen it, but as a result of one's being-in-the-world.

As I have pointed out, authentic anxiety as Heidegger envisages it in *Sein und Zeit* bears resemblance to the unhomelike attunement of illness, which we have focused on above. Authentic anxiety seems to mean a breakdown of all ordinary understanding and activity. Is this not equivalent to the defective transcendence of the ill way of being-in-the-world? Is it not in a way similar to a total lack of transcendence (death, coma)? This would indeed be true if the authentic anxiety lasted for a long time, but according to Heidegger (1986, p. 344) it does not. The total withdrawal from the activities of the world only lasts for a moment (*Augenblick*). After this rewarding experience the phenomenologist has to return to everyday doings – but with new insights, of course, which could change his everyday life. The unhomelike attunement of the ill existence would, in contrast to the attunement of authenticity, be lasting. It could, however, in a similar way, in some cases, offer new perspectives on the worldliness of human being.[18]

Certainly not every breakdown in the ongoing activity of being-in-the-world is equivalent to the anxiety of authentic understanding. Firstly, many moments of anxiety do not result in authentic understanding. The reason for this is that the person tends to identify the source of anxiety with a thing in the world and not with the world-structure itself. But in addition to this, very few breakdowns are total, in the sense of making our *entire* existence senseless. Most are merely partial, interrupting our ongoing activity on account of the failure, malfunctioning or obtrusiveness of a certain tool in the world.

Heidegger's (1986, pp. 69, 84) famous example is the hammer. While engaged in building a house, we do not, in the act of hammering, explicitly focus our attention upon the hammer in an intentional way. We are absorbed in the activity. Suddenly, the head of the hammer flies off, and the tool can no longer be used for striking nails. The activity is now interrupted, and we are forced to focus upon the hammer, to become conscious of it as a broken tool which must be repaired or replaced, if we are to be able to go on building the house. Heidegger's (1986, pp. 73 ff.) point here is that through breakdowns in the activity, it becomes possible to grasp the hammer as a piece of equipment – a tool – and not just as a thing. Similar to authentic understanding, certain breakdowns of activity (which, of course, must also be attuned but not necessarily in the mood of anxiety) make it possible to explicate *parts* of the world-structure as tools.

In *Sein und Zeit* Heidegger neglects to highlight the part played by the body in activity and understanding. The lived body (*Leib*) is merely a part of Dasein's spatiality (Heidegger, 1986, p. 108). He never extensively discusses the body as a set of tools or part of the meaning-structure of the world.[19] There is no valid argument, however, for restricting the concept of tool (*Zeug*) as it is used in *Sein und Zeit* to things outside the human body.[20] To place the limit of the meaning-patterns of the world at the surface of the biological organism would be to proceed from a materialist perspective, and not from the phenomenological point of view that Heidegger adopts in this work. Once the positions of dualism and reductionism have been given up in the phenomenological attitude they cannot be brought in again in order to draw a line between the hammer and the hand in the phenomenological analysis. If the hammer and not the hand can be said to constitute a tool in the phenomenological sense, this line of demarcation must be drawn from the phenomenological position. I will now try to show how this line can be drawn – not at the surface of the biological organism, however, and not by denying the hand the position of a certain type of equipment – but rather by stressing the distinction between *self* and *world*.[21]

If the hand instead of the hammer breaks, the activity will likewise come to an end. The hand consequently is also a sort of tool for activity and understanding constituting a nodal point in the meaning-pattern of the world. The hand, like the hammer, can be repaired or maybe even replaced by a prosthesis that would adequately perform the functions of a hand. But the situation of the broken hand is also different from the situation of the broken hammer since it is very differently attuned. The confusion and annoyance of the broken hammer can hardly be compared to the pain we experience having broken a wrist. This pain, if continuous, is typically an attunement of illness.[22] The attunement of illness is linked to a defective understanding, since the meaning-patterns of the world are disturbed as a result – in this case – of the body-tools breaking down. The understanding is limited since a vital part of the totality of relevance has been removed. Still, in a way, it would be wrong to call the body parts tools since they are also part of *Dasein* as a self. They are not only a part of the totality of tools, but also, as lived (*leibliche*), they belong to the projective power of the self. The hand and the hammer are, thus, rightly regarded as *different* forms of tools; but the line of demarcation is not drawn with reference to the surface of the biological organism. It is rather determined through an appeal to the importance the tool plays in the totality of relevance for the human being in question. If the tool belongs to the region we would identify with the self rather than the world in the person's being-in-the-world, then it would consequently be more likely to result in illness if broken.

The self must, however, always be understood as acquiring its meaning and identity by way of its being-in-the-world.[23] The loss of a beloved could lead to an attunement of illness if he formed an irreplaceable part of the person's self as a being-in-the-world; whereas the loss of some body parts – such as the appendix – could take place without us afterwards experiencing any pain or illness, and in this case it could actually mean a recovery from illness. The loss of all hair, or the loss of an irreplaceable toupee that had made one's being-in-the-world homelike again after one had gone bald, would be interesting borderline examples. Could the loss of either of these two 'tools' constitute illness? The question would have to be answered through a careful analysis of the person in question and his being-in-the-world.

The body as lived (*Leib*) would, in any case, just like language, form a vital part of our transcendence into the world. It would indeed seem appropriate to accord the lived body the status of a fourth existential of our being-in-the-world, alongside the other existentials of understanding, attunement and language. Heidegger (1994, p. 293) says to Medard Boss in 1972:

Everything that we refer to as our lived body (unsere Leiblichkeit), including the most minute muscle fibre and the most imperceptible hormone molecule, belongs essentially to our mode of existence. This body is consequently not to be understood as lifeless matter, but is part of that domain that cannot be objectified or seen, a being able to encounter significance (Vernehmen-könnens von Bedeutsamkeiten), which our entire being-there (Da-sein) consists in. This lived body (diseses Leibliche) forms itself in a way appropriate for using the lifeless and living material objects that it encounters. In contrast to a tool (Werkzeug) the living domains of existence cannot, however, be released from the human being. They cannot be stored separately in a tool-box. Rather they remain pervaded by human being, kept in a human being, belonging to a human being, as long as he lives.[24]

VII. THE INTERTWINING OF THE EXISTENTIALS

That the body is pervaded (*durchwaltet*) by human being (*Mensch-sein*) must certainly mean that it is attuned and transcending. As Merleau-Ponty (1962, p. 139) writes, the body 'understands' and 'inhabits' the world. Heidegger (1994, p. 113) expresses this notion through the neologism 'Das Leiben des Leibes.' Existence is a 'bodying forth' in the meaning-structures of the world. These meaning-structures are consequently not confined to language or consciousness. Indeed, as early as in *Sein und Zeit,* Heidegger had given priority to everyday *actions* in his analysis.

Interestingly enough, in the *Zollikoner Seminare* Heidegger (1994, p.

118) refers to the 'bodying forth' of human being as 'Gebärde' – gestures – a term connected to language, the language of the body:

> In philosophy we must not confine the meaning of the term gesture to 'expression' ('Ausdruck'). This term we should instead use to characterize all behaviour of human being (alles Sich-Betragen des Menschen) as a being-in-the-world determined by the life of the lived body (Leiben des Leibes). Every movement of my body (Leib) as a gesture and behaviour enables me not only to take place in physical space. Rather the behaviour is always already situated in a region (Gegend) that has been opened up by the thing that I am occupied with, as for instance when I take something in my hand

Body and language are inter-nested just like attunement and understanding. To reach out with the hand and to speak both belong to the transcendence (*Entwurf*) of human being. They are not identical, but rather, just like attunement and understanding, are different aspects of our being-in-the-world. The four existentials all permeate and support each other. Illness would mean an unhomelikeness of this structure of being-in, which can be thematized as a peculiar kind of attuned understanding, or as a breakdown in the tool-structure related to the self.

In many cases of illness the basic experience of being ill is tied to the body. In *The Context of Self* Richard Zaner (1981, pp. 48-55) investigates different ways in which the body announces itself as uncanny in illness. My body is not just a dwelling I live in (this is the basic mistake of dualism) – it is *me* (failing to recognize the significance of this 'mineness' is the basic mistake of reductionism). I *am* my own body. Yet this body, as Zaner remarks, also has a life of its own. As we have come to realize through our reading of Heidegger, I belong to the world just as much as the world belongs to me. In the same way I also belong to my body.

The body is alien, yet, at the same time, myself. It involves biological processes beyond my control, but these processes still belong to me as lived by me. The body seems to be the very form of finitude, in the sense of referring to our being born as well as our having to die. Our finitude is so to say incarnated in the being of the body as our form of existence. Illness is an uncanny (unhomelike) experience since the otherness of the body then tends to present itself in an obtrusive, merciless way. In illness the body often has to be surveyed as something other than oneself, something that has its own ways and must be regulated if one is to survive. The behavior of the body in illness is often no longer under control; the emptying of bowel and bladder, for instance, takes place without the ill person being able to regulate it. Of course the body has a life of its own even in health – certain needs must be satisfied; but the inconvenience, experienced in health, of urgently having to heed the call of nature, is very different from the inconvenience of urinating or defecating without noticing it until it is too late. The change in outer appearance and the loss of mastery over one's own body, which take place in many cases of illness, can in themselves be stigmatizing and lead to problems in relationships with other people.[25]

The unhomelikeness of illness would never consist in one or several of the existentials being totally absent. This is indeed not even a possibility, since the existentials are not merely added to each other as things in the world but permeate each other in an unravellable way, forming the necessary structure that makes being-in-the-world possible at all. The *entire* being-in-the-world of the ill person is always involved in illness. Even if one cannot speak (dumbness) or indeed possesses no 'language' in the usual sense (severe disturbance in mental development), one's body would still 'express' things in its bodying forth in the world. Analogously, tetraplegics who are able to speak or express themselves through writing (perhaps through some sort of computerized device) would still in some restricted sense be present in their own bodies.[26] Consequently, no transcendence and no illness is entirely somatic or entirely mental. On the contrary, these terms are applicable only on the psychological and physiological level – to diseases – not on the meaning-level of health and illness.

VIII. FINAL REMARKS – HEALTH, HOMELIKENESS AND HEALTH CARE

The point of my phenomenological approach has been to give a *general characteristic* of health and illness and provide a vocabulary with the aid of which one can talk about different illnesses and different ill persons and, in each case, be able to understand why and how *this* person is ill. I think the general scheme I give, with four existentials (understanding, attunement, language, lived body) structuring our being-in-the-world in a homelike or unhomelike way, provides a promising ground for adequate descriptions of different illnesses with different forms of breakdowns of meaning-structures accompanied by unhomelike understanding.

I am not saying, however, that all that phenomenology can provide is a good description of what different individuals experience. As I have stressed before, phenomenology is not a part of descriptive psychology. The phenomenological description aims at uncovering necessary features of the meaning-patterns structuring our life. The explication of phenomena – in this case, ill persons' experiences – can, therefore, also help us understand in which sense they are *ill* (and not, for example, merely unhappy). *Unhomelike being-in-the-world* is the feature I have suggested to fill this illuminating role in my analysis. The structures that break down in illness are the meaning-patterns of being-*in*-the-world, the way we inhabit the tool-structures we transcend in our homelike being-in-the-world. The unhomelikeness of illness is consequently a certain form of senselessness, an attunement of, for instance, disorientedness, helplessness, resistance, and despair.

My claim has been that all forms of attunements of illness are best understood as unhomelike ways of being-in-the-world. But is not unhomelikeness a more general aspect of human existence that is experienced in many different types of situations, and not exclusively in illness? The question is important since a characterization of illness that encompasses too many conditions and situations (normally referred to as suffering and unhappiness but not as ill health) threatens to expand the territory of medicine in an unrealistic and even dangerous way. It has not been my intention in this paper to provide more fuel for the ongoing process of a medicalization of life, but rather to supply clinical medicine with the possibility of a more human approach when it encounters ill persons. I will, therefore, by way of conclusion devote some time to possible counterexamples to my phenomenological proposal for a characterization of health and illness.

What about the unhomelikeness one experiences in getting lost in the woods for instance? This attunement is surely unhomelike and uncanny and yet the lost person can hardly be referred to as ill. I think this counterexample can be met with a reference to the lasting character of an illness mood. Feeling disoriented in one's being-in-the-world happens regularly in exploring new territories (geographical as well as thematic) and is indeed a part of a *healthy* being-in-the-world, the person being interested in learning new things that he does not initially control. As I wrote above, the being-in-the-world of the person can best be understood as characterized by encounters with both familiar and unfamiliar phenomena: the world does not only belong to me, it also has a sense of alienness and is thus basically both homelike *and* unhomelike for me, even when the former phenomenon is predominant. To be permanently lost 'in the woods' – that is, in-the-world – would, however, in most cases, mean illness. We can hardly specify a precise time-lag necessary for the unhomelikeness of being-in-the-world to become ill, in contrast to other types of homelessness. Time in phenomenology always means lived and meaningful time – a structure of human existence – not the objective time of physics.[27]

Stressing the time aspects of illness, while contrasting it to other forms of unhomelikeness, is indeed a start, but it hardly answers all our questions. An existential crisis (after losing a loved one, or losing one's faith in God, man or life in general) is clearly a paradigm example of unhomelikeness, but it is not always synonymous with depression – that is, with illness. Interestingly enough, we here seem to have a kind of mixture of Heidegger's authentic unhomelikeness and the unhomelikeness of illness. The moment of authentic unhomelikeness is mitigated enough to be extended and allow a process of lasting attunement which does not mean total paralysis. Thus, life goes on, but is still fundamentally questioned as a foundation for meaning. Existential crisis can be transformed into an illness mood, like in depression, and in this form it is no exception to my analysis. I admit, though, that the permanent unhomelikeness, in which a person questions his place in life, but is not looked upon as ill either by himself or others, seems to be a possible scenario. This situation, however, always borders on illness if it goes on for a long time. The reason for this seems to be that the unhomelikeness in the latter case is tied to a permanent deformation of the meaning-structure of the self, which the person is unable to change. To see this more clearly let us turn to yet another possibility of unhomelike being-in-the-world without illness.

To be locked up in prison for years, exposed to horrible conditions, is another example of unhomelikeness without illness. First, what is important to stress here is that it is the being-*in*-the-world of the person that counts as homelike or unhomelike in the phenomenological theory. To live in an environment means to interpret it and assign it meaning through feelings, thoughts and actions. Thus the lifeworld of phenomenology is not identical to *physical* surroundings, but is a meaning-pattern partly created by the person himself. Whether being locked up in prison results in unhomelike being-in-the-world depends partly on the world that the person is thrown into (its factual and cultural characteristics) and partly on the way he projects this given world of necessities and possibilities in his life. The prisoner might in some cases be able to adjust to a homelike existence behind bars, although the conditions of imprisonment in most cases would offer too much resistance to allow this homelike reinterpretation of the subject's life project.

Second, as I stressed above, the unhomelike being-in-the-world of illness is characterized by a fatal change in the meaning-structures, not only of the world, but of the *self*. Although self and world are always interconnected in a synthetic way through the being-*in*-the-world of the self, it is still possible to make a distinction between the person and the world he inhabits. In this way it is possible to distinguish between a more general homelessness of being-in-the-world, which is merely due to breakdowns in the 'external' tool-pattern, and the unhomelikeness of illness, which is always accompanied by a fatal change in the meaning-structure of the self. As we have seen, the parts of the meaning-pattern that are hardest to replace or reinterpret are the ones tied to our own embodiment. The lived body forms the center of the self. The 'body-tools' are most fundamental for our transcendence, while the surrounding ones often are replaceable to a greater extent. The self, however, is not identical with the lived body, but also has other 'mental' characteristics, which are made lucid by the existentials of understanding and language. This distinction between breakdowns in the meaning-patterns of self versus world should be sufficient to counter the argument that the characterization of illness as unhomelike being-in-the-world threatens to encompass too many conditions that we would normally refer to as unhappiness rather than illness.

Given this distinction between different forms of unhomelikeness, which is important since the mission of health care can hardly be to provide solutions for every kind of unhomelike being-in-the-world, clinical medicine still encompasses not only the sciences of biology and pathology, but also the art of providing of a way home for the patient. Human life (*bios*) suffers (*pathe*) from a basic homesickness that is let loose in illness. If this unhomelikeness is taken as the essence of illness, the mission of health-care professionals must consequently be not only to cure diseases, but actually, through devoting attention to the being-in-the-world of the patient, also to open up possible paths back to homelikeness. This mission can only be carried out if medicine acknowledges the basic importance of the meaning-realm of the patient's life - his life-world characteristics.

University of Linköping
Linköping
Sweden

NOTES

[1] This paper is a revised, shortened version of Part 2 of my book, *The Hermeneutics of Medicine and the Phenomenology of Health* (2000). I refer the reader to this work for a more exhaustive account of my proposal for a phenomenology of health.

[2] The first philosopher to have shown this in a systematic manner seems to have been Georges Canguilhem (1991 [1966]), who refutes in a convincing manner the biological-statistical health concept, which he shows to have been present in medicine ever since Claude Bernard and Auguste Comte. He also develops a view according to which health should be approached as an evaluative concept studied at the level of personal, clinical experience.

[3] Other attempts which have been made to develop 'holistic' theories of health - that is, theories taking account of the whole person in his environment, and not exclusively of his biology - have relied on analytic action-theory rather than on phenomenology. See Nordenfelt (1995) for such an attempt, which also includes critical investigations of other proponents of holistic theories of health. The most well-known proponent of a biological-statistical theory of health - that is, an account of health which is made on the basis of the concept of disease as biological malfunctioning - is Christopher Boorse (1997). For an extensive discussion of different health-theories and their relation to the phenomenological alternative I present in this paper see Part 2 of Svenaeus (2000).

[4] For a comprehensive survey of the phenomenological movement, see the books by Herbert

Spiegelberg (1972, 1982).

[5] See Part 1 of Svenaeus (2000).

[6] In addition to 'disease' and 'illness' a third term, 'sickness', is often used in talking about health and its opposites. When contrasted to illness it is often used in the sense of a social role, a 'sick-role', ascribed to a person by other people, rather than being something experienced by the person himself. However, the term is also often used interchangeably with 'illness', especially in American English.

[7] See Drew Leder (1995) for a survey of the fragmentary but interesting attempts that have been made, mainly by Kay Toombs (1992a), Richard Zaner (1981, 1988) and Leder himself (1990). Psychiatrists and psychoanalysts inspired by the philosophy of Martin Heidegger have in a (at least partly) phenomenological manner developed theories of *mental* health and illness. Three well-known and influential examples are the theories of Ludwig Binswanger, Medard Boss, and Jacques Lacan. See Richardson (1993) and Spiegelberg (1972).

[8] I abstain from giving any substantial account of Husserl's philosophy in this paper, since I will not turn to him, but to his follower Martin Heidegger, in developing a phenomenology of illness. The authoritative guide to Husserl's phenomenology, examining the genesis of his philosophy and its main themes, is Bernet et al. (1989; translated into English in 1993). Although I choose to go with Heidegger rather than with Husserl, I do not want to claim here that it would be impossible to work out a theory of health working within the scaffold of Husserl's phenomenology. Two interesting recent attempts to approach questions of normality and medicine on the basis of Husserlian phenomenology are Steinbock (1995) and Waldenfels (1998).

[9] This short presentation limits its scope to Heidegger's early philosophy and first main work, *Sein und Zeit* from 1927: (1986, 16th ed.) The excellent new translation by Joan Stambaugh from 1996 has been a great help in transferring Heidegger's vocabulary into English.

[10] This is Heidegger's own famous example (1986, p. 84); one could think of countless others.

[11] The only exception is the *Zollikoner Seminare* (1994 [1987]), which Heidegger conducted together with the psychiatrist Medard Boss during the sixties. These seminars appear to be one of the very few places where Heidegger addresses not only health and illness, but also embodiment (*Leiblichkeit*). Heidegger, otherwise reluctant to discuss the *specific* activities of everydayness, is here forced to address these themes in the presentation of his philosophy. The encounter between the famous philosopher and the doctors offers very stimulating reading since Heidegger (even more than in his lecture courses) has to mobilize all his pedagogical skills in the face of the questions of the philosophically untrained audience. We will return to these seminars in what follows.

[12] I am thinking here primarily on Medard Boss's attempt to develop a phenomenology of health on the basis of Heidegger's philosophy (1975). Nietzsche's notion of a 'große Gesundheit' seems to be a forerunner to such interpretations (1973, pp. 15-17, 317-319). That is, 'the great health' of Nietzsche could be understood as a form of authenticity – the mode of existence of the 'fitter', stronger, braver individual or culture, which can endure and learn from pain and illness – rather than as health in the usual sense. On this subject see Raymond (1999). The remarks by Nietzsche on *Gesundheit der Seele* in *Die fröhliche Wissenschaft* could also, however, be interpreted as promoting an individualistic, holistic health-concept: 'Es kommt auf dein Ziel, deinen Horizont, deine Kräfte, deine Antriebe, deine Irrthümer und namentlich auf die Ideale und Phantasmen deiner Seele an, um zu bestimmen, *was* selbst für deinen *Leib* Gesundheit zu bedeuten habe' (1973, p. 105).

[13] The double meaning of the German word 'unheimlich' – it means both 'uncanny' and 'not at home' or 'unhomelike' ('unheimisch') – cannot be translated directly into English. For the etymology of the word 'unheimlich' and the relation between the phenomenon of *Unheimlichkeit* and mental illness, see Freud (1919).

[14] Erwin Straus's idea of an *I-Allon* (I-other) relation as the basis of our being-in-the-world has

been very helpful here in my reading of Heidegger. See Straus (1966a).

[15] This thought is inspired by Straus's essay *The Upright Posture* (1966b), in which, however, the author also tries to give this phenomenon a moral significance which is absent in my analysis.

[16] See Plügge (1962, p. 97).

[17] To transcend in Heidegger's phenomenology does not mean to go beyond the world. This is the way the concept has usually been employed in philosophy (or religion), sometimes to indicate a mystical relationship – a relationship to something outside the world (God). In contrast to this, transcendence in *Sein und Zeit* means the way human beings (*Dasein*) are thrown *into* the world (1986, pp. 363-364). This transcendence, which belongs to the being-in-the-world of all human understanding, means that we always find ourselves already in the world, given over to some meaning-structures that we have not chosen. These meaning-structures (*Bewandnisganzheiten*) are intersubjective, historical and cultural; that is, they change gradually between different times and places, but they are nevertheless always there in some form, as a necessary 'facticity' (*Faktizität*) which one must make one's own in order to develop a human understanding and self (1986, p. 135).

[18] Merleau-Ponty (1962 [1945]) offers an analysis of the meaning-structures of perception performed mainly with the aid of different defects. It is also noteworthy here that to return to health from severe illness generally seems to provide people with new ideas and feelings about the structure and meaning of life.

[19] Many philosophers, like Sartre for instance, reproached him for this exclusion. To deem Heidegger an idealist on these grounds would, however, clearly be wrong, since his phenomenology is intended to transcend both naive realism and idealism (1986, pp. 207-208). Merleau-Ponty's *Phénoménologie de la perception* from 1945, a work highly dependent not only upon Husserl's phenomenology, but also upon the analysis of worldliness in *Sein und Zeit*, makes up for this shortcoming of Heidegger by focusing upon the role of the lived, phenomenal body in perceptual activity. Interestingly enough, the *Zollikoner Seminare* provides evidence that Heidegger was aware of this shortcoming already when he wrote *Sein und Zeit* and that the body as lived – *Leib* – forms a more important part of *Dasein's* being-in-the-world than is evident from the book: 'Sartres Vorwurf kann ich nur mit der Feststellung begegnen, daß das Leibliche das Schwierigste ist und daß ich damals eben noch nicht mehr zu sagen wußte' (1994, p. 292). It is highly likely that Heidegger's analysis of the body in the *Zollikoner Seminare* was inspired by a reading of Merleau-Ponty, though he is never mentioned there.

[20] As Straus has pointed out the original meaning of the Greek *organon* is indeed tool (1966b, p. 150).

[21] Some philosophers before me have suggested the broken tool example in Heidegger as a promising way to a phenomenology of illness. They have, however, never actually carried out the analysis. See Leder (1990, pp. 19, 33, 83-84), Rawlinson (1982, p. 75), and Toombs (1992b, p. 136 ff.).

[22] The phenomenology of pain is without doubt the best explored area of the phenomenology of illness. A classic case is Buytendijk (1962). I would also like to mention the excellent book by Elaine Scarry, *The Body in Pain* (1985), which explores the effects of torture as severe damage to the person's being-in-the-world. Language, for example, as a way of inhabiting the world, is destroyed through pain. Pain is certainly not restricted to a sensation in the body, but consists in an atmosphere, a 'pain-mood' disrupting understanding. See Leder (1984-85); and Schrag (1982). See also Heidegger (1989, pp. 118-119): 'Eine Magen "verstimmung" kann eine Verdüsterung über alle Dinge legen.'

[23] For phenomenological studies of the genesis of the self, see Zaner (1981), and Ricoeur (1992).

[24] This quotation inevitably provokes thoughts about the era of organ transplantation which was still in its infancy when Heidegger said this in 1972. There still remain important limits, however, regarding which organs it is possible to store in the 'tool-box' and for how long. Nevertheless, given the emergence of cyberspace and artificial organs, the difference between

hand and hammer certainly seems even more diffuse today than in Heidegger's time.
[25] For a study of the stigmatizing effects of illness and handicap, see the classic book by Goffman (1974). See also Toombs (1992a, 1992b).
[26] On the phenomenology of the lived brain see Leder (1990, p. 111 ff.).
[27] On the subject of lived time and the illness experience, see Toombs (1990).

BIBLIOGRAPHY

Bernet, R., Kern, I. and Marbach, E.: 1989, *Edmund Husserl: Darstellung seines Denkens*, Felix Meiner Verlag, Hamburg.
Boorse, C.: 1997, 'A Rebuttal on Health', in J. Humber and R. Almeder (eds.), *What is Disease?* Humana Press, New Jersey, pp. 1-137.
Boss, M.: 1975, *Grundriss der Medizin und der Psychologie*, Hans Huber Verlag, Bern.
Buytendijk, F. J. J.: 1974, *Prolegomena to an Anthropological Physiology*, Duquesne University Press, Pittsburgh.
Buytendijk, F. J. J.: 1962, *Pain: Its Modes and Functions*, University of Chicago Press, Chicago.
Canguilhem, G.: 1991, *The Normal and the Pathological*, Zone Books, New York.
Freud, S.: 1919, *Das Unheimliche*, Vol. 12. *Gesammelte Werke*, Imago Publishing, London.
Gadamer, H.-G.: 1993, *Über die Verborgenheit der Gesundheit*, Suhrkamp Verlag, Frankfurt am Main.
Goffman, E.: 1974, *Stigma: Notes on the Management of Spoiled Identity*, J. Aronson, New York.
Heidegger, M.: 1994, *Zollikoner Seminare*, V. Klostermann, Frankfurt am Main.
Heidegger, M.: 1989, *Nietzsche*, Vol. 1. G. Neske, Pfullingen.
Heidegger, M.: 1986, *Sein und Zeit*, 16th ed., Max Niemeyer Verlag, Tübingen.
Held, K.: 1993, 'Fundamental Moods and Heidegger's Critique of Contemporary Culture', in J. Sallis (ed.), *Reading Heidegger: Commemorations*, Indiana University Press, Bloomington, pp. 286-303.
Leder, D.: 1995, 'Health and Disease: The Experience of Health and Illness', in W. T. Reich (ed.), Vol. 2. *Encyclopedia of Bioethics*, Simon & Schuster Macmillan, New York.
Leder, D.: 1990, *The Absent Body*, University of Chicago Press, Chicago.
Leder, D.: 1984-85, 'Toward a Phenomenology of Pain', *Review of Existential Psychology and Psychiatry* **19**, 255-266.
Merleau-Ponty, M.: 1962, *Phenomenology of Perception*, Routledge, London.
Nietzsche, F.: 1973, *Die fröhliche Wissenschaft*, Div. 5. Vol. 2. *Kritische Gesamtausgabe*, Walter de Gruyter, Berlin.
Nordenfelt, L.: 1995, *On the Nature of Health: An Action-Theoretic Approach*, Reidel Publishing, Dordrecht.
Plügge, H.: 1962, *Wohlbefinden und Missbefinden*, Max Niemeyer Verlag, Tübingen.
Rawlinson, M. C.: 1982, 'Medicine's Discourse and the Practice of Medicine', in V. Kestenbaum (ed.), *The Humanity of the Ill: Phenomenological Perspectives*, University of Tennessee Press, Knoxville, pp. 69-85.
Raymond, D. (ed.): 1999, *Nietzsche ou la grande sante*, Éditions L'Harmattan, Paris.
Richardson, W. J.: 1993, 'Heidegger Among the Doctors', in J. Sallis (ed.), *Reading Heidegger: Commemorations*, Indiana University Press, Bloomington, pp. 49-63.
Ricoeur, P.: 1992, *Oneself as Another*, University of Chicago Press, Chicago.
Scarry, E.: 1985, *The Body in Pain*, Oxford University Press, Oxford.
Schrag, C.: 1982, 'Being in Pain', in V. Kestenbaum (ed.), *The Humanity of the Ill: Phenomenological Perspectives*, University of Tennessee Press, Knoxville, pp. 101-24.

Spiegelberg, H.: 1982, *The Phenomenological Movement: A Historical Introduction*, 3rd rev. ed., M. Nijhoff, The Hague.
Spiegelberg, H.: 1972, *Phenomenology in Psychology and Psychiatry*, Northwestern University Press, Evanston, Illinois.
Steinbock, A. J.: 1995, *Home and Beyond: Generative Phenomenology after Husserl*, Northwestern University Press, Evanston, Illinois.
Straus, E.: 1966a, 'Norm and Pathology of I-World Relations', in *Phenomenological Psychology*, Basic Books, New York.
Straus, E.: 1966b, 'The Upright Posture', in *Phenomenological Psychology*, Basic Books, New York.
Svenaeus, F: 2000, *The Hermeneutics of Medicine and the Phenomenology of Health: Steps Towards a Philosophy of Medical Practice*, Kluwer Academic Publishers, Dordrecht, The Netherlands.
Toombs, S. K.: 1992a, *The Meaning of Illness: A Phenomenological Account of the Different Perspectives of Physician and Patient*, Kluwer, Dordrecht.
Toombs, S. K.: 1992b, 'The Body in Multiple Sclerosis', in D. Leder (ed.), *The Body in Medical Thought and Practice*, Kluwer, Dordrecht, pp. 127-37.
Toombs, S. K.: 1990, 'The Temporality of Illness: Four Levels of Experience', *Theoretical Medicine* **11**, 227-241.
Waldenfels, B.: 1998, *Grenzen der Normalisierung: Studien zur Phänomenologie des Fremden 2*, Suhrkamp Verlag, Frankfurt am Main.
Zaner, R. M.: 1988, *Ethics and the Clinical Encounter*, Prentice Hall, New Jersey.
Zaner, R. M.: 1981, *The Context of Self: A Phenomenological Inquiry Using Medicine as a Clue*, Ohio University Press, Athens, Ohio.

PER SUNDSTRÖM

DISEASE: THE PHENOMENOLOGICAL AND CONCEPTUAL CENTER OF PRACTICAL-CLINICAL MEDICINE

I. INTRODUCTION

In an ordinary English dictionary, chosen at random, you may read: "Someone who is ill is suffering from a disease or health problem which makes them unable to work or to live normally" (Collins, 1987, p. 721).

Such a down-to-earth explication of illness says little or nothing about how it *feels* to fall, and to be, ill. And it does not tell you anything about that forbidding existential fact which infects any serious illness: being ill, seriously ill, is being lost and trapped in your self, your own weak self, which finds itself bereft of a world to care for and inhabit, and so unable to care for anything or anyone except itself, i.e., your self, which confines you to itself – your petty self.

Instead, the dictionary resolutely locates – indeed, relocates – the primitive, experiential and existential illness phenomenon in two contexts – the medical and the social – which we find only at some remove from the feeling, existential self. And not without reason, because these are the contexts in which – daily and indefinitely – illness is spoken of, discoursed upon, and thematized; even as the wretched, suffering self – muted in spite of itself – turns on itself, its hapless self.

Now, leaving the naked and muted self behind, and entering the medical and social contexts to which illness has been relocated, we must immediately recognize that the causal order here is such that the medical context takes precedence over the social one: the decisive event is the advent of "a disease or health problem" – which, subsequently, makes its bearer or sufferer "unable to work or to live normally." This essay will be concerned with illness in a medical context – or, more specifically, in a practical-clinical medical context – where it is often turned into, regarded as, or simply equated with, *disease*.

There is illness without disease; and there is disease without illness. But most of the time, the two go together, disease/illness presenting itself as a mixed subjective/objective phenomenon, which is disturbing, disabling, or even calamitous to the individual patient, *and* – darkly or clearly, vaguely or readily – recognizable by the medically trained professional as a deviation from the normal functioning, or shape, of the body/organism.

Now, the main concern of medicine and health care is precisely this mixed subjective/objective phenomenon, which, in clinical practice, is divided, almost endlessly, into several diseases or diagnoses. These, in turn, are pluridimensional conceptions that account for conglomerates of subjective and objective phenomena – reflecting the general, mixed phenomenon of disease/illness (Sundström, 1987). The present essay deals with disease and diseases in this – i.e., the clinical – sense. Later on, we shall consider the fact that sometimes illness occurs in the absence of disease – and sometimes, conversely, disease in the absence of illness.

II. THE VAGUENESS AND POWER OF 'DISEASE'

The meaning of the word 'disease' is vague, elusive, and unstable. You may reach for it, but you won't grasp it, except in pieces and fragments. It is prone to changes, permutations, and shatterings – according to the circumstances, or irrespective of them. And yet, you can use the word, and you may feel confident that it carries some meaning – at least most of the time.

The semantic instability and vagueness of 'disease' is a general phenomenon; it holds for common and professional medical usage alike. And it is a great asset in both – though first and foremost for the physician, whose powers to cure, to ease, and to comfort, rest on it.

If you are a medical doctor, you do feel confident that 'disease' carries meaning. In spite of the instability of its meaning, you make use of the word as if its meaning could be grasped; you just grab it and do momentous things with it. There is, potentially and in the hands of a medical professional, great power in 'disease'. Unblushingly, a doctor takes advantage of the word in the economy of finding out what there is to know, to do, and to say vis-à-vis a given, individual patient – or a certain category of patients. 'Disease' is malleable, and it works powerfully in this particular context by dint of the order, ritual, and discipline of the medical enterprise.

The medical enterprise begins and ends with the dyad physician-patient, in encounter and dialogue. This encounter – the clinical encounter – is, in most cases, and most of the time, conspicuously asymmetric as regards power. The physician is the doer and the wielder of the power invested in the word 'disease'. This turns 'disease' – as used in the clinical encounter between physician and patient – into a most powerful, and morally ambiguous, tool. It can be used to probe and to dissect a state of mind and body – as a first step in the restoration of their integrity, and so of health *tout court*. And it can be used to put a name on that which is inexplicably, purely, and mysteriously, distressing – and thus to soothe a troubled mind. But it can also be used to insert a hurting and damaging needle into an unsuspecting mind. And it can be used to bring almost any patient – who is then just anybody, any body – low. Finally, it can be used to conjure up a spurious self-identity, especially when there is only a weak one to compete with. It can even be used simply to awe the credulous.

In professional medicine there is a long history of attempts to discipline 'disease' by forging classifications or taxonomies: the various manifestations of disease – which are then called diseases or diagnoses – are arranged and ordered in accordance with perceived or imagined similarities between them (Faber, 1923; King, 1963). There is power in this too: the power of lining things up and of holding a line once established; and the power to have people assigned and confined to a prefabricated locus, cognitive as well as spatial.

But whatever order – real or imagined – and power there may be in various classifications of diseases, they will consistently feed on the instability and vagueness of 'disease' – vagueness being subdued, but not vanquished, by a spirit of discipline and order. In other words, 'disease' is supple enough to pull – or to be made to pull – the various, often quite diverse, states, processes and events which come under the purview of a physician together; 'disease' is the flexible cement of the physician's universe.

Now, by recognizing the semantic vagueness and instability of 'disease', we may learn that quibbles about the true meaning, or the definition, of 'disease' are precisely that: quibbles. Nevertheless, the word cannot be made to mean just anything; the frivolous saying 'anything goes' is geared to fake reasoning – i.e., reasoning that promotes nothing but the would-be reasoner herself – rather than to genuine inquiry into the reality of disease.[1] Two elementary clues as to the meaning of 'disease' may be disentangled – and should indeed be recognized, lest all serious, responsible communication of shared meanings disintegrate.

The first clue derives from the etymology of the word 'disease': someone who suffers from a disease – of body, mind, or spirit; mental or physical – is *not at ease* with herself, and so in a state of uneasiness (OED, 1989).[2]

The second clue derives from the privileged status that professional medicine – theoretical and practical – enjoys vis-à-vis other meaning-conferring social or linguistic contexts with regard to the meaning of 'disease.' That is, 'disease' is primarily a professional medical word, the proper use of which, no matter how metaphorical or poetic, should never radically cut the semantic moorings to its medical sense or senses.

III. THE LINGUISTIC DOMINANCE OF PROFESSIONAL MEDICINE AND BIOMEDICAL SCIENCE

Whether we like it or not, there is not much talk about illness or disease that is not influenced, indeed colonized, by the peculiar language, interests, and goals of professional medicine.

There may have been a time – prehistoric time – when the presence of illness was communicated by gestures, groans and complaints, which might have been recognized – by an anachronistic observer overhearing the scene – as instances of a primitive, quasi-medical language. If so, that would have been before the entrance of the physician, i.e., the medical professional, on the historical scene, and so before the emergence of that peculiar species of language – integrating empirical knowledge, sound judgment, and benevolent practice – which we recognize as medical. Any primitive, quasi-medical language that may have preceded professional medicine is lost in the dusk of prehistory, and today we have to acknowledge that – with the possible exception of lingering quasi-medical ingredients in the magic and religion of so-called primitive cultures – there is almost no discourse on illness/disease available to us that is independent of the medical profession and its peculiar attitudes, practices, and modes of linguistic expression.

There are several aspects to the linguistic dominance of medicine over all current disease-talk:

First, we have the language of biomedical science: atoms, molecules, genes, cells, tissues, receptors, antibodies, electrolytes, counts, graphs, etc. – words, phrases and meanings which provide the ontological baseline, as it were, of any serious talk about disease or illness. Anything aspiring to be considered real with regard to disease ought to be founded on, or at least linked to, the basic strategies, methods, and entities, of biomedical science. This is an unavoidable cultural imperative today.

Secondly, we have the ingrained idea that the human body, any body (equality reigns on this level), is an instantiation, an infinitely complex instantiation, to be sure, of the entities, processes, and events that science charts.

Finally, we have the practical-clinical medical perspective, which focuses on individual patients, i.e., human beings whose stake in medicine is their own health and wellbeing, as well as their hopes and prospects of becoming its beneficiaries, on a strictly individual basis. The ultimate aim and purpose of medicine is the health of individual human beings. And clinical medicine typically helps by putting the powerful arsenals of science, and all the ingenious paraphernalia of a scientific technology, to work for this humane purpose. This is medicine as a through-and-through ethical project –

which is a real possibility (and no mere ideal), in spite of its being turned, frequently and continually, into something else, such as a quest for biomedical information and knowledge.

Now, the remarkable thing about modern medicine, on all three counts, is that it *works* – on analogy with other modern things that, more often than not, work, such as cars and refrigerators. Or, at least, it is believed to work. It is noteworthy, that those who profess this belief – and indeed consider it to be more than a belief, viz., a fact – are not only, or even typically, those who have a professional stake in medicine, but also, and manifestly, the general public, i.e., its supposed beneficiaries, who, naturally, *want* it to work. Occasional disappointments are largely powerless to trouble this slightly idolizing faith in the accomplishments of a scientifically and technologically equipped medical profession. On the contrary, even disappointments feed on a deep trust in the worldview of science, in the accurateness of a scientific view of human being, and in the knowledge, powers, and reliability – even wisdom – of those who are educated and trained in science. The mistakes of modern medicine – whether constitutional or occasional – will prompt few, if any, calls for other things than science. And so, the remedy for science gone awry is not less but more, and better, science – whether basic, applied, or one that is specifically geared to the peculiar humane purpose of clinical practice. Healing efficacy – which is what everyone demands in the end – rests with science and the votaries of science.

The upshot of all this is that modern medicine – inspired by, and indeed impregnated with, science – will shape and steer all discourse, whether spoken or written, on disease or illness, and that the human life-world, in all matters pertaining to illness or disease, will be suffused with medicine's peculiar ways of perceiving, processing, and putting things.

IV. A PHENOMENOLOGICAL PHYSIOGNOMY OF DISEASE

Some species of philosophical prejudice have it that you must *define* things before you may begin a meaningful, informed, and knowledge-generating discussion of them. Needless to say, that is not the way a phenomenologist would begin and proceed. In fact, a strict definition may vitiate a promising inquiry – by pre-empting it and prejudging its outcome (Cf. Peirce, 1931-1935, 5.447, 5.505-508; Potter, 1996, pp. 65-77). It is no disaster, then, if it soon turns out that a sufficiently broad, as well as distinct and concise, definition of 'disease' is hard to find; definition is mere pastime and make-believe anyway.

In fact, we have more than one reason to abandon the quest for a strict definition of disease. The words 'disease' and 'illness' figure in various linguistic contexts, spoken and written, lay and professional, where they are made to mean various things to various people. So, whatever meaning they may have, and whatever understanding they may convey, it is to be sought and found in the contexts in which the words are used. In ordinary as well as professional usage, there are lots of notions, meanings, ideas and thoughts that relate, one way or another, to 'disease'; there is a spectrum of meanings which are related so as to make up a family, a semantic family, a family of meanings which – like the members of any family – resemble each other while being indubitably distinct individuals. Wittgenstein's metaphor 'family resemblance' catches the essence of this semantic relatedness in spite of considerable variability (Wittgenstein, 1953, §66).

There is a continuous exchange of disease-related words, phrases, and meanings between medical professionals and patients, and between ordinary and professional usage. Physicians and other professionals take the lead by introducing new words and expressions which stem from the peculiar vocabularies and linguistic habits of biomedical science and technology. The professional side to the linguistic and semantic exchange stands for novelty, innovation and fresh discovery. The linguistic and semantic contribution of patients, on the other hand, derives mostly from the age-old fact that disease or illness has to do with life, individual life, gone awry; with the body and its self sliding out of joint; with pain and suffering, and with death as the ultimate threat. And this ever recurring contribution to the meaning, or rather meanings, of disease would be no less weighty than that of the professional propagators of biomedical science and technology.

Now, given the insight that the formulation, indeed fabrication, of a strict definition of 'disease' is neither feasible nor desirable, we may still approach a reasonably adequate linguistic-semantic characterization of 'disease' by making a short, admittedly rather cursory, inventory of meanings, or species of meaning, which make up the semantic family 'disease'. I will here consider three traits that contribute to the phenomenological physiognomy of disease:

(1) Disease is closely related to the organism, the living and lived organism, the organism as an integrated whole, which in its diseased state does not function, move and work, as it should, or as it is expected to do, or as it did before. Disease is one aspect of this impaired functioning, this change for the worse, this failure – just as, seen from another angle, the impaired functioning is one aspect of disease. In other words, the functional integrity of the organism proves to be deficient, wanting in resilience and strength, broken.

Since the organism is the junction where the human as body, mind and spirit; as feeling, reaction and thought, and as past, present and future history, meet, we can expect that its impaired functioning and broken integrity will surface and become manifest in various ways: in the appearance, moves and comportment of the body; in the cognitive moves and maneuvers of the mind; in feeling and emotion, whether intensified or depressed; etc. (Jonas, 1966).

The change for the worse, the impaired or broken integrity of the organism, relates to both an imagined, ideal normality, which would be the norm that holds for each and everyone, *and* the individual's own previous level of organismic functioning and wellbeing. Disease means newness, undesired newness: a new state of things; a new situation; often even a new way of life and being. And this undesired newness typically affects both the individual herself *and* her family – as well as, trivially, physicians, nurses, and other members of the medical staff.

(2) In most instances, disease means diminished wellbeing for the affected; he or she has lost the relative ease of ordinary, everyday life. The loss may be temporary, it may last indefinitely, or it may be a loss for life. It may strike now, without respite, or in the future, or today in anticipation of a dreaded tomorrow.

For a patient in pain, the recollected past, as well as any prospective future, is swallowed up by the dreadful pit of the present. But in most instances – where severe pain is not an all-pervading presence – the future, the individual's prospective, anticipated, and feared, future, emerges as vitally important; disease typically means a threat to the affected individual's future life. The individual patient's physical, mental, or social abilities may become impaired – or the disease may be perceived as a threat of such impairment and disability.

Furthermore, disease may threaten, indeed ruin, the existential integrity of the individual. That is, life – i.e., human existence, or *Dasein* – may begin to appear ever less meaningful to the self – the self perceiving itself bereft of a proper place to inhabit and a task to perform in the world.

Ultimately, disease means a threat to life itself; it demonstrates death as a real possibility, to fear and to reckon with; disease may be a reminder of death, sometimes healthy, sometimes devastating.

(3) To some extent disease can be visualized, literally visualized. Skin affections can often be seen by the naked eye, though their visibility is regularly enhanced if mediated by a microscope. Moreover, aided by microscopes, endoscopes, ultrasound scanners, X-rays, and other, technically even more advanced, devices, the eye may reach far and wide below the surface of the body, penetrating into its remotest cavities and recesses (Reiser, 1978). In most diseases the inner landscape of the body and its

geography undergo changes that may be searched and ascertained under the gaze of the technologically equipped and enhanced eye. Indeed, such changes make up one aspect of disease itself. Surgical practice has always meant the literal unveiling, at times even the extirpation, or at least resection, of disease. And thanks to the unremitting progress of scientific technology, internal medicine – for which the interior of the body used to be uninspected territory – has been able to take advantage of another sort of unveiling, not a literal unveiling this time, but an unveiling by other, knife-less, means.

Just as disease may be a matter of enlargement or shrinking, widening or narrowing, lengthening or shortening, shape or shapelessness, it may also be a matter of too much or too little, of excess or shortage, sufficiency or deficiency. In other words, disease may, at times and to some extent, be subjected to quantitative analysis, where more, or less, of a chemical substance means more or less of disease – and similarly with the flow of fluids, the propagation of electrical currents, the speed of chemical reactions, etc. And so, the mean between extremes – being one way of spelling out normality – emerges as an important notion in the phenomenology of disease; disease is then a deviation from the mean and the norm – an ideal, general norm, or a norm relative to the individual's own, previous life.

V. DOCTORS' EXPERIENCES AND PATIENTS' EXPERIENCES

Diseases are not – unlike physical bodies and organisms – parts of nature. Diseases emerge only with suffering individuals seeking the help of other individuals, who care for them and try to help them – a help which is, ideally or typically, founded on knowledge, accumulated experience, and sound judgment.

In other words, nature as perceived by science does not contain – and indeed, is absolutely ignorant of – disease as a human concern and challenge. Yet, as we have seen, science will have a lot to say about diseases. But whatever it has to say, it will say it not directly, as items of pure science, but indirectly, by way of the medical enterprise. And at the phenomenological center of that enterprise we find the primitive medical situation: a patient, who has nothing but his or her weakness and need to offer; and a medical professional, who has, or professes to have, the power to meet the need of the other.

The fundamental asymmetry of the doctor-patient relationship – as regards both power and the motive for entering into such a relationship – is bound to have a profound influence on, and to be reflected by, the way disease emerges and is conceived of in practical-clinical medicine. Any tenable phenomenology of disease must take a wide range of experiences into account, doctors' experiences as well as patients' experiences – and experiences shared by both parties. It must demonstrate some awareness of the dread suffered by the patient's self facing decay, disability, and ultimately death; and it must be familiar with the dedicated, and possibly fairly ambitious, doctor's eager pursuit of data, facts, evidence (physical, chemical, physiological, etc.) – the more, and the more exact, the better. Disease is a territory colonized by science, with death lurking in the background, lining the horizon. And between these experiential extremes – that of the mortal self, and that of the neutral eye of science – innumerable items of experience are dispersed.

Obviously, some species of experience are, typically, the patient's, rather than the doctor's, concern, whereas others typically belong to the physician's domain. But they all do belong – actually or potentially, as real possibilities – in the clinical encounter and will have some bearing on how diseases are perceived, conceived, and apprehended. The wide range of potential experiences provides a suitable emotional and conceptual horizon for the clinical encounter between physician and patient.

It is a common mistake, and a well-worn cliché, to believe that a physician is merely a scientist who has adopted the task of maintaining, or mending, as the case may be, human bodies. It is true that a physician's training is, or ought to be, scientific; he or she ought to assimilate the general scientific spirit of critical, evidence-based inquiry, as well as acquire a fair knowledge of the current methods of science. But if scientific detachment is an asset, indeed a virtue, to the physician, so is proximity and empathy. A physician needs both detachment and empathy in order to understand what sort of a creature a patient is. The absence of either would seriously disable him precisely *qua* physician; he would become a pitiable counterfeit of the real thing, as unfit for his task as a one-legged runner. And he would soon drop out of business altogether.

Medicine is a science and a technological art intent on helping individual human beings – in various ways, but principally by way of their bodies or organisms, the morphological and functional integrity of which may, through the interference of medicine, be supported, defended, or restored. Human bodies and organisms are, to a large extent, similar between individuals, and they may be similarly treated, to good effect. Standardization is therefore a medical stock in trade. But that is not the whole story, because "all humans have / prejudices of their own / which can't be foreseen" (Auden, 1972, pp. 13-15), which is a fact that calls for an attentive, considerate, empathic, and, not least, diversified approach.

Thus, the proper clinical approach is one of standardization and diversification at once. And this, in turn, calls for, indeed presupposes, the wide range of experiences – patients' as well as doctors' – with which we are here concerned. In other words, proper clinical action issues from the acknowledgment of a wide and open experiential horizon, which saves the physician from the narrowmindedness of misplaced standardization, while allowing him to profit from the blessings of routine and standardization. A wide experiential horizon – which corresponds to a broad fore-understanding (*Vorverständnis*) on the physician's part – would provide the only intellectual and emotional ambience in which practical-clinical medicine can flourish.[3]

Considered as a form of human experience, science distinguishes itself by being at once subtle (in its cognitive content) and impressively powerful (in its workings). Its prestige rests on both properties, but the latter, in particular, makes for its being the largely uncontested backbone of medicine. Nevertheless, scientific experience must be supplemented by a host of other sorts of experience, which yield understanding if not explanations, and insight if not generalizable knowledge.

This basic truth about medicine – viz., that it is a scientific, yet more than scientific, enterprise – is reflected in the conceptions of disease entertained by good, i.e., truly professional, physicians.

VI. THE CLINICAL CONCEPTIONS OF DISEASE

In practical medicine, it is taken for granted that disease – or rather, this or that particular disease – ought to be dealt with, one way or another. You may think of any disease, and you will be thinking of something that, as a matter of course and received principle, ought to be dealt with, treated, ultimately disposed of. It is the existence of diseases that provides a tenable *raison d'être* for physicians, nurses, and other members of the medical staff – and precisely as doers, professional doers.

The world of practical medicine is emotionally plagued, and cognitively colonized, by diseases; it is oriented towards them, its intentionality is structured by them. This may sound trivial, and indeed, it is trivial in the sense that it happens every day, over and over again, and simultaneously in countless places and surroundings – even as it is the steady state of affairs wherever medicine is practiced. As a physician, one is constantly on the watch for diseases. To pursue diseases – in order eventually to bring them down – is to be on the right track for medical men and women. And the signposts of that pursuit emerge from a huge set of standard conceptions of disease – which correspond to so many diagnoses and disease names.

The disease-orientation of practical medicine is not trivial, however, if by 'trivial' is meant self-explanatory; it is not trivial as crossing the street, drinking a glass of water, or going to bed in the evening, are trivial acts; it is not a simple, straightforward fact, the meaning of which is transparent.

A patient, or prospective patient, who suspects, fears, or squarely realizes that she is seriously ill, senses an "intuitive awareness [...] that the symptoms are part of a larger whole. That is, the various isolated bodily disturbances point to, or signify, a more complex entity of which they are simply one facet" (Toombs, 1992, p. 35). In other words, finding herself ever less at ease with herself, with her *Dasein*, the patient apprehends that beyond the discrete sensations that stir, and disturb, her body (as well as her mind and self), there is "a synthetic totality which transcends them all" (Toombs, 1992, p. 35). This "synthetic totality" is disease, or rather, some disease or other, which infects the *Dasein* of its contractor, thwarting its plans and projects, as well as its sense of freedom and hope.

The patient's intuition, which has been formed, at least partly, by the medical way of perceiving and talking, tells her that disease is a dreadfully down-to-earth, yet also eerily abstract, sort of reality. Furthermore, a disease once contracted is intuited as something that resides in the body, though it is also distinct from it – which suggests its being both a threat within, attached to body and self, and a detached object of cognition, manipulation, and control. Now, this double intuition – disease as existential threat and manipulable object – corresponds to, and reflects, two structuring aspects of the standard clinical conceptions of disease, which provide the cognitive, and to some extent the emotional, horizon of clinical perception and judgment, as well as provide guidelines for professional clinical conduct (Sundström, 1987).

Disease is a multi-faceted reality, though always with a human being as its carrier and vehicle, and regularly with a professional helper as its interpreter and knower. The interaction between the two is complex and difficult to elucidate. When disease is dealt with in a clinical context, several things – of a biological, emotional, social, moral and spiritual nature – happen simultaneously, or in rapid sequences of events. Disease as a reality in the encounter and interaction between patient and professional helper unleashes actions, reactions, thoughts, emotions and events which are difficult to account for in a straightforward manner. However, some order is brought into this process by the fact that the professional helper employs, and is guided by, a set of pluridimensional conceptions of disease, each of which provides a general picture of that which establishes the presence of a particular disease (the diagnosis); of the overall management and treatment of the disease; and of its likely long-term outcome (prognosis) (Sundström, 1987).

The pluridimensional, clinical conceptions of disease – which correspond to disease names such as acute myocardial infarction, diabetes mellitus, and asthma – provide professional helpers with a composite, yet integrated, picture of clinical reality as structured and directed by the presence of diseases. Integrating several dimensions and phenomena, while being integral to the exigencies and challenges of the clinical encounter, these conceptions may be dubbed the *integral clinical* conceptions of disease.

In medicine as science, art, and praxis, such integral clinical conceptions of disease continually emerge and evolve, wax and wane; they are encoded in the language of clinical medicine, spoken and written, thereby conveyed to others, in ever wider circles, and disseminated throughout the medical enterprise and beyond.

Highlighting the essentials from the practicing physician's point of view of, and being attended to, if not venerated, by physicians and other members of the medical staff, the integral clinical conceptions of disease provide more than general overviews of the phenomena of disease; they direct the professional's attention, inspire a helping attitude, and induce a readiness for action.[4]

Now, the art of diagnosis is the art of finding the integral clinical conception of disease that corresponds to the state and predicament of the help-seeking carrier of, as yet unspecified, disease; or, in other words, it is the art of finding the conception of disease that induces action that helps, cures, restores health. Each diagnosis corresponds to its own integral clinical conception of disease, which suggests, in general terms, how the patient suffering from the disease in question should be handled. The integral clinical conception does not define disease, or anyone's disease, but it directs and focuses professional attention with a view to proper, i.e., helping, medical action (Sundström, 1987).

VII. ON THE RELATION BETWEEN REDUCTIONIST AND INTEGRAL CLINICAL CONCEPTIONS OF DISEASE

It might be argued: Disease is an elusive thing, not least in clinical medical practice; a disease may have its signs, symptoms, prodromes, pathogenesis, physiology, natural history, prognosis, etc., but all that is not yet the disease itself, which we need to know in order to be able to understand and to help. This argument is interesting and vague – and interesting because of its vagueness. It is susceptible of two (seemingly) divergent interpretations:

According to one interpretation, the disease itself is that which emerges as the end product of a scientific, explanatory reduction of the multifarious evidence gathered clinically; according to the other, the disease itself is the integral conception of all these features assembled (in various patterns in various diseases). Now, where will reasoning along each of these two lines lead us?

The first, reductionist interpretation carries a risk that reasoning along that line will end up with empty hands, simply because, at the end of the road, no satisfactory, simple and powerful, scientific explanation has been found. The disease itself then is and remains a postulated something. But notice that this very postulate rests on the existence of a clinical state of things that was recognized as disease in the first place, viz., the undefined state of disease with which the reductionist quest started. But then, again, we seem to be led on to the second interpretation, where disease is taken to be an integral conception of all that may be found and figured out in the clinical encounter between doctor and patient. However, we should also consider the possibility that the reductionist quest has been successful.

Let us assume that the reduction has been successful on its own criteria; a simple, manipulable physiological mechanism has been found, which entails both an explanation and a potential undoing of the undesirable condition which was the clinical starting-point, and which triggered the scientific, reductionist inquiry. The disease has now been redefined, or rather, defined for the very first time; the simple mechanism which has been found by way of the reductionist move now defines disease. But again, here we must ask ourselves: What about the initial conception of disease, the imprecise, clinical one with which the physician and medical scientist was dissatisfied in the first place, dissatisfied presumably because he could do so little to cure or otherwise help his patient? Has that conception been rendered redundant by the successful reductionist definition? No, it does not seem so, because what ultimately confirms the new, reductionist definition of disease is the disappearance of all those features which justified the initial, overall judgment of the patient as, in fact, diseased. If the integral clinical conception of disease – which is wedded to the physician's overall clinical judgment – corresponds to the reductionist definition, then the measures taken to correct the faulty physiological mechanism, and thus to counter disease in the reduced, pathophysiological sense, will work out the relief of the *clinical* disease as well. That is, the patient will be relieved in the process. Everything is then all right, both 'diseases' – clinical and reductionist – being on the verge of being efficiently cured. What has happened in the process, by way of the reductionist move, is that the integral clinical conception of disease has been enriched, or even metamorphosed, in some of its constituent features – reductionism in the service of integral clinical reality.

To sum up: The reductionist journey toward scientific explanation, and a well-defined disease mechanism, leads back to the integral clinical conception of disease, from which it once departed, and to which it finally returns with the empty hands of failure or the gift of a fresh scientific discovery. Certainly, the reductionist route does not render any clinical conception of disease redundant.

Our argument has established an irreducible link between reductionist and integral clinical conceptions of disease; in the end it is the latter that provide the meaning and the *raison d'être* of the former. But, as we have seen, there is a transfer of meaning in the other direction too, the reductionist quest – when successful – qualifying and enriching the integral clinical conceptions of disease.

It turns out that the two interpretations that triggered the present train of thought are linked and in fact – in a clinical culture that is both humane and science-oriented – mutually dependent. However, though by no means necessarily opposed, they are not equals from a phenomenological, and ontological, point of view. It is the integral clinical conceptions of disease that provide the most authentic, as well as the most encompassing and unified, view of disease. The integral conceptions of disease are *sui generis* in the clinical life-world, which is simply the human life-world as seen from one specific angle. Arguably, this is the center where all considerations pertaining to disease – phenomenological, conceptual, ontological, etc. – converge.

VIII. THE DISTINCTION DISEASE–ILLNESS: REVEALING *AND* MISLEADING

It is often claimed that you will understand more of that age-old phenomenon which in (Classical) Greek was called *nosos*, in German is labeled *Krankheit*, and in Swedish *sjukdom*, if you make use of the two English words *illness* and *disease* – and make a strict distinction between them (Feinstein, 1967; Reading, 1977; Kleinman, 1980; Nordenfelt & Twaddle, 1993; Fábrega, 1997, p. 104). *Illness* is then taken to designate an individual's (subjective) feeling that all is not well, with her body and her self, whereas *disease* is taken to be a neutral and objective scientific description of events and processes in the body which are considered abnormal. Such a distinction is perfectly feasible; it is reasonably clear, and either of its two parts corresponds to something indubitably real. We have all experienced illness;

and disease is the subject-matter of the medical subdiscipline pathology. Yet, the distinction is a double-edged sword when it comes to understanding what this intriguing and disturbing thing disease-illness is in a clinical medical context.

To begin with, we may note that in ordinary English usage there is no strict distinction between illness as subjective experience and disease as objective description. This goes for ordinary lay usage, as well as for ordinary, everyday clinical usage – two species of ordinary usage which, inevitably, continuously exchange words, phrases, sentences, meanings. True, there is a linguistic-semantic tendency to invest comparatively more subjectivity in 'illness' than in 'disease,' and, conversely, more objectivity in 'disease' than in 'illness.' But this fact does not detract from the wider truth that the two words overlap semantically and so, to a large extent, may be used interchangeably – to which should be added, perhaps, *sickness*, another member of the same semantic family. Now, these linguistic-semantic facts obliges anyone who wishes to make use of a strict distinction between illness and disease to clarify this particular usage by stipulating what each word is supposed to mean in a given context. If he does not do this, misunderstandings are likely to ensue.

There are two rather special, though not too infrequent, situations in current health care where a strict distinction between disease and illness is both adequate and useful. On the one hand, we have that patient who clearly feels that all is not well with his body and himself, who complains and feels afflicted, and no doubt has a reason to complain, but whose ailment – and this is the crux of the matter – will not, or indeed cannot, be diagnosed by a medical professional as some known disease or other. The professional is here wont to declare, sadly: "we didn't find anything," which is her way of saying: "I acknowledge your illness, but unfortunately there is little I can do to help you, since your illness corresponds to no disease that I know of." In other words: this is a case of illness without disease, which confirms that a strict distinction between illness and disease may be useful and enlightening.

On the other hand, we have that patient who is unwilling to recognize himself as a patient, simply because he feels perfectly well, only that he has now come under the purview of medical professionals, who have examined his organism – anatomically, physiologically, chemically, immunologically, genetically, etc. – and found some abnormality or other, possibly amounting to the diagnosis of a readily recognizable disease. This is then a case of disease without illness, and it confirms – no less than does the inverse case (illness without disease) – that a strict distinction between illness and disease may be useful and enlightening.

It should be carefully noticed, however, that the two sorts of cases and situations here considered – illness without disease, and disease without illness – do not fail to emerge as anomalies in the clinical encounter between physician and patient. If the patient does not present *both* disease and illness, the physician is likely to lose confidence about what to do, or not to do, vis-à-vis his patient, and their relationship may become infected with feelings of unease, disappointment and distrust, or even hostility and rage, outrage – feelings which, no doubt, emanate mainly from the patient's reactions and perceptions of the situation, before they, gradually, infect the whole relationship. Fortunately, this is by no means what usually happens in the clinical encounter. In most clinical cases, both illness and disease are present, often corresponding to each other in ways that facilitate and further a physician-patient interaction and dialogue characterized by mutual trust and understanding. Which is to say that, in most cases, a strict distinction between illness and disease is misplaced, indeed a misleading fabrication.

The distinction turns out to be particularly misleading if it is taken to coincide with the distinction between the physician and the patient, the former supposedly being concerned exclusively with disease (= neutral and objective scientific description), the latter exclusively with illness (= subjective experience). Used in this way the distinction is no more than a well-worn, simplistic cliché, which conceals the real situation and obscures the important facts that disease/illness is a mixture, or alloy, of subjective and objective considerations: facts, feelings, perceptions, attitudes, reactions, ideas, events, etc., and that this is so for both parties, for doctors and patients alike, indeed not in quite the same way, or in identical proportions, but still for both a mixture that makes any strict distinction between illness and disease inadequate and misleading.

Fortunately, few patients and, I hope, no doctor would be quite as one-dimensional in their outlook as the strict distinction between illness and disease suggests. Doctors and patients would, from the very beginning, find it extremely difficult to communicate in a meaningful manner if they did not, to a considerable extent, *share* conceptions of what disease/illness is. And the eventual outcome of such a clinical encounter, based on a radical lack of dialogue and mutual understanding, would be poor, or disastrous – at least for the patient.

But the truth is that there is a continuous exchange of meanings related to disease/illness between common usage and professional medical usage. And so, each and everyone who is concerned about disease/illness as something that affects, disturbs, disables, hurts, tortures, and ultimately destroys people, will look upon it as a mixture of subjective 'illness' and objective 'disease' – thus healing the spurious distinction-dichotomy between illness and disease.

Center for Medical Ethics
University of Oslo
Norway

NOTES

[1] I have borrowed the expression 'fake reasoning' from Haack (1998, pp. 189-91).
[2] I have tried to achieve a statistically gender-neutral text by a fifty-fifty ratio of female and male gender. Furthermore, when 'he or she' does not sound too awkward, I have used that expression instead.
[3] On 'horizon' and 'fore-understanding' as phenomenological and hermeneutical concepts, see Gadamer (1989).
[4] In my book *Icons of Disease: A Philosophical Inquiry into the Semantics, Phenomenology and Ontology of the Clinical Conceptions of Disease* (1987) I introduce the term 'icon' as short for 'integral clinical conception'.

BIBLIOGRAPHY

Auden, W.H.: 1972, *Epistle to a Godson & Other Poems*, Faber & Faber, London.
Collins Cobuild English Language Dictionary: 1987, Collins, Glasgow.
Faber, K.: 1923, *Nosography in Modern Internal Medicine*, Humphrey Milford, London.
Fábrega, H. Jr.: 1997, *Evolution of Sickness and Healing*, University of California Press, Berkeley.
Feinstein, A.R.: 1967, *Clinical Judgment*, Williams & Wilkins, Baltimore.
Gadamer, H-G.: 1989, *Truth and Method*, Second, revised edition, Crossroad, New York.
Haack, S.: 1998, *Manifesto of a Passionate Moderate: Unfashionable Essays*, University of Chicago Press, Chicago.
Jonas, H.: 1966, *The Phenomenon of Life: Toward a Philosophical Biology*, University of Chicago Press, Chicago.
King, L.S.: 1963, *The Growth of Medical Thought*, Chicago University Press, Chicago.
Kleinman, A.: 1980, *Patients and Healers in the Context of Culture*, University of California Press, Berkeley.
Nordenfelt, L. and Twaddle, A.: 1993, *Disease, Illness and Sickness: Three Central Concepts in the Theory of Health – A Dialogue*, TEMA: Department of Health and Society, Linköping.
Peirce, C.S.: 1931-1935, *Collected Papers of Charles Sanders Peirce*, 6 vols. The Belknap Press of Harvard University Press, Cambridge, Mass.
Potter, V.G.: 1996, *Peirce's Philosophical Perspectives*, Fordham University Press, New York.
Reading, A.: 1977, 'Illness and Disease', *Medical Clinics of North America* 61, 703-10.
Reiser, S.J.: 1978, *Medicine and the Reign of Technology*, Cambridge University Press, New York.
Simpson, J.A. et al. (eds.): 1989, *The Oxford English Dictionary*, Second edition, Clarendon Press, Oxford.
Sinclair, J. et al. (eds.): 1987, *Collins Cobuild English Language Dictionary*, Collins, Glasgow.

Sundström, P.: 1987, *Icons of Disease: A Philosophical Inquiry into the Semantics, Phenomenology and Ontology of the Clinical Conceptions of Disease*, TEMA: Department of Health and Society, Linköping.

Toombs, S.K.: 1992, *The Meaning of Illness: A Phenomenological Account of the Different Perspectives of Physician and Patient*, Kluwer Academic Publishers, Dordrecht.

Wittgenstein, L.: 1953, *Philosophical Investigations*, Basil Blackwell, Oxford.

RICHARD M. ZANER

THINKING ABOUT MEDICINE

I. INTO THE MIDST OF THINGS

The acceptance of many new medical procedures is fueled by their being promoted as 'treatments' promising to correct or ameliorate some 'condition,' rendered as both awful and 'medical.' In vitro fertilization is a clear instance.

Insisting from the very beginning that 'infertility' is a 'disease,' the swiftly developing industry became pervaded with the idea that IVF is a 'treatment' (Edwards, 1974; Kass, 1972a). It was soon clear that IVF was not going away. Indeed, largely unregulated, programs quickly expanded into clinics, hospitals, and private offices in almost every sector. The accepted idea was: if you're 'infertile,' go to a fertility clinic where we'll finesse 'nature' – with infertility circumvented and pregnancy just around the corner. Beneath all that is something usually unspoken, but quite potent: namely, a silent, 'moralized' sense that being infertile is unfortunate and unnatural since it does not conform to what 'normal' people are supposed to be and do.

Much the same tacit connection between silent moral feeling and overt advertisement of 'therapeutic promise' plays out in disputes over genetics and especially cloning. It seems rather clear that the public's rapid acceptance of the Genome Project is due to the prospect not simply of new knowledge about genetic flaws and anomalies, but far more the always sincere promises of 'treatment.'

Still, it is evident that there is something deeper at play here. On the one hand, it is now obvious that the development of genetic therapy is not only perplexingly difficult but probably is further in the future than usually stated. On the other hand, a key part of this is the idea that certain human traits are deeply negative and should be corrected – violence and sexual perversity are very high on that list. Suggestively, there has been a good deal of public éclat to the idea that with the new genetics it may finally be possible to 'do something' about human proclivities to stupidity, violence and perversity.

And just here that other, darker side of genetics creeps in: the ability to diagnose, and perhaps someday to treat, also harbors ineluctably the ability to enhance and improve – according to *someone's* notion of 'norm,' 'good,' and 'evil.' The eugenic motif is a serious, if often hidden, motivation for the mapping and sequencing of the human genome (*Time*, 1999). We are, in a word, on the horns of a dilemma stemming from the increased presence of a wholly new kind of power: the ability to make and re-make human beings, without yet having the essential guide of responsible distrust at hand. The eugenic 'breeding of people' (for which 'cloning' is a key component) – often regarded as scarily inevitable (Kass and Wilson, 1998) – is here to stay.

II. THE 'NEW PARADIGM': MOLECULAR MEDICINE

Hans-Jorg Rheinberger (1995, p. 251) has argued that the traditional restorative character of Western medicine harbors an epistemological paradox, inasmuch as it is "a cultural endeavor set in place, not to conquer, replace, or transcend nature, as most other cultural endeavors have been, but to maintain the integrity of the body as a natural entity in its own right and for its own sake." One of its key traits is that the body "has been treated as a given and as a gift," a view that has not been successfully challenged until our times. Now, however, Rheinberger (1995, p. 251) suggests, there is a widespread, possibly irreversible "transformation of living beings toward deliberately engineered beings, including the human body," which signifies the "end of 'natural' evolution" – a vision already urged by Sir John Eccles (1979), Sir Macfarlane Burnett (1978) and more recently, Walter Gilbert (1992).

Key to this stunning development is the "molecularization" of medicine whose character and aims are rapidly being changed by recombinant DNA techniques – restriction, transcription, replication and ligation enzymes, plasmids, and other vectors, as well as bits and pieces of DNA and RNA. Prior to this, molecular techniques were intended to construct experimental models in an environment where the milieu of the living cell could be replicated in the laboratory, then studied and explained. Extracellular models represented intracellular configurations.

The fundamental change that occurred with the introduction of these rDNA techniques essentially reversed that relationship, for with these and other technologies, molecules became conceived as "informational." Constructed within extracellular conditions, these molecules are subsequently implanted into intracellular environs where the organism itself transposes, reproduces, and 'tests' them. The organism has thus become a *"locus technicus,"* a representational space in which new genotypic and phenotypic patterns are probed, tested and articulated. A technique of potentially unlimited medical significance and application, now for the first time "metabolic processes are becoming susceptible to manipulation" at the cellular level, the "level of *instruction*. Until that point was reached, medical intervention, even in its most intrusive forms, was restricted to the level of metabolic *performance"* (Rheinberger, 1995, p. 252).

The extracellular, therefore, no longer 'represents' or 'explains' the intracellular; just the reverse: the intracellular now represents some specific extracellular project. What is 'inside' (genome) is now taken to explain what is 'outside' (behavior). In this way, Rheinberger (1995, p. 253) concludes, "the deliberate 'rewriting' of life" has occurred. The organism itself has become a "laboratory" in which bits and pieces of itself, or related organic materials, are designed and implanted by molecular engineers in bodily cells where they are "tested" and eventually found to "work" (or not). A "new medical paradigm: molecular medicine" has thus emerged (Rheinberger, 1995, p. 254) – a remarkably accurate fulfillment of what early science fiction writers like James Blish had postulated for the new biology.

Clearly, too, not only has a new, highly profitable industry emerged within the medical-industrial complex, with all the usual venture-capital accoutrements (Hubbard & Wald, 1997; Kevles & Hood, 1992; Annas & Elias, 1992), but profound social changes in expectations and perceptions will follow inevitably in their wake. What Gilbert and others regard as the "holy grail" is precisely this new molecular medicine. As has already been suggested in many quarters, the answer to the ultimate questions of our nature, duty, and destiny will then be postulated either to *be* the genes or to be *in* the genes – and not in the familiar metaphysical Easter-egg hunts of philosophers and theologians.

There is, of course, even more on the agenda. As the 'molecularizing' of diseases and their possible treatments gains even firmer grip on modern medicine, we will doubtless witness the very same transformation of the common understanding of 'human nature' – subjecting it to deliberate technical blue-print projects designed by scientists and technologists (or by those who fund their efforts) that are intended to alter that 'nature,' be it to get rid of genetic disease or to enhance the human genome (Rheinberger, 1995, p.254).

Ingredient to the newly emerging molecular medicine is the notion that "human diseases are essentially nature's occasional misprints, to be corrected in the next edition" (Rheinberger, 1995, p. 254). With this notion the lines have inevitably begun to blur between what used to be seen as 'basic' research on the one hand, and medical 'applications' on the other. As Jonas well understood, it has now become perfectly obvious that modern science is technological by nature. Correlatively, the same blurring can be detected within one of the cornerstones of modern self-understanding generally and of modern science in particular – the assumed ontological difference between nature and culture. Having gained access to the corporeal core of human 'nature,' the consequent disappearance of 'evolution' in its usual sense is also at hand, as Eccles urged two decades ago: "natural evolution has come to an end. Molecular biology will come to invent biological reality" (Rheinberger, 1995, p. 256).

Reflecting on these matters, Rheinberger suggests there is something unspeakable at the root of this paradigm shift, a kind of 'scandal' in the sense first given this notion by Claude Lévi-Strauss when he identified the incest taboo as a 'scandal.' Trying to account for the historical emergence of savage communities, Lévi-Straus (1967, p. 10, cited in Rheinberger, 1995, p. 258) formulated his well-known principle that "...everything that is universal in man belongs to the order of nature and is characterized by spontaneity, and that everything bound to norms belongs to culture and has the characteristics of the relative and of the particular."

Yet, having said this, Lévi-Strauss (Rheinberger, 1995, p. 258) then noted that the incest taboo is exceptional, for it is neither culture nor nature. Although it "grounds a rule, [it is] a rule which, uniquely among all social rules, at the same time possesses the character of universality." That 'rule' cannot be 'natural' since incest can be chosen; but neither is it 'cultural' for it is universal. There is thus a "scandal" here: while incest *can* be chosen, it *ought not* be chosen – hence, the "taboo." Incest is banned for it lies outside the limits of social definitions and norms. Yet, the taboo implies acts of incest, however abhorrent; therefore, the taboo leaves "in the realm of the unthinkable what has made it possible" (Rheinberger, 1995, p. 258).

As Jacques Derrida (1978, p. 283) pointed out (although he has far more subtle issues in view which I must ignore here) the taboo is a taboo solely within a context that accepted the opposition between nature and culture, yet the 'scandal' is "something which no longer tolerates the nature/culture opposition he has accepted." It is 'unthinkable' in the sense that it makes possible both the nature/culture distinction and opposition, and in that sense grounds the very possibility of philosophy and knowledge (Rheinberger, pp. 283-84).

Rheinberger is convinced that, in just the way that incest taboo is a scandal, there is a 'scandal' at the heart of the 'new medicine'. What is it? We may catch a glimpse if we think about a key feature of biomedical science: the requirement of informed consent. If medicine's very point is to help sick, compromised people who cannot help themselves – people who for these very reasons are multiply disadvantaged and at their most vulnerable (Zaner, 1988) – why would there ever be any question at all about informing people and ensuring no one takes advantage of them? Yet, just this has become a centerpiece of medicine and biomedicine in our times.

There is an intriguing ambiguity, both with what Lévi-Strauss studied and with the well-understood need for informed consent when using human subjects as experimental subjects. In both cases, there would be no need either for taboo in the case of incest, or legal requirement in the case of informed consent, if it were not first of all true that the need for both taboo and legal demand pre-exists. If vulnerable patient-subjects were not abused in some manner in the first place, obtaining informed consent would be pointless – as would a taboo on incest, if no parent or sibling engaged in sexual activity with child or sibling. Yet both occur often enough that it is not unreasonable to wonder which – the action of incest and the abuse of patients; or the taboo on incest and the need for the requirement for informed consent – constitutes a real 'scandal'. And, there is something unspoken here as well: though people can be and have been abused in experimental settings, this is rarely talked about – or if so only indirectly, by emphasizing, indeed requiring, informed consent.

How do we make sense of these remarkable developments which were only barely suspected in the 1960s when some of us began working within medicine, became clearer by the early 1980s, and sensationally obvious since the genome project got under way, even more since the advent of animal cloning in 1997. To help grapple with the many complex issues here, I suggest a somewhat circuitous route through a wonderfully fascinating novel, *Mendel's Dwarf*, by the British molecular biologist, Simon Mawer (1998).

III. ON BEING A DWARF

The principal character is Dr. Benedict Lambert, a world-renowned geneticist, the great-great-great nephew of Gregor Mendel, and a dwarf (achondroplasia). Whether delivering a scientific lecture, sitting in one of his college classes, or simply walking the streets, Ben is invariably gawked at and thereby made acutely aware of himself over against the "phenotypically normal" folks who gape at him. He is on the outside socially; but, with superb irony, Mawer constructs his character as a scientist who possesses the awesome power made available by the new genetics: after years of research, he discovers the achondroplasia gene.

After the lecture he delivers to the Mendel Symposium, commemorating his great-great-great uncle Gregor in the city where Mendel's work was done, a crescendo of applause still ringing in his ears, Ben is greeted by the secretary of the association, a "large and quivering mountain of concerned flesh." She purrs, "Gee, Ben, that's wonderful. So brave, so brave...." Ben thinks to himself: "Brave. That was the word of the moment. But I'd told

Jean [his lady-love] often enough. In order to be brave, you've got to have a choice" (Mawer, 1998, p. 5). Precisely, he thinks, what he never had, merely instead that "tyranny of chance" when just one of the innumerable sperm erupted from his father's orgasm and impregnated his mother's ovum, that magic moment of conception from which Ben is born. As the headmaster of his first school later remarks, after the typical round of teasing from classmates ("Mendel, Mendel, Mendel's dwarf"), "it's a problem you have to live with." And Ben silently rages: being a dwarf is not like premature baldness or a stammer, "it is me. There is no other" (Mawer, 1998, p. 21).

I doubt that even having a choice is enough to support such moral notions as "brave" or "courageous." Choice requires an awareness of options (if only 'yes' or 'no' to one of them) and, for that very reason, a reflexive awareness of oneself as the one who chooses and is responsible for the choice. Obviously, Ben never chose to be a dwarf, and comes to awareness of himself as dwarf through others – those who are phenotypically *different*. For example, while in college he meets Dinah, the first girl he ever kissed. He helps her get through a genetics class and she is beside herself when she learns she's passed. She suddenly kisses him, then promptly regrets it, but is unable to say why out loud. Befuddled yet on fire, riveted by "*she* kissed *me*!", Ben tells her he loves her. Her response? "I knew you'd do this…can't you see it's impossible?" To which Ben shouts: "Of course it's impossible. It's the impossible that attracts me. When you're like I am, who gives a toss about the possible? …I'll say it for you: you can't love me because I'm hideous and deformed, a freak of nature, and people would stare" (Mawer, 1998, p. 52).

His discovery of the achondroplasia gene leads to the invitation by the Mendel Symposium. His sister telephones him about a newspaper's report of his discovery. She reads the headline: "Dwarf Biologist Discovers Himself," and then the text:

[S]uper geneticist Ben Lambert has finished his search of a lifetime. Genetic engineering techniques and years of patience have finally led him to discover the gene that has ruled his own existence, for Ben, thirty-eight and a researcher at one of the world's leading genetics laboratories, is…a dwarf. Little in body but big in spirit…." (Mawer, 1998, pp. 242-43).

Later, wandering through the small village after his lecture, Ben winds up in the convent garden standing before the statue of Gregor Mendel. He reflects:

This acre of space was where it all started, where the stubborn friar lit a fuse that burned unnoticed for thirty-five years until they discovered his work in 1900 and the bomb finally exploded. The explosion is going on still. It engulfed me from the moment of my conception. Perhaps it will engulf us all eventually (Mawer, 1998, p. 10).

At one point Jean, a librarian with whom he falls in love, tells Ben they really cannot be together any longer. They had sexual relations, resulting in his making her pregnant; but she had aborted. However, she later asks him to help her become pregnant again using IVF, since her husband (whom she just cannot leave) is infertile. Ben, who had become accomplished in reproductive techniques, not only uses his own sperm but later observes the fertilized embryos and gets ready to do the embryo transfer to Jean's uterus. Thus, Ben, whose dwarfness was not chosen, finds himself with the chance and the power to choose whether *any* babies will be born at all and, if so, whether they will be like *him* or like *her*. What should he do? Like Gyges in Book Two of Plato's *Republic*, having found the ring of power, what now? (Zaner, 1994)

Ben is a dwarf without having chosen that; moreover, as his dwarfism is unable to be 'restored', what, exactly, *is* he for conventional medicine? As he himself knows keenly, he is not only socially but also medically outside, he is phenotypically *non-* or *ab*normal. As he says what his college friend, Dinah, could not bring herself to say, in conventional terms he is "hideous, deformed, a freak of nature...a shrunken monster." 'Deformed', yes, but he is not medically 'sick' or 'injured'. Geeks and freaks are socially constructed as fit only for carnivals or circuses. Thus, Dinah cannot see herself as 'in love' with, nor even 'as loved' by Ben. And, despite her love for him, Jean cannot see herself married to Ben. Dwarfs and freaks are scandals, it would seem. One dwarf may be an imp or a good fellow, another a rogue, but all stand outside due to that "tyranny of chance" both of birth and from its social construction. The social world is not designed for, nor especially friendly to, geeks and freaks.

Still, I continue to oscillate, for there are the many layers of Mawer's cutting irony: Ben is a brilliant geneticist, a descendant of Mendel the great geneticist. Ben identifies the 'dwarf gene' and later has the choice whether to splice that gene into or out of his and Jean's embryo. He feels the power that he has, his ability to choose about that very issue over which neither he nor restorative medicine had any control whatever. He seems, indeed, to sense himself beyond the power of chance. He has the power to cancel the accident of birth for his progeny. He lives and knows the sharpest edges of the new genetics. He has the kind of choice that very few of us ever will have.

Unlike in restorative medicine, then, the new paradigm – like most other cultural projects – is centrally concerned to conquer, replace and transcend nature. Thus, on the new agenda is not merely curing but nipping heritable diseases in their genetic bud, correcting otherwise debilitating anomalies, and altering the human condition itself so as to ensure (so the promissory note reads) immunity against those 'genetic' diseases and anomalies. The membrane separating nature and culture is porous. Still, we have to ask, where is the scandal, in the sense that the incest taboo was regarded by Lévi-Strauss, or in the manner of the requirement of informed consent?

IV. CLOSING IN ON THE UNSPEAKABLE

I want to begin an ever so cautious turn into the unforgiving unthinkable. Several movements are at hand.

First movement. There may be nothing, or not much, that can be done to restore Ben's body, although it is not beyond the pale as it once was: bone-lengthening, nerve-regeneration, and other novel interventions are announced daily. To be sure, we could all work to change social attitudes toward the disabled. Echoing an element in the IVF debate, we would still have to wonder whether it is the business specifically of medicine to be involved with, or even concerned about, the grievous social mistreatment Ben regularly receives (as do most of the disabled). On other side of it, Ben is a famous scientist appointed to a famous institute, is doing quite well in many of the ways that count – what need does he have for anything from medicine?

But even if the bodies of dwarfs are not, or not yet, on the agenda of the new medicine, their progeny most surely are on it – witness Ben's example – and for that reason, if for no other, the new paradigm is unavoidably eugenic. Everything, in short, is up for review, new exploratory designs and inventive blueprints, and open for possible (if not yet probable) reversal, correction, even replacement if need be (already implicit to transplant medicine). In avid pursuit of geneticist Gilbert's "holy grail," nothing can be ruled out! Indeed, quite literally, everything 'human' is very much 'in bounds': mind, body, self, emotion, action, the works.

Still, to speak of something unspeakable here seems at best ill mannered if not irreverent and irrelevant. Yet, in that very hubris is a whiff of dread, however faint.

Second movement. Mawer's novel is deeply instructive. Why is Dinah so disturbed by Ben who is very nice to her and goes out of his way to help her get through her class? At the very moment she kisses him, there is something unspoken. Or, it is spoken only because the unspeakable one, Ben, says *for* her what she cannot bring herself to say: that he is "hideous and deformed, a freak of nature, and people would stare." But why is it so hard for her to say this to Ben? Why is that so difficult, even insulting, for anyone to say to someone like Ben?

That whiff of the absurd and scandalous seems more definite, more like spoor than road maps and we, avid hunting dogs, hot on its trail even while things remain foggy.

Third movement. When discussing using people as research subjects, Claude Bernard (1865) did not mention anything like 'informed consent' (it probably never occurred to him). His comment is well-known:

> It is our duty and our right to perform an experiment on man whenever it can save his life, cure him, or gain him some personal benefit. The principle of medical and surgical morality, therefore, consists in never performing on man an experiment which might be harmful to him to any extent, even though the results might be highly advantageous to science, i.e. to the health of others (Cited in Katz, 1992, p. 229).

Commenting on this passage, Jay Katz seems taken aback not only by Bernard's not thinking about informed consent, but also that he mentions only "our duty" and "our right." There is, Katz says, one fundamental question that has "not been thoroughly analyzed to this day: When may investigators in any way expose human beings to harm in order to seek benefits for them, for others, or for society as a whole?" The fact is, he contends, one cannot find any "general justifications" for using human beings as research subjects – not that they cannot be found at all. Rather, Katz (1992, p. 231) remarks, "I only wish to point to the pervasive silence...." But why, we might wonder, should Katz be even slightly disturbed? Aren't such physician-scientists rigorously governed by the Hippocratic imperative, *nolo nocere*? Is it that he or we cannot trust such scientists never to cause harm? Can't we count on researchers wanting subjects to understand and not be coerced? In any case, Katz exaggerates a bit when he alleges that these issues have never been "thoroughly analyzed," for he apparently forgot that it is precisely these matters that a troubled Jonas (1974, p. 105, n.1) analyzed many years ago – specifically bringing his concerns to the attention of the scientists who had invited his "philosophical" thoughts about using human beings as experimental subjects.

In any event, Katz (1992, p. 231) thinks that "the pervasive silence" leads directly to "a slippery slope of engineering consent" which in turn leads "inexorably to Tuskegee, the Jewish Chronic Disease Hospital in Brooklyn, LSD experiments in Manhattan, DES experiments in Chicago" which are "done in the belief that physician-scientists can be trusted to safeguard the physical integrity of their subjects."

It might be said, to be sure, that the physician-scientists who conducted those experiments, like those who did the experiments in the Nazi camps, the

Gulag Archipelago, Willowbrook, or others, were simply perverse people. Science and medicine are, on this view, 'value-neutral'; perversity stems not from science but from misuses and abuses by individuals who are misled or evil. Yet, since these scientists chose to engage in such research, and since people ought not be treated that way when they are already terribly vulnerable, is this not as unspeakable as engaging in an act of forbidden incest? The smell of scandal seems more definite.

Fourth movement. If the possibility is attractive nowadays to 'do something' as much for the dwarf, the freak, the geek, as for those crushed by genetic disease and other anomalies, does this give a sort of license for physicians in the newly emerging era to do whatever they deem necessary, to replace and transcend both 'natural evolution' and unfortunate 'products' like Ben – precisely as was urged by Burnett, Eccles, and others? Is not being a dwarf deeply abhorrent, in precisely the way that the 'feebleminded' were regarded by Darwin as "highly injurious to the race of man" (Darwin, 1974/1874, pp. 130-31)?

There is that 'slippery slope of engineering consent' which Katz believes leads 'inexorably' to the dreadful perversions as Tuskegee, Willowbrook, and others. How could any physician in the Hippocratic, restorative tradition ever be caught up in such corruption of that tradition's noble aims? (see Jonas, 1974, pp. 113-17)

V. TRYING TO ENTER THE UNSPEAKABLE

Rheinberger (1995, p. 259) suggests: "Just as the incest prohibition became the scandal of anthropology, so has the commandment of truth become the scandal of the sciences of natural things," which includes medicine precisely because of its focus on the human body. A basic aim of the new genetics is the "deliberate 're-writing' of life," and thereby a kind of power is introduced into human society that is unprecedented and exhibits a kind of universalization: it inexorably sweeps up everything in its path, introducing a sense that nothing is beyond the pale. Scientists now do as Ben did: choose which sperm to use, which ovum and which, if any, embryo to implant. Some, though probably few, even may have had the chance to choose their own children. In any case, is this power to 're-write' life in the hands of finite human beings something 'scandalous' – whether it is moved merely by the passing fancy of current fads or by something more noble? Or is it merely an instance of hubris, not unheard of in human affairs?

Rheinberger and others have noted, too, that the wider public generally does not, often cannot, and even doesn't care to understand the nature and implications of reproductive technologies, much less such exotic undertakings as the genome project. In this people have apparently ceded their choice to the very scientists who engage in such experiments and the engineers who produce the goods for our use. Consumers have no need to understand in order to consume, and therein lies an intriguing prospect: they are voters, and politicians perforce heed (even seek to create) their wants. If, as Gerald Weissmann (1982, p. 110) notes, society "gets not only the government, but the science, it deserves," then scientists are unavoidably in the midst of the political arena competing for funding and, therefore, in the game of explaining publicly what few want, need, or can understand. Rheinberger (1995, p. 154) argues nothing "could describe the political moves of James Watson, Walter Gilbert and their combatants better...." Put a bit crudely, scientists have to 'sell' what they're up to, an act of persuasion that tells the public (and those holding the purse-strings) what they want to hear or what they will find attractive. We thus witness the spectacle of political finesse by scientists seeking immense funding for their scientific projects. More subtly, is the 'scandal' that frankly eugenic designs are cleverly concealed by the charade of promised therapies – promising, at some point even perchance delivering, what is merely part of the real story anyway? Is the 'scandal' that only those with the skill and knowledge to practice genetics will actually use that power, while never letting on about the 'truth' of what goes on 'behind closed doors'?

Are greed, deceit and the passion for epistemic and political power the 'scandal' within the new medicine, occurring in 'pervasive silence' not unlike the radiation experiments over the past fifty years, thus giving ample reason to distrust science since 'the truth' seems unspeakable? As Rheinberger (1995, p. 259) suggests, "just as the incest prohibition became the scandal of anthropology, so has the commandment of truth become the scandal of the sciences of natural things." The work of science would then be a sham, and medicine on sale to the highest bidder. Still, there seems something still deeper here.

What comes to mind is the powerful Grand Inquisitor scene in Dostoevsky's, *The Brothers Karamasov*, when, after hearing Ivan's tale of the return of Christ (whose name is unspoken, perhaps unspeakable) and the Inquisitor's objections to His return, Alyosha cries out with his riveting question: Is anything then permitted? Is nothing forbidden? Are we free to do anything, simply because at any moment we happen to want it? I am also reminded of Husserl's (1965; 1970) abiding concern for what he analyzed as the "crisis" of Western science (and humanity), along with his conviction that the "naturalization of consciousness" means that the very idea of standards or norms of reason is threatened.

VI. THINKING ABOUT BIRTH, AND BEYOND

Those are strong words. Yet, my sense is that there is still something else lurking the darker corners of those labs. At stake in the Genome Project is a fundamental philosophical-anthropological issue. It concerns not only how 'self' is at all known or experienced, but whether there is 'self' at all, or instead, merely genetic information encoded in and on strands of DNA/RNA nestled within each individual cell. This, of course, is the reason for Gilbert's over-heated praise for the genome project, the "holy grail" that has him so enthralled. " Here is a human being; it's me!" he announces, and it can be etched on a CD – more likely, I should think, in the tiniest of our tiny body-parts, tucked away in a cell like some neo-modernist spy, a mole whose code is known only by those who implanted it. Here is a challenge no philosopher can ignore. But does it pose the same question of 'scandal' that Rheinberger dares us to face?

To help make my way through these brambles, I need to mention several of the odder passages in the work of Alfred Schutz, and return to Ben the dwarf.

After insisting in his critique of Husserl that intersubjectivity is a "given" and not a "problem," Schutz (1975, p. 82) wrote: "As long as man is born of woman, intersubjectivity and the we-relationship will be the foundation for all other categories of human existence." Precisely because of this everything in human life is "founded on the primal experience of the we-relationship" which, though Schutz didn't explicitly say so, must surely be the experience of *being born* – or, in his phrase, being "born of woman." Since *all* "other categories of human existence" are founded on this primal experience, Schutz (1975, p. 82) calls intersubjectivity "the fundamental ontological category of human existence in the world and therefore of all philosophical anthropology."

In one of his essays on Max Scheler, Schutz's (1967, p. 168) pronouncement is just as fascinating. He first pointed out that there is one taken-for-granted assumption which no one, not even the most ornery skeptic, for a moment doubts: "we are simply born into a world of Others.... As long as human beings are not concocted like homunculi in retorts but are born and brought up by mothers, the sphere of the 'We' will be naively presupposed." Here, too, it is reasonable to surmise that what is "naively presupposed" is precisely that "primal experience" of being "born of woman" and (he added here) being raised by mothers, as opposed to being "concocted...in retorts."

Note first the idea that "being born of woman" constitutes *the* (not even 'a') "fundamental" ontological and anthropological category of human existence. It is true that it has occurred to almost no philosophers to consider this phenomenon of "having been born of woman." Reflections on death and dying are plentiful, while those on birth and being born are oddly absent. Still, if we think about it – as we must, if we would understand our own humanity, even if we can get at this only indirectly through other people (Schutz and Luckmann,1973, p. 46) – my having been born is surely as constitutive of 'my life' as is my 'going to die.'

Schutz himself did not probe this phenomenon beyond these scant references. Still, I have to take it quite seriously. In a compelling way it is the primal experience of being (or having been) born that constitutes the crucial other side of the central fact of our temporal condition, our being-with-one-another as we grow old together. We could not experience ourselves as growing older if we did not *begin to be* – so to speak, come at some always-already-ongoing time in our lives to find ourselves as having-already-been-thrust-into life: birthed and thereby 'worlded.'

I want to suggest briefly and tentatively that being human is something that *happens* to each of us; being human is being *given*, even *gifted* with my life and the sense of myself as having to be achieved. The primal Other is my mother; the primal body is her body; the primitive *occurrence* and *place* of embodiment is her body, her womb. She 'gives' me *myself*, she 'gifts' me with *herself*. Later, she 'gifts' me with history, culture, my and the *world* (mainly through daily discourse and its stories). I am not only 'being-toward-death,' therefore, but just as fundamentally 'being-from-birth.'

In every significant sense, being-born at all with whatever initial biological wherewithal is as unchosen, as radically unfair, as being born a dwarf. Yet being born inevitably brings with it a radical debt that seems to be at least one crucial root of responsibility – 'radical' in the precise etymological sense, 'of or going to the roots.' I am gifted not simply with life but with my self-conscious life, which I must then go on to accomplish. What and who I am, is what and who I in multiple ways become, and this is first set in motion in the essentially mysterious and accidental ways of every birth.

How is it that 'I', this person now asking the question, got here at all, became this specific person, me? How is it that just one specific sperm made its way into that one specific ovum produced at just one specific time by my mother, and came to be 'me'? What is this, that each of us is born as 'me'? Even were a couple, like Ben and Jean, to undergo an *in vitro* infusion of a pre-selected sperm – 'get that one there, Shirley...no, not him, *that* one' – how account for, how make understandable, just that unique 'life' which, if all goes well, becomes, say, Ben Lambert, brilliant geneticist, monstrous dwarf?

What these reflections evoke is not so much that the new medical paradigm puts this awesome process at risk; nor that it threatens my 'who I am' and our foundational social relationships (father, mother, son, daughter, etc.), as Leon Kass (1998) alleges. Rather, it is *my being at all* that is forced into a radically new light, and in this constitutes a fundamental *mystery*: *that* I am *at all*, *that* I have come to be or been brought into being (not just 'life' but *my* 'life') neither through my own nor anyone else's action or choice. The 'accident of birth' harbors ontological significance (see Spiegelberg, 1986). Being a self is being *my*self, and how can that be accounted for? Is this not, choosing words carefully, an essential mystery just because it is essentially fortuitous?

VII. BACK TO BEN

What is 'scandalous' about that? At one point in Mawer's (1998, pp. 197-98) complexly ironic novel, Ben succeeds in sequencing the likely genes which, incorporating a trivial error in a single base pair in "this enigmatic, molecular world," eventually resulted in that "genetic error" – a "simple transition at nucleotide 1138 of the *FGFR3* gene." In the dark recesses of his mother's womb this somehow mutated into him, thanks to one mistake in the 3.3×10^9 base pairs in Ben's genome, a single substitution of guanine for adenine, in the transmembrane domain of the protein. Of all the millions of pairings along those sinewy, snaky threads of deoxyribonucleic acid busily replicating the building blocks of life, proteins, a single letter error occurred, somehow – and there's Ben, the achondroplasic dwarf, that hideous, misshapen "monster" who, despite everything, loves Jean. Jean, likewise an accident but without, so far as we know, misspellings in her genome. Jean, who tries mightily to love Ben, too, but in the end, "it just isn't right, Ben, it isn't right...." Can 'right' or 'not right' bear the brunt of good sense?

Two things: one common, the other novel, and for that reason very uncertain as to its implications. The first? Just this, picking up on Spiegelberg's (1944) insight: *that*, having no choice in his birth, not even *that* he will be born, this unique Ben gradually emerges from that globally undifferentiated entity we call 'baby,' and from then on bears the brunt of being himself. *That*, in other words, his birth is happenstance, chance, an utter accident resulting from those many other uncanny 'accidental' happenings. But then, there is this: was he *born* 'Ben'? Or, did 'he' *become* 'Ben' at some point? Or, within that accidental substitution of guanine for adenine, was that one 'error' also 'Ben'? I mean, is his '*what* I am' – all the accoutrements and wherewithal implied by being born with just this body, just this mind, just these circumstances, just this time and place, etc. – all the sheerest of 'chances'? 'Ben' thus could just as readily and by the same array and course of 'chances,' *not* have been born, hence not *be* at all; or if born, then born without that chance mutation, and for any number of 'reasons' just as incalculable and accidental as that all the 'right' processes and timings managed to eventuate in his birth. And is this not the essential sense of the 'accident of birth' – what Spiegelberg acutely observed is the purest kind of "moral chance" and, therefore, utterly undeserved? 'I' am, just as 'Ben' is, and I am no more entitled by birth to 'what' I am than 'Ben' is entitled to 'what' he is.

And second? Of course, the choice Ben faces when Jean asks him to use his sperm for her *in vitro* fertilization. He agrees, checks the fertilized eggs and himself does the PCR amplification, finds that embryos 2, 5, 6 and 7 are unaffected (no misspelling of adenine by guanine) but 1,2, 4, and 8 show the mutation. Now what? By chance, four 'normals' and four mutations – dwarfs-to-be if allowed to be at all. It is his choice. What should he do? Note well: he *can* actually choose whether to do anything, and also choose which embryo to allow to grow to become a baby. This, Mawer reflectively observes about the scene, is unlike the way God works:

Benedict Lambert is sitting in his laboratory playing God. He has eight embryos in eight little tubes. Four of the embryos are proto-Benedicts, proto-dwarfs; the other four are, for want of a better word, normal. How should he choose?

Of course, we all know that God has opted for the easy way out. He has decided on chance as the way to select one combination of genes from another. If you want to shun euphemisms, then God allows pure luck to decide whether a mutant child or a normal child shall be born. But Benedict Lambert has the possibility of beating God's proxy and overturning the tables of chance. He can choose. Wasn't choice what betrayed Adam and Eve? (Mawer, 1998, p. 238)

When the deed is done and the baby on its way, Jean telephones to ask him what he did: "Will it be 'all right'?" The interchange heats up as Ben adroitly evades and dodges. Jean pleads with him, at which point he's had it, and grows angry "at the docile stupidity of her, at the pleading, whining kindness of her, at her naiveté. 'Well, you'll have to wait and see, won't you?' I said to her." Then he hangs up. What did he do? Since Ben won't say, should she abort? Why did he choose not to tell her? Why did he choose to implant an embryo in her uterus? Which one did he implant? Why did he choose it?

For all our questions, once Ben chooses which embryo to implant, is there any way for him to justify whatever decision he makes, whether 'normal' or 'dwarf' embryo? If his taking matters into his own hands, making it so that he had to choose, was not 'playing God' *what was it*? But what is it that Ben actually did, whichever embryo was implanted? Choosing the embryo is not choosing the person – that occurs, despite any and all planning, by its own permutation of chance. What happens after the brute fact of the accidents of conception and birth, that's something else, something with respect to which this or that course of life may or may not ensue with responsibility properly meted out for these as for all other facets of a person's life. Even if we acknowledge that Ben chose 'this' and not 'that' embryo, that of itself does not get rid of the bald, brutal fact of the chance of initial biological conception and birth with its own particular, utterly 'unchoosable' wherewithal. And *that it is not and cannot be chosen* seems utterly mysterious. It would, however, be quite wrong-headed to say, for oneself (if born as Ben) or for another (Ben's mother), 'God did it' and is responsible for the deed, natural or unnatural! Ben reflects, after using his sperm:

But what is natural? Nature is what nature does. Am I natural? Is superovulation followed by transvaginal ultrasound-guided oocyte retrieval natural? Is *in vitro* fertilization and the growth of multiple embryos in culture, is all that natural? Two months later ... I watched shivering spermatozoa clustering around eggs, *my* spermatozoa clustering around *her* eggs. Consummation beneath the microscope. Is that natural? (Mawer, 1998, pp. 214-15)

The scandal, if there be any: is it within that essential contingency of human life? Or, perhaps, is it that we cannot stand being contingent and, when we think about it at all, we try so hard – as Camus (1955) understood – to conceal it through wily postulations of transcendence?

Here, square on, is a fundamental mystery: each of us without exception is an *accident;* we find ourselves at some point after that rudimentary fact on this earth, and the sheer fact of being here at all – how and in the way we each happen to be – occurred with a plain throw of the dice, the sheerest of chances. I think of Gilbert, Watson, or Eccles; and I think of their vision of having control in a world profoundly governed, contrary to that dream, by the genius of chance.

Vanderbilt University Medical Center
Nashville, Tennessee
U.S.A.

BIBLIOGRAPHY

Annas, G.J. and Grodin, M.A. (eds): 1992, *The Nazi Doctors and the Nuremberg Code: Human Rights in Human Experimentation*, Oxford University Press, New York.
Annas, G.J. and Elias, S.: 1992, *Gene Mapping: Using Law and Ethics as Guides*, Oxford University Press, New York.

Bayertz, K.: 1994, *GenEthics: Technological Intervention in Human Reproduction as a Philosophical Problem*, Cambridge University Press, Cambridge.
Beecher, H.K.: 1966, 'Ethics and clinical research', *New England Journal of Medicine* 74,1354-60.
Burnet, M.: 1978, *Endurance of Life: The Implications of Genetics for Human Life*, Cambridge University Press, London.
Camus, A.: 1955/1983, *The Myth of Sisyphus, and Other Essays*, Vintage International, Random House, Inc., New York.
Cassell, E.J.: 1985, *The Healer's Art: A New Approach to the Doctor Patient Relationship*, MIT Press, Cambridge.
Clark, K.B.: 1973, 'Psychotechnology and the pathos of power', in F. W. Matson (ed.), *Within/Without: Behaviorism and Humanism*, Brooks/Cole Pub. Co., Monterrey, CA, pp. 93-99.
Darwin, C.: 1974/1874, *The Descent of Man and Selection in Relation to Sex*, Revised edition, Rand, McNally & Co., Chicago.
Derrida, J.: 1978, 'Structure, sign and play in the discourse of the human sciences', in *Writing and Difference*, University of Chicago Press, Chicago, pp. 112-135.
Dostoevsky, F.: 1961, *The Brothers Karamazov*, C. Garrett (trans.), W. Heinemann, London.
Edwards, R.G.: 1974, 'Fertilization of human eggs in vitro: Morals, ethics and the law,' *Quarterly Review of Biology*, 3-26.
Eccles, J.: 1979, *The Human Mystery* (The Gifford Lectures, 1977-78), Springer-Verlag International, Berlin/Heidelberg/New York.
Ehrenreich, B.: 1993, 'The economics of cloning,' *Time* (November 22), 86.
Gilbert, W.: 1992, 'A vision of the grail,' in Daniel J. Kevles and Leroy Hood (eds.), *The Code of Codes: Scientific and Social Issues in the Human Genome Project*, Harvard University Press, Cambridge, pp. 83-97.
Hubbard, R. and Wald, E:. 1997, *Exploding the Gene Myth*, Beacon Press, Boston.
Husserl, E.: 1965a, 'Philosophy and the crisis of European man,' in Q. Lauer (trans.), *Phenomenology and the Crisis of Philosophy*, Harper & Row, New York.
Husserl, E.: 1965b, 'Philosophy as rigorous science' in Q. Lauer (trans.), *Phenomenology and the Crisis of Philosophy*, Harper & Row, New York.
Husserl, E.: 1970, *The Crisis of European Sciences and Transcendental Phenomenology*, D. Carr (trans), Northwestern University Press, Evanston.
Illich, I.: 1976, *Medical Nemesis: The Expropriation of Health*, Bantam Books, New York.
Jonas, H.: 1966, 'The practical uses of theory,' in *The Phenomenon of Life: Toward a Philosophical Biology*, University of Chicago Press, Chicago.
Jonas, H.: 1974, *Philosophical Essays: From Ancient Creed to Technological Man*, Prentice-Hall, Inc., Englewood Cliffs, NJ.
Kass, L.R.: 1972a, 'Babies by means of in vitro fertilization: Unethical experiments on the unborn?' *New England Journal of Medicine* **285**, 1174-79.
Kass, L.R.: 1972b, 'New beginnings in life,' in M. P. Hamilton (ed.), *The New Genetics and Future of Man*, William B. Eerdmans, Grand Rapids, MI.
Kass, L.R. and Wilson, J.Q.: 1998, *The Ethics of Human Cloning*, The AEI Press, Washington, D.C.
Katz, J.: 1992, 'The consent principle of the Nuremburg Code: Its significance then and now', in G.J. Annas. and M.A. Grodin (eds), *The Nazi Doctors and the Nuremberg Code: Human Rights in Human Experimentation*, Oxford University Press, New York, pp. 227-39.
Kevles, D.J. and Hood, E.: 1992, *The Code of Codes: Scientific and Social Issues in the Human Genome Project*, Harvard University Press, Cambridge.
Kleinman, A.: 1988, *The Illness Narratives*, Basic Books, Inc., New York.
Laing, R.D.: 1967, *The Politics of Experience*, Pantheon Press, New York.

Latour, B.: 1988, *The Pasteurization of France*, Harvard University Press, Cambridge.
Lévi-Strauss, C.: 1967, *Les structures Élementaires de la parente*, 2nd ed., Mouton, The Hague.
Mawer, S.: 1998, *Mendel's Dwarf*, Penguin Books, New York.
Pellegrino, E.D.: 1982, 'Being ill and being healed: Some reflections on the grounding of medical morality', in V. Kestenbaum (ed.), *The Humanity of the Ill: Phenomenological Perspectives*, University of Tennessee Press, Knoxville, pp. 157-66.
Pellegrino, E.D. and Thomasma, D.C.: 1993, *The Virtues in Medical Practice*. Oxford University Press, New York and Oxford.
Rheinberger, H-J.: 1995, 'Beyond nature and culture: A note on medicine in the age of molecular biology', in *Medicine as a Cultural System*, M. Heyd and H. Rheinberger (eds.), *Science in Context* **8**:1 (Spring), 249-63.
Schutz, A: 1967, 'Scheler's theory of intersubjectivity and the general thesis of the Alter Ego', in M. Natanson (ed.), *The Problem of Social Reality*, Vol. I, *Alfred Schutz, Collected Papers*, Martinus Nijhoff, The Hague, pp. 150-179.
Schutz, A. and Luckmann, T.: 1973, *Structures of the Life-World*, R.M. Zaner and H.T. Engelhardt, Jr. (trans.), Vol. I, Northwestern University Press, Evanston, Illinois.
Schutz, A.: 1975. 'The problem of transcendental inter-subjectivity in Husserl', in I. Schutz (ed.), *Studies in Phenomenological Philosophy*, Vol. 3, *Alfred Schutz, Collected Papers*, Martinus Nijhoff, The Hague, pp. 51-91.
Spiegelberg, H.: 1944, 'A defense of human equality', *Philosophical Review* **53**, 101-24.
Spiegelberg, H.: 1986, *Steppingstones Toward an Ethics for Fellow Existers: Essays 1944-1983*, Martinus Nijhoff, The Hague.
Time: 1999, 'Smart Genes?', September 13, 154.
Tulipan, N. and Bruner, J. P.: 1998a, 'Myelomeningocele repair in utero: A report of three cases, *Pediatric Neurosurgery* 28, 177-80.
Tulipan, N., Hernanz-Schulman, M. and Bruner J. P.: 1998b, 'Reduced hindbrain herniation after intrauterine myelomeningocele repair: A report of four cases', *Pediatric Neurosurgery* **29**, 274-78.
Van den Berg, J. H.: 1978, *Medical Power and Medical Ethics*, W. W. Norton and Co., Inc., New York.
Weissmann, G. A.: 1982, 'The need to know: Utilitarian and esthetic values of biomedical science,' in W.B. Bondeson, H.T. Engelhardt, Jr., S.F. Spicker, and J. White (eds.), *New Medical Knowledge in the Biomedical Sciences*, D. Reidel Pub. Co., Dordrecht, Holland/Boston, pp. 105-112.
Zaner, R.M.: 1994, 'Encountering the other', in C.S. Campbell & A. Lustig (eds.), *Duties to Others*, Theology and Medicine Series. Kluwer Academic Publishers, Boston, pp. 17-38.
Zaner, R.M.: 1991, 'The phenomenon of trust in the patient-physician relationship,' in E. D. Pellegrino (ed.), *Ethics, Trust, and the Professions: Philosophical and Cultural Aspects*. Georgetown University Press, Washington, D.C., pp. 45-67.
Zaner, R.M.: 1988, *Ethics and the Clinical Encounter*, Prentice-Hall, Inc., Englewood Cliffs, N.J.

SECTION TWO

THE BODY

SHAUN GALLAGHER

DIMENSIONS OF EMBODIMENT: BODY IMAGE AND BODY SCHEMA IN MEDICAL CONTEXTS

I. INTRODUCTION

Perhaps the most general and most obvious fact about medicine is that it concerns the body. If one eliminates the body one eliminates the subject and object of medical science and practice. In contrast, philosophy has traditionally been dismissive of the body. One need not return to Plato, Plotinus, or Descartes to find representatives of this view. More recently, in the branch of philosophy known as philosophy of mind, and in the context of writing about personal identity, the philosopher Sidney Shoemaker (1999, 1976) distinguishes between three different concepts of embodiment: biological embodiment, sensory embodiment, and volitional embodiment. Shoemaker readily dismisses biological embodiment as being of any feasible importance to the issue of personal identity (he calls it a "radical" supposition to maintain that the biological body has relevance to this issue). The other two concepts of embodiment are analyzed in the context of functionalist thought experiments (involving such logical possibilities as body exchanges and brain transplants), putatively showing how unessential the body is to self-identity, understood as psychological continuity. Shoemaker, and others who frame their theories of personal identity in terms of psychological continuity, are not alone in this regard. Even theorists who reject approaches based on psychological continuity, and defend a biological approach to the question of personal identity, may dismiss the body as having relevance. Eric Olson (1997, p. 150), for example, captures the sentiment of numerous contemporary philosophers when he suggests that "the notion of a human body is best left out of philosophy, or at least out of discussions of personal identity."

In contrast to philosophers who dismiss the body, certain phenomenological philosophers recognize the ineliminable role of the body in the constitution of human subjectivity. Merleau-Ponty is well known for this emphasis. His work borrowed heavily from a tradition that included not only phenomenologists such as Husserl, but physicians, neurologists, and psychologists like Ravaisson, Goldstein, Head, and Schilder. In this essay I focus on and extend several of the concepts that informed Merleau-Ponty's analysis of the lived body. Specifically I want to make clear a distinction between the concepts of *body image* and *body schema*. These are concepts that continue to play a role in medical research and treatment, and that are, I will argue, essential to the philosophical analyses of questions pertaining to cognition and personal identity.

Merleau-Ponty (1945) was careful to distinguish between kinaesthetic or perceptual impressions of one's own body (body percept) and what he termed the *schema corporel*. Following Head (1920, 1926), he understood this latter term to signify a dynamic sensory-motor functioning of the body in its environment. As a system of dynamic motor equivalents belonging to the realm of habit rather than conscious choice, the body schema works in an integrated way with a "marginal awareness" of the body. Despite the fact that authors familiar to Merleau-Ponty often referred to the body schema as a "body image" (e.g., Schilder, Wallon), and despite the unfortunate mistranslation of his term '*schema corporel*' as 'body image' (in the English translation of his work, *The Phenomenology of Perception* [1962]), Merleau-Ponty was careful not to confuse these terms.

The translation of '*schema corporel*' as 'body image' is both problematic and a symptom of a long tradition of terminological and conceptual confusion. The concepts of *body image* and *body schema* have a long history of usage dating back to the 1890s (e.g., Munk, 1890; Bonnier 1905). I could cite numerous examples of how these terms have been, and continue to be, defined and used in a confused way throughout the literature (see Gallagher, 1986a and 1995). Instead I will refer to an exhaustive survey by Tiemersma (1989). His review reflects both the extensive employment of these concepts across numerous disciplines, (including developmental studies, medicine, psychology, psychopathology, phenomenology, and philosophy) and their ambiguous nature. In brief, in the extant literature the schema or image of the body is alternately characterized as a physiological function, a conscious model or mental representation, an unconscious image, a manner of organizing bodily experiences, an artificially induced reflection, a collection of thoughts, feelings, or memories, a set of objectively defined physical positions, a neuronal "map" or cortical representation, and an ideational/conceptual activity. Although it is a fact that the body schema (or image) is conceptually complex it is unlikely that it is at once or even at different times all of these things. The variety of usage suggests that, rather than one complex concept with a plurality of names, there is, over and above the terminological ambiguity, a basic conceptual confusion.

This confusion has motivated some authors to suggest that we ought to give these terms up, abandon them to history, and formulate alternative descriptions of embodiment. Poeck and Orgass (1971, p. 254), for example, argue that these concepts are ill defined, "difficult to reconcile with modern theories of central nervous functions," unhelpful for explaining pathologies, and are thus simply inadequate for scientific explanation (see De Renzi, 1991 and Straus, 1967 for similar criticisms). Such objections, however, are not objections "*in principle*." For the most part they express a large and justifiable complaint about conceptual misuse in the history of psychology and related disciplines. I argue here that it is important to retain the terms *body image* and *body schema*, and to formulate a clear conceptual distinction between them. It will be apparent in the following discussion that this particular distinction cuts across a number of other distinctions, such as conscious/nonconscious, personal/subpersonal, explicit/tacit and particularly willed/automatic. The body image/body schema distinction, however, is not reducible to any one of these, and no other distinction is able to carve up the intellectual space in quite the right way.

I have argued elsewhere that the objections raised by Poeck and Orgass can be answered by addressing four areas of concern: first, and most importantly, distinguish between the concepts of body image and body schema and clear away the confusions found in the literature (Gallagher, 1986a, 1995); second, provide an account of the ontogenetic development of the body schema and body image (Gallagher and Meltzoff, 1996); third, relate these concepts to recent developments in neuroscience (Gallagher, Butterworth, Lew, and Cole, 1998); and fourth, show how these concepts can be usefully employed in systematic explanations of relevant pathological conditions (Gallagher, 2000; Gallagher and Cole, 1995).

In the remainder of this essay I will address each of these points and suggest how the concepts of body image and body schema can play an essential role in understanding self-consciousness and personal identity. Finally, I will return to the context of medicine and indicate some recent uses of these concepts in experimental therapies.

II. PHENOMENOLOGICAL AND CONCEPTUAL CLARIFICATIONS

I begin with an initial characterization: A *body image* is an intentional content of consciousness that consists of a system of perceptions, attitudes, and beliefs pertaining to one's own body. A *body schema*[1] is a pre-noetic (automatic) system of processes that constantly regulates posture and movement – a system of sensory-motor capacities and actualities that function without the necessity of perceptual monitoring.

A body image involves more than occurrent perceptions. It can include mental representations, beliefs, and attitudes where the object of such intentional states (that object or matter of fact towards which they are directed, or that which they are about) is or concerns one's own body. The body schema, in contrast, involves certain motor capacities, abilities, and habits that enable movement and the maintenance of posture. It continues to operate, and in many respects operates best, when the intentional object of perception is something other than one's own body. So the difference between body image and body schema is like the difference between a *perception* (or conscious monitoring) of movement and the actual *accomplishment* of movement. Obviously, however, a perception of one's own movement (or indeed someone else's movement) can be complexly interrelated with the accomplishment of one's own movement, although not all movement requires a body percept.

The body image, consisting of a complex set of intentional states – perceptions, mental representations, beliefs, and attitudes – in which the intentional object of such states is one's own body, involves a form of reflexive or self-referential intentionality. Studies involving body image (e.g., Cash & Brown 1987; Gardner and Moncrieff 1988; Powers, et al, 1987) frequently distinguish among three of these intentional elements: (a) the subject's *perceptual* experience of his/her own body; (b) the subject's *conceptual* understanding (including folk and/or scientific knowledge) of the body in general; and (c) the subject's *emotional* attitude toward his/her own body.

Although (b) and (c) do not necessarily involve an occurrent conscious awareness, they are maintained as sets of beliefs or attitudes, and in that sense form part of an intentional system. Conceptual and emotional aspects of body image are no doubt affected by various cultural and interpersonal factors, but in many respects their content originates in perceptual experience.

In contrast to the body image, a body schema is not the result of a perception, a belief, or an attitude. Rather, it is a system of motor functions or motor programs that operate "below" the level of self-referential intentionality. It involves a set of tacit performances – subpersonal processes that play a dynamic role in governing posture and movement. In most instances, movement and the maintenance of posture are accomplished by the *close to automatic* performances of a body schema, and for this reason the normal adult subject, in order to move around the world, neither needs nor has a constant body percept. In this sense the body tends to efface itself in most normal activities that are geared into external goals. To the extent that one does become aware of one's own body in terms of monitoring or directing perceptual attention to limb position, movement, posture, pleasure, pain, kinaesthetic experience, and so on, then such awareness helps to constitute the perceptual aspect of a body image. Such awareness may then interact with a body schema in complex ways, but it is not equivalent to a body schema itself.

Although a body schema operates in a *close to automatic* way, its operations are not a matter of mere reflex. Movements controlled by a body schema can be precisely shaped by the intentional experience or goal-directed behavior of the subject. If I reach for a glass of water with the intention of drinking from it, my hand, completely outside of my awareness, shapes itself in a precise way for picking up the glass. It takes on a certain form in conformity with my intention. Thus, it is important to note that although a body schema itself is neither a form of consciousness nor a cognitive operation, it can enter into and support (or in some cases, undermine) intentional activity, including cognition. Motor action is not completely automatic; it is often part of a voluntary, intentional project. When I jump to catch a ball in the context of a game, or when I walk across the room to greet someone, my actions may be explicitly willed, and governed by my perception of objects or persons in the environment. My attention, however, and even my complete awareness in such cases, is centered on the ball or on the other person, and not on the precise accomplishment of locomotion. In such cases the body moves smoothly and in a coordinated fashion, not because I have an image (a perception) of my bodily movement, but because of the coordinated functioning of a body schema. In such movements a body schema contributes to and supports intentional action.

The body-schema system, as the result of a variety of perceptual and non-perceptual inputs, "pre-noetically" governs the postures that are taken up by the body in its environment. That a body schema operates in a pre-noetic way means that it does not depend on a consciousness that targets or monitors bodily movement. This is not to say that it does not depend on consciousness at all. For certain motor programs to work properly, I need information about the environment, and this is most easily received by means of conscious perception.

The case of eyestrain illustrates how the body deals with its environment in a pre-noetic way. In becoming conscious of eyestrain, a subject is first intentionally directed towards the environment rather than towards the body.

<small>When the eyes become tired in reading, the reader does not perceive his fatigue first, but that the light is too weak or that the book is really boring or incomprehensible.... Patients do not primarily establish *which* bodily functions are disturbed, but they complain about the fact that "nothing works right anymore," "the work does not succeed," that the environment is "irritating," "fatiguing" (Buytendijk, 1974, p. 62).</small>

As a reader in this situation, I am not at first conscious of my posture or of my eyes as they scan the pages. Rather, totally absorbed in my project of reading, I begin to experience eyestrain as a series of changes in the things and states of affairs around me. Gradually the perceived environment begins to revise itself; the text seems more difficult, the lighting seems too dim, etc. Without the agent's conscious decision or perceptual awareness, the body maneuvers to position itself closer to the text, the fingers guide the reading process over the difficult path of words, the eyes begin to squint, etc. In the end I discover the true problem – the fatigue, the headache. The eyes that have been reading have been "anonymous" eyes, doing their work without my reflective awareness of them. I was not conscious of my eyes at all. Now, however, my attention is directed to my eyes. They suddenly emerge out of pre-noetic anonymity and become explicitly owned. My pain now becomes my present concern, and my body in general gets in the way of my reading comprehension.

When I am engaged in the world I do not perceive such body-schematic performances. Such operations do not become apparent to consciousness until there is a reflection on my bodily situation brought on by pain, discomfort, fatigue, etc. When the body is "in tune" with the environment, when events are smoothly ordered, when the subject is engaged in a task that holds conscious attention, then the body pre-noetically performs in a way that is in excess of that of which I am conscious.

It is also the case, as I mentioned, that in some situations a body image or percept contributes to the controlled production of movement. The visual, tactile, and proprioceptive attentiveness that I have of my body may help me to learn a new dance step, improve my tennis game, or imitate the novel movements of others. The dancer or athlete who practices long and hard to make deliberate movements proficient so that movement is finally accomplished by the body without conscious reflection, uses a consciousness of bodily movement to train body-schematic performance. Various experiments show that visual awareness of one's own body can correct or even override body-schematic functions (Gurfinkel and Levick, 1991). For example, visual perception of wrist, elbow, or shoulder can recalibrate motor performance distorted by the effects of prismatic goggles (Paillard, 1991). Focused attention, or the lack of it, on specific parts of the body may alter postural or motor performance (e.g., Fisher, 1970; Winer, 1975). In such cases the contribution made to the control of movement by my perceptual awareness of my body finds its complement in capacities that are defined by the operations of a body schema that continues to function to maintain balance and enable movement. Such operations are always in excess of what I can be aware of. Thus, a body schema is not reducible to a perception of the body; it is never equivalent to a body image.

Importantly, it is possible to identify cases that involve a double dissociation between body image and body schema. For example, one finds evidence of a functioning body schema in the absence of a completely intact body image in some cases of unilateral neglect. Denny-Brown and his colleagues report that a stroke patient fails to notice the left side of her body; excludes it from her body percept. She fails to dress her left side or comb the hair on the left side of her head. Yet there is no motor weakness on that side. Her gait is normal, although if her slipper comes off while walking she fails to notice. Her left hand is held in a natural posture most of the time, and is used quite normally in movements that require the use of both hands, for example, buttoning a garment or tying a knot. For instance, she uses her left hand, and thus the motor ability of the neglected side, to dress the right side of her body (Denny-Brown, Meyer, and Horenstein, 1952). Thus her body schema system is intact despite her problems with body image on the neglected side.[2]

Dissociation of the opposite kind can be found in cases of deafferentation (loss of peripheral sensory input). A subject who has lost proprioceptive input from the neck down is deprived of essential aspects of the body schema. The motor system is not automatically updated in regard to limb position or posture, and he can control his movement only by cognitive intervention and visual guidance of his limbs. In effect he employs his body image (primarily a visual perception of his body) in a unique way to compensate for the impairment of his body schema (see Cole, 1995; Gallagher and Cole, 1996).

A more complex dissociation is cited by Paillard and Stelmach (1983). A patient who suffers from a deafferented forearm following a parietal lesion, when blindfolded and thereby deprived of a body percept for that arm, is unable to detect the presence or location of tactile stimulation on the forearm. She is, however, able automatically to point to the place of stimulation with her intact hand. Information available in the body-schema system allows for an automatic identification of body location without perceptual awareness, that is, without the ability to locate the position in terms of a conscious body percept. One might think of this as the motor equivalent of blindsight, as Paillard suggests (also see Rossetti et al., in press). Paillard (1997, 1999), following the distinction made in Gallagher (1986) and Gallagher and Cole (1995), suggests that this case represents one half of a double dissociation between body image and body schema. The other half is to be found in the cases of deafferentation mentioned above (see Paillard, 1991).

Such dissociations suggest that establishing a conceptual distinction between body image and body schema is only the beginning of an explication of the role played by the body in action and cognition. Nonpathological relations between body image and body schema can be worked out in detail, however, only if the conceptual distinctions between them are first made clear.

III. DEVELOPMENT

The traditional and strongly empiricist view assumes that the newborn infant has no body image or body schema, and that such mechanisms are acquired through prolonged experience in infancy and early childhood. This view has been worked out in a number of ways and in a variety of contexts in scientific and philosophical discussions. Until thirty years ago this position was the consensus among developmental theorists. At that time, however, on several fronts, new evidence surfaced in support of a more nativist position. The idea that body schemas may in fact be "innate" was put forward in studies of phantom limb in cases of congenital absence of limb (Weinstein and Sersen, 1961). In the 1970s, in studies of neonate imitation (Meltzoff and Moore, 1977), further evidence was provided to show that certain elements of what previously were understood to be learned motor actions were in fact already present in the newborn.

A. *The Traditional View*

Given the confusions in the literature about body image and body schema, we need to be careful when interpreting claims about either of these concepts. Nevertheless, on the issue of whether the body schema/body image is acquired or innate, the traditional view is the same for either of these concepts: the body schema/body image is an acquired phenomenon, constituted through experience, the product of development.

Marianne Simmel (1958, 1962) one of the few psychologists who recognizes a theoretical distinction between body schema and body image, contends that there is no question that the body image is something which comes relatively late in development, through organized tactile and visual experience. Furthermore, concerning the body schema, Simmel (1958, p. 499) claims that it is

> built up as a function of the individual's experience, i.e., it owes its existence to the individual's capacity and opportunity to learn. This means that at some early time in the development of the human organism the schema has not yet been formed, while later . . . it is present and is characterized by considerable differentiation and stability.

Simmel is a good representative of traditional developmental psychology in this regard. Merleau-Ponty (1962), who was greatly influenced by his study of developmental psychology, including the work of Piaget, Wallon, Guillaume, and Lhermitte, also took the traditional view that the body schema is a product of prolonged experience of movement in early infancy. Although he conceives of the body schema as an anterior condition of possibility, a dynamic force of integration that cannot be reduced to a sum of experienced sensations, still, in terms of development, the operations of the body schema are "'learnt' from the time of global reactions of the whole body to tactile stimuli in the baby . . ." (Merleau-Ponty, 1962, pp. 101, 122n). The body schema functions *as if* it were an "innate complex" (Merleau-Ponty, 1962, p. 84), that is, as strongly and pervasively as if it were innate, yet, it is in fact not innate but acquired through developmental history.

For Merleau-Ponty (1964), motor experience and perceptual experience are dialectically or reciprocally linked. This means, first, that the mature operation of a body schema depends on a developed perceptual knowledge of one's own body; and, second, the organized perception of one's own body and of the external world, depends on a proper functioning of the body schema.

According to Merleau-Ponty, the infant does not yet have a body schema because of the immaturity of its neurological development, especially in regard to myelination.[3] The development of the body schema can happen only after certain neurological conditions are met and thus, only in a fragmentary way at first. Motor schemas are gradually integrated into a reciprocal system with external perception and sensory inputs, so that they become "precise, restructured, and mature little by little" (Merleau-Ponty, 1964, pp. 123).

Merleau-Ponty also links the emergence of an organized body schema with the development of a body image, that is, the development of a conscious awareness in which the "child takes notice of his own body." At a later stage, the mirror or specular image, which involves a conscious objectification of the body, provides a way for further developing the body image (Merleau-Ponty, 1964, pp. 121-25).

The idea that body image and body schema were products of relatively late development implied, for many traditional psychologists, that the conscious experience of the infant is a disorganized, "blooming, buzzing confusion," as James (1890) famously put it. It also implied certain limitations about what is possible for the infant. Consider, for example, the capacity for imitation, an important capacity directly related to questions about perception, knowledge of other persons, and the origins of a sense of self. Here we can let Piaget represent the traditional view. Piaget defines invisible imitation as the child's imitation of another person's movements using parts of the child's body that are invisible to the child. For example, if a child does not see its own face, is it possible for the child to imitate the gesture that appears on another person's face? Piaget's answer is that at a certain point in development it is possible; but in early infancy it is not, because invisible imitation requires the operation of a relatively mature body schema. Thus, according to Piaget (and most other classical theorists of development), invisible imitation is not possible prior to 8 to 12 months of age (Piaget, 1962). His reasoning on this question involves two issues that are essentially related. First, for invisible imitation to be possible, the child must have developed the proper body schemas that define her specific motor capacities and to which she can relate or co-ordinate the visible gesture of the other person. Piaget's view on invisible imitation is completely consistent with the idea that body schemas are not sufficiently developed in early infancy, from which it follows that imitation dependent on such schemas would not be possible until their acquisition at 8 to 12 months of age. Thus, invisible imitation is absolutely out of the question for a neonate. Second, there must be some mechanism that allows for the translation between what the infant sees and what movements the infant is capable of. The traditional view on this issue, as far back as John Locke's original statements on sense perception (1690), is that different sense modalities (vision, touch, and proprioception, for instance) are not naturally intermodal, and that prolonged experience is required to be able to translate between vision and tactilo-kinesthetic sensation. Piaget claims that the young child's intellectual mechanisms are not sufficiently developed to relate sense experiences (vision and proprioception) that are not naturally related.

Merleau-Ponty's (1964, pp. 116-17) view is consistent with Piaget in regard to these issues. To imitate, he states,

[I]t would be necessary for me to translate my visual image of the other's [gesture] into a motor language. The child would have to set his facial muscles in motion in such a way as to reproduce [the visible gesture of the other] .... If my body is to appropriate the conducts given to me visually and make them its own, it must itself be given to me not as a mass of utterly private sensations but instead by what has been called a "postural" or "corporeal schema."

Even beyond three months, when the infant could distinguish facial features of the other person, it has little or no familiarity with its own face, and has little or no control over moving that face, since motor schemas are not yet in place. As a result, the young infant could not even begin to imitate the facial gesture of the other person, although, as Merleau-Ponty (1962, p. 352) observes, by 15 months, this kind of imitation is possible.

B. Neonate Imitation And The "Innate" Body Schema

In complete contrast to this traditional view, studies of imitation in infants conducted by Meltzoff and Moore (1977, 1983) show that invisible imitation does occur in newborns. Their experiments demonstrated that normal and alert newborn infants systematically imitate adult gestures of mouth opening and tongue protrusion. Others have replicated and extended their results to include smile, frown, angular tongue protrusion, and other gestures. (Meltzoff and Moore, 1983; for a summary see Meltzoff and Moore, 1994; Gallagher and Meltzoff, 1996). Infants between the ages of 16-21 days also imitate facial gestures after a delay. In one experiment a pacifier is placed in an infant's mouth and then a facial gesture is presented. After the presentation is complete, the pacifier is removed and the infant proceeds to imitate the gesture. Thus, the presented gesture is remembered and imitative responses are delayed. Even in circumstances of longer delays (of 24 hours) 6-week-old infants clearly remember a gesture and imitate it (Meltzoff and Moore, 1994).

Imitation of this sort is not a matter of "reflexes" (or release mechanisms). Reflexes are highly specific – that is, narrowly circumscribed to limited stimuli. One cannot have a reflex for imitation in general. As a result, the range of behaviors displayed by infants would require the unlikely postulate of distinct reflexes for each kind of imitative behavior: tongue protrusion, tongue protrusion to one side, mouth openings, smile, frown, etc. Furthermore, neonate imitative behavior involves memory and representation, since imitation can happen even after a delay. It is also the case that infants can improve or correct their imitative response over time. Neither delayed response nor improvement in response is compatible with a simple reflex or release mechanism.

If we follow the logic expressed by the proponents of the traditional view, namely, that imitation requires the existence of a body schema, then the studies on newborn imitation suggest that the body schema is innate, in the sense that it is already developed and operational at the time of birth. It may be a primitive schema, but one sufficiently developed at birth to account for the ability to move one's body in appropriate ways in response to environmental stimuli.

Is it also possible to speak of an innate body image? Gallagher and Meltzoff (1996) suggested that some primitive perceptual element of the body image seems to be required for the infant to have awareness of its own face. For example, without awareness of its face it would be difficult to explain how the infant could improve its imitative performance. Newborns, however, do not have a visual perception of their own face, nor are they in a position to have conceptual or emotional aspects of a developed body image. The capacity for proprioception, however, develops prenatally. It seems likely that the neonate has a proprioceptive awareness of its own face. This pro- prioceptive awareness is a tacit, non-perceptual awareness (Gallagher, in press), and constitutes the very beginning of a primitive body image. It is in the intermodal interaction between proprioception and the vision of the other's face that one's body image originates. To be proprioceptively aware of one's face (or more generally, one's body) does not involve making one's face an object of perception. The infant does not have to "consider" which of its body parts to move in order to accomplish the imitation. It does not have to identify its mouth or tongue *as* a mouth or tongue, or even as the correct body part to move. Proprioception is what Merleau-Ponty would call a pre-reflective awareness that allows the body to remain *experientially transparent* to the agent who is acting.

The newborn infant's ability to imitate others, and its ability to correct its movement, which implies a recognition of the difference between its own gesture and the other's gesture, indicates a rudimentary differentiation between self and non-self. This fact is not unimportant for issues related to having a sense of self and the role that plays in the constitution of personal identity. When, in both philosophical and psychological traditions, the sense of self is conceived as developing in a relatively later time-frame, it is frequently discussed in terms that are explicitly related to cognitive development. Such cognitive models of the self clearly imply that personal identity or a sense of self may be primarily and for the most part a psychological phenomenon. In contrast, if a sense of self is operative earlier, and specifically, if a sense of self is already involved in neonatal behavior, the concept of self starts out closer to an embodied sense than to a cognitive or psychological understanding.

This issue is further clarified in considerations of the notion of intermodal communication between vision and proprioception. Vision seems to direct the child outward toward the world and other people; proprioception, as the word itself signifies, involves information about the self. Traditionally, intermodal connections between these two sense modalities had been thought to develop late in infancy, on the basis of experience. In contrast, the studies on newborn imitation indicate that these intermodal connections are operative from the very beginning. Infants already apprehend, with quickly improving precision, the equivalencies between the visible body transformations of others and their own invisible body transformations which they experience proprioceptively.

From early infancy, one's visual experience of the other person communicates in a code that is related to the embodied self. What one sees of the other's motor behavior is reflected and played out in terms of one's own motor possibilities. There exists in the neonate a natural, unconfused, intermodal coupling between self and other. Rather than confusion, from the very start there is a self-organizing collaboration between visual perception and proprioception, between sensory and motor systems, and between the self and the other. Body schemas, working systematically with proprioceptive awareness, constitute a proprioceptive sense of self from the very beginning.

IV. NEUROPHYSIOLOGY

What neurological mechanisms account for these possibilities in the neonate? By asking this question we are also asking about the mechanisms that underlie body-schema and body-image systems in adults, as well as certain aspects of embodied self-consciousness. In part, the answer involves the notion of an intermodal sensory system. This idea is supported by experiments showing an early relation between different sense modalities, including vision, touch, and proprioception, as well as between the auditory perception of speech and the visual perception of particular lip movements that cause the speech (see Meltzoff, 1993).

In an intermodal system, proprioception and vision are already in communication with each other. In certain cases, what I see automatically gets translated into a proprioceptive sense of how to move. On this basis, visual perception can help to inform and coordinate movement and can correct proprioceptive distortions or illusions.[4] Proprioception itself has a dual nature. It consists of both non-conscious, physiological information (PI) that informs the body schema, and a non-perceptual awareness (PA) that contributes to the body percept. This dual nature helps to explain how intermodal communication between vision and proprioception is at the same time a communication between sensory and motor aspects of behavior. On the physiological level PI and PA depend on the same proprioceptors, and in some cases the same central neural structures. These mechanisms generate the information necessary for both the automatic governing of movement and the sensation of one's own movements (Phillips, 1985). Since PI and PA depend on and are constituted by the same physiological mechanisms it would not be unreasonable to suggest an immediate connection or close interactive coordination between proprioceptive information which updates motor action at the level of the body schema, and proprioceptive awareness, as a pre-reflective, performative accompaniment to that action.

Proprioception and vision are also integrated with vestibular information about head motion and orientation. The vestibular nucleus, a relatively large midbrain structure, serves as a complicated integrative site where first-order information about head position is integrated with wholebody proprioceptive information from joint receptors and oculo-motor information about eye movement. This integrated, multimodal information projects to the thalamus, informing connections that project to cortical areas responsible for control of head movement. Vestibular neurons in the parietal lobe respond to vestibular stimulation, but also to somatosensory and optokinetic stimuli. More generally, in structures that are mature at birth, there is cortical integration of information concerning self-motion, spatial orientation, and visuo-motor functions (Guldin, Akbarian, and Grüsser, 1992; Jouen and Gapenne, 1995).

In the case of neonate imitation, the imitating subject depends on a complex background of embodied processes, and in particular a bodyschema system specified by visual, proprioceptive and vestibular information. In the foreground, what the infant sees gets translated into a proprioceptive awareness of her own relevant body parts; and proprioceptive information allows her to move just those parts so that her proprioceptive awareness matches up to what she sees.

Recent studies in neuroscience suggest that there are specific neurophysiological mechanisms that can account for the intermodal connections between visual perception and motor behavior. These are mechanisms that operate pre-noetically, as general conditions of possibility for motor stability and control. They are also directly related to the possibility of imitation. Indeed, these studies suggest that the "mirror stage" in later childhood, considered to be important for the development of a mature body image and the onset of self-recognition, may be prefigured in the earlier onset of an interior mirror process. I refer here to what neuroscientists now describe as processes that involve mirror neurons (Gallese et al., 1996; Rizzolatti et al., 1996). Mirror neurons link up motor processes with visual ones in ways that are directly relevant to the possibility of imitation.

Mirror neurons were discovered in the premotor cortex of the macaque monkey. There is now good evidence that they also exist in area F5 of the ventral premotor cortex and Broca's area in the human (Fadiga et al., 1995; Rizzolatti et al., 1996; Grafton et al., 1996). Mirror neurons respond *both* when a subject *performs* a particular (goal-directed) action with arm, hand, or mouth *and* when the subject *observes* such actions being done by another subject. They constitute an intermodal link between the visual perception of action or dynamic expression, and the *intra*subjective, proprioceptive sense of one's own capabilities. Their functioning clearly helps to account for the communication between proprioception and vision, and between specific movements and the visual perception of the same movements made by others.[5] In this regard, the following hypothesis seems reasonable: when the neonate sees another person perform a specific motor act, for instance a tongue protrusion, the visual stimulus initiates the firing of the same neurons that are involved in the infant's own performance of that motor act. Although precise experiments that would verify this hypothesis have not been carried out, this kind of mechanism would go a long way toward accounting for neonatal imitation capability.

The neurological picture involving intermodal and mirror neurons correlates well with phenomenological observations made by Husserl (1973; "D" manuscripts of the early 1930s). He described the complex integration of alterations in kinesthetic and proprioceptive senses with alterations in the perceptual field. The movement and posture of the body is fully integrated with visual perception, for example, and motivates the organization of tactile exploration. One's own movement conditions perceptual experience and there is a tight correspondence between the stream of kinaesthetic sensation and the movement of the perceptual world (Gallagher, 1986b). Husserl extends this idea to the empathic perception of others. Empathy (*Einfühlung*) allows one to experience, as in the other, what one is able to experience kinaesthetically in oneself. Jean-Luc Petit (1999) has suggested that these Husserlian insights on empathy and the internal imitation of the other's movement are consistent with the recent discovery of mirror neurons.

I have focused on neurophysiological structures that may account for the possibility of the intermodal communication required for neonate imitation. The neurological picture can be expanded in two directions. First, one can explore in more precise ontogenetic terms how something like the basis for an innate body schema is hard-wired into the nervous system. Second, one can ask about the neurophysiology of normal adult movement. In both cases we can get a better understanding of how these systems work by examining pathological cases.

V. PATHOLOGIES

A. Phantoms and the Innate Body Schema

We can understand how the neurological basis for a body schema develops through prenatal life, by considering instances of phantom limb in cases of congenital absence of limb (*aplasia*). The traditional view concerning this issue followed the same logic that shaped the traditional answer about the possibility of neonate imitation. If a body schema is something that is acquired only over the course of experience (in the first 8-12 months of life), then an aplasic phantom is just as impossible as neonate imitation. In contrast to this view, if there is an innate body schema, then it should be quite possible to find cases of aplasic phantoms – and in fact such cases have been found (Melzack 1989; Poeck 1963, 1964; Scatena 1990; Vetter & Weinstein 1967; Weinstein & Sersen, 1961; Weinstein, Sersen & Vetter 1964).[6]

Although much brain development involves a genetic blueprint that predetermines the pattern of neuronal growth, self-organizing movement plays an important role in stimulating and promoting such growth. Reflex movement in the embryo begins around week seven gestational age and grows in complexity in the eighth week (Flower, 1985). The actual development of embryonic neural tissue depends, in part, on fetal movement, and on components that are important for the attainment of postural balance. Proprioceptors in the muscles (muscle spindles) which will ultimately be responsible for a sense of position and movement first appear at 9 weeks gestational age (Humphrey, 1964); spontaneous and repetitious movements follow shortly (de Vries *et al.*, 1982). The development of semicircular ear canals that, as part of the vestibular system, later provide a sense of balance begins as early as the fourth month of gestation (Jouen and Gapenne, 1995). Even as the cortical plate begins to form early connections are made with the brainstem (Flower, 1985), and later neural development of the motor cortex and other cortical and subcortical areas continue to be stimulated by fetal movement (Sheets-Johnstone, 1998).

In early fetal development a specific kind of movement emerges. Ultrasonic scanning of fetuses shows that movement of the hand to the mouth occurs between 50 to 100 times an hour from 12 to 15 weeks gestational age (DeVries, Visser & Prechtl, 1984). This suggests that hand-to-mouth movement may be an aspect of an early, centrally organized coordination that eventually comes to be controlled proprioceptively. This kind of prenatal movement may be precisely the movement that helps to generate or facilitate the development of specific body schemas. That is, quite consistent with the traditional hypothesis, the formation of body schemas may depend on motor experience. The only difference is that this movement occurs much earlier, and by implication, body schemas develop much earlier than the traditional account permits, that is, in the fetus rather than in the 8-12 month old infant.

There is an important continuity between this very early movement and postnatal movement. Spontaneous movements such as whole-body flexions or more localized limb movements occur in human neonates until approximately the third month of life. These movements are very similar to fetal spontaneous movements (de Vries et al., 1982) and are generally thought to reflect the relative motor immaturity of humans at birth compared to other mammals (Hopkins and Prechtl, 1984; Prechtl and Hopkins, 1986). Butterworth and his colleagues, however, have discovered relatively organized movements of hand to mouth (similar to the fetal movement mentioned above) embedded within these spontaneous movements. They suggest this as evidence for an innate coordination between the hand and the perioral region (Butterworth and Hopkins, 1988; Lew and Butterworth, 1995). Important here is the fact that in such movements the mouth "anticipates" arrival of the hand. When the infant is hungry, the mouth more frequently opens in anticipation of arrival of the hand than when the infant has been fed. This suggests that hand-mouth coordination in the fetus and the neonate may be an early form of orally-targeted reaching linked to the appetitive system.

This behavioral evidence of an early link between the hand, mouth and appetitive system is supported by neuroanatomical and single-cell recording evidence in primates. There is a network of neurological interconnections in the primate between regions in the orbital and medial prefrontal cortex which receive gustatory inputs from cortical, and subcortical areas, as well as inputs from regions of the somatosensory cortex that represent the hand, arm and face. There are also projections from the ventral premotor cortex to these same prefrontal areas (Carmichael and Price, 1995).[7] The complexity of this network makes it difficult to pin down all of the neuronal details. It seems clear however, that such a network can constitute the innate neural basis for a specific body schema, the behavioral evidence for which we see in fetal and neonate hand-mouth coordination.

The notion of an innate motor schema in connection with hand-mouth coordination can help to explain the aplasic phantom as a product of innate mechanisms. Where a functional system is disrupted by failure of objective limb formation, the missing limb may nevertheless manifest itself as a phantom part of the lived body, because a specific *movement coordination* is represented within a neural matrix that remains sufficiently intact (Gallagher, Butterworth, Lew, and Cole, 1998). The hand-mouth coordination has two related elements: the hand movement and the mouth movement. The mouth opens to anticipate the arrival of the hand. In the case of a missing limb, it is not just that the intact body part involved in the coordination (the mouth and perioral region in the hand-mouth coordination) is neurally represented. Insofar as the motor coordination or schema itself is represented, there must be some implicit representation of the "other end" of the coordination. Even if the arm and hand are not there, a circuit or a defined schema (involving some definite tendency of arm movement) would require that both sides of the circuit be neurally defined.

If this is the case, then, under certain conditions, a stimulation of the mouth would be sufficient to activate the joint mouth-limb neural system. The virtual limb (the phantom) comes into existence when the coordination is activated. A singular representation of the coordinated schema or action pattern incorporates both ends of the movement, the mouth and the hand, even when, objectively, there is no hand. That this is a part of an action pattern rather than a conscious image is important. It corresponds to what Merleau-Ponty and others have termed the phenomenon of "forgetting" found in cases of post-amputation phantoms (Simmel 1966; Poeck 1964; Melzack 1990; Merleau-Ponty 1962). In such cases, although the subject clearly knows about and acknowledges the loss of limb, in certain instances of motor behavior he seems to be unaware of the loss and relies on the phantom as one would on the real limb. For example, an amputee who attempts to walk with his phantom leg is surprised when he falls. The close-to-automatic functioning of his body schema leads him to "forget" that he does not have the leg to walk on. In the case of the aplasic hand, the mouth, opening in anticipation of the hand, "forgets" that there is no hand.[8]

This account depends on an innate body schema, involving in this case a specific coordination or motor capacity. Obviously this does not require that the 12-15 week-old fetus be conscious of the movement. Still, it is likely that at some point in development a proprioceptive accompaniment develops along with this movement. The enactment of the motor schema would, at some later ontogenetic point, produce a proprioceptive sense of movement. In this case, one might say, the movement precedes the awareness of movement but motivates that awareness (in the form of proprioception) when the system is sufficiently developed to allow for consciousness. This specific proprioceptive sense of hand-mouth coordinated movement would then form the initial aspect of an experienced phantom.[9]

B. Schizophrenia and the Sense of Agency

I have suggested that body schemas play a role in the earliest sense of a distinction between self and non-self. Processes that involve body-schemas and body images are also implicated in experience that contributes to more developed aspects of self-consciousness. This can be seen in pathological cases where those aspects come into question. Specifically, in certain positive symptoms of schizophrenia – for example, delusions of control in regard to bodily movements, thought insertion, and auditory hallucinations – the sense of agency, one important aspect of self-awareness, is disrupted.

In the normal phenomenology of intentional action, the sense of agency involves an awareness of generating or being the willful initiator of action, where action can mean either a motor or a cognitive activity. This sense seems to be indistinguishable from a sense that the movement or thought is the movement of my own body or a thought in my own stream of consciousness. In the case of *involuntary* action, however, it is quite possible to distinguish between the sense of agency and the sense of ownership. I may acknowledge ownership of a movement – for example, I have a sense that I am the one who is moving or is being moved. I can thus self-ascribe it as my movement. I may rightly claim, however, that I am not the author of the movement, because I do not have a sense of causing or controlling the movement – I have no sense of agency for it. The agent of the movement is the person who pushed me from behind, or, for example, the physician who is manipulating my limb in a medical examination. Likewise, in the case of involuntary *cognitive* processes, I may acknowledge that I am the one who is thinking, but claim that the thoughts are not willfully generated by me. For example, certain unbidden thoughts or memories (or frequently melodies) may impinge on my consciousness, even if I do not intend for them to do so, or even if I resist them (see Frankfurt, 1976). In such cases my claim of ownership (my self-ascription that I am the one who is undergoing such experiences) may be consistent with my lack of a sense of agency.

For schizophrenics who suffer delusions of control in regard to bodily movements (as well as thought insertion and auditory hallucinations) the sense of ownership remains intact, but the sense of agency is disrupted. Such patients make mistakes about the agency of various bodily movements, and may report that their movements are made or caused by someone or something else. The motor action responsible for the speech is in fact the patient's own motor action, and the patient acknowledges that his own lips are moving, but he makes an error of identification concerning who produced this motion. An example provided by Mellor (1970, p. 17; cited by Spence, 1996, p. 82) makes this clear: "A 29 year old shorthand typist described her actions as follows: 'When I reach my hand for the comb it is my hand and arm which move, and my fingers pick up the pen, but I don't control them'."

How can one explain this disruption in the sense of agency? The classic theory in this regard is based on the notion of efference copy. Normally, when a motor instruction to move is sent to a set of muscles, a copy of that instruction, the efference copy, is also sent to a comparator or self-monitoring system. Held (1961) suggested that efference copy sent to a comparator is stored there, and then compared to the reafferent (proprioceptive or visual) information about what movement is actually made. In the case of the schizophrenic's motor experience, the absence of a match between movement and efference copy at the comparator explains his lack of a sense of self-agency. On this explanation, the sense of agency would seemingly have to wait for perceptual feedback and would come after the fact as a kind of verification that it was I who did the moving (see, e.g., Campbell, 1999).

This account in terms of *sensory-feedback* corresponds in part to ecological explanations of motor action and proprioceptive awareness. The control of motor action depends in part on feedback from proprioceptive and visual proprioceptive processes, and more generally on an ecological sense of one's own self-movement (Gibson, 1987). If something seems to be going wrong with the action, it is quite possible to correct for it on the basis of this ecological proprioceptive awareness.

Another important component to the control of movement can also be interpreted in terms of a comparator model. In this respect, the comparator is understood to be part of a pre-motor system responsible for the control of movement, operating *prior to* the actual execution of movement and *prior to* sensory feedback. On this *"forward* model," motor control does not depend on sensory feedback. Rather, the comparator registers the efference copy of motor commands as correctly or incorrectly matching motor *intentions* and makes automatic corrections to movement prior to any sensory feedback. Consistent with this idea, experimental research on normal subjects suggests that an agentive awareness of action is based on central motor processes that precede action and that translate intention into movement, rather than on actual feedback from movement or from peripheral effort associated with such movement.[10] That is, although the content of experience is a sense of agency for the action, its source is in fact what lies between intention and performance.

Both the ecological, sensory-feedback model and the forward model are descriptions of body-schematic processes involved in motor control, and in generating the senses of ownership and agency for action. To be more precise, the distinction between ecological (sensory-feedback) control and the forward, pre-action control corresponds to the phenomenological distinction between sense of ownership and sense of agency, respectively (Gallagher, 2000). The evidence for this correspondence is to be found in involuntary movement, and in the case of schizophrenic delusions of control. In the normal experience of involuntary action, the sensory feedback system tells the subject that it is he who is moving or being moved (providing a sense of ownership for the movement). Without efference copy at the forward comparator, that is, absent any pre-action preparatory processes at the neurological level, the body-schematic system fails to register a sense of agency, as it should be in the case of involuntary movement.

Schizophrenics who suffer from delusions of control have problems with precisely the forward control of movement, but not with motor control based on sensory feedback (Frith, 1992). Control subjects in experimental situations correct their movement either relatively slowly using sensory feedback or quickly by means of the forward (pre-action) mechanism. In contrast, schizophrenics are unable to control their movement using the forward mechanism (Malenka et al., 1982; Frith and Done, 1988). These findings are consistent with controlled studies that show abnormal pre-movement brain potentials in schizophrenia, which Singh associates with elements of a neural network involving supplementary motor, premotor, and prefrontal cortexes (Singh, et al., 1992).

In delusional experience of control the schizophrenic feels under the influence of others. How close is this to normal experience of involuntary movement? In both cases there is an absence of a sense of agency, reflecting an absence (in involuntary movement) or problem (in schizophrenia) in regard to the forward, pre-action aspect of the body-schema system, but not an absence of a sense of ownership. In the case of the schizophrenic's delusion of control, however, since he is in fact the agent, the disruption of the sense of agency is caused by something going wrong with efference copy or the comparator. The experimental data indicate that the schizophrenic is still able to correct movement errors if sensory feedback is available. Likewise, in the case of involuntary movement, the ecological, sensory feedback system continues to operate, informing the subject that he is moving (Gibson, 1986). Thus, both in the schizophrenic's delusional experience of control and in the normal experience of involuntary action, the ecological, sensory feedback system tells the subject that it is he who is moving or being moved (providing a sense of ownership for the movement). The absence or disruption of efference copy at the forward comparator (the absence of any pre-action preparatory processes at the neurological level) deprives the subject of a sense of agency, a sense that it is the subject himself who is the willful generator of the movement.

VI. RECENT CLINICAL APPLICATIONS

I have argued that: (1) properly distinguished concepts of body schema and body image can be usefully employed in systematic explanations of certain pathological conditions (including aplasic phantoms and schizophrenia, as well as unilateral neglect and deafferentation); (2) that explanations of movement in terms of body schema and body image are supported by and can be integrated with current ontogenetic and (3) neurological explanations; and (4) that various elements of the body-schema system are responsible for important aspects of self-reference and a sense of personal identity, specifically involving a basic phenomenological differentiation between self and non-self, and the senses of agency and ownership.

Much more could be said concerning self-reference and personal identity in relation to a variety of other pathologies. The complexity of cases that involve disordered relations between body image and body schema often extends to associated cognitive problems, including problems of self-reference. Rather than pursue philosophical implications concerning personal identity, however, I want to conclude by reviewing some recently developed experimental medical treatments that depend on a clear distinction between body image and body schema. In particular I focus on the innovative use of Virtual Reality (VR) technology in clinical settings to address a variety of pathologies that can be classified in relation to problems involving either body image or body schema.

A number of medical scientists have designed therapies that employ VR as a way to address pathological distortions involving body image. Specifically, subjects who suffer from anorexia nervosa, bulimia nervosa, obesity, or some form of somatoform disorder (body dysmorphic disorder) have problems that involve self-perception and body awareness. The use of virtual environments effects two changes in these patients that may help to address these problems. First, self-perception of one's own body (the body image) undergoes profound changes (Cioffi, 1993).[11] Second, movement is distorted, and this affects body schema operations. Because changes in the body schema can affect the body percept (Gallagher, 1995), the manipulation of motor performance can have an effect on body perception (for review, see Riva, 1998; Riva and Melis, 1997; Riva *et al.*, 2000). In VR environments, the subject's normal sensorimotor loops, including proprioception, are altered. Mismatches between the proprioceptive signals and the visual signals of a virtual environment can alter body percepts, sometimes causing discomfort or simulator sickness (Sadowsky and Massof, 1994). Insofar as the VR experience violates the information that currently informs the body image and body schema, the subject is forced to become conscious of that information, to recalibrate her movement, and to adapt to a reorganized bodily experience. This process of adaptation to the new circumstances motivates a greater awareness of perceptual and sensory-motor processes and, together with other therapeutic procedures, facilitates a modification of how the patient represents her body.

Unilateral neglect and hemispatial neglect also involve problems with body image that result from neurological damage following stroke. Robertson, et al. (1993) reviewed strategies for rehabilitation of neglect that involve manipulation of visual, vestibular, and proprioceptive variables. These strategies also involve attempts to manipulate the body-schema system to correct for the problematic body percept. For example, optokinetic stimulation increases awareness of body movements in the neglected field (Vallar et al., 1993). A dramatic example of this approach can be found in Rossetti's use of wedge-prism glasses to shift the visual field to the left. The effect of this is to introduce adjustments to the egocentric spatial framework of perception by recalibrating proprioception and the body-schematic system. In patients who manifest hemispatial neglect this recalibration happens quickly and has relatively long-lasting effects that reduce the degree of neglect (Rossetti, et al., 1998). It has also been suggested that interventions of this sort might be made through the use of VR technology (Wann et al., 1998).

Virtual reality environments may also have application to pathologies that involve the body schema and motor control. Because vision and visual proprioception contribute to the control of movement, virtual environments have also been used to experiment with motor disorders, e.g., in some diseases of the basal ganglia – hyperkinesia, akinesia, particularly Parkinson's disease (Riess, 1998). The characteristics of movement in the virtual environment are manipulated to create a conflict between the perceived visual space and proprioceptive space. Visual cues are added to help the subject correct movement or modify their representation of movement.

The theoretical justifications for such experimental treatments depend on a clear distinction between body image and body schema (see, e.g., Riva 1994, 1995). The actual therapies, however, also depend on an understanding of how these systems interact. In both of these respects the phenomenology and the neuroscience that help to make the distinctions and the interconnections more precise are directly relevant to medical practice.

Canisius College
Buffalo, New York
U.S.A.

NOTES

[1] One can refer to *body schemas* in the plural. Head, for example, used the plural as well as the singular form of the noun. In the plural, body *schemas* (or schemata) refer to a collection of motor programs or motor habits that individually may be defined by a specific movement or posture, for example, the movement of hand to mouth. Such specific motor schemas are more complex than they first might seem. In the case of the hand-to-mouth movement, not only are the anatomical parts of hand and face involved, but a large number of muscle systems throughout the body are also activated for purposes of maintaining balance. One might refer to this entire complex organization of movement as 'a body schema,' and consider that a subject is capable of many such complex patterns. In this essay, the phrase 'body schema' may refer to a particular schema, or, more generally, to the larger system or collection of schemas.

[2] Similar cases are reported by Ogden (1996) and Pribram (1999). Further observations about intact body schemas have been made in tests of hand grip, conducted with unilateral neglect patients. Despite neglect of the left hand, for instance, a subject made to pick up an object with the left hand shows completely normal hand grip for the task (Milner, 1998). That is, the hand reaches and shapes itself in the appropriate fashion for picking up the particular object – the details of such motor action being quite different for different objects, a puck rather than a cup, for instance. Normal hand grip relative to different objects is not the result of conscious decision nor is it controlled by mechanisms that involve body awareness. Thus, even in cases where the hand is not integrated into the subject's body image, motor programs are sufficiently intact to accomplish the action.

[3] The myelination of nerve fibers responsible for proprioception, which occurs 3-6 months (postnatally) and is later in some limbs than in others (feet vs. hand; left hand vs. right hand), was thought to be required for the full and proper functioning of relevant parts of the nervous system. Windle (1971) notes: "The theory that initiation and maturation of function of the nervous system depends upon formation of myelin sheaths has had proponents for many years .... Nevertheless, much well-organized activity of animal fetuses is present before there is any myelin" (p. 71).

[4] Adjustments made to the visual system (for example, by wearing wedge-prism glasses), can recalibrate proprioception and the body schematic system, as well as the egocentric spatial framework of perception. This recalibration in hemispatial neglect patients happens quickly and has relatively lasting effects (Rossetti, et al., 1998). One can also correct proprioceptive and nocioceptive sensations associated with a phantom by creating a visual illusion of the phantom using mirrors (Ramachandran and Rogers-Ramachandran, 1996).

[5] See Iacoboni et al. (1999) for a recent attempt to identify the neurological substrate for imitation.

[6] Unfortunately, much of the evidence cited for an innate body schema in discussions of aplasic phantoms is complicated by the traditional confusion concerning the distinction between body image and body schema (see Gallagher and Meltzoff, 1996). Indeed, the evidence that is usually cited to support the idea of an innate body *schema* in this context misses the mark because it actually pertains to the notion of body *image*. For purposes of this essay, however, I set this issue aside and focus on what we know about the innate body schema.

[7] A good number of such neural convergences motivate Carmichael and Price (1995) to postulate a network in the prefrontal cortex that involves hand-mouth coordination and is dedicated to feeding behavior. Rizzolatti, et al. (1988) identified neurons in the prefrontal cortex (ventral area 6) that fire in relation to movements that can be described as "grasping with the hand and/or mouth," and specifically when these kinds of movements are directed at food items. Stimulation of ventral premotor cortex elicits both oral and hand movements in owl monkeys (Preuss, Stepniewska and Kaas, 1996). This is consistent with the observation that the area of ventral premotor cortex plays a role in hand-to-mouth movements (for review see Jackson and Husain, 1996).

[8] The point is that in neither case is there a literal "forgetting." Rather, there is simply the normal functioning of the body schema which tends to carry on without perceptual monitoring on our part.

[9] Of course, things are much more complicated. To account for the fact that only 17% of subjects born without limbs experience a phantom (in contrast to close to 85% in amputee patients) one needs a complex account of neural platicity (see Gallagher, Butterworth, Lew, and Cole, 1998). It

has also been suggested that mirror neurons may play a role in the generation of aplasic phantoms (Peter Brugger, private correspondence; also see Brugger *et al.*, 2000). Since the same mirror neurons fire either when the subject *sees* a specific action such as grasping or movement of hand to mouth performed by another person or when the subject *performs* the action herself, observation of an other's actions may activate the body-schematic representation of one's own limb-action. If, in such circumstances, the limb involved in that representation is congenitally missing, neural circuits may be activated nonetheless, thereby causing the phantom.

[10] See Marcel (in press), and Fourneret and Jeannerod, (1998). In addition, research which correlates initial awareness of action to recordings of the lateralised readiness potential and with transcranial magnetic stimulation of the supplementary motor area, strongly indicates that one's initial awareness of a spontaneous voluntary action is based on the anticipatory or pre-movement motor commands relating to relevant effectors (Haggard and Eimer, 1999; Haggard and Magno, 1999).

[11] Approximately 40% of subjects "feel as if they had 'dematerialised' or as if they were in the absence of gravity; 44% of the men and 60% of the women claimed not to feel their bodies. Perceptual distortions, leading to a few seconds of instability and a mild sense of confusion, were also observed in the period immediately following the virtual experience" (Riva, 1995).

BIBLIOGRAPHY

Berlucchi G. and Aglioti S.: 1997, 'The body in the brain: neural bases of corporeal awareness', *Trends in Neuroscience* **20**, 560-64.
Bonnier, P.: 1905, 'L'aschematie', *Rev. Neurologie* **13**, 604-609.
Brugger, P. et al.: 2000 'Beyond remembering: Phantom sensations of congenitally absent limbs', *Proceedings of the National Academy of Science*, **97**, 6167-72.
Butterworth, G. and Hopkins, B.:1988, 'Hand-mouth coordination in the new-born baby', *British Journal of Developmental Psychology* **6**, 303-14.
Buytendijk, F.J.J.: 1974, *Prolegomena to an Anthropological Physiology*, A. I. Orr (trans.), Duquesne University Press, Pittsburgh, PA.
Campbell, J.: 1999, 'Schizophrenia, the space of reasons and thinking as a motor process', *The Monist* **82**, 609-25.
Campbell, J.: 1995, 'The body image and self-consciousness', in J. Bermúdez, A. Marcel, & N. Eilan (eds.), *The Body and the Self*, MIT/Bradford Press, Cambridge, MA, pp. 29-42.
Carmichael, S.T. and Price, J. L.: 1995, 'Sensory and premotor connections of the orbital and medial prefrontal cortex of macaque monkeys', *The Journal of Comparative Neurology* **363**, 642-64.
Cash T.F. and Brown, T.A.: 1987, 'Body image in anorexia nervosa and bulimia nervosa: A review of the literature', *Behavior Modification* **11**, 487-521.
Cioffi, G.: 1993, 'Le variabili psicologiche implicate in un'esperienza virtuale [Psychological variables that influence a virtual experience]', in G. Belotti (ed.), *Del Virtuale*, Il Rostro, Milano, Italy.
Cole, J. : 1995, *Pride and a Daily Marathon*. MIT Press, Cambridge, MA.
Denny-Brown, D., Meyer, J.S. and Horenstein, S.: 1952, 'The significance of perceptual rivalry resulting from parietal lesion', *Brain* **75**, 433-471.
De Renzi, E.: 1991, 'Spatial disorders', in M. Swash and J. Oxbury (eds.), *Clinical Neurology, vol. I*, Churchill Livingstone, Edinburgh, pp. 44-53.
De Vries, J.I.P., Visser, G.H.A., and Prechtl, H.F.R.: 1984, 'Fetal motility in the first half of pregnancy', in H.F.R. Prechtl (ed.), *Continuity of neural functions from prenatal to postnatal life*, Spastics International Medical Publications, pp. 46- 64.
De Vries, J.I.P., Visser, G.H.A., and Prechtl, H.F.R.: 1982. 'The emergence of fetal behaviour: I. Qualitative aspects', *Early Human Development* **7**, 301-22.

Fadiga, L. et al.: 1995, 'Motor facilitation during action observation: a magnetic stimulation study', *Journal of Neurophysiology* **73**, 2608-11.
Fisher, S.: 1970, *Body Experience in Fantasy and Behavior*, Appleton-Century-Crofts, New York.
Flower, M.J.: 1985, 'Neuromaturation of the human fetus', *Journal of Medicine and Philosophy* **10**, 237-51.
Fourneret, P. and Jeannerod, M.: 1998, 'Limited conscious monitoring of motor performance in normal subjects', *Neuropsychologia* **36**, 1133-40.
Frankfurt, H.: 1976. 'Identification and externality', in A.O. Rorty (ed.), *The Identities of Persons*, University of California Press, Berkeley, pp. 239-51.
Frith, C.D.: 1992, *The Cognitive Neuropsychology of Schizophrenia*, Lawrence Erlbaum Associates, Hillsdale, NJ.
Frith, C.D. and Done, D.J.: 1988, 'Towards a neuropsychology of schizophrenia', *British Journal of Psychiatry* **153**, 437- 43.
Gallagher, S.: (in press), 'Non-perceptual self-awareness ', *Theoria et Historia Scientiarum: International Journal for Interdisciplinary Studies* (Poland).
Gallagher, S.: 2000, 'Self-Reference and schizophrenia: A cognitive model of immunity to error through misidentification', in D. Zahavi (ed.), *Exploring the Self: Philosophical and Psychopathological Perspectives on Self-experience*, John Benjamins, Amsterdam & Philadelphia, pp. 203-239.
Gallagher, S.: 1995, 'Body schema and intentionality', in J. Bermúdez, A. Marcel, and N. Eilan (eds.), *The Body and the Self*, MIT/Bradford Press, Cambridge, MA, pp. 225-244.
Gallagher, S.: 1986a, 'Body image and body schema: A conceptual clarification', *Journal of Mind and Behavior* **7**, 541-54.
Gallagher, S.: 1986b, 'Hyletic experience and the lived body', *Husserl Studies* **3**, 131-66.
Gallagher, S., Butterworth, G., Lew, A. and Cole, J.: 1998, 'Hand-mouth coordination, congenital absence of limb, and evidence for innate body schemas', *Brain and Cognition* **38**, 53-65.
Gallagher, S. and Meltzoff, A.: 1996, 'The earliest sense of self and others: Merleau-Ponty and recent developmental studies', *Philosophical Psychology* **9**, 213-36.
Gallagher, S. and Cole, J.: 1995, 'Body schema and body image in a deafferented subject,' *Journal of Mind and Behavior* **16**, 369-90.
Gallese, V., Fadiga L., Fogassi, L., et al.: 1996, 'Action recognition in the premotor cortex', *Brain* **119**, 593-609.
Gardner R.M. and Moncrieff, C.: 1988, 'Body image distortion in anorexics as a non-sensory phenomenon: A signal detection approach', *Journal of Clinical Psychology* **44**, 101-107.
Gibson, J. J.: 1987, 'A note on what exists at the ecological level of reality', in E. Reed and R. Jones (eds.), *Reasons for Realism: Selected Essays of James J. Gibson*, Erlbaum, Hillsdale, NJ.
Gibson, J.J.: 1986, *The Ecological Approach to Visual Perception*, Erlbaum, Hillsdale, NJ.
Grafton, S.T. et al.: 1996, 'Localization of grasp representations in humans by PET: II. Observation compared with imagination', *Experimental Brain Research* **112**, 103-111.
Guldin, W.O., Akbarian, S., and Grüsser, O.J.: 1992, 'Cortico-cortical connections and cytoarchitectonics of the primate vestibular cortex: A study in squirrel monkeys (*Saimiri sciureus*)', *Journal of Comparative Neurology* **326**, 375-401.
Gurfinkel, V.S. and Levick, Y.S.: 1991, 'Perceptual and automatic aspects of the postural body scheme', in J. Paillard (ed.), *Brain and Space*, Oxford University Press, Oxford, pp. 147-62.
Haggard, P. and Eimer, M.: 1999, 'On the relation between brain potentials and the awareness of voluntary movements', *Experimental Brain Research* **126**, 128-33.
Haggard, P. & Magno, E.: 1999, 'Localising awareness of action with transcranial magnetic stimulation', *Experimental Brain Research* **127**, 102-107.
Head, H.: 1920, *Studies in Neurology*, Vol 2, Oxford University Press, London.
Head, H.: 1926, *Aphasia and Kindred Disorders of Speech*. Vol 1, Cambridge University Press, Cambridge.

Held, R.: 1961, 'Exposure-history as a factor in maintaining stability of perception and coordination', *Journal of Nervous and Mental Diseases* **132**, 26-32.
Hopkins, B. and Prechtl, H.F.R.: 1984, 'A qualitative approach to the development of movements during early infancy', in H.F.R. Prechtl (ed.), *Continuity of Neural Functions from Prenatal to Postnatal Life*, Blackwell, Oxford.
Humphrey, T.: 1964, 'Some correlations between the appearance of human fetal reflexes and the development of the nervous system', *Progress in Brain Research* **4**, 93-135.
Husserl, E.: 1973, *Ding und Raum: Vorlesungen (1907)*. Husserliana 16, U. Claesges (ed). M. Nijhoff, The Hague.
Iacoboni, M., Woods, R. P., Mazziotta, et al.: 1999, 'Cortical mechanisms of human imitation', *Science* **286**, 2526-28.
Jackson, S.R. and Husain, M.: 1996, 'Visuomotor functions of the lateral premotor cortex', *Current Opinion in Neurobiology* **6**, 788-95.
James, W.: 1890/1950, *The Principles of Psychology*. Dover, New York.
Jouen, F. and Gapenne, O.: 1995, 'Interactions between the vestibular and visual systems in the neonate', in P. Rochat (ed.), *The Self in Infancy: Theory and Research*, Elsevier Science B.V., pp. 277-301.
Kass, J.H.: 1991, 'Plasticity of sensory and motor maps in adult mammals', *Annual Review of Neuroscience* **14**, 137-67.
Lew, A. and Butterworth, G.E.: 1995, 'Hand-mouth contact in newborn babies before and after feeding', *Developmental Psychology* **31**, 456-63.
Locke, J.: 1690/1959, *An Essay Concerning Human Understanding*, A.C. Fraser (ed.), Dover, New York.
Malenka, R.C., Angel, R.W., Hampton, B., at al.: 1982, 'Impaired central error correcting behaviour in schizophrenia', *Archives of General Psychiatry* **39**, 101-107.
Marcel, A.J.: (in press), 'The sense of agency: Awareness and ownership of actions and intentions', in J. Roessler and N. Eilan (eds.), *Agency and Self-Awareness*, Oxford University Press, Oxford.
Mellor, C.S.: 1970, 'First rank symptoms of schizophrenia', *British Journal of Psychiatry* **117**, 15-23.
Meltzoff, A.: 1990, 'Foundations for developing a concept of self: The role of imitation in relating self to other and the value of social mirroring, social modeling, and self practice in infancy', in D. Cicchetti and M. Beeghly (eds.), *The Self in Transition: Infancy to Childhood*, University of Chicago Press, Chicago, pp. 139-64.
Meltzoff, A.: 1993, 'Molyneux's babies: Cross-modal perception, imitation, and the mind of the preverbal infant', in N. Eilan, R. Mccarthy, and B. Brewer (eds.), *Spatial Representation: Problems in Philosophy and Psychology*, Basil Blackwell, Oxford, pp. 219-235.
Meltzoff, A. and Moore, M.K.: 1977, 'Imitation of facial and manual gestures by human neonates', *Science* **198**, 75-78.
Meltzoff, A. and Moore, M.K.: 1994, 'Imitation, memory, and the representation of persons', *Infant Behavior and Development* **17**, 83-99.
Melzack, R.: 1989, 'Phantom limbs, the self and the brain', *Canadian Psychology* **30**, 1-16.
Merleau-Ponty, M.: 1964, *The Primacy of Perception*, W. Cobb (trans.), Northwestern University Press, Evanston.
Merleau-Ponty, M.: 1962, *Phenomenology of Perception*, C. Smith (trans.), Routledge and Kegan Paul, London.
Merleau-Ponty, M.: 1945. *Phenomenologie de la perception*, Gallimard, Paris
Milner, D.: 1998, 'Unconscious visual processing for action: Neuropsychological evidence', paper presented at Towards a Science of Consciousness, Third Conference, Tucson, April 27, 1998.
Munk, H.: 1890, *Über die Functionen der Grosshirnrinde*, 2 Aufl., Hirchwald, Berlin.
Ogden, J.A.: 1996, *Fractured Minds: A Case-Study Approach to Clinical Neuropsychology*, Oxford University Press, Oxford.
Olson, E.: 1997, *The Human Animal: Personal Identity without Psychology*, Oxford University Press, Oxford.

Paillard, J.: 1999, 'Body schema and body image: A double dissociation in deafferented patients', in G. N. Gantchev, S. Mori, and J. Massion (eds.), *Motor Control, Today and Tomorrow*, Academic Publishing House, Sofia.

Paillard, J.: 1997, 'Divided body schema and body image in peripherally and centrally deafferented patients', in V. S. Gurfinkel and Yu. S. Levik (eds.), *Brain and Movement*, Institute for Information Transmission Problems RAS, Moscow.

Paillard, J.: 1991, 'Knowing where and knowing how to get there', in J. Paillard (ed.), *Brain and Space*, Oxford University Press, Oxford, pp. 461-81.

Paillard, J. and Stelmach, G.: 1983, 'Localization without content: a tactile analogue of "blind sight"', *Archives of Neurology* **40**, 548-51.

Petit, J-L.: 1999, 'Constitution by movement: Husserl in light of recent neurobiological findings', in J. Petitot, et al. (ed.), *Naturalizing Phenomenology: Issues in Contemporary Phenomenology and Cognitive Science*, Stanford University Press, Stanford, pp. 220-44.

Phillips, C.: 1985, *Movements of the Hand*, Sherrington Lectures, XVII, Liverpool.

Piaget, J.: 1962, *Play, Dreams, and Imitation in Childhood*, Norton, New York.

Pick, A.: 1915a. 'Störung der orientierung am eigenen körper: Beitrag zur lehre vom bewusstsein des eigenen körpers', *Psychologische Forschung* **1**, 303-18.

Pick, A.: 1915b. 'Zur pathologie des bewusstseins vom eigenen korper: Ein beitrag aus der kriegsmedizin', *Neurologisches Centralblatt* **34**, 257-65.

Poeck, K.: 1964, 'Phantoms following amputation in early childhood and in congenital absence of limbs', *Cortex* **1**, 269-75.

Poeck, K.: 1963, 'Zur psychophysiologie der phantomerlebnisse', *Nervenarzt* **34**, 241-56.

Poeck, K. and B. Orgass.: 1971, 'The concept of the body schema: A critical review and some experimental results', *Cortex*, 7, 254-77.

Powers, P.S., Schulman, R.G., Gleghorn, A.A, et al.: 1987, 'Perceptual and cognitive abnormalities in bulimia', *American Journal of Psychiatry* **144**, 1456-60.

Prechtl, H.F.R. and Hopkins, B.: 1986, 'Developmental transformations of spontaneous movements in early infancy', *Early Human Development* **14**, 233-83.

Preuss, T.M., Stepniewska, I. and Kaas, J.H.: 1996, 'Movement representation in the dorsal and ventral premotor areas of owl monkeys: A microstimulation study', *The Journal of Comparative Neurology* **371**, 649-76.

Pribram, K.H.: 1999, 'Brain and the composition of conscious experience', *Journal of Consciousness Studies* **6**, 19-42.

Ramachandran, V. S., Rogers-Ramachandran, D. and Stewart, M.: 1992, 'Perceptual correlates of massive cortical reorganization', *Science* **258**, 1159-60.

Riess, T.J.: 1998, 'Gait and Parkinson's Disease: A conceptual model for an augmented-reality based therapeutic device', in G. Riva, B. Wiederhold, and E. Molinari (eds.), *Virtual Environments In Clinical Psychology And Neuroscience: Methods and Techniques in Advanced Patient-Therapist Interaction*, IOS Press, Amsterdam.

Riva, G.: 1998, 'Modifications of body image induced by virtual reality', *Perceptual and Motor Skills* **86**, 163-170.

Riva, G.: 1994, 'What is body image?', *European Commission VREPAR Virtual Body Project*, (http://www.ehto.org/ht_projects/vrepar/whatbody.htm).

Riva, G.: 1995. Virtual Reality in the Assessment and Treatment of Body Image. Paper presented at Center for Disabilities 1995 Virtual Reality Conf.(http://www.csun.edu/cod/95virt/0015.html).

Riva, G., Bacchetta, M., Baruffi , M. et al.: 2000, 'Virtual reality based experiential cognitive treatment of obesity and binge-eating disorders', *Clinical Psychology and Psychotherapy* **7** (2).

Riva, G. and Melis, L.: 1997, 'Virtual teality for the treatment of body image disturbances', in G. Riva (ed.), *Virtual reality in neuro-psycho-physiology: Cognitive, Clinical and methodological issues in assessment and rehabilitation*, IOS Press, Amsterdam, pp. 95-111.

Rizzolatti, G., Camarda, R., Fogassi, L., et al.: 1988, 'Functional organization of inferior area 6 in the macaque monkey. II. Area F5 and the control of distal movements', *Experimental Brain Research* **71**, 491-507.
Rizzolatti, G., Fadiga, L., Gallese, V., et al.: 1996, 'Premotor cortex and the recognition of motor actions', *Cognitive Brain Research* **3**, 131-41.
Robertson I.H., Halligan, P. and Marshall, J.C.: 1993, 'Prospects for the rehabilitation of unilateral neglect', in I.H. Robertson and J.C. Marshall (eds.), *Unilateral Neglect: Clinical and Experimental Studies*, Erlbaum, Hillsdale, NJ.
Rosenfield, I.: 1992, *The Strange, Familiar, and Forgotten : An Anatomy of Consciousness*, Knopf, NY.
Rossetti, Y., Rode, G. and Boisson, D.: (In press), 'Numbsense: a case study and implications', in De Gelder, De Haan and Heywood (eds.), *Varieties of Non-conscious Processes*.
Rossetti, Y., Rode, G., Pisella, L., et al.: 1998, 'Prism adaptation to a rightward optical deviation rehabilitates left hemispatial neglect', *Nature* **395**, 166-69.
Sadowsky, J. and Massof, R.W.: 1994, 'Sensory engineering: the science of synthetic environments', *John Hopkins APL Technical Digest* **15**, 99-110.
Scatena, P.: 1990, 'Phantom representations of congenitally absent limbs', *Perceptual and Motor Skills* **70**, 1227-32.
Sheets-Johnston, M.: 1998, 'Consciousness: A natural history', *Journal of Consciousness Studies* **5**, 260-94
Shoemaker, S.: 1976, 'Embodiment and behavior', in A. Rorty (ed.), *The Identities of Persons*, University of California Press, Berkeley.
Shoemaker, S.: 1999, 'Self, body, and coincidence', *Proceedings of the Aristotelian Society* Supplementary Volume **73**, 287-306.
Simmel, M.L.: 1962, 'Phantoms--experiences following amputation in childhood', *Journal of Neurology, Neurosurgery and Psychiatry* **25**, 69-78.
Simmel, M.L.: 1958, 'The conditions of occurrence of phantom limbs', *Proceedings of the American Philosophical Society* **102**, 492-500.
Singh, J. R., Knight, T., Rosenlicht, N., et al.: 1992, 'Abnormal premovement brain potentials in schizophrenia', *Schizophrenia Research* **8**, 31-41.
Spence, S.: 1996, 'Free will in the light of neuropsychiatry', *Philosophy, Psychiatry, and Psychology* **3**, 75-90.
Strauss, E.: 1967, 'On anosognosia', in E. Straus and D. Griffith (eds.), *Phenomenology of Will and Action*, Duquesne University Press, Pittsburgh, pp.103-125.
Tiemersma, D.: 1989, *Body Schema and Body Image: An Interdisciplinary and Philosophical Study*, Swets and Zeitlinger, Amsterdam.
Vallar, G., Antonucci, G., Guariglia, C., et al.: 1993, 'Deficits of position sense, unilateral neglect and optokinetic stimulation', *Neuropsychologia* **31**, 1191-1200.
Vetter, R.J. and Weinstein, S.: 1967, 'The history of the phantom in congenitally absent limbs', *Neuropsychologia* **5**, 335-38.
Wann, J. P., Simon, K., Rushton, M.S., et al.: 1998, 'Virtual environments for the rehabilitation of disorders of attention and movement', in G. Riva (ed.), *Virtual Reality In Neuro-Psycho-Physiology: Cognitive, Clinical and Methodological Issues in Assessment and Rehabilitation*, IOS Press, Amsterdam, (http://www.psicologia.net/pages/book1.htm).
Weinstein, S., Sersen, E.A. and Vetter, R.J.: 1964, 'Phantoms and somatic sensation in cases of congenital aplasia', *Cortex* **1**, 276-90.
Weinstein S. and Sersen, E.A.: 1961, 'Phantoms in cases of congenital absence of limbs', *Neurology* **11**, 905-11.
Windle, W.F.: 1971, *Physiology of the Fetus*, Charles C. Thomas, Springfield, IL.
Winer, G.A.: 1975, 'Children's preference for body or external object on a task requiring transposition and discrimination of right-left relations', *Perceptual and Motor Skills* **41**, 291-8.

MAUREEN CONNOLLY

FEMALE EMBODIMENT AND CLINICAL PRACTICE

I. INTRODUCTION

During my most recent visit with my physician, we had an argument. This is not unusual; we often argue or, rather, we engage in mature and respectful professional disagreement. This recent "discussion" stemmed from a much earlier point of contention between us: Several years ago, I took myself off the ERT (Estrogen Replacement Therapy) prescribed for my bodily manifestations of precocious menopause. My physician had warned me at that time that all the physical distress controlled by the ERT would return; I assured her that the physical distress was a small price to pay for the return of my personality and my felt sense of my body.

And so the saga continues in the recent visit. She asks me what symptoms I am experiencing; I tell her I am experiencing *NO* symptoms since menopause is not a pathology. She tells me that this is an unfortunate and unrealistic perspective for a woman of my education to take on my condition; I suggest to her that anyone who believes she can be in four rooms at the same time is not in a position to speak to me about realism, and so it goes.

Eventually we arrive at an impasse; we regard each other evenly and make a silent truce not to move into open hostility and disrespect. This visit – as have most others – ended courteously. It is not so much that we have an active dislike of each other; it is more the case that we have radically differing orientations to the lived body. It is also the case that these orientations cannot co-exist with dignity and integrity.

Another brief anecdote will illustrate this incompatibility. I was recently involved in interviewing candidates for an academic position in Disability Studies. One candidate, a brilliant young scholar in neuroscience, gave a presentation on Parkinson's Disease, and in that presentation showed a videotape of clinical encounters between physicians and Parkinson's clients. The clinical encounters depicted formulaic, objectifying, remarkably ungentle touch-based encounters with a series of clients exhibiting advanced Parkinson's profiles. The young scholar, perhaps noticing the look of horror on my face, quickly commented, "I know it looks really brutal, but they're used to it." I imagine that this was meant as some sort of consolation and/or rationalization. He continued to discuss the clients as "afflicted with a debilitating illness," "confined to a wheelchair"; he described the treatment interventions as "turning them on and turning them off, like a switch, as necessary"; he described the medical efforts as "attacking" and "killing" the root of the problem. At the conclusion of this presentation, he asked me for my feedback and responses. I explained that Disability Studies is based primarily in post-structural ideology critique, grounded in a semiotic phenomenological sensibility which locates meaning in the lived, contingent, recalcitrant body, and that the biomedical objectifying of these people was beyond unacceptable. Further, since the language perpetuates valorization of an upright, stable, productive, predictable body, it is inconceivable that any other way of living in the world might be meaningful or valuable in and of itself.

The use of the term "confined to a wheelchair" rather than "wheelchair user" justifies the heroic efforts taken to release the "afflicted person" from that confinement. Unfortunately, once the harsh reality of chronicity is accepted, the heroics are abandoned, and the day-to-day management of a complex condition is much less attractive in terms of heroics or status (Greenwood & Nunn, 1992; Wendell, 1996).

I begin this paper with these anecdotes for several reasons. The first is my firm belief that first person, first hand, insider accounts are the touchstone of phenomenological work. Mundane activities are sites for meaning and theorizing. Second, I wish to illustrate the profound incompatibility between traditional or typical biomedical approaches to the body, and lived body approaches to the body. Third, I want to suggest that these approaches cannot co-exist, and that the thrust of most clinical encounters is in the direction of eliminating – by various means – the lived body of the person seeking medical care.

While I will weave and intersect these three foci in the unfolding of the paper, at least my early emphasis will be on exploring and interrogating the clinical elimination of the lived body. The paper will unfold in four sections: there will be a section which articulates what I believe to be the strategies and tactics of clinical elimination of the lived body; there will be a brief overview of phenomenological and feminist work on female embodiment; this section will be followed by a more lengthy discussion of how female embodiment is constructed in clinical practice; I will conclude with a section on re-imagining otherwise in clinical practice.

II. KILLING THE LIVED BODY

Foucault (1980) has argued against the widespread notion that a Western scientistic allegiance to dualism has neglected the body in preference to the intellect. He suggests, on the contrary, that the control and repression of the body have played a fundamental part in the establishment and maintenance of power required for the growth of industrial capitalism since the 18th century. Foucault suggests that since the 1950's, the means of subjugating the body have evolved with the technology, so that the regimes of corporeal control, while no less insidious, are much more "invisible," largely due to their unquestioned familiarity. Foucault's central point is that, since at least the 18th century, far from neglecting the body, societal (read capitalist-productivist) interests have recognized the crucial significance of the body and movement in relation to the exercise of power. The importance of Foucault's work in relation to the body and power is that it shows the physical dimension of our beings to be infused with social and cultural significance (Kirk and Tinning, 1990).

The aforementioned points from Foucault frame both the argument for an (albeit implicit) intentionality of subjugation toward the lived body by practitioners within productivist-infused social and cultural codes (e.g., medicine, finance, education, industry) and the necessity of deconstructing these codes from a lived body-friendly theoretical framework. Richard Lanigan's semiotic phenomenology provides a helpful theoretical frame for experiencing embodied contingency by combining phenomenological explication of the freedom of individual expression with semiotic analysis of the field of culturally sedimented perception (i.e., what Bowers, 1984, presents as a dialectic between (a) peoples' self-understandings and efforts to create an enabling context to question taken for granted beliefs, and (b) the authority culture has over us. In clinical settings we need to discover the necessary conditions that free people to engage in critique given the often unconscious hold of illusion or habit). The advantage of working through semiotic phenomenology of "the experience of the body coupled with the consciousness of choice" is that it provides a critical theoretical frame to balance (interrogate and critique) the normative logics and inscriptions, the phenomena of the lived body as sign (of political discourse) and the sign systems that hold them together (Lanigan, 1988, 105; Craig, 1997, 52 and 54).

A dramatic example of Foucault's thematic of the regimes of corporeal control is the healthist, largely biomedical notion of the "healthy" body as unproblematically achieved. Neither contingency nor social intersections of race, class, gender, or embodiedness are considered relevant as potential barriers to the severely able body (Craig, 1997). When bodies present as less than ideal, they are "othered" within the existing structures of biomedical and capitalist agendas. Wendell (1996) proposes an idealized body as being illusory and unattainable yet strangely normative, whose criteria include being young, male, productive, strong, homogeneous, and unproblematic. "Othering" of bodies which deviate from this "ideal" involves projecting rejected aspects of ourselves on to groups of people who are then (necessarily) designated "the other".

When we make people other, we group them together as the objects of our experience instead of regarding them as subjects of experience with whom we might identify and we see them primarily as symbolic of something else – usually but not always, something we reject and fear and project onto them (Wendell, 1996, p. 60).

This notion of creating "the other" is akin to Julia Kristeva's (1982, p. 3) process of abjection: "an act of expulsion, a self-purging whereby with the food I vomit I also 'expel myself,' I spit myself out, I abject myself within the same motion through which 'I' claim to establish myself". Commenting on Kristeva, Elizabeth Grosz (1990, p. 89) elaborates:

Expelling (parts of) myself to establish myself as a member of the symbolic order, I create corporeal boundaries between myself and what is not myself, and, in so doing, actively constitute myself as an idiosyncratic entity.

Gail Weiss (1999, p. 47), consolidating and interrogating both Kristeva and Grosz, posits Kristeva's notion of abjection as a law which necessitates the purging of the abject from the culture. This desire to move beyond all limits (Kristeva, 1982, p. 180) is a fantasy of death, the death of the abject other that is viewed as necessary to maintain the corporeal integrity of the self. However, like Foucault, for Kristeva this desire reinforces rather than denies the power the abject wields within the Symbolic Order's own carefully defined boundaries, hence the non-ideal body is the ultimate threat to the sign system of predictability, productivity, and control.

Following this logic, then, the lived body, which will not present in discrete categories of illness or symptomology, which resists tidy diagnosis, and which goes on to resist heroic attempts to return "it" to its pre-trauma state, is abject for biomedical approaches to healing and treatment. Women's bodies or feminized bodies are even more abject because of how they evade the explanation, prediction, and control of rationalist science.

That which is abject – for no other reason than the threat it presents within the Symbolic Order – must be cast out, othered, discredited, killed.

What are the clinically acceptable modalities of elimination? Wendell's (1996) list includes epistemic invalidation, alienation, social abandonment, failures of communication and gaps in knowledge. Kum-Kum Bhavnani (1993) suggests that invisibility, erasure, denial, and tokenism are the most effective ways to wipe out a culture. Certainly, clinical applications of these, as exemplified in silencing, over-medicating, over-testing, psychiatric diagnosis for physical ailment, underestimating and trivializing pain and stress, using procedures which perpetuate relations of dominance and subordination, are effective ways of eliminating the lived body as a legitimate partner in the clinical encounter.

From the previous section we can discern several normative logics and inscriptions regarding clinical encounters:
➤ the lived body of the person seeking help from the health care system is not the most reliable source of information about itself;
➤ the control of information about the body creates and perpetuates a power differential based in the assumption that medical knowledge is privileged and mysterious;
➤ illness is recognizable, treatable and curable given the appropriate context and variables;
➤ bodies are the problem, medicine is the solution.

We can also discern how the lived body presents as a transgressive sign. In *The Absent Body* Drew Leder (1987) suggests that the cultural body, or the body of the natural attitude, is largely absent to itself until it begins to make itself known in illness or dysfunction. Once present in dysfunction, it presents as a threat to the disembodied status quo. Such a dysfunctional body must be treated, cured and returned to its normative docility, or it must be made silent and/or invisible, and/or shockingly, pathetically visible so as not to act as incentive for other bodies which might be contemplating outing themselves as other than indestructible, stable, productive, and predictable.

The sign systems which hold sway over the normative logics and inscriptions, and which keep the transgressive lived body at bay, are any political agendas which benefit from the perpetuation of the myth of control and the normativity of the idealized body (i.e., any social institution whose continued success and status depend upon production, prediction, and control). Perceived threats to the ongoing success and status of these sign systems must be eliminated.

It should be obvious that "killing" the lived body need not – indeed seldom does – involve an actual death. Rather, the lived body is invalidated, denied, patronized, mystified, and problematized until its abjection is complete. Only then it can be managed at various levels and degrees within, or tangential to, clinical encounters in particular, and health care systems in general.

III. FEMALE EMBODIMENT

Discussing female embodiment is tricky business. I live in a culture which is simultaneously obsessed with and ignorant about the body. The bodies of athletes, models, performers, and public figures are analyzed, criticized and fetishized. In the commercial, media-soaked societies of North America I experience the idealization and objectification of the body, and the concomitant demand that people control and attempt to perfect their bodies. Susan Wendell (1996, p. 9), a feminist philosopher whose writing takes up issues of embodiment and disability, proposes that

[R]efusal to come to terms with the full reality of bodily life, including these aspects of it that are rejected culturally, leads people to embrace the myth of control, whose essence is the belief that it is possible by means of human actions, to have the bodies we want and to avoid illness, disability, and death.

This culture also valorizes productivity, so the body must not only maintain standards of appearance, but also standards of performance and pace. There is a proliferation of products designed to improve aspects of appearance as well as internal functioning, which, in turn, contribute to external demonstrations of competence. The demand for self-improvement products is high, even though the majority of people have little or no knowledge about the structures, functions, and processes of the human body or how these products interact with bodily structures, functions, and processes. Through its effects on the flow of communication about bodily experience the authority of medicine shapes and limits what the culture knows about the human body (Wendell, 1996, p.9).

The body as object – observed, studied, measured, cosmetized, dissected – is neither a subjective nor intersubjective body; studying the objectified body does not constitute work in the study of embodiment. Embodiment as subject matter must necessarily include a deep consideration of the lived experience of the body as mediated by the normative disciplines of a culture. It must take into account the intersections of body, experience, and culture, and the role and/or place of the body in the experience and construction of meaning and consciousness.

Maurice Merleau-Ponty's (1962) bold and courageous work in *Phenomenology of Perception* (preceded by early progressive interrogations in *The Structure of Behavior*) compelled a focus on the problem of embodiment for phenomenology. Yet, Merleau-Ponty compelled more than neutral involvement. His examination of perception as embodied meaning for a consciousness attempts the daunting task of theorizing the transformation of sensation into meaning structures within humans. Merleau-Ponty also models (and articulates) phenomenological method as he works through a recursive, and embedded process of description, reduction, and interpretation in the careful consideration of bodily-based phenomena in high functioning, typical, atypical, and low functioning humans. Merleau-Ponty's students – Michel Foucault and Pierre Bourdieu – continued the emphasis on the body within culture, moving scholars from diverse locations to a more critical engagement with the disciplinary practices that produce the "docile bodies" required by modern social institutions (Foucault, 1979).

In spite of its legitimacy within philosophical study – largely phenomenological and post-modern in orientation – embodiment remains a nebulous and, for some, a highly suspect focus of study. Moreover, even when taken seriously as the subject of study, it is seldom taken up methodologically, both within and outside of phenomenological and postmodern approaches to inquiry.

Feminist analyses of female embodiment have moved through a progression similar in their unfolding to the study of embodiment within phenomenological and postmodern orientations. "Female" as a concept and a lived reality has been problematized through postmodern analysis and critique of the notions of fixed, unified, unambiguous, categories of identity. Other feminist analyses cast a jaundiced eye on a philosophical orientation which de-stabilizes the female subject just as she has found her voice and some access to the structures of power.

Postmodern feminist theorists Judith Butler (1989), Luce Irigaray (1977, 1993) and Susan Bordo (1993) have contributed to some reconciliation between the two aforementioned positions, and the critical, sociological and historical works of theorists Iris Young (1990), Maxine Sheets-Johnstone (1992), Linda Alcoff (1988), Page du Bois (1988), and Marianne Paget (1993) (among others) have also enabled powerful critique of modernist (and largely, sexist) norms of behavior and power distribution, while preserving a viable female subject. Kristeva's (1984) notion of a "subject on trial/in process" is also helpful in the re-imagining of a female subject within an ongoing building up and breaking down of normative structures and fixed categories. Of course, much present day feminist scholarship is indebted to Simone de Beauvoir's (1952) early articulations and theorizing on *The Second Sex*.

An important area of clinical practice, the discipline of nursing, is also indebted to the extensive and ground-breaking phenomenological and politically sensitized work of Patricia Benner (1999, 1984). Benner's work has demonstrated the power of the mundane not only in theorizing, but also in the theoretical impact on practitioner preparation and ongoing professional development in nursing. Janice Morse's (1991) work is similarly located in nursing practices and the power of story and lived relation in the development of a body of compassion and comportment.

Sandra Lee Bartky's work is especially illuminating in its emphasis on the social construction of femininity through idealization, objectification, and demands for control of the female body. Bartky (1990, p. 65) has examined "those disciplinary practices that produce a body which in gesture and appearance is recognizably feminine". Especially poignant and alarming in her analysis is the insight that these norms and inscriptions of appearance and behavior are not forced upon women by anyone in particular, and appear to be voluntary or even "natural" while at the same time exerting tremendous power in women's lives.

... the disciplinary power that is increasingly charged with the production of a properly embodied femininity is dispersed and anonymous; there are no individuals formally empowered to wield it; it is...invested in everyone and in no one in particular. This disciplinary power is peculiarly modern: It does not rely upon violent or public sanctions, nor does it seek to restrain the freedom of the female body to move from place to place. For all that, its invasion of the body is well-nigh total: The female body enters "a machinery of power that explores it, breaks it down, and rearranges it." The disciplinary techniques through which the "docile bodies" of women are constructed aim at a regulation which is perpetual and exhaustive – a regulation of the body's size and contours, its appetite, posture, gestures, and general comportment in space and the appearance of each of its visible parts (Bartky, 1990, p. 80).

Emily Martin (1987) proposes reproduction (and everything associated with it – menarche, pregnancy, child birth, menopause) as a metaphor of social and medical control. Through the medicalization of typical healthy physiological processes in developing and aging females, physicians have wrested control and knowledge of the body from those living the embodied experience. Further, the mystification, folklore, and pathology constructed around pre-menstrual syndrome, problem births, and menopause (described in some medical textbooks as "a design flaw") function to perpetuate women's alienation from their bodies.

Cosmetic surgery, homophobia, compulsory heterosexuality, and ageism are other significant normative influences on the social construction of women's bodies. Social constructions of reality are intensely political processes, reflecting competing groups' divergent vested interests and differential access to power (Weitz, 1998). Female embodiment – and women's authorship of it – remains a contested landscape. Powerful, passionate and politicized prose and poetry [notables here include Adrienne Rich (1986), Audre Lorde (1984), Nancy Mairs (1994), and bell hooks (1996)] blend with philosophical scholarship and academic writing until the lines are blurred, yet most within the fields associated with feminist scholarship would agree that anecdotal and autobiographical work, while necessary, is still insufficient as the philosophical ground for effective theorizing in the politics of (female) embodiment. (By this I mean theorizing that would have a significant impact on the power structures that define and control women's bodies.)

As this section concludes, I wish to draw attention to the norms and inscriptions within Western contemporary culture that hold sway over women's (and men's) "choices" regarding appearance, behavior, and awareness of embodied consciousness. Wendell (1996) and Bartky (1990), refer to these norms and inscriptions or "normative logics" (Lanigan, 1988) as disciplinary practices of physical normality and/or femininity. Wendell (1996, p. 88) maintains that "the disciplines of normality are pre-conditions of participation in every aspect of social life, yet they are unnoticed by most adults who can conform to them without any conscious effort."

These norms and inscriptions include, but are not limited to, standards of pace, stamina, and energy; standards of compliance and adherence (and their requisite behavioral and appearance performances); pathologizing of excess (e.g., excessive fat, excessive size of particular, non-sexualized body parts); associations between particular bodily presentation and culturally approved markers of success; fear of subordination and/or loss of control; fear of loneliness, shame and exclusion – in a phrase, the horror of unspoken awareness. By this I mean that persons have a felt sense of how they will be treated if they do not live up to the aforementioned disciplines of normality, hence are kept in check (and lock step) by their embodied fear of what their own culture and peer group will do to them if they make visible those self-same oppressive, unrealistic, and disembodied norms.

Given these norms and inscriptions, the lived body – even the mythologized "healthy body" – presents as a sign of problematic political discourse insofar as it cannot adhere to these norms and inscriptions; i.e., at some point bodies sleep, get sick, get injured, and die (Craig, 1997, 1998), hence is always hovering as a transgressive and threatening presence. This in turn means that if the norms and inscriptions are to continue, the body – and person's awareness of it – must be devalued, objectified, controlled, fetishized, commodified, and pathologized. In this way the larger sign systems which benefit from the disciplines of normality (heterosexism, ageism, sexism, healthism, productivism, capitalism, ableism) can perpetuate themselves as structures of power within and across culture. Cultural and societal institutions – medicine, education, commerce, and social services, to name a few – function within the larger sign systems of productivism, capitalism, and heterosexism, and enforce the disciplines of normality via ageist, sexist, healthist, and ableist normative logics (insidiously within the "natural attitude" of a given culture or cohort).

Norms and inscriptions, and unconscious allegiance to sign systems, are significant contributors to the natural attitude, the bane of the epoché in phenomenological method, wherein the researcher attempts to suspend all influences of culture, society, family, ethnicity, and embodiment (as if this were completely or even remotely possible). The deep subjectivity required to recognize and subvert the disciplines of normality instead depends on the researcher's ability to embrace the always, already present, contingent, recalcitrant body as a starting point for the crisis of cultural transgression necessary to move beyond our (albeit unacknowledged) certainties. This same courage to suspend one's certainties presents immense risk in the clinical context. Physicians who are unable or unwilling to entertain alternative or even preposterous explanations are not only lacking in imagination and resourcefulness – aspects of a clinical profile one might find admirable – but are unable or unwilling to acknowledge their own limits. This desire to exceed limits is, as Kristeva has reminded us earlier, a fantasy of death – the death of the abject that embodies uncertainty and hence, unpardonable transgression.

IV. FEMALE EMBODIMENT AND CLINICAL PRACTICE

I will begin this section with what I feel to be necessary reference to, and commentary on, Elaine Scarry's (1985) *The Body in Pain*. In that it dramatically points to the issue of epistemic invalidation and the denial of lived experience, Scarry's analysis of the inexpressibility of physical pain poignantly describes the context in which most clinical encounters occur:

> Thus, when one speaks about "one's own physical pain and about "another person's physical pain," one might almost appear to be speaking about two wholly distinct orders of events. For the person whose pain it is, it is "effortlessly" grasped (that is, even with the most heroic effort, it cannot *not* be grasped); while for the person outside the sufferer's body, what is "effortless" is *not* grasping it (it is easy to remain wholly unaware of its existence; even with effort one may remain in doubt about its existence or may retain the astonishing freedom of denying its existence; and, finally, if with the best effort of sustained attention one successfully apprehends it, the aversiveness of the "it" one apprehends will only be a shadowy fraction of the actual "it"). So, for the person in pain, so incontestably and unnegotiably present is that "having pain" may come to be thought of as the most vibrant example of what it is to "have certainty", while for the other person it is so elusive that "hearing about pain" may exist as the primary model of what it is "to have doubt". Thus pain comes unsharably into our midst as at once that which cannot be denied and that which cannot be confirmed (Scarry, 1985, p.4).

Scarry's (1985, p.3) project in her book is threefold. She wishes to examine and interrogate (a) the difficulty of expressing physical pain; (b) the political and perceptual complications that arise as a result of that difficulty, (c) the nature of both material and verbal expressibility. Scarry utilizes several contexts as sites for examples and anlaysis – medical practice is one of those sites. Medical practice is seen as a site wherein the person experiencing pain may be regarded as an unreliable narrator, particularly when the language used to describe pain is deemed naive, insufficient and/or inaccurate by the medical professional not experiencing it, or if the symptoms or features expressed do not add up to a discrete, unproblematic pathology. Here we have an intense example of the phenomenon described earlier: to have great pain is to have certainty; to hear that another person has pain is to have doubt. As Scarry (1985, p. 7) notes, "the doubt of the other person, here as elsewhere, amplifies the suffering of those already in pain." (Note: Medical practice can also provide helpful and hopeful responses, such as the development of the Gate Control Theory of Pain and the McGill Pain Questionnaire, both of which will be addressed in the recommendations section.)

Wendell (1996, p. 120) believes that:

> [T]he cognitive authority of medicine in the doctor-client encounter gives far more weight to the doctor's metaphysical stance, undermining the confidence of clients in the importance of their bodily experiences. When that happens, clients cease to expect acknowledgement of the subjective suffering or help in living with it. This can leave them not only isolated with their experience but feeling obligated to discount or ignore it, alienating them further from their own bodies.

For persons living with profound and ever-present pain, a diminished sense of the pain's intensity may occur by virtue of that person's adjustment to an ongoing pain state. This pain need not always be "certain" even for those who experience it. The pursuit of certainty in the face of the ineffability of pain or the elusiveness of bodily sensation is yet another example of how the lived body, having presented in ways that cannot be accounted for, must now be "eliminated" (i.e., invalidated as a legitimate source of information) by a sign system invested in the precision of its own rationality (i.e., its ability to "account for" human physiological structure and function).

More often than not, the typical clinical encounter, is one in which the person in pain, seeking treatment, or seeking information is in a subordinate position to a physician who has been socialized into sexist and elitist attitudes (Fugh-Berman, 1992; Wendell, 1996; Sherwin, 1992). Further, because medical knowledge has been mystified through various means of information control, people who are not trained as medical insiders do not have familiarity with, or access to, this discourse which controls their behavior within the clinical encounter. Even more insidious is the potential threat of psychiatric diagnosis for clients with the audacity to arrive at a clinical encounter with their own research preparation done or with specific questions to discuss with their physician. Apparently the least significant variable in the clinical encounter is an embodied speaking subject.

Gender further complicates the already problematic embodied speaking subject (Craig and Connolly, 1997). Given the unacknowledged cultural power of binary constructions of gender, it is no surprise that the gender to which attributions of non-valued behavioral or expressive features have been attached will be seen as transgressive and threatening to a system whose currency is based in the attributional opposite. If stoicism and strength are seen as valued personal traits in a productivist system, then the persons with more power will have those traits attributed to them (whether they possess them or not). If those without power exhibit these same traits, the traits will be re-interpreted in order to maintain the unequal distribution of power. This is an example of both denial and invisibility as proposed by Bhavnani (1993) cited earlier in this paper.

Women, manifesting the same (or similar) symptoms, as men – such as angina, heart attack, chronic obstructive pulmonary disease, cancer, diabetes, to name a few examples – will receive different, and in some cases, dismissive treatment. This differential treatment is based in gender stereotyping – women cannot be susceptible to the same intensities of trauma as men because women's lives are less stressful – and in genuine ignorance: symptomatology in women may indeed be different for particular conditions, but the research is so sparse that there is very little definite information in this regard (Armstrong and Armstrong, 1993).

How the medical establishment constructs women is a powerful influence controlling nearly every facet of the women's health care system. This control exercised by the medical industry is implicated in the training of medical students and new physicians, and in men's and women's attitudes toward women clients (Albino et al, 1990; Theriot, 1993). Women's physical complaints are devalued, interpreted as "psychological" or "emotional" in etiology, and dismissed as trivial. Even gender harassment is prevalent. Further, there are tendencies to categorize particular kinds of conditions or ailments as "women's diseases." Conditions which present as non-discrete, ambiguous, inconsistent or episodic, chronic, recalcitrant, conditions such as fibromyalgia, lupus, myalgic encephalo-myelitis, depression, to name a few, are often "feminized"; that is, the condition itself is devalued as a "real" medical condition by virtue of its being associated with women. The characteristics of the persons in the client role are also categorized as feminine by virtue of their choice of language and expressive repertoire, i.e., if a client presents as confused, upset, anxious, curious, frightened, vulnerable – or a whole host of other emotions appropriate to a traumatic or untenable situation – then she or he is feminized (Craig & Connolly, 1997; Findlay, 1993). The feminization of disease serves to control and silence women and men in clinical encounters and the (usually) concomitant medicalization of the body ensures unquestioning client compliance and reinforces alienation from the lived body.

Wendell (1996, p. 122) claims that the cognitive and social power of medicine is so pervasive and influential that a person's subjective descriptions of her bodily experience need the confirmation of medical descriptions to be accepted as accurate and truthful. She goes on to state that "since modern medical science does not exercise much modesty about the extent of its knowledge, it has a tendency to ignore, minimize the importance of, or deny outright any bodily experiences it cannot explain." This epistemic invalidation of lived bodily experience can be a reality shattering experience. Being told "there is nothing wrong with you," regardless of how acute or debilitating a condition may be, can completely undermine a person's confidence in her lived experience and relationship to reality (Wendell, p.123).

When a woman – or anyone – in a clinical encounter is placed in a situation of having to deny and distrust her felt sense of her own embodiment, then there is no ground for lived relation. When the lived body has no validity, then the object body becomes the viable site for commodity exchange within ostensible and paradoxical health "care" contexts. From there, it is a short distance to the abject and eventually, absent body, a body which one can manage, ignore, and forget.

V. RE-IMAGINING OTHERWISE IN CLINICAL PRACTICE

How might a phenomenological sensibility inform re-imagined praxis in health care and clinical practice? How might it present alternatives to Young's (1990, p. 153) chilling statement that "women in sexist society are physically handicapped"?

First person, thick description accounts of pain, illness, diagnosis, treatment and client-physician interactions give voice to insider stories of lived experience. These accounts – which ought to embrace as heterogeneous a sampling as possible (i.e., accounts from clients, physicians, nurses, family and friends) – must continue to be seen as legitimate, necessary, and sufficient capta in phenomenologically oriented research (as well as in other orientations to research). The ground breaking and courageous work of authors like Kay Toombs (1992), Arthur Frank (1991), Arthur Kleinman (1988), Marianne (Tracy) Paget (1993), Tom Craig (1997, 1998, 2000), Nancy Mairs (1994) (among others) is rich not only in the power of the narratives and startling transparency of the expressive word, but also in its unsparing detail and unsettling political impact. Work like this – writing which has the potential for robbing one of all one's certainties – must be actively cultivated and supported in the face of capitalist and productivist agendas which would characterize it as threatening and irrational. Make no mistake about it – pain, trauma, chronicity, stressed embodiment (Craig, 2000) *ARE* irrational. We must move outside of rationalist models and norms if we are to get any insight into the complexities and power of these lived experiences.

Toni Jeffreys (1982, P. 183) points out that in the modern scientific world an illness does not exist until the medical industry says it does. When people took to their beds in other centuries, their families, the community, assumed they had a good reason for doing so. They did what was necessary for them. The lives of Theresa of Spain, Florence Nightingale, and Charles

Darwin are examples of people chronically ill for many years but never "diagnosed". They were just ill, and that was enough for the rest of the world. But in the twentieth century, if one takes to one's bed and does not seek the medical "seal of approval" for illness – medical diagnosis and treatment – one is suspect (p.183).

Wendell (1996, p. 128), commenting on Jeffrey's work, suggests a solution to at least some of the epistemic invalidation that modern medicine inflicts on clients: "Acknowledge the possible ignorance of individual practitioners and the incompleteness of medical science, and assume the reality of clients' symptoms unless there is overwhelming positive evidence that they are imagined or pretended." One need only read the work of medical sociologist Marianne Paget (1993), whose ongoing research on error in medical practice and posthumously published memoir on the last year of her life offer startling testimony to the imperfect arts of diagnosis and treatment. Other powerful work in this genre includes Deric Longden's (1989) *Diana's Story* (later produced as the movie "The Wedding Gift"). I am not suggesting that medical practice be held to unreachable standards of perfection; I am suggesting that the prevailing myth that medicine can do everything is perpetuated inside the field as much as it is assumed outside. Further, if more modesty, disclosure, and willingness to admit to limitations were features of clinical encounters (in traditional and alternative models), then the subject of discourse could indeed be the embodied speaking subject, and the unfolding consultation and treatment could be authentic communication within reciprocally respectful lived relation.

Scarry (1985) describes the work of Ronald Melzack who, along with his colleague Patrick Wall, authored the widely respected "Gate-Control Theory of Pain", and with his colleague W.S. Torgerson also developed the "McGill Pain Questionnaire." The theory and the questionnaire were developed out of the first person accounts of clients describing their lived experience of pain. Melzack and Torgerson propose descriptive categories generated from the insider language of persons in pain. The "temporal," "thermal" and "constrictive" groups are among those that together express the sensory content of pain, while certain other word groupings express pain's affective content, and still others its evaluative or cognitive content. Greenwood and Nunn (1992) also acknowledge the narrative, mythological, and emotional character of pain, and how these fuse with somatic and myofascial memory. Greenwood and Nunn's treatment protocols work with their conceptualization of pain as a layered, fused and embedded experience. The development of diagnostic and treatment protocols based in the expressions of the lived experience of the embodied speaking subject grant legitimacy and dignity to the speakable and the unspeakable in the clinical encounter.

In a recent publication, I describe how it feels to do touch therapy on a chronic pain client (who happens to be my life partner); the co-author of the paper describes his lived experience of treatment – mine, and other practitioners'. What remains remarkable to me is this person's amazement at being believed. In my own practice, I find it inconceivable that the experiencer would not be believed – especially when I can feel with my hands the presentation of pain in his joints, soft tissue, and fascia. Given the increasing absence of human contact in our culture, I would encourage touch practices ("touch" as distinguished from "palpation") in diagnosis and treatment, and, necessarily, in medical education. In this regard, medical education could take a page from the book on disability education whereby the pre-service preparation includes vicarious experience (empathy sessions), direct contact, prolonged engagement, and persuasive, meaningful messages/subject matter. I realize that in medical education direct contact and prolonged engagement are programmed into the preparation of practitioners, but the empathy and meaningfulness work is lacking and is necessary for attitudinal change (Connolly, 1994; Fugh-Berman, 1992).

Other insights that arose in the writing of this aforementioned paper (Craig and Connolly, 1999) include the realization that sometimes pain wants and will have its voice and its time, and that the best response is to be present with the person and her/his pain. Deep intersubjectivity – sharing silence or verbal expression – is often a more effective treatment than invasive, relentless, neutralizing, or deadening interventions. And, often, negotiated management of pain, illness, chronicity, or other modalities of stressed embodiment are more realistic and honorable than tortured efforts at fixing and/or curing. When clinical practice is reflectively constructed as a commitment to honorable lived relation with embodied speaking subjects, the clinical encounter can be imbued with a sensibility of compassion and an acknowledgment of the textured and intertwined character of the body, in time, and space, with others.

Page du Bois (1988, pp. 195-86) insists:

> ... our feminist work will be limited if we cannot learn to historicize our own theories of subjectivity and those theories' construction. Feminists interested in transforming gender, class, and race relations must take account of subjectivity, of the ways in which the myth of the ego, the isolated self, is produced psychically.... To try to will away the cultural experience of a gendered subjectivity without understanding its production is to court failure by appealing always to the rational, to the adult ego, a fiction which resists being acknowledged as a fiction.

Du Bois' injunction applies equally well to phenomenologically oriented work in medicine and embodiment – female and otherwise.

Maurice Merleau-Ponty (1962, p.408) critiques an ideology of rationality as being unable to go beyond itself, by virtue of having to use its own criteria to continue to validate its own concordances. According to Merleau-Ponty (1962, p. 409) all one has to do to get out of this loop is to "recognize these phenomena which are the ground of all our certainties. The belief in an absolute mind, or in a world in itself detached from us is no more than a rationalization of this primordial faith."

If the subject *is* in a situation, even if he is no more than a possibility of situations, this is because he forces his ipseity into reality only by actually being a body, and entering the world through the body. In so far as, when I reflect on the essence of subjectivity, I find it bound up with that of the body and that of the world, this is because my existence as subjectivity is merely one with my existence as a body and with the existence of the world, and because the subject that I am, when taken concretely, is inseparable from this body and this world. The ontological world and body which we find at the core of the subject are not the world or body as idea, but on the one hand, the world itself, contracted into a comprehensive grasp, and on the other hand, the body itself as a knowing body (Merleau-Ponty, 1962, p.408).

The knowing, world-disclosing, reality-shaping body of the embodied speaking subject is the body to be acknowledged and consulted in clinical encounters. Is this a naive and utopian re-imagining of practitioner-client interactions? Utopian, perhaps, in its expectations of a deconstruction of the dominance-subordination script, one in which those benefiting from the arrangement are loathe to surrender those privileges; and naive only in so far as it might be formulated in an ahistoricist political vacuum. Neither of these is presumed here. Our journey to encountering Merleau-Ponty's knowing body will be agonizingly slow and fraught with difficulty, but I believe the encounter is unavoidable. Stressed embodiment (and gendered embodiment) will neither recede nor compliantly disappear, and our engagement with them will define us as a culture.

Kristeva's (1984) notion of the subject on trial/in process is a helpful frame for all of us working in, and living, the impossible dialectic of semiotic and symbolic, the project of examining and overturning our certainties. When both clinician and client acknowledge the self as a divided subject, who is constantly negotiating the pulls and expectations of unconscious and consciousness awareness, as well as negotiating the needs and rights of the body with the responsibilities and oppressions of symbolic (usually productivist) codes, then each makes a step towards a consciousness of choice. That is, in acknowledging their own internal tensions between affiliating with cultural codes of disembodiedness and honoring the body that is, clinicians can no longer engage in clinical conduct as if they did not have choices about the lived quality of that encounter; how they dwell in the encounter is now a choice, not a cultural habit, and they have a consciousness of that choice, of how and what they choose. Likewise, clients can choose to participate in cultural replays of dominance and subordination, or they can claim and name their lived bodies in the clinical encounter. Making this shift to a consciousness of choice is risky. As a practitioner and/or as a client I can choose the trial and the process, the rupture of heterogeneity, or I can choose another haven which allows me to live as if I were not a body in intersubjective relation.

It is my hope that practitioners, clients, clients, educators, indeed, anyone with the potential for living as an embodied speaking subject, will make the journey to a consciousness of choice and continue the joyful, ongoing task of dwelling in the heterogeneous body with unsparing self-reflexivity and attentive wonder.

Brock University
St. Catherine's, Ontario
Canada

BIBLIOGRAPHY

Albino, J., Tedesco, L., and Shenkle, C.: 1990, 'Images of women: Reflections from the medical care system', in M. Paludi and G. Steuernagel (eds.), *Foundations for a Feminist Restructuring of the Academic Disciplines*, Haworth, New York, pp. 220-40.

Alcoff, L.: 1988, 'Cultural feminism versus post structuralism: The identity crisis in feminist theory', *Signs*, **13**, 405-36.

Armstrong, P. and Armstrong, H.: 1983, *A Working Majority. What Women Must Do For Pay*, Supply and Services Canada for the Canadian Advisory Council on the Status of Women, Ottawa.

Bartky, S.L.: 1990, *Femininity and Domination: Studies in the Phenomenology of Oppression*, Routledge, New York.

Beauvoir, S. de: 1952, *The Second Sex*, Alfred A. Knopf, New York.

Benner, P.: 1999, *Clinical Wisdom and Intervention in Critical Care*, Saunders, Philadelphia.

Benner, P.: 1984, *From Novice to Expert – Excellence and Power in Clinical Nursing Practice*, Addison Wesley Publishing Company, Reading, MA.

Bhavnani, K.K.: 1993, 'Talking racism and the editing of women's studies', in D. Richardson and V. Robinson (eds.), *Thinking Feminist: Key Concepts in Women's Studies*, The Guilford Press, New York, pp. 27-48.

Bordo, S.: 1993, *Unbearable Weight:Feminism, Western Culture, and the Body*, University of California Press, Berkeley.

Bowers, C.A.: 1984, *The Promise of Theory: Education and the Politics of Cultural Change*, Longman, New York.
Butler, J.: 1989, *Gender Trouble: Feminism and the Subversion of Identity*, Routledge, New York.
Connolly, M.: 1994, 'Practicum experiences and journal writing in adapted education: Implication for teacher education', *Adapted Physical Activity Quarterly* **11**, pp. 306-28.
Craig, T.: 2000, 'A funny thing happened on the way up to renown: On the concrete essence of chronic illness and the intercorporeal weight of human suffering', Paper presented at the Society for Phenomenology and the Human Sciences, Pennsylvania State University, State College, PA.
Craig, T.: 1998, 'Liminal bodies, medical codes', *Semiotics 1997*, Peter Lang Publishing, New York.
Craig, T.: 1997, *Disrupting the Disembodied Status Quo: Communicology in Chronic Disabling Conditions*, Unpublished doctoral dissertation, Southern Illinois University at Carbondale.
Craig, T. and M. Connolly: 1999, 'Your body tells me stories: Living pain, flirting madness, transforming care', *Narratives of Professional Helping*, **5**, 16-26.
Craig, T. and M. Connolly: 1997, 'Medicalization of the body, feminization of disease – developing regimes of silence', in *Semiotics 1996*, Peter Lang, New York.
Du Bois, P.: 1988, *Sowing the Body*, University of Chicago Press, Chicago.
Findlay, D.: 1993, 'The good, the normal, and the healthy: The social construction of medical knowledge about women', *The Canadian Journal of Sociology* **18**, 115-36.
Foucault, M.: 1980, *Power/Knowledge: Selected Interviews and Other Writings, 1972-1977*, Colin Gordon (ed.), Harvester Press, Brighton.
Foucault, M.: 1979, *Discipline and Punish*, Vintage, New York.
Frank, A.: 1991, *At the Will of the Body: Reflections on Illness*, Houghton Mifflin, Boston.
Fugh-Berman, N.: 1992, 'Tales out of medical school', in Estelle Pisch (ed.), *Reconstructing Gender - A Multicultural Anthology*, Mayfield, Mountainview, CA, pp. 351-55.
Greenwood, M. and P. Nunn: 1992, *Paradox and Healing*, Paradox Publishers, Victoria, BC.
Grosz, E.: 1990, 'The body of signification', in John Fletcher and Andrew Benjamin (eds.), *Abjection, Melancholia, and Love: The Work of Julia Kristeva*, Routledge, New York.
hooks, b.: 1996, *Outlaw Culture*, Routledge, New York.
hooks, b.: 1994, *Teaching to Transgress*, Routledge, New York.
Irigaray, L.: 1977/1985, *This Sex Which is Not One*, Cornell University Press, Ithaca, NY.
Irigaray, L.: 1993, *An Ethics of Sexual Difference*, Cornell University Press, Ithaca, NY.
Jeffreys, T.: 1982, *The Mile-High Staircase*, Hodder and Stoughton, Sydney.
Kirk, D. and R. Tinning: 1990, 'Physical education, curriculum, and culture', in D.Kirk and R. Tinning (eds.), *Physical Education, Curriculum and Culture: Critical Issues in the Contemporary Crisis*, The Falmer Press, London, pp. 1-21.
Kleinman, A.: 1988, *The Illness Narratives: Suffering, Healing, and the Human Condition*, Basic Books, New York.
Kristeva, J.: 1986, "The system and the speaking subject", in Toril Moi (ed.), *The Kristeva Reader*, Columbia University Press, New York, pp. 24-33.
Kristeva, J.: 1984, *Revolution in Poetic Language*, Margaret Waller (trans.), Columbia University Press, New York.
Kristeva, J.: 1982, *Powers of Horror: An Essay on Abjection*, Leon S. Roudiez (trans.), Columbia University Press, New York.
Lanigan, R.: 1988: *Phenomenology of Communication*, Duquesne Press, Pittsburgh.
Leder, D.: 1987, *The Absent Body*, The University of Chicago Press, Chicago.
Longden, D.: 1989, *Diana's Story*, Corgi Books, London, UK.
Lorde, A.: 1984, *Sister Outsider: Essays and Speeches*, The Crossing Press, Freedom, CA.
Mairs, N.: 1994, *Voice Lessons: On Becoming a (Woman) Writer*, Beacon Press, Boston.

Martin, E.: 1987, *The Woman in the Body*, Beacon Press, Boston.
Merleau-Ponty, M.: 1962, *The Phenomenology of Perception*, Colin Smith (trans.), The Humanities Press, New Jersey.
Morse, J. (ed.): 1991, *Qualitative Nursing Research: A Contemporary Dialogue*, Sage, Newbury Park.
Morse, J. and J. Johnson (eds.): 1991, *The Illness Experience – Dimensions of Suffering*, Sage, Newbury Park.
Paget, M.: 1993, *A Complex Sorrow*, Temple University Press, Philadelphia.
Rich, A.: 1986, *Of Woman Born: Motherhood as Experience and Institution*, W.W. Norton, New York.
Scarry, E.: 1985, *The Body in Pain*, Oxford University Press, New York.
Sheets-Johnstone, M.: 1992, *Giving the Body Its Due*, State of New York Press, Albany.
Sherwin, S.: 1992, *No Longer Client: Feminist Ethics and Health Care*, Temple University Press, pp. 41-60, pp. 41-60, Philadelphia.
Theriot, N.: 1993, 'Women's voices in 19th century medical discourse: A step toward deconstructing science', *Signs* **19**, 1-31.
Toombs, S.K.: 1993, *The Meaning of Illness: A Phenomenological Account of the Different Perspectives of Physician and Client*, Kluwer Academic Publishers, Dordrecht, The Netherlands.
Weiss, G.: 1999, 'The abject borders of the body image', in G. Weiss and H. Haber (eds.), *Perspectives on Embodiment - The Intersections of Nature and Culture*, Routledge, New York, pp. 41-60.
Weitz, R. (ed.): 1998, *The Politics of Women's Bodies*, Oxford University Press, New York.
Wendell, S.: 1996, *The Rejected Body*, Routledge, New York.
Young, I.M.: 1990, *Throwing Like a Girl and Other Essays in Feminist Philosophy and Social Theory*, Indiana University Press, Bloomington.

GLEN A. MAZIS

EMOTION AND EMBODIMENT WITHIN THE MEDICAL WORLD

I. THE EMOTIONAL BODY AS RELATIONAL MATRIX EXCEEDS CALCULATIONS

The medical world with its extended chains of instrumental appropriation of the body is perhaps the site in our postmodern world that seems most oppressive to letting the emotions have their voice and sway over the course of our lives. This had seemed true even without the growing managerial cost-accounting that has also come to transform medicine into a further reifying practice. Yet, paradoxically, most medical practice is located between the most routine kind of maintenance and the acute reaction to trauma, i.e. situations that involve a longer term coping with crises and chronic medical conditions that are obvious openings where authentic revelation through the power of emotion is legitimized. Medicine works at the boundaries of many of the most culturally charged terms of identity and meaning, intervening in regions that dramatically change our sense of who we are and what our life is about. Yet it does so blindly and awkwardly, because it intervenes in areas in which the emotional dimensions are central to these identities and meanings without an ontology that allows it fully to comprehend this. Insofar as regaining health entails regaining an emotionally capacious understanding and relationship to changes in embodiment, the biomedical model of body is unable to facilitate full healing.

Whether it is through the removal of a breast, disruptions in movement capacities, or damage to cognitive faculties, to name a few of the myriad examples, the changes in embodiment with which medicine must deal undercut simple exclusionary definitions of mind-body, self-other, nature-technology, freedom-necessity, or gender difference. For example, the biomedical model assumes: (1) that "repairs" to the biochemistry or physiology of the client's "body" exhaust the healing task of the physician, while emotional "adjustments" are a separate matter for other professionals or family members to facilitate in the "emotional" or "mental" realm; (2) that the client is "alone" within the "subjectivity" of his/her emotional struggles, instead of recognizing phenomenologically that emotions are co-constituted with all those involved in the client's world – especially physicians and other healthcare professionals; (3) that a prosthesis is just a "mechanical part" added on to a functioning organism, not realizing that the client now has a new sense of embodiment which calls for an emotional realignment of relationship if the prosthesis is to become integrated into the "lived body"; and (4) that the client's freedom in an abstract matter of "will" or "mind" confronting a physical world, instead of comprehending how choices are "situated" in a context in which emotional clarification is part of a process of making possible new dimensions of life in working through ambiguous situations with others. Certainly, it is not surprising that, in a cultural context in which gender assumptions are often unconscious, helping clients make emotional shifts in their sense of being gendered beings with differing bodily capacities or appearances is often overlooked.

In all these ways the biomedical model of body keeps apart dimensions of meaning that are treated more holistically through focusing on the emotional sense of the client's embodiment. It is understandable how medical practice can easily slip into these mistakes, since the sudden, acute breakdown of health in which medicine is most spectacularly called into action seems to marginalize the kind of sense to which the emotions are most attuned. Furthermore, as a profession, medical practice has often sought to stabilize its power of intervention in a reduction of the body to an objective map using the emergency as paradigm. However, every emergency takes place within a larger context of the client's life-world.

When assisting healthcare clients in their ability to move beyond the crisis, it is often the case that the more objective indicators of the body's physiological systems fail to fathom the overall well-being of the affected person. Their referents can be perfectly verified and their coherence may seem to have reached closure within the prescribed medical systems of signification. Yet, the state of affairs of client recovery they seek to describe and capture in its truth may utterly elude them. What is most likely to be missed in the exclusive focus on the rationally representable world of the body-as-mechanism is the world of emotional understanding, the dimension of affective feeling that wells up from the body. Certainly the body is partially mechanical, but not solely so, nor are all its dimensions rationally delineable. Vital rhythms of attunement and identity within the client's world are primarily emotionally grasped and articulated. This becomes more obvious when medicine deals with treatment that is not abrupt and invasive, but that helps clients with ways of structuring their world to alter behaviors, diet, activities, expectations, meanings of events, etc. Regardless of whether it is the scope of activity, the kinds of food the client is able to eat, the way s/he can enter into relationships with others in terms of energy expenditure, excitement levels, nature of sexual options, and so forth, all these dimensions have emotional significance that directly affect how the person fathoms his/her identity and the meaning of his/her life within her/his embodied being. Thus, if healthcare professionals are to assist clients in dealing with traumatic or gradually learned alterations in these dimensions, they must attend to the client's way of affectively embodying herself or himself.

However, even the most traumatic and abrupt medical emergency reveals the unity of embodiment with emotion. Oliver Sacks (1984) relates a striking example of such a moment in which all the medical signs fail to signify. In this instance, Sacks, the noted physician, is the healthcare client at the center of the narrative. Hiking up to a mountain summit at 9,000 feet, above the Hardanger fjord in Norway, he had ignored the sign warning him of the presence of a massive bull until, on his way down, the bull appeared to Sacks in far too close a proximity. In his panicked flight from the bull, Sacks fell and broke his leg and severed the connection to muscles. After surgery, medical interventions, and a period of healing, his surgeon – who is quite pleased with the surgery – examines Sacks. The pain has subsided to a bearable level, the swelling has gone down, and Sacks agrees with the surgeon, "It seems fine, Sir," Sacks responds, "surgically speaking." However, Sacks still cannot even contract his muscles, move his leg, or even feel where his leg is located. He is panicked and discouraged. The surgeon replies to Sacks (1984, p. 105):

"Nonsense, Sacks," he said sharply and decisively. "There's nothing the matter. Nothing at all. Nothing to be worried about. Nothing at all."
"But…"
He held up his hand, like a policeman halting traffic. "You're completely mistaken," he said with finality. "There's nothing wrong with the leg. You understand that, don't you?"
With a brusque and, it seemed to me, irritable movement, he made for the door, his Juniors departing deferentially before him.

Sacks' physicians were frustrated. They were adamant that the medical findings indicated that his leg had been repaired. All the relevant significations, whether x-rays or lab values, have declared the truth and the rest is some sort of problem with Sacks, not with "the leg." Sacks, of course, is correct that his leg is not "cured," for a leg that is fully functional is one that feels like part of his body, one that feels like it is part of his world, one that not only moves in ways he wishes, but also fits in with the rhythm of his existence. About these things, the doctor's calculations have little to say.

Being the holistic investigator that he is, Sacks finds a meaningful way to rediscover and reintegrate his leg. His solution highlights how the emotions and the body are inseparable, how the emotions of the client permeate his/her world, and how emotional well-being and the health of the client are the same matter. We will use Sack's case as a touchstone in this essay as we unravel the place of the emotions in medical care.

Medicine seems to construct a context that maximizes the extent to which the body can become submerged in discursive practices. The exhaustive batteries of tests circumscribe the meaning of illness into arithmetic values, reified structures, and efficient causal relations. The procedures proliferate to manipulate the body as a mechanism. There is endless analysis, interpretation, and prognostication in diagnosis. The body in its most material reality has been subsumed into a series of readouts and signs. The flesh is seen as biophysiological functioning. Yet, the reality of the client's "condition" is an ever-anxious waiting for how it feels and what it means within the horizon of experience of that person. In excess of the numbers and physiological observations, there is the silence of how the body is present for that unique individual within his/her sense of the world. This is the realm to which discursive practices point and try to map, but they cannot capture or exhaust the pertinent level of meaning. Experiencing this lack in the medical model, clients (and families) often become frustrated as they try to cope with the fact that the well-being of the client is about the quality of his/her relationships that comprise a unique world – a shifting, fragile, and nuanced sense of how its *feels* welling up from the visceral and perceptual depths of experience.

Within the confines of a massive scanner, under the knife of a surgeon, or as part of a mechanically assisted process, the subjectivity of the client is submerged into a cyborg existence of biotechnical machinery. The way in which the client's subjectivity stands uniquely with his/her world – which is the sense that vitalizes emotion – is subordinated to engineering dilemmas of immediate intervention. In the midst of this domain of practical action, the determinate teleology of achieving physical "health" reigns supreme and the emotional experience of the client within this world seems the least determinative factor for assessing the client's well-being. Yet, it is precisely at this moment of breakdown, of life crisis of the body, that emotion erupts – and mere matter and mere discursive formulation miss exactly what is happening to the patient. Thus, at the very moment when clinicians, family, friends, and the client forsake emotional reality as insignificant, they discover that it is the primary reality in several key senses. For, ultimately, what has been affected is the quality of the relationships of the person to his/her world and this is precisely what the client's deeper sense of the body as "flesh" registers in emotion.

Suffering a medical crisis – and the sudden vulnerability and need for help, not only in dealing with physiological symptoms but also in gaining some rational and emotional understanding of what one is undergoing – reveals there is no isolable subjectivity, or detached realm of the mental, but rather there is the world, others, and self, lived through the perceptual matrix of the body. As Richard Zaner (1981, p. 246) has eloquently stated, this is the realization that illness forces in upon us: "Thus the variety of responses by family, medical or non-medical observers – of facing an 'abyss,' a glass 'wall,' etc. – can be seen as modes of self and other recognition, evoking an awareness of self in respect of its crucially constitutive moment: the other self." Given the Cartesian tradition, it is difficult for many of us to experience this interweavement of persons (rather than being an isolated "I"), and recognize that this is true because we are fleshly creatures whose emotions are not peripheral but at the heart of who we are. This is where medicine can draw upon phenomenology's rich descriptions of how we are co-presences who feel *and* think our way together through the world.

Phenomenology has also revealed the body as "embodying. We are not only interwoven on the bodily, affective realm. The world itself is structured by the body, and the body takes its sense from the demands of the world. Again, the Cartesian sense of an isolated "I" and one that is detached in essence from the world, is not only undermined by the work of phenomenology, it is revealed through the personal experience of illness. The medical crisis is more powerful than a philosophical deconstruction of the objective body: "It is thus not so much that 'the' organic body signals the 'latent crisis of every known ontology.' It is rather that '*my*' organic body does" (Zaner, 1981, p. 21). Merleau-Ponty's (1968) idea of the "flesh" allows us to see the body as a dialogical exchange of significance through the depths of perception that is fluid and revelatory of vital aspects of the world of the client outside the tests of bioscience. As this essay will explore, to be forced to confront through illness that I am a body and am in the world through my body also means that I (and my loved ones) will discover that the primary registration of how I am – and how I am in relation with the world – occurs within emotional revelation of a more complex sense of embodiment.

II. THE NEED FOR MEDICINE TO SEE THE BODY'S EMOTIONAL ENGAGEMENT IN THE WORLD

Using Merleau-Ponty's (1962, 1968) perspective, we can describe embodiment as an ongoing process of inhabiting situations in a relational network of perception as motivated discovery and expression. "Motivated discovery" means that the body encounters its world through *solicitations* from that which is perceived. The body is open to "appeals" from the perceptual field in its ongoing attempt to find meaning in perceptual experience (Merleau-Ponty, 1962, pp. 403-34). So, for example, as we will see in discussing Sacks' (1984) case further, the body does not move by fiat, but because it is drawn into action by its context – whether it is to avoid being hit by an oncoming car, to reach the table where a meal is waiting, or in response to the myriad "appeals" made to us by our environmental context. These appeals will vary in how much they strike our emotional marrow. Avoiding an oncoming car will strike a much deeper chord of panic and will, therefore, be a stronger solicitation to our body than walking towards an unappetizing bowl of mush. It is as seeking, or as intentional – rather than reacting to its environment impassively and mechanically through efficient causal chains of events – that perception occurs. In other words, the body moves into areas of concern, into contexts that have an emotional meaning for us that the body translates into activity. For instance, if someone has a need and skill to type, she may be able to move her hands in dexterous ways when sitting down at a keyboard, yet be unable to perform such motions in response to an abstract command from a physician.

Embodiment is dialogical or responsive in this sense, not mechanically assertive or self-propelled. Embodiment occurs within a context of our personal histories. The context is filled with areas that have emotional significance for us. One person may love to walk on the beach, another to dance, and yet a third person may live with their greatest vitality when working in a printing shop. Arms, legs, and even perceptual organs, are "geared" into these settings. To understand each person's body fully is to have some sense of embodiment in this context of richly emotionally colored areas of concern for that person. How vital this may be for healthcare professionals in treating clients will become most clear when looking at Sacks' attempt to regain the use of his leg. The ambiguities of perception are presented as problems or tasks to be further resolved, ultimately inexhaustibly so. So, for example, we do not merely use our legs to see how strong or coordinated they are, we live them as ways to get up that steep hill to a daughter's school; and we try to decipher an oncoming road sign, in order to know whether to slow down or not, rather than simply to test our eyesight. Thus, as an "ongoing process," embodiment is not "located" in the body as an isolated object, but is rather "dispersed throughout its lived context (Merleau-Ponty, 1962, p. 248). This context is one marked out by our emotional commitments. So, the body is where my emotional investments are located.

The "relational network" of the perceived situation refers to the fact that each perceived has its identity only within a field where each member of the field takes its meaning from the vantage points of every other member of the field *as if* they all were perceivers (not literally so). So, for example, the sick person lying in a hospital bed can see her husband's body from the vantage of the wall behind him, of the ceiling above their heads, and of her friend standing on the other side of the bedside. This is what "her" vision is about at any moment. "Her" vantage draws on all these other perspectives in their co-presence. The same is true for other dimensions of apprehension: they are all "given" as drawn together from donations of meaning throughout the encompassing "phenomenal field" or "life-world." As an "ongoing process," embodiment is not located in the body as an object we use as a vehicle, but occurs as an evolving, fluctuating, and transforming site of interaction, or interweaving, of perception (and its layers of other concomitant apprehensions) and the world that we *are* (Merleau-Ponty, 1968, pp. 132, 161-62, 229). This notion of the interweaving of perceiver and world, and of perception with all levels of apprehension, undermines the dualistic separation of body from world and of body from mind, imagination, emotion, memory and intuition (Merleau-Ponty, 1968, p. 239).

Merleau-Ponty details how the perceiving body structures what it apprehends through a context and with a habit body of embedded personal, cultural and historical significance, but only does so by constantly shifting itself in relation to how the perceived beckons to it. This renders perceiving-perceived as a system or a relational ebb and flow. The body is an open question to its environment which itself questions back the body for its possible ways of joining in with its unfolding of meaning. In terms of his later formulations, both are of a "flesh" that folds back on itself, enveloping the perceiving-perceived (Merleau-Ponty, 1968, pp. 132, 229). To cite an example used by Merleau-Ponty: it is true that we see a Cezanne painting only within a history of regarding paintings in a certain way which is embedded in a particular culture, history, personal set of experiences, etc. – so one person is apt to approach it reverentially and another dismissively, and a third might have no idea of what they are looking at – but the canvas itself demands a certain optimal distance from which to be viewed, a certain lighting, a certain posture, a certain rhythm of attention and time of investigation, etc. A similar equilibrating of body and world occurs with all perception and action. Our bodies are shaping and even apprehend the environment, because of what is in the environment itself. This means that, as an embodied being, a healthcare client only has a part of himself (or herself) examined when doctors probe the body in isolation. The relational context of which that body is a node of circulating sense and motivation is missing, or has been replaced by a very impoverished environment. In some sense, the body carries this field or world, or larger sense of the flesh in which it is enfolded, everywhere. So a client may transpose some of the meaning of his/her old context onto the limited hospital context. Perhaps, for example, when confined to the hospital, an avid hiker and bird photographer has a measure of that excitement and love focused on the view from a window at the end of the corridor – a place he may visit once a day. However, this co-presence becomes more and more attenuated as medical examination/intervention becomes prolonged and the body is treated as an isolated physiological object.

A famous line from Merleau-Ponty's (1962, p. xi.) introduction to the *Phenomenology of Perception* is "there is no inner man." In other words, the person is only found in relation to an environment, a world. The same thought is registered in his later formulation by saying that human being is a "dehiscence" or a "deflagration" with the world, a fission with itself into the world and also with the world, from which we return to ourselves at the end of circulating through this world and its significance (Merleau-Ponty, 1964, p. 180). This insight is key to our consideration of emotion, *because emotion is the registration of how it stands with this relationship between the embodying process and the surrounding environment* (Mazis, 1993, p. 22). We could paraphrase Merleau-Ponty's denial of the "inner" person also to say there is no "contained body" or no "inner body mechanism," insofar as this problem of where the body is located arises in medicine. Using emotion as the general term that includes more momentary flashes of affect, the visceral dimension of feelings, and the more longer term passions and emotional attunements to the world, we can say emotion is the lining of bodily perception with a sense of how it qualitatively stands with the web of relationships with the world. As such, the emotions are key resources for medicine to find the "larger" body of the client as interwoven with the world.

However, to consider this more fully, it is important to add to Merleau-Ponty's notion of embodiment by considering Drew Leder's (1990) analysis in *The Absent Body*. Leder emphasizes that this relational interweavement between our embodiment and the world remains largely absent to our explicit awareness. Our embodiment, in general, is unobtrusive, and especially "the hiddenness of vital organs ... is essential to healthy functioning" (Leder, 1990, p. 44). There is a sense in which our perceptions normally take place with the body's presence withdrawn from attention and with the viscera doubly withdrawn into what Leder (1990, p. 53) calls a "depth disappearance" in which they are only very indirectly ever perceivable. Yet, at the moment when medicine intervenes, it is usually when there is what Leder terms "an affective call" of the body in pain of some sort. Pain announces something wrong with the body, but not from the body as an isolated mechanism, but rather as at the heart of all the relationships of the person to the activities, things, and people within his/her world. As Leder (1990, p. 73) puts it, "As such, pain reorganizes our lived space and time, our relations with others and with ourselves." It is only because the body is embodying, forging all these relationships, that the body's disrupted functioning now disrupts them. Yet, the pain itself does not indicate the past relationships, but is an intense focus on the body withdrawn from its normal sphere of interconnection. Leder focuses on the call of pain, but, of course, there are other experiences, such as fatigue or dizziness, etc., with their own affective charge that also disrupt our normal time, space, and context of relationships.

As Leder describes, pain has a compelling quality that constricts our accustomed context of life. The pained body seems an alien body, no longer me. The body that I am normally takes me into the world, or as Merleau-Ponty (1962, p. 235) phrased it, "My body is the fabric into which all objects are woven." This is true of the weakened body, the dizzy body, the disoriented body, etc. All of these conditions are registered in a variety of disruptive emotions, such as psychic pain, fear, frustration, panic, sadness, and other affective registrations of these shifts. With illness, the world that was the body's natural extension seems far away. There is a shift in our embodiment: "We are no longer dispersed out *there* in the world, but suddenly congeal right *here*" (Leder, 1990, p. 75). The world of food, for example, which marked my rhythms of the day and my immersion in certain neighborhoods, my anticipation of certain activities, like cooking and dining with others, etc., is gone in the searing pain of the colon cancer which prevents eating. Similarly, the temporal context is disrupted: "As it pulls us back to the *here*, so severe pain summons us to the *now*" (Leder, 1990, p. 75). All the projects that were echoes of what my body could do without any thought – that constituted my past and future horizons of engagement – have been put out of play, as the disorientation or weakness or pain absorbs me and undercuts these avenues of involvement. Especially, when symptoms reveal serious illness and disability, the world in space and time sinks away. The medical objectification of the body within the biotechnological focus of practice exacerbates the same sort of de-contextualizing of the body, the same sort of alienation from the world of the body, with which these disruptive feelings and illness already menace the client.

In this sense, medical practice itself is dis-abling. If each part of the body embodies "I can" relationships to the world – such as my arm being a reference to the writing desk in my study, or to the car I can drive, or to the shovel and flower bed I can dig, etc. – and pain and illness force me to live my body as "the no longer can" (Leder, p. 81), this isolates me from all around me. Then, insofar as medicine views the body as an isolated mechanism, it prevents the body from forging new relationships to the world or to renewing the old relationships in new ways. If emotion throws into relief the ways in which person and world are interrelated, then medicine needs to pay attention to the emotions of healthcare clients if it is to assist them to re-achieve health or a sense of well-being. This does not mean necessarily restoring a client's previous physiological functioning (although it may include that) but rather facilitating a restored emotional attunement with the world, no matter how the client's physiological functioning may have become altered. The emotions can allow the world to "come back" into relationship with embodiment and achieve a new sense of meaningfulness. Otherwise, there is no restoration of embodiment and health.

This is what Sacks' doctor didn't understand. For him, the body was a mechanism to repair. However, the living of the body as a way of forging meaningful relationships with the world is inseparable from this physiological functioning. If either alters, so does the other. In Sacks' case, his leg would not function mechanically unless some sense of the emotional excitement and investment in contexts that involved his leg was restored. For other clients, it might be the reverse, i.e. they might need to see how differently functioning aspects of embodiment can lead to new types of relationship with the world. For most clients, as for Sacks, it is probably some of both. Surgically "fixing" the parts of, say, the leg without restoring the leg's – or rather, the person's "legged relationships" – with the world does not allow the person to rediscover well-being, or to achieve health defined as the ability to lead a meaningful life through making the most fruitful use of the sort of embodiment one has.

If pain, panic, or depression, etc., take the client's embodiment away from its matrix in the world, then other emotions that can restore the diminished connections of orientation, significance and intention should be cultivated as a part of treatment. An important part of the healing process is to catalyze and nurture emotions that restore the interweavement of body and world. Even if pain did not obscure these connections of body and world, and even if medicine did not exacerbate this disconnection with its approach to the isolated body as mechanism to be repaired, the mundane, everyday, absence of awareness of how the body relates to its world probably needs to be undone if the client is to discover ways to restore her/his body's potential connections to the world. Medicine is not solely about "curing" in the sense of replacing body "parts" so that they become "like new." We all age, suffer various illnesses, and eventually die. Medicine as an achievement of well-being is about maximizing our embodiment as a rich matrix for meaning with the world.

To see what this might mean, let's return to Sacks' and his discouragement that his surgically perfect leg has remained "entirely motionless, toneless, senseless, beneath its white sepulcher of chalk." Sacks no longer felt "at one" with his leg, nor did his leg feel like it "fit" into his world. This condition had remained for weeks, although there had been involuntary twitches of "the leg" but, as Sacks eloquently describes it, they had no personal quality, no sense that they were actions, just a functioning of parts. Then, a visitor brought Sacks a tape of Mendelssohn's Violin Concerto. Suddenly, Sacks (1984, p. 119) felt a "decisive effect":

From the moment the tape started, from the first bars of the Concerto, something happened, something of the sort I had been panting and thirsting for, something that that I had been seeking more and more frenziedly with each passing day, but which had eluded me. Suddenly, wonderfully, I was moved by the music. The music seemed passionately, wonderfully, quiveringly alive and conveyed to me a sweet feeling of life. I felt with the first bars of music, a hope and an intimation that life would return to my leg—that it would be stirred, and stir, with original movement, and recollect or recreate its forgotten motor melody.

In hearing the music, Sacks felt as though he were experiencing the "animating and creative principle of the whole world." For Sacks, music was a vital part of how his body was connected to his world. This sense that "life itself was music" had always profoundly moved him and it was a source of hope and vitality. With the music, he could feel the sparks of life in his healing leg as being hopeful signs of recovery. However, even though Sacks felt inspired by the music, which was important to his healing, he did not at first realize its importance.

As time passed Sacks' leg gradually resumed more function and he could even walk somewhat, but the leg always felt artificial, awkward, and as if it were not really *his* leg. One day, when he was trying to walk, something else happened:

> And suddenly – into the silence, the silent twittering of motionless frozen images – came music, glorious music, Mendelssohn, fortissimo! Joy, life, intoxicating movement! And as suddenly, without thinking, without intending whatever, I found myself walking, easily, joyfully, *with* the music. ... Suddenly, with no warning, no transition whatever, the leg felt alive, and real, and mine, its moment of actualization precisely consonant with the spontaneous quickening, walking and music (Sacks, 1984, p. 144).

What Sacks discovered was that his leg became *his* when it found its place within the context of part of his world that had great emotional meaning for him. Given its insertion within part of the world of heightened emotional significance for Sacks, the leg became him again, rather than a mere mechanical part to be manipulated. Sacks realized that, before this, all his therapy had been impersonal. It had been rational and cold, but then "came music, warm, vivid, alive, moving personal" (Sacks, 1984, p. 148).

We will return to Sacks' case, because as it turns out, another breakthrough was needed in order for his leg really, fully to become "him" again. This breakthrough has to do with further refinements we must articulate about the role of emotions in healing damaged embodiment. However, it is important to note that we are related to aspects of our world emotionally and for each of us this is a different fabric of connection. If medicine is to heal, it is necessary for some members of the team of caregivers to help clients explore how they are related to the world, especially in reference to the parts (or dimensions) of their embodiment that have become injured or compromised in illness.

III. POTENTIAL DIMENSIONS OF EMOTIONAL HEALING IN MEDICAL CARE

The experience of Oliver Sacks is not unique, but points towards shared structural features of the lived experience of serious illness or physical trauma. If, in the experience of threatening emotions and impaired functioning, every illness involves an alienation from the body, and if medical practice contributes even further to deepening this uncanny experience (Zaner, 1981, pp. 54-55), then the clients of healthcare require a way to reawaken their bodies to the power of embodying. As Kay Toombs (1993, p. 87) states it boldly: "the objectification of the body results in the loss of embodiment. That is, as an object the body is no longer embodying." Like Sacks, sick and injured persons find themselves displaced from their usual contexts of engagement, of concerns, loves, hates, tasks – lost somewhere in a time and space that have been altered by illness. "Illness is a state of disharmony, disequilibrium, dis-ability, and dis-ease which incorporates *a loss of the familiar world.*" (Toombs, 1993, p. 96). If the sick person is to transform the object body that has been "worked upon" by medical science into a power of taking up creatively the myriad relationships s/he has with the world, then s/he must find a way of becoming connected, directed, and responsive again to her/his world. These powers, however, are the powers of emotional engagement with the world. Healing care requires facilitating this emotional discovery since re-embodying the client is world engagement through emotion.

The body as embodying is an ongoing dynamic process of becoming enmeshed in the world and being transformed by its efforts to transform the world through its expression. The emotions are the body's immediate registration of how this interrelation is going: how it has been met and towards where it is heading. However, the emotions themselves are dynamic, always changing, revealing new aspects of the unfolding meshing with the world and others: "within the dawning of emotional sense, the past is ever coming to new significance, is ever becoming, which has always been recognized as the dangerous fluidity of e-motion" (Mazis, 1993, p. 106-107). As such, the emotional sense of the world is fragile, changeable, and in motion. This has threatening aspects and exciting aspects, both of which are central to the experience of those who suffer illness and dis-ease. The threatening aspect of suddenly feeling a loss of security in the body, helplessness, fear, pain, uncertainty, aloneness, and sadness through serious illness, is obvious and unavoidable when one's life or capacities are threatened or permanently compromised. On the other hand, it is the same dynamic and fragile dimension of emotions that has revitalizing potential for all those who suffer illness and injury. However, the healing potential of emotion within the healthcare context will not just happen unless we focus on facilitating its discovery for each client. Learning to forge new emotionally meaningful relationships with aspects of one's life-world through one's altered sense of embodiment is not a separate issue from achieving health. Thus, medical practice needs to facilitate the emotional side of re-embodiment.

The need for others to enter the emotional world of the healthcare client is crucial in the context of serious illness. One of the shocks of serious illness is this: that this body which opens me to others and to our shared emotional life together seems to have taken me away from others in a way that perhaps none of us can correct. This may be true in a purely physiological way, but perhaps not in a fuller sense. Yet, this is what first strikes the seriously ill person emotionally. As Toombs (1993, p. 96) puts it, "Existential aloneness is necessarily a part of serious illness." Sacks has a very powerful chapter describing how he fell into a limbo and nothingness, into what he called "a hole in reality itself, a hole in time no less than in space, and therefore cannot be conceived as having as term or ending" (Sacks, 1984, pp. 108-109.) The person who becomes ill or injured inevitably feels these moments of being torn away from the rest of humanity; yet, if they are to engage on the journey of coping and healing, it is through others that sick persons are most likely to discover who they are becoming and what their new world is like.

This need for others to enter the void of aloneness is often misunderstood as a need for sympathy or cheerfulness or other solicitous emotional display. Although these emotional interactions may have some constructive role in dire moments of facing serious illness and disease, what the sick person needs is for others to enter the search for greater emotional understanding of who s/he has become in this new phase of embodiment brought about by illness or injury (which also means within a new world in some sense). This does not mean the ill person needs psychotherapy, but it does mean that healthcare professionals who are experienced with the effects of various sorts of illness and trauma need to pay attention to what these effects mean for the everyday lives of their clients. Using the acquired knowledge from past clients, healthcare professionals can then enter into dialogue with new clients concerning these dimensions of illness and injury. It also means that physicians, nurses and other healthcare professionals must be open to exploring with clients the specific emotional impact of illness. When Toombs (1993, p. 111) states, "What the patient seeks is not simply a scientific explanation of the physical symptoms, but also some measure of understanding of the personal impact of the experience of the lived body disruption," she is pointing to a lived meaning that has a major emotional component.

It is on an emotional level that we first succeed in addressing the need of the ill person "to make sense of this particular experience of illness" (Toombs, 1993, p. 111). Healthcare clients need information and intellectual understanding, but what medicine often fails to provide is facilitation of the emotional registration of the client's situation. This is achieved by helping the client to focus on his/her emotional relationships to the world and by exploring the transformations and revelations that have taken place.

Specifically, there are at least three dimensions of emotional understanding that should emphasized in good medical care. One aspect of emotions is their *exploratory* nature: "Through 'feeling' in its emotional sense, the body moves forward gropingly into the world, not as self-sufficient, not as holding a meaning already to be signified ... but rather as touching things in order to be touched back" (Mazis, 1994, p. 30). In making sense of illness and injury, being encouraged to follow up on feelings as they arise and lead to further feelings, images, expressions, and finally insight, is a process that can be supported, encouraged and facilitated by healthcare professionals. Unlike many aspects of medicine that abstract the client from his/her personal situation in order to understand the "case" as one instance among many, emotional clarification requires a movement toward appreciating the client's *existential uniqueness*: "the emotions do not help us move beyond our embodied situation in which we are caught up in particulars, but rather keep us moving back to that point at which we marvel at the singularity of each particular as if it were experienced for the first time" (Mazis, 1994, p. 31). Again, this kind of emotional clarification – from initially vague emotional feelings to those that resonate with who the particular individual is in his/her unique circumstances – is a process of focusing in which a caring, but professional, interlocutor can be tellingly helpful. Finally, there is the *transformative* power of emotion to open the situation: "If one allows emotion to open up surges of sense, new dawnings of relation, then the world, no matter how familiar, can take on radically new identities" (Mazis, 1994, p. 107). Like the other two aspects of emotion, a person with the distance of a professional – but with genuine concern for the particular individual – may be better able to help the client feel new possibilities for his/her transformed embodiment and world than can those who are closest to him/her. Even if the family does help, medical caregivers should be ready to attend to these dimensions of emotional understanding.

To find new ways of understanding the world emotionally, to become emotionally grounded in the particularity of the unique medical and life situation, and to feel new possibilities, are life and death tasks for coping and healing from dis-ease. A close relative of mine had open-heart surgery that was "very successful" in repairing the damaged biomechanical parts. However, he never achieved any emotional understanding of how he might become a changed person, might understand his personal habits, diet, relations with others, activities, hobbies, etc., in new ways, or come to see new possibilities for activities in life that would help him heal. He was emotionally devastated, felt as if his life was over, felt cut off from the world, could not fathom a new identity, and fell into despair. Not only was the quality of his life horrible for his two remaining years, but his doctors were mystified as to why he died after "successful surgery."

Every serious illness and injury harbors a possibility for falling into despair. The moment of despair is the moment when one feels as though the meaning of life is gone and one no longer has a place in the world. Most illnesses change who we are and what our world is like. If we are not helped to embrace our emotional capacities for forming new relationships with others, the world, and ourselves, we may never heal the breach that has opened with our old life and world. We cannot just retrieve them. Healthcare clients need help to go forward into transformation. This is an emotional task. In a culture that is obsessed with being in control versus the world and others, that structures life according to future goals and accomplishments, that has little emphasis on self-knowledge and meaningful dialogue, that has little tolerance for pain without feeling persecuted by it, that promotes an obsession with cosmetic beauty, and does little to teach people to improvise creatively in order to cope with untoward situations, the emotional demands for healing from serious illness are made much more difficult. To promote capacities for dealing creatively with situations beyond personal immediate control, to experience pain and struggle as meaningful aspects of existence, to find hope in a commitment to shape the life one has been given within restraints so that it is expressive and personally significant, and to experience wonder, appreciation, and joy for whatever level of activity one can carry out with altered capacities, are emotional tasks that healthcare professionals must address in promoting authentic healing. Healthcare is about achieving well-being through embodiment. An example I sometimes recall is a Navajo story of a healing ceremony for an old woman with advanced cancer. She was "healed" even though she died three months later. What this meant was that she again discovered her wonder with life, her enjoyment of her grandchildren, her equanimity in the face of pain, and her gratitude for each day of being alive. This is the kind of promoting of the feeling of vitality that is essential to well-being and healing. Healing can be augmented by maximum biological functioning (which is the prime goal of the medical model) but is not at all determined by it. Rather, it is promoted by a care for the emotional attunement of the client with their embodied relation with the world.

Coming full circle from his time immersed in isolated nothingness, towards the end of his rehabilitation Sacks realized that, although he had to perform the acts that would allow him the highest level of health he could now achieve, it was others who were as much a part of his emotional ability to go forward:

> But I could not do this by will-power, or on my own steam, alone. The initiation, the impulse, had to come from without. I had to *do* it, give birth to the New Act, but others were needed to deliver me, and *say*, 'Do it!' They were the permitters, the prescribers, the midwives of the act – and of course, its supporters and encouragers (Sacks, 1984, p. 182).

If Sacks was to make the journey back emotionally – which was the same journey as physical rehabilitation – other people had to become part of his psychic structure. Healthcare is not just from "the outside" (concerned with the objective functioning of organic "parts"), it must also infiltrate the world as it is structured by the client's emotional network of relationships. This network is not separable from the client's embodiment. It is its "other side" – a side that requires medical care to achieve the goal of health (i.e. the most meaningful use of one's embodiment to live a meaningful life).

However, Sacks still had one more barrier to cross. He had "forgotten" how to use his knee properly, even after he found his leg's overall rhythm and got back the sensation that it was *his* leg. He went to see another doctor who asked him what activities he truly loved. When Sacks answered that he used to love to swim, the doctor sent him down to a nearby pool in a taxi. Unknown to Sacks, the lifeguard there was the doctor's conspirator. He shoved Sacks into the water, so he would be tricked into swimming. After that the knee "worked," since it had become reintegrated with the rest of the body through a beloved activity. This reintegration could not be rationally planned and executed. It had to be spontaneous – the hallmark of emotional experience. The final answer to Sacks' healing was to be thrown back into the parts of the world that he loved that would spontaneously solicit the actions his body could perform. This kind of "treatment" would not be right for most patients – it was tailored specifically for Sacks. The point is, however, that such a process takes creative thinking and a focus on the person's unique situation. Medicine has to discover, and then creatively use, what each client cares about – what inspires his/her passion. The world, other people, and our bodies are one circulation of emotional meaning that must be tapped for healing.

Soka University of America
Aliso Viejo, California
U.S.A.

BIBLIOGRAPHY

Leder, D.: 1990, *The Absent Body*, The University of Chicago Press, Chicago.
Mazis, G.: 1993, *Emotion and Embodiment: Fragile Ontology*, Peter Lang Publishing, New York.
Merleau-Ponty, M.: 1968, *The Visible and Invisible*, A. Lingis (trans.), Northwestern University Press, Evanston, Ill.
Merleau-Ponty, M.: 1964, *The Primacy of Perception*, Northwestern University Press, Evanston, Illinois.
Merleau-Ponty, M.: 1962, Phenomenology of Perception, C. Smith (trans.), Routledge and Kegan Paul, London.
Sacks, O.: 1984, *A Leg to Stand On*, Harper and Row, New York.
Toombs, S. K.: 1993, *The Meaning Of Illness: A Phenomenological Account of the Different Perspectives of Physician and Patient*. Kluwer Academic Publishers, Boston.
Zaner, R.: 1981, The Context of Self: A Phenomenological Inquiry Using Medicine as a Clue, Ohio University Press, Athens, Ohio.

BRUCE WILSHIRE

THE BODY, MUSIC, AND HEALING

Music is the sound of life – Carl Nielsen[1]

As true stillness comes upon us, we hear, we hear, and we learn that our whole lives may have the character of finding that anthem which would be native to our own tongue, and which alone can be the true answer to the questioning, the calling, the demand for ultimate reckoning which devolves upon us – Henry Bugbee[2]

Nothing is more ironical than strengths that generate weaknesses. I think particularly of the ability of Western philosophy and science to objectify the world. To so sharply demarcate things and pin them down that they become objects. Ob-jecta – the Latin roots connote things thrown in front of us. So framed and constituted in experience, things transformed into objects "out there" are subjected to scientific observations and experimentations that tease out, or force out, some of their hidden features.

But this strength comes only with a price that is typically hidden. Acts of objectifying the world and turning things into objects become greatly effective within the sphere of scientific and technological control. So effective they become automatic. The result? We tend to lose touch with things in the full amplitude and impact of their immediate sensuous presence; that is, things as they appear to us before they are objectified for the purposes of science, technological control, or for any short ranged gain.

But within things' immediate sensuous presence our species has grown up and taken shape over millions of years of pre-human and human evolution. Indeed, our own immediately lived personal selves tend to be occluded! What we gain in the right hand, we tend to lose from the left.

Probably most who have introduced undergraduate students to philosophy have encountered the first obstacle. Most students come in as "naive realists." "How do we know the world? Well, there are objects out there that stimulate our senses. We open our eyes and see them, and can choose to examine them closely." The students are assuming – without knowing they are assuming – that a prior question has been asked and answered: What makes objectification possible? How have we come to regard things as objects "out there" to be inspected and controlled, rather than as presences that are, perhaps, spirits that control us?

And one's own body, naively-realistically understood? It is another object, just one that is always with us somehow. The price paid for this detachment from one's body is great. We lose touch with ourselves immediately involved in the world. Capitally, we lose touch with the fundamental roles that both art-making and religion-making play in constituting who we are.

On one essential level of analysis, what I immediately sense visually (I don't say perceive) are electro-chemical events in my brain. And what I immediately sense auditorially are also electro-chemical events in my brain. Each of us must learn to construct a world-experienced – a world of things that sound and show themselves – things that can later be treated as objects. A necessary condition for this being possible at all, is that we are human bodies that become mimetically engulfed in other human bodies that have already learned to construct a shared world-experienced.

* * * * *

My topic is the role of sound and music in healing. Any remnant of naive realism fatally poisons what we can learn about this. Since sight is the distancing and detaching sense, par excellence, and since we today tend automatically to objectify the world and to detach ourselves, it is all too easy to privilege the visual and to eclipse, dampen down, or occlude the auditory. This of course would defeat my project.

We look around and find ourselves in a family of snares. The quicksilver movement of detachment is endemic, affecting all that we think before we think that it does. For example: Why do we facilely assume that each of our lives begins with our birth, begins with detachment and separation from the mother? This is a fatal patriarchal prejudice.

Our eyes were closed in the womb, but after a certain point we could hear. The muffled commotions and sounds of the day alternated rhythmically with the relative quietness of the night. But always foggy sounds, basal rhythms and novelties coming from outside the mother's body interweaving silkenly with basal rhythms and sounds from inside her. And both these with sounds from within the unborn infant's own body, the beating of its own heart and the gurglings of its fluids. And all in the darkness! This was – and still is, I think – the primal generative and regenerative matrix for each of us. For instance, Dylan Thomas cites "the close and holy darkness" in his "A Child's Christmas in Wales."

Here is the first clue to why music can heal. The past is not simply left behind. One moment in our lives is not simply severed from those that precede it – or that follow it. The past has left a living, vibrant residuum in our nervous and glandular systems. Artfully formed sounds can excite the residuum's regenerative powers. As R. W. Emerson (1982, p. 85) knew, the whole universe pours back regeneratively through each node of itself – as long as each node can last at all. The whole universe is "circular power returning into itself," "the inexplicable continuity of this web of god." Whatever prompts each life to sweep back through its sources, and to connect fully again with its roots and commingle in its sap, tends to renew it.

The fetal infant in the womb can turn its head only slightly. Sheathed in fluids, its ability to cock its ears and locate a source of a sound stereophonically is greatly limited. Probably things sound all around and through the infant. Apparently it is this phenomenon that is relived by the Native American holy man, shaman, and thinker, Black Elk (Neihardt, 1988). After a great vision experienced during a coma at age nine, Black Elk is finally ready at eighteen to share its power with his struggling and suffering people. A father has asked him to minister to a very sick child. Black Elk prepares himself by going out in the open country and finding what he senses "he will have need of."

It was a clear evening with no wind, and it seemed that everything was listening hard to hear something. While I was looking over there I felt that somebody wanted to talk to me. So I stood up and began to sing the first song of my vision, the one that the two spirits had sung to me.
"Behold! A sacred voice is calling you!
All over the sky a sacred voice is calling!" (Neihardt, 1988, p. 178)

Non-locality in space. "All over the sky a sacred voice is calling!" And it is but "somebody" who wants to talk to him about "something" that is to be heard. Vibrating, pulsing plenipotentiality permeates from all sides the whole situation and the person engulfed in it.

Moreover, the purport of Black Elk's experiencing is not localized in time either:

Then the daybreak star came slowly, very beautiful and still; and all around it there were clouds of baby faces smiling at me, the faces of the people not yet born...and birds were singing somewhere yonder and there were horses nickering and blowing as they do when they are happy, and somewhere deer were whistling and there were bison mooing too. What I could not see of this, I heard (Neihardt, 1988, pp. 186-87.)

Non-locality in time as well as in space. Evoked, I believe – become available again – is the primal sustaining and regenerating experience of floating in the womb. And, to be sure, that experience is not of something detached or isolated! For indeed it is the fetus's experience of floating in the womb of a woman who is at home and sustained every minute in the larger body of her people. And in turn, the people are sustained and nourished within "the great Hoop of the World," in which Wakan Tanka – the Great Mystery – had provided for the Oglala Lakota to brood and raise their children (Neihardt, 1988, p. 43).

This gift made available, Black Elk brings to his people. Neither time nor space is exclusively linear, diced up, segmented. All flows together with everything else in circular power returning into itself. Time, sacred time, is also cyclical. The gift is a kind of eternal present: "energy new as on the first day of creation" is now available to flow in full course through the people.[3]

Having prepared himself with what "he will have need of," Black Elk enters the tepee where the sick boy lies (Neihardt, 1988, pp. 200 ff.). In fact, he lies at the edge of the inner space of the tepee next to its wall and at the northeast point of its circumference. All these directions are immensely significant for these people. For each of the six directions – the four cardinals and the upper and the lower worlds – hold funded within them the nesting of energy and visceral meaning built up immemorially in the nervous systems of these people. The shaman healer prepares a space and time where "the much at once" (William James's phrase), or the eternal present, can pour in from all sides. Black Elk enters the tepee from the south, "the source of heat and life."

He enters with four virginal young women – virginal meaning the pulsing possibility of new life. He then drums. He also smokes the sacred pipe at each station, addressing the Grandfather officiating in that direction and making it sacred. He does this at each station that leads to the boy lying in the northeast sector of the tepee.

The drum "arouses the mind," Black Elk says, "and makes men feel the mystery and power of things." But the sound is not isolated, cut off from other events: there is no sound "in itself." Drumming "arouses the mind" because it feeds into the total arousal of the situation, which feeds back into the drumming, and everything else involved. Circular power returns into itself.

After the smoking and drumming with which Black Elk addresses the South, he turns to the West, where the sun sets each day and dies, as each of us must do some day. Thus quickened is his own near death experience at age nine, which is necessary for him to heal the nearly dying ones now.

He then proceeds to the north "where the white hairs are" – the Grandfather there – and from whence comes "the cold wind of the north that teaches endurance." He then approaches the boy, who smiles wanly at him, and Black Elk feels the energy of the earth and lower world come up through his feet and legs into his mid-section. He is irradiated and emboldened by Mother Earth.

He then drinks from a wooden cup of water with flakes of red-willow bark in it, and gives the cup to one of the virgins to give to the boy to drink. He approaches the boy and stamps four times on the earthen floor. He then puts his mouth on the boy's abdomen and sucks the wind of the north that teaches endurance through his body.

Black Elk arises and addresses the east and the Grandfather there – the east, the source of renewed light and understanding each day. After instructing one of the virgins to help the boy up, he then exits to the south where he first entered, having completed the cycle through the Four Directions and the Six Powers, and not waiting to monitor progress by the boy. The elderly Black Elk told Neihardt that the boy lived until thirty.

* * * * *

Now, we who derive from Europe or the West (quite a different West!) are primed to dismiss all this out of hand. For probably before we know it, we have abstracted ourselves from immediate involvement in the world and have construed things in the world as objects to be studied and controlled. These things include our own bodies. (Kierkegaard called this the lunatic postulate: to abstract and objectify and to forget what we've done). To have taken a third-person attitude toward our own bodies is to find Black Elk's cure incomprehensible. Standing at a distance from bodily selves, we see only physical objects "out there." But we know that there must be – or probably are – minds "in them" too: "unobservable mental realities." Then the poisoned and prejudiced question arises: How could a non-physical idea or mental image of "the north wind that teaches endurance" possibly affect the physical body and cure it? The alleged cure seems laughably dismissible, a fraud.

And we do all this in the names of objectivity, concreteness, and realism. But, to regard the real north wind as what it is minus its impact on a people's aware and involved bodies over the generations, that is the abstract, partial, prejudiced account. The actual north wind includes its effects on the nervous, glandular, and muscular systems of the people who have traditionally and habitually experienced it in a certain definite way. A particular experience is impossible without a particular bodily state. And a bodily state that correlates with an experience of "the cold wind of the north that teaches endurance" may very well activate the immunological and regenerative capacities of the body that bring about healing. The point is to find connections between centers in the engaged brain, not between a fantasized mental domain and a physical one.

To think that a real north wind can only be movements of air from the north measurable by a meteorologist's instruments epitomizes the desiccated, atomized, hollowed-out scientism and nihilism of our own times (Wilshire, 2002).

* * * * *

Immediately lived, in first person involvements in actual situations, one's body is one's self. What is needed to grasp how artfully created sounds – music – can heal, is a particularly acute and receptive description of just how the whole engaged body makes and hears music. We need a suitable phenomenology. If we succeed in avoiding hasty and unwitting objectification, we will not need to dismantle dualistic construals and dissolve artificial problems such as: Music is spiritual and the body is physical. How then can the body make and hear music?

Valuable contributions to this end have been made by the philosopher and polymath, Rudolf Steiner (one of his achievements being to inspire Waldorf schools).[4] Well-known phenomenologists have already given us valuable phenomenological descriptions of the engaged body: see for example Maurice Merleau-Ponty's (1962) *The Phenomenology of Perception*. Merleau-Ponty and William James have described how the body must be when it perceives, then remembers, or anticipates. What Steiner gives us specifically is the brilliant beginnings of a phenomenology of music-making.

Briefly, though the sense modalities work as an ensemble, auditory perceiving, for instance, differs essentially from visual: In unaided visual perception, we can only see directly in front of us; the face seeing must be facing the thing seen. Whereas in dominantly auditory perception, we hear what's happening all around us, even in us (psychologists point out that deaf persons tend more to paranoia than do the blind for they cannot hear and detect who may be approaching from behind, nor can they hear the reassuring sounds of their own inner bodies). Music's vibrations animate the body with a peculiar force, directness, and pervasiveness. Particularly so with electronic rock music: communication is tympanic – walls of blasting air embrace and shake us. Our bodily existence is confirmed in an especially vivid and compelling way.

A few more words on salient and general features disclosed in phenomenologies of the body; these can help us mark Steiner's distinctive phenomenology of the body as music maker and hearer. To perceive the present and actual situation through any of the sense modalities that may be dominant at the moment requires the body to be fully – or nearly fully – engaged in the situation (even sounds or smells that just happen to come to us present themselves as encroaching or attractive, as exacting some regard for the present and actual situation). Whereas in remembering, the body cannot be involved totally and focally with the present and actual situation. It must be at least partially disengaged from it. Likewise it must be partially disengaged when anticipating – or when fantasizing – about what might happen. The difference is that it is "turned forward" in time in the latter cases, and "backward" in the case of remembering.

Now for Steiner's distinctive contributions, and how they help us understand the healing properties of music. I concentrate on his "Eurythmy as Visible Singing", lectures given just before his death in 1925 ((Steiner 1996/1927). Right off the bat, in his personal notes included in this volume, he hits hard. He contrasts speech sound with musical sound: speech sound is created on the periphery of the bodily self, whereas musical sound is created in its center (Steiner, 1996, p. 105). Then, "The perfect octave gives the human being his very self out of the world" (Steiner, 1996, p.107). Notable is the immediacy and totality of involvement that Steiner's descriptions convey, their ring of truth. "Yes, that's the way it really feels – in that octave I come home to myself." Likewise again: "The keynote is always inner unrest – stepping – the fifth is always calming (and) forming (and) fulfilling" (Steiner, 1996. p. 107). Heard again is the ring of truth for immediate, first-person, actual experiencing – the instantaneous and self-forgetful moment of abstraction has been blocked from happening.

Equally arresting are his phenomenologies of the body that is making-hearing or just hearing (incipiently making) music in major keys in contrast to minor. [Sketches of the different bodily attitudes are provided (Steiner, 1996, pp. 4-5)]. The major key is a distinctive instance of active engagement in the world. And because we speak of music, an art, and not just everyday life, the engagement can equally well be in the past or future world as in the present and actual one.

Or, the world can be felt to be nearly timeless, an eternal present. In the major key the person is depicted striding forward with the right foot and holding out nearly levelly the right arm. Steiner says, "The human being is (recreated) outside himself, he steps forward – movement directed outwards"; whereas in minor keys, "the human being is concealed within himself," and the right foot withdraws inward toward the left, in (relative) disengagement from the world.

The implications for healing are not far to seek. As is the case with Steiner, and as is also typical in indigenous or primal ways of regarding sickness, disease is dis-ease, lack of balance and flow. The treatment of disease aims at restoring this. So a person fixated in melancholy and self-doubt may be given exercises and/or dances in the major musical key. And, logically enough, persons fixated in impulsivity and willfulness – fragmented, wayward – may be given work to do in the minor. These latter persons may learn in time to trust their own inwardness, and to achieve an abiding, quiet self-regard. If they learn this, they will not gain merely cerebral knowledge, but foundational knowing involving their whole bodies, through and through. It will be "experiential" learning.

This is a beginning. The implications for healing are practically limitless because balance can be destroyed – and possibly restored – in probably countless ways. In famous passages in Ecclesiastes, Solomon details how there is a time for this and a time for that, and a time for this other thing and that other thing, and, again, this still other thing and that still other thing. Every activity has its season. Lack of balance may be a lack of coordination with the cycling seasons of ever-regenerating Nature in which we were formed over countless millennia. There is spring, summer, fall, and winter music, and in vital cultures each is played in synchrony with the seasons. Even today with the jumbling of times and places, we can care for the seasons, and be cared for by them.

Or lack of balance may take the form of simple lack of variety. Rats, for example, find boredom excruciating, and will run over an electrically charged grid: the pain breaks the monotony. The symphonies of Carl Nielsen often juxtapose abruptly the major and minor keys; they shake us out of deadening habit and monotony. For best therapeutic effect, the symphonies could be danced or mimed.

* * * * *

Above all, it is music's moving presence that ministers to whatever ails the listener or player. Music that either complements one's mood, or which might conceivably counteract a negative state, is a welcome gift. One is not alone. For he or she is supported and understood at the heart of themselves. Many today have no diagnosable illness. But their lives are emotionally flat, dull, and empty. They drift in a fog of loneliness. Lacking a vivid perceiving, sensing, and urging life, they often seek to substitute passing sensations – riveting only temporarily – for missing enlivening and reliable emotions and moods. They may be addicted to sensations that must ever be repeated on pain of increased emptiness and withdrawal agonies (Wilshire, 1998).

Centering himself in our reality as bodily selves – and similarly to Black Elk – Steiner rediscovers the several meanings and motivations in fast/slow, syncopated/steady, up/down, and in right/left, forward/backward. Up, for instance, he calls "spiritual." I would prefer "fiery" and "spirited." Or, we are drawn as powerfully to what he calls "the physical."[4] I prefer Black Elk's terms: "the earth and lower world," that presence that is stable, supporting, equilibrating – steadily energizing. Mother Earth.

A perpetuum mobile, for example, is music that summons and gathers all the energies that surround and potentially permeate us in any of our actions. Whether occurring in Schumann, Beethoven, Bruckner, or in an acid-rock repeating, pounding cadence, the effect is always the same: power and empowerment. It is momentous emotion – emotion as momentum – that can drive us through the shocks and disappointments, the black holes, of time. It is universal therapy. The perpetuum mobile taps the energies of the ever-regenerating universe that formed our forebears and that can support us now. Circular power that returns into itself in majestic periods: daily, monthly, yearly, generational.

Of more than passing note is what is becoming known as neurotheology. Around 1997 brain scientists found that repetitive rhythms, incantatory hymns, and the rapturous dancing of Sufi mystics excite the hypothalamus. Participants feel as if they are channeling the energies of the universe. Using ever-improving brain scans (PET and SPECT), scientists have recently studied the brains of people in deep states of religious meditation and prayer. Studying these people's brains, scientists find that "the orientation area" of the brain is suppressed. That is, the experience of where the "self stops" and the "rest of the world begins" is greatly blurred. So persons tend to feel that they have "merged with God" (Newberg, et al., 2001; Newberg and D'Aquili, 1999). This prompts some scientists to conclude that the way the brain is wired genetically may explain the origin and power of religious beliefs.

The scientific discoveries are significant. But are the inferences some scientists draw from them valid? It's all well and good to talk about genetic influence on the formation of our brains and on our experiencing. But the genetic origin of all this? Do the genes drop down from heaven? The genes themselves have evolved over millions of years of pre-human and human evolution. Our forebears evolved as whole human bodies living a rigorous hunting-gathering life in wilderness environments. The whole inter-circulating energic universe – God? – is the origin and sustainer of our brains and of our experiencing! And properly describing this experiencing tends to lead us back to our roots, toward our true origin.

I believe nothing better epitomizes how the strengths of science generate, insidiously and ironically, the great weaknesses and limitations of our times. So easily are we thrown out of touch with our sensuous depths and our substantive, spiritual selves. In the guise of expansion of spirit and of objective knowledge, we get reduction of self and spirit. Tracing something to its easily imagined proximate cause or "origin," its "material basis," reduces its immediately felt significance and its power in our lives.

Of course, this reduction occurs also to all that we create, for example, our music. There may well be a large genetic component in our music-making. But thinking that genetics alone reveals the origin of it radically limits its power to reveal our place and nature in the whole universe. It is this universe that has poured itself through us and our kind, and that has generated the fullness of our experience.

* * * * *

There never has been a vital culture without music. Music guarded, cherished, licensed for performance at inviolable times. These are sacred cusps of transition, times triggering individual and group growth. It is understandable that certain analytic thinkers who labor under the influence of abstraction, detachment, and mind/matter dualism will never understand the cogency and power of music. And essential to their benightedness is that they think they are being "scientific." Steven Pinker, for instance, in *How the Mind Works* (1997, pp. 534-37), thinks of the study of mind as "reverse engineering." We see what proves adaptive for organisms in environments, and then try to discover what has produced it. But he finds music unamenable to his procedures, anomalous.

This must mean that his methods are gravely limited and incomplete. For music is perhaps the most adaptive thing we do. It can sustain us through grief, arouse and embolden us when depressed or fearful, and quiet and center us when flustered or desperate. We are body-selves. Lacking music we would not be left as pure, transcendent intellects, but as less than human.

Let the penultimate words be Emerson's in "Poetry and Imagination" (1990):

> We are lovers of rhyme and return, period and musical reflection. The babe is lulled to sleep by the nurse's song. Sailors can work better for their yo-heave-o....Metre begins with pulse beat, and the length of lines in songs and poems is determined by the inhalation and exhalation of the lungs. If you hum or whistle the rhythm of common English metres...you can easily believe these metres to be organic, derived from the human pulse, and to be therefore not proper to one nation, but to mankind. I think you will be set on searching for the words that can rightly fill these vacant beats.

Such music is the basal order of meaning: it cannot be paraphrased, translated, or put into any other terms. The sailors' yo-heave-o is more sounds than words, and no mere expression, pressing out, of an internal "mental state." Nor is it the mere expression of the genes. Rather it is a releasing of sighs and sobs and groans into the ripping and rumbling waves against which the sailors bend their oars. When this release and the water's response take the form of repeatable articulation, we have art's truth, which is intimate with ritual (Wilshire, 1998).

Rutgers, The State University of New Jersey
New Brunswick, New Jersey
U.S.A.

NOTES

[1] Nielsen's words are quoted in Robert Simpson's extensive notes to *Carl Nielsen: The Complete Symphonies* (LP's), Unicorn Records, RHS 324-330. Simpson adduces recorded examples of abrupt key juxtapositions.

[2] Bugbee's words are taken from his *The Inward Morning: A Philosophical Exploration in Journal Form*, Univ. of Georgia Press, Athens, Georgia, 1999 [1958], as quoted in my *The Primal Roots of American Philosophy: Pragmatism, Phenomenology, Native American Thought*, Penn State University Press, Univ. Park, 2000.

[3] Hocking, W.E., *The Self: Its Body and Freedom*, Yale Univ. Press, New Haven, 1928. Quoted as epigraph in my *The Primal Roots of American Philosophy: Pragmatism, Phenomenology, Native American Thought*, Penn State Univ. Press, Univ. Park, 2000.

[4] I think that Steiner slips far too often into versions of psycho/physical dualism. He speaks of the "astral body" and of the "spiritual" in a way that tends inevitably, given our Western history, to oppose them to the actual body each of us is. I delete all this in the attempt to unblock the full force of his fine phenomenological descriptions of the body in action, in music-making-hearing. I can imagine a further critique aimed at Steiner's "ethnocentrism": "Major and minor keys are not absolutes and do not occur in every culture." I grant the point but doubt its force. We are – in our bodies – enculturated creatures. That western music does not speak equally to all peoples does not invalidate its value for westerners. Finally, I am aware that there is a whole professionalized field of music therapy that I do not address in this paper. It should not be inferred from this that I find no value in such efforts.

BIBLIOGRAPHY

Emerson, R.W.: 1982, 'The American Scholar', in L. Ziff (ed.), *Ralph Waldo Emerson: Selected Essays*, Penguin Books, New York.

Emerson, R. W.: 1990, 'Poetry and Imagination', in R. Poirier (ed.), *Ralph Waldo Emerson*, New York, NY.

Hocking, W.E.: 1928, *The Self: Its Body and Freedom*, Yale Univ. Press, New Haven, CT.

Neihardt, J.: 1988/1932, *Black Elk Speaks*, Univ. of Nebraska Press, Lincoln, NE.

Newberg, A., D'Aquili, E. and V. Rause: 2001, *Why God Won't Go Away: Brain Science and the Biology of Belief*, Ballantine Books, New York.

Newberg, A. and E. D'Aquili.: 1999, *The Mystical Mind: Probing the Biology of Religious Experience*, Fortress Press, Chicago.

Merleau-Ponty, M.:1962, *Phenomenology of Perception*, C. Smith (trans.), Routledge and Kegan Paul, London.

Pinker, S.: 1997, *How the Mind Works*, W. W. Norton, New York.

Steiner, R.:1996/1927, *Eurythmy as Visible Singing*, A. Stott (ed.), Third English Edition, The Anderida Music Trust, Stourbridge, England. Consult this edition to learn of all the hands that have brought the book into being through many stages.

Wilshire, B.: 2002, *Nihilistic Consequences of Analytic Philosophy*, SUNY Press, Albany, New York.

Wilshire, B.: 2000, *The Primal Roots of American Philosophy: Pragmatism, Phenomenololgy, Native American Thought*, Pennsylvania State University Press, University Park, PA.

Wilshire, B.: 1998, *Wild Hunger: The Primal Roots of Modern Addiction*, Rowman and Littlefield, Lanham, MD.

SECTION THREE

LIVED EXPERIENCE

ARTHUR W. FRANK

EXPERIENCING ILLNESS THROUGH STORYTELLING

My initial experience of illness was a series of disconnected shocks, and my first instinct was to try to bring it under control by turning it into a narrative. Always in emergencies we invent narratives. We describe what is happening, as if to confine the catastrophe. When people heard that I was ill, they inundated me with stories of their own illnesses, as well as the cases of friends. Storytelling seems to be a natural reaction to illness (Broyard, 1992, pp. 19-20).

Filmmaker, Gerry Rogers is a natural storyteller. When she was diagnosed with invasive infiltrating ductal cancer in her left breast, it was only natural that she make a film about it. The resulting documentary, *My Left Breast*, was premiered at the Newfoundland Film Festival in October, and was aired on CBC Newsworld in the same month.

"When you find out you have breast cancer, you feel like you no longer have control over your life," she says. "Making the film has helped me feel like I had a little bit of control, and it's helped me be a little less afraid. The film isn't an educational video about breast cancer or the latest treatment, it's about what I've been through, and it's full of ambiguity" (From "in the pink", Winter 2000, official newsletter of the Canadian Breast Cancer Foundation).

These two statements span almost a decade, concern two very different illnesses (Broyard was dying of prostate cancer), and express two different sensibilities who create in different media. Yet their similarities override these differences. Each presents illness as a loss of "control" and storytelling as a process of regaining a new sense of control. Broyard's statement that "Storytelling seems to be a natural reaction to illness" is echoed in the editorial description of Gerry Rogers as "a natural storyteller" for whom it was "only natural" to make a film about her experience.

Phenomenologists can take a variety of interests in first-person narratives of illness. Merleau-Ponty's philosophy might direct attention to how intentionality is shaped in and through the body; Ricoeur's work opens up dimensions of temporality and narrative construction; Heidegger could lead us to consider not only illness as what is to be disclosed but, more significant to narrative, being ill as a mode of disclosure; Heidegger also directs attention to death as a horizon of finitude; Levinas raises issues of suffering and mutual obligation, refocusing Heidegger's theme of care. This list of authors and interests could certainly be expanded.

Recognizing the finitude of this essay, I focus on the repetition of the word "natural" in the quotations above and will use the work of Alfred Schutz to pose the question of what is "natural" in the illness narrative. Those who turn their illnesses into stories take for granted much that is "natural" both about illness and about narrative form, yet they also question what society takes for granted about illness, and they question what other ill people take for granted about themselves and their lives. This interplay of taking for granted and questioning deconstructs Schutz's concept of the natural attitude, even as illness narratives are written both within and against this attitude.

I present this essay as a work of applied phenomenology. An articulate benchmark of what might be "pure" phenomenology can be taken from Richard Zaner's early work:

> Still, the 'pure' science of consciousness which Husserl sees as urgently demanded by the very character of philosophy, not to say by the current crisis, *does not purport to be empirical or autobiographical, but eidetic*. What is at issue are those features by virtue of what any possible process of consciousness is what it is, or without which it would not be that which it is. It seeks to ascertain the *generic nature of any possible* mental process, and to delineate the generic types, interconnections, and foundations. In view of that, empirical evidence concerning the *de facto* cases is neither relevant nor sought for (Zaner, 1970, p. 132; original emphases, here and in quotations below).

My choice to quote Zaner is strategic since his recent work (e.g., 1994, 2001) exemplifies an applied phenomenology with an empirical basis and an inherently "natural" interest: alleviating the suffering of the ill by assisting them in moral deliberations that reflect their past autobiographies and form their futures. "*De facto* cases" are not only sought for, such cases seek out the applied phenomenologist. Beyond being the inescapable horizon of thought, "cases" present a Levinasian call to respond (see Zaner, 1993). "How should I respond?" is the constant question of the applied phenomenologist, and this essay moves toward that question.

Yet if applied phenomenology is empirical and autobiographical, a Husserlian core of concern remains. The applied phenomenologist is no less committed to disclosing "those features by virtue of what any possible process of consciousness is what it is, or without which it would not be that which it is." If applied phenomenology defers the search for the "*generic nature of any possible* mental process," what still counts is how a particular form of consciousness is what it is. Moreover, what counts is how the world experienced by that consciousness is what it is as a result of the activity of conscious experiencing.

My applied phenomenological concern is neither to summarize what certain narratives disclose about illness as lived experience, nor to elaborate the variety of narrative forms of these disclosures (see Couser, 1997; Frank, 1995; Hawkins, 1993; Mattingly and Garro, 2000). Rather I seek to analyze how such narrative disclosures originate in an attitude of consciousness that I can best call *telling-illness*, comprising both speaking and hearing as reciprocal aspects of shared telling. Telling-illness is more than a consciousness of being ill. It is rather a consciousness – reflected in the quotations above from Broyard and Rogers – of living illness as a story in which one is both narrator and narrated, organizing consciousness and consciousness that is organized. The consciousness of telling-illness is the recursive elaboration of this loop of narrating and being narrated.

I. TELLING-ILLNESS AS FINITE PROVINCE OF MEANING

Schutz's 1955 essay "Symbol, Reality, and Society" begins with William James's argument "that there are several, probably an infinite number of orders of realities, each with its special and separate style of existence" (Schutz, 1971, p. 340). Schutz marks his shift from James's usage of "subuniverses" to his own preferred "finite provinces of meaning," to emphasize "that it is the meaning of our experiences, and not the ontological structure of the objects, which constitutes reality" (1971, p. 341). Yet Schutz, as a sociologist as well as a philosopher, is acutely aware that "the meaning of our experiences" is not something we invent individually. The experiencing subject is always already an intersubjective subject and a member of a society. As a member of a society the subject creates meaning using resources of common-sense knowledge and taken-for-granted assumptions; she or he draws on a social stock of knowledge and thinks within certain horizons of relevance.

The experience of the social world begins, on Schutz's (1971, p. 341) account, with what he calls "the paramount reality" which comprises "the reality of our everyday life which our common-sense thinking takes for granted" including "physical objects, facts, and events." The capacity of consciousness to take common sense for granted seems, however, short lived. Schutz describes how consciousness is forced to depart from the paramount reality and enter finite provinces of meaning. His account provides a commentary on Broyard's description, quoted above, of his transition from experiencing illness "as a series of disconnected shocks" to his increasing ability to turn these shocks into a coherent narrative.

The world of everyday life is taken for granted by our common-sense thinking and thus receives the accent of reality as long as our practical experiences prove the unity and congruity of this world as valid. Even more, this reality seems to us to be the natural one, and we are not ready to abandon our attitude toward it without having experienced a specific shock which compels us to break through the limits of these 'finite' provinces of meaning and to shift the accent of reality to another one (Schutz, 1971, pp. 343-344; cf. 231).

Schutz elaborates several of these provinces of meaning, including worlds of play, art, dreams, and scientific contemplation. His idea of "specific shock" may describe entry into the world of illness experience better than it fits some of these other worlds.

Even a finite province of meaning remains, however, socially recognized and includes socially recognizable features: a "province of meaning – the paramount world of real objects and events into which we can gear our actions" (Schutz, 1971, p. 341). Each province of meaning is, for consciousness within it, still "paramount." In provinces of meaning the experiencing subject finds "real objects and events" and can rely on socially available knowledge of how to gear action to these features. Yet each province is "finite" since within it any individual's experience "has its particular cognitive style" (Schutz, 1971, p. 341), presumably developed to deal with the specific shock that shifted the accent of reality and forced abandonment of unquestioned reliance on common sense. Recognizing what Schutz (1971, p. 342) adds almost immediately, I would amend his phrase to read that each province of meaning has its own *embodied*-cognitive style, since cognition is an embodied process even as "the body" is constantly being apprehended cognitively.

The experiencing subject must develop a particular cognitive-embodied style appropriate to her or his unique experience in the province of meaning. This particular style consists, "among other things," of four characteristic features:

[1] a specific tension of consciousness (from full awakeness in the reality of everyday life to sleep in the world of dreams), [2] by a specific time-perspective, [3] by a specific form of experiencing oneself, and, finally, [4] by a specific form of sociality (Schutz, 1971, p. 341; numbers inserted).

These four features offer a structure to describe telling-illness as a province of meaning, and I will apply this structure below.

Illness as storytelling – the attitude of telling-illness – begins when the taken for grantedness of what is intersubjectively shared within paramount reality no longer applies. In paramount reality, "we can communicate with our fellow-men [sic, here and below] and thus establish a 'common comprehensive environment' in the sense of Husserl" (Schutz, 1971, p. 342), or what Schutz (1971, p. 12, emphases omitted) calls "the general thesis of reciprocal perspectives." Within finite provinces of meaning, however, the reciprocity of perspectives breaks down. Some aspects of experience "cannot be intersubjectively shared" (Schutz, 1971, p. 342). Schutz's examples of what cannot be intersubjectively shared include dreams, daydreams, and "the lonely vision of the mystic." In these examples, both the quality of self-absorption (e.g., being asleep while dreaming) and the content of the experience work against intersubjective sharing. Dreamers do write about their dreams, and mystics about their visions, but only after returning to paramount reality. And then they write with a consciousness of the gap between the experience itself and its narrative recollection.

Illness seems to be a province of meaning that can be shared while it is being experienced, but such sharing can never be taken for granted. As in other finite provinces of meaning the subject is aware of being unable to assume a reciprocity of perspectives with others. However much of one's own perspective can be communicated, what counts is that the other's understanding does not, literally, go without saying.[1] The development of some degree of reciprocity of perspectives about the experience of the finite province of meaning must be a self-conscious and limited achievement.

Thus to summarize, a "specific shock" dislodges consciousness from paramount reality and shifts "the accent of reality" to a finite province of meaning. My opening quotations from Broyard and Rogers express their moments of recognition that the natural attitude of taken-for-grantedness no longer applies; in these moments consciousness experiences what can be called a *narrative imperative*. Consciousness must tell its experience, both to itself as a means of reestablishing coherence in a milieu where routine expectations no longer hold, and to others as a means of seeking new terms of intersubjectivity.

Finite provinces of meaning are, on the account I have tried to develop, domains that require and privilege storytelling. A crucial part of being shocked out of the natural attitude of paramount reality is the recognition that this newly discovered (if not personally invented) province of meaning exists only insofar as its members *perpetually recreate it*, and storytelling – from literary stories to physical enactments (Frank, 1997a) – is the medium of that recreation. Finite provinces of meaning are not private worlds; intersubjectivity remains not only possible but imperative, but the reciprocity of perspectives cannot be taken for granted. Ethnomethodology (Garfinkel, 1967) may have demonstrated that the intersubjective is always an on-going achievement, but in the finite province of meaning this achievement is atypically self-conscious; people know they have to work to achieve intersubjectivity.

Being told, as Broyard was, that you have prostate cancer, or as Rogers was, that you have invasive, infiltrating ductal cancer is the "specific shock" that initiates entry into illness as a finite province of meaning. If paramount reality has its natural attitude, this finite province of meaning has a characteristic attitude that I am calling illness-telling. This attitude can now be described, returning to Schutz's four characteristics of a finite province of meaning.

First, telling-illness is an attitude of self-conscious awareness of the tension between its accent of reality and the reality of the attitude in which health is taken for granted (Leder, 1990). This tension is expressed beautifully in Nancy Mairs's (1996) title, *Waist-High in the World: A Life Among the Nondisabled*; Kay Toombs (1995) offers of phenomenology of experiencing the world from a wheelchair. The late Robert Murphy, anthropologist and quadriplegic, reads Merleau-Ponty's discussion of phantom-limb syndrome through the lens of his own experience to conclude that: "The amputee is missing more than a limb: He is also missing his conceptual links to the world, an anchor of his very existence" (Murphy, 1987, p.99). This unmoored condition is not a function of the disability itself but of being disabled in "the same society that sustains the body beyond its normal limits [but] shies away from the results" (Murphy, 1987, p. 138).

Second, this tension generates the time-perspective of illness in which at least three temporalities conflict: the "natural" time of the healthy and their projects; the "treatment time" of organized medicine and hospitals (Zerubavel, 1979); and the temporality of being ill with its delays, waits, and changed horizons of expectation (Charmaz, 1991), including awareness of the imminence of death (Elias, 1985).

Third, the experience of oneself takes place in tension between the terms of valuing and being valued that are commonsensical within the world of health and new terms of value that the ill must discover as applicable to themselves. Any illness narrative provides testimony that being diseased or disabled is a condition of embodiment placing the person outside the horizons of value that are natural when the body's health can be taken for granted. Diseased or disabled embodiment precipitates a subjective isolation (see Murphy, 1987; Robillard, 1999) that can be opened by telling-illness but never resolved by that telling.

Telling-illness is Schutz's fourth characteristic, "a specific form of sociality." This form of sociality is distinguished by communication that cannot take for granted the reciprocity of perspectives. Like the dreamer or the mystic, the ill person is caught in the paradox of being moved to communicate precisely because what has been experienced can never be fully communicated. Storytelling becomes imperative when the sharing of the story can no longer be taken for granted, and the story confronts this limit even as it seeks to transcend it.

These four characteristics (tension, time, experience of self, and mode of sociality) are obviously interdependent, but my particular concern below will be how the experience of oneself is created dialogically in the "sociality" of storytelling.

II. STORYTELLING: CREATING COMMUNITIES OF MEANING

Anatole Broyard and Gerry Rogers, with whom this essay begins, do not communicate within the natural attitude, although any communication depends on sharing what Schutz calls "real objects and events into which we can gear our actions" and our understandings. My argument above is that ill people like Broyard and Rogers tell stories in response to being shocked out of the taken-for-grantedness of the natural attitude. This shock may first happen at a particular moment – for example, the experience of some symptom or being told a diagnosis – but shock becomes the *continuing* and perpetually elaborating consciousness of one's difference; this distinction is particularly important in chronic illness and disability.

How on-going shock sets consciousness outside the natural attitude is evoked in one of the earliest of the contemporary genre of literary illness narratives, written by the political journalist, Stewart Alsop, who died of leukemia soon after completing his memoir. Near the end of his book Alsop reflects on his physician's assertion that he has been living "a normal life."

> But it has not been altogether normal. It is not normal to wake up every night just before dawn, with a fever of 101 or so, take a couple of pills, and settle down to sweat like a hog for four or five hours. It is not normal to feel so weak you can't play tennis or go trout fishing. And it is not normal either to feel a sort of creeping weariness and a sense of being terribly dependent, like a vampire, on the blood of others [matched relatives who provided platelets]. After eight weeks of this kind of "normal" life, the thought of death loses some of its terror (Alsop, 1973, p. 289).

The taken-for-grantedness of health includes sleeping until the alarm goes off, being able to play tennis and trout fish, and going through life without a sense of "creeping weariness" and vampire-like dependence. In the natural attitude of health people relate to each other on the assumption of their mutual capacities to perform these activities; such capacities define their horizons of "normal" (Robillard, 1999, pp. 63, 72). Alsop's story calls the reader to leave this natural attitude and engage an embodied reality in which "normal" is a possibility for others but not for one's self. Like every illness story, his is about the tension between his embodiment and the embodiment that is taken for granted as "normal."

The climax of Alsop's story concerns his changed attitude toward his imminent death. Death occupies a privileged place in Schutz's understanding of the natural attitude:

> But in a word, we want to state that the whole system of relevances which governs us within the natural attitude is founded upon the basic experience of each of us: I know that I shall die and I fear to die. This basic experience we suggest calling the *fundamental anxiety*. It is the primordial anticipation from which all the others originate. From the fundamental anxiety spring the many

interrelated systems of hopes and fears, of wants and satisfactions, of chances and risks which incite man within the natural attitude to attempt the mastery of the world, to overcome obstacles, to draft projects, and to realize them (Schutz, 1971, p. 228).

The natural attitude is built up around the epoché of this fundamental anxiety; the denial of death (Becker, 1973) is enacted by everyday consciousness bracketing death's eventuality out of consideration. Alsop's daily encounters with death's imminence place him in a wholly different relation to the fundamental anxiety and thus precipitate a different attitude of his consciousness.

Alsop describes one night when he felt so sick he was certain his death was imminent, but he felt none of the panic that he had experienced earlier in his disease. His reflection on why he no longer feels panic exemplifies the particular epoché that grounds telling-illness. He has finally bracketed the natural attitude's bracketing of death:

But the most important reason why I felt no panic fear last Saturday was, I think, the strange, unconscious, indescribable process which I have tried to describe in this book – the process of adjustment whereby one comes to terms with death. A dying man needs to die, as a sleepy man needs to sleep, and there comes a time when it is wrong, as well as useless, to resist (Alsop, 1973, p. 289).

The wit of Alsop's "indescribable process which I have tried to describe" derives from the experience of dying being foremost among Schutz's "finite provinces of meaning which cannot be intersubjectively shared" – if there is consciousness in death itself, that would be the ultimate finite province of meaning that cannot be shared. Alsop testifies that from within his finite province of meaning, the anxiety of death is, if not entirely dissipated, less than fundamental, and therein lies a distinctive quality of writing like his. To write is still to be engaged in a project, and projects require that death be displaced; as projects enact this displacement, death's fundamental anxiety is deferred (Bauman, 1992). The difference of Alsop's writing is not only that he makes thematic the anxiety of death, but that he no longer imagines attempting, as Schutz puts it in the quotation above, "the mastery of the world." This illusion of mastery is the denial of death.

Alsop does not write as a philosopher, but neither does he write within "the natural attitude of man living within the world he accepts, be it reality or mere appearance" (Schutz, 1971, p. 104). Alsop's meditation on his increasingly permeable boundary between life and death is also a consideration of reality and mere appearance. He questions the reality of the natural attitude from which his physician can evaluate the mere appearance of his life as "normal." Deeper still, he evokes how familiar the reality of death becomes to someone who has been living with leukemia – or a number of other chronically critical diseases – for years. His book may be a shock to some readers, allowing them limited entry into its finite province of meaning. Yet the text can only suggest an experience that remains indescribable in its embodied consciousness and consciousness of embodiment.

These passages from Alsop illustrate being ill as a finite province of meaning, but they suggest less about telling-illness. Alsop may disclose much about the province of meaning called illness – in particular his words convey the fatigue of his body and how this embodiment affects his thinking about life and death – but illness remains a province of meaning that he occupies by himself (albeit with a great deal of support from others, including their blood). As a storyteller he graciously pulls back the curtain of his solitary world, but he has limited expectations how far his readers can enter.

The writing of Audre Lorde, whose *The Cancer Journals* (1980) is one of the best-known and most often cited illness narratives, exemplifies stories that self-consciously seek to gather others into their world. Alsop organizes his memoir around the progress of his illness, but in other respects he writes the traditional autobiography of a public person offering the reading public a backstage look at a notable life. Alsop writes about politics; Lorde's writing is political. For Lorde, to understand illness as a finite province of meaning is a political observation. Illness is rendered "finite" by a politics of exclusion; what is "taken for granted" is the isolation – the accomplishment of invisibility – not only of the ill but also of those who share other identities she claims: an African American, a feminist, and a lesbian. Lorde writes in order to transform what has been an excluded province of meaning – a collection of silenced, invisible solitudes – into a community of mutually supportive voices demanding recognition. Thus she begins:

I do not wish my anger and pain and fear about cancer to fossilize into yet another silence, nor to rob me of whatever strength can lie at the core of this experience, openly acknowledged and examined. For other women of all ages, colors, and sexual identities who recognize that imposed silence about any area of our lives is a tool for separation and powerlessness, and for myself, I have tried to voice some of my feelings and thoughts about the travesty of prosthesis, the pain of amputation, the function of cancer in a profit economy, my confrontation with mortality, the strength of women loving, and the power and rewards of self-conscious living (Lorde, 1980, pp. 9-10).

Lorde tells her story to shock readers out of the paramount reality of their racist/sexist/healthist silences (silences legitimated by common-sense typifications), and to provide a rallying point for women who want to recognize – literally to be able to see and greet – other women who share their illness. She seeks to transform a finite province of meaning into a generalized *community* of meaning. Her autobiography is intended to turn her personal pain into a public dialogue over the causes and treatment of cancer. Experience is used to self-conscious political effect.

Lorde's emblem of society's "natural attitude" is the prosthesis that social convention and medicine want her to use to disguise her mastectomy. The prosthesis sustains the mere appearance of the expected female body by disguising the reality of cancer. At two central moments in her story she is coerced to use a prosthesis. The first is a pleasant visit from a Reach for Recovery volunteer who presents the prosthesis as a taken-for-granted, desirable way to conceal the results of surgery (Lorde, 1980, pp. 42-44). In the second scene a nurse tells Lorde that her failure to wear a prosthesis is "bad for the morale of the office." "I could hardly believe my ears," Lorde (1980, p. 59) writes, "I was too outraged to speak then, but this was to be only the first such assault on my right to define and to claim my own body."

Lorde's 1980 counter-assault on society's prosthetic natural attitude has evolved, twenty years later, into a movement represented by the newsletter from which I quoted the comments about Gerry Rogers at the beginning of this essay. The same newsletter reports what might seem to be Lorde's dream come true: in the annual Run for the Cure people across North America take to the streets to proclaim the effect of breast cancer in their lives; in Canada alone, 100,000 participants in 29 communities. On their bibs participants write the names of loved ones and friends who have breast cancer, as well as their own cancers. Yet is this what Lorde imagined? Instead of "an army of one-breasted women descend[ing] on Congress and demand[ing] that the use of carcinogenic, fat-stored hormones in beef-feed be outlawed" (Lorde, 1980, p. 16), the Run for the Cure is depoliticized; mainstream medical research is supported, and the carcinogenic and discriminatory societal practices that outraged Lorde are not questioned.[2] Lorde's (1980, p.66) imagination of "the design and marketing of items for wear for one-breasted women" has lost to her nightmare: support-group newsletters routinely feature (and are financed through revenue from) prosthetics and cosmetics advertisements.

Critics can readily claim that Lorde's radical gesture of writing explicitly about her changed body has become routinized, and that routinization is the inevitable result of seeking political goals through individual storytelling.[3] Each personal account may be authentic in itself, but what is the effect of their aggregation? Personal accounts proliferate: Gerry Rogers's video about her experience is one of two such videos advertised in one newsletter, along with a notice about a performance of *Wit*, Margaret Edson's Pulitzer Prize-winning play about a woman dying of cancer. Newspapers and magazines routinely include stories by people with cancer. An open question is whether these works collectively challenge society's common sense about health, illness, medicine and mortality, or whether their sum publicity institutionalizes storytelling as another normal expectation within a natural attitude of being ill that is not really new but only a more visible and vocal version of old relations and tensions of illness.

Schutz (1971, p. 347) asked, "To what extent are significative and symbolic appresentations dependent upon the sociocultural environment?" The contemporary answer seems to be, more and more so. As I noted above, people discover stories – their own and others – but they do not invent the forms of storytelling. The advantage of making some forms of storytelling more readily available is equally the risk that telling these stories can become routinized as expectations, their power of critique thus being dissipated.

Unquestionably the increasing proliferation of illness stories makes these stories recognizable within the natural attitude; their "specific shock" becomes lost. The significant question is whether the natural attitude is thus changed and illness is typified differently, or whether illness becomes told in terms that impose an existing natural attitude on experience. In most stories both possibilities are realized. If stories of illness suspend some aspects of society's natural attitude, they rely on other aspects. Stories can create a moment of suspension when the paramount reality of the natural attitude loses its capacity to "prove the unity and congruity of this world as valid" (Schutz, 1971, p. 343; quoted above). Stories not only testify to the shock of these moments but can enact this shock; to read many illness stories is to experience the unity and congruity of the world as healthy being suspended. But these stories can only go so far in creating their own world; the limit of telling-illness reaffirms Merleau-Ponty's (1962, p. xiv) observation, "The most important lesson which the reduction teaches us is the impossibility of a complete reduction." The pure reduction or epoché is a vanishing point; just as the natural attitude is suspended, it reasserts itself in consciousness. Just as the natural assumption of health is suspended and the fundamental anxiety of death is seen for what it is, these aspects of the natural attitude reassert themselves: what illness stories tell become new typifications; their narrative forms become new rhetorical expectations.

Stories of illness are told within this dialectic of suspension and reassertion of the common sense, everyday world. They are told because the natural attitude has broken down for the teller, but their telling is built out of other aspects of the natural attitude. Illness stories critique everyday typifications, even as they themselves become typified. The best answer to the risk of these stories becoming routinized seems to be what they can achieve during the process of their routinization.

III. APPLYING PHENOMENOLOGY: "HOW SHOULD I RESPOND?"

What do storytellers of illness want? What does this attitude of telling-illness have as its project? The idea of the story having a project situates it within the natural attitude, even as the grounds of telling the story – the shift in fundamental anxiety toward death – move it outside the natural attitude because the specific content of the story questions what society takes for granted. But stories are told to some purpose. "Thus," Schutz (1971, p. 209) wrote, "it may be correctly said that a pragmatic motive governs our natural attitude toward the world of daily life. World, in this sense, is something that we have to modify by our actions or that modifies our actions."

Stories are actions that modify the world in at least three overlapping, empirically inseparable domains. The first is the intrasubjective reformulation of who the ill person has become in the wake of the disruption of illness (G. Becker, 1997). Lorde (1980, p. 16) provides one of the earliest, most explicit statements of this need: "Any amputation is a physical and psychic reality that must be integrated into a new sense of self." Her writing is the medium of that integration, and at this level of her project her writing still reflects a *finite province* of meaning. Stories of illness work much like other narratives in which identity is established and reaffirmed.[4] The story, in brief, recollects meanings that make the teller's finite province of meaning coherent, if never entirely unified, for him or her.

The second domain is intersubjective, in which the story builds relationships and communities. Here the story proposes a distinctive stock of knowledge that can be shared within what becomes a *relationship* of meaning, and that relationship in turn changes the intrasubjective sense of self. This relationship may initially be finite to the immediate co-presence of the storytelling encounter, but it expands quickly into the third domain, the social. Here stories are told to affect others whom the storyteller does not know immediately. At the social level the interpersonal relationship of meaning makes a claim on the generalized *community*, offering or demanding a shift in its common sense. Lorde proposed to change consciousness about breast cancer and while that has happened (as evidenced, for example, by touring exhibits of art depicting breast cancer experiences and body images), activism has probably had its greatest successes with respect to AIDS and disabilities. From a phenomenological perspective, the fundamental "right" proclaimed by these groups is have the natural attitude shift. Activists seek to change taken-for-granted language to recognize the normality of what has previously been typified as abnormal. Thus "AIDS victims" have become, in frequent if not exclusive common usage, "persons with AIDS" and "the handicapped" have become "the disabled", even as activists press to become the "differently abled."[5]

But how, across these three intersecting levels, is storytelling an action that seeks to modify its world? The loss of the taken-for-granted world – being wrenched out of the natural attitude and facing the fundamental anxiety of death – induces panic, in the mythic sense of unexpectedly encountering the terrifying god who screams in despair. "The patient has to start by treating his illness not as a disaster, an occasion for depression or panic, but as a narrative, a story," Broyard writes (1992, p. 20). For all the multidisciplinary literature – psychotherapeutic and neurological, anthropological and literary – that has attempted to explain storytelling, perhaps storytelling is too primal – too foundational to what it is to be human – to be explained. Perhaps we can only say of storytelling what Schutz, among other phenomenologists, says of intersubjectivity: that it simply *is* as a ground of all else. "We stated before that the world of daily life into which we are born is from the outset an intersubjective world" (Schutz, 1971, p. 218). Storytelling is the medium of this intersubjectivity, and it too is "from the outset." In the story I and the other become "*we*", sharing "*our* vivid present" (Schutz, 1971, p. 219), including its fundamental anxieties, bracketed or not.

This creation of a "we" who share a vivid present is – without developing the full Levinasian (1998) argument – a moral relationship; each is in the storytelling relation *for* the other, in Levinas's sense. Schutz (1971, p. 220) writes, "We grow older together"; illness and disability are, for many people if not all, inevitable aspects of that growing old. Sharing illness stories affirms our continuing intersubjectivity through these crises and confronts, together, the fundamental anxiety of death. Illness stories also disrupt society's natural attitude in which the typification of illness as abnormal is one strategy of silencing and displacing the fundamental anxiety about death.

At its core the illness story claims a "we" relationship that represents a shared commitment to the fullest sensory awareness of each other's suffering, from physical pain to existential distress. If the natural attitude is in many respects closed to the apprehension of suffering, the attitude of telling-illness initiates an openness. This openness to suffering as an inevitable part of growing old together is the feature that makes the consciousness of telling-illness *what it is* and, in Zaner's (1970, p. 132, quoted above) terms, "without which it would not be that which it is."

Stories are the media of what I have called illness as moral occasion (Frank, 1997b). On a day before I wrote this essay I spent a couple of hours at the bedside of a stranger who has cancer. Our talk was not a research interview and the content of his story has no place in this essay. The process of our storytelling, however, is my experience as much as his. It was the same process described by theologian and bioethicist, William May (1991, pp. 4, 131) as "rising to the occasion." Richard Zaner (1993), Barry Hoffmaster (1992), and Daniel Callahan (1993) are some of those who have noted people's intense desire to rise to the occasion of illness. This young man's story looped back and forth between the various crises in his life that illness had set in motion or intensified; it had multiple plot lines and no particular end.

The socio-linguistic narrative was a medium for what I can best describe as *the formation of a moral subject*. At the basis of all his contingent dilemmas were two foundational questions: what do I value in my own life, and how do I and others express to each other what we value about each other? Illness complicated those questions, but it was also – and he realized this – the only reason he was able to pose the questions at all. His talk was a means of coming to the realization that these are the questions any human life revolves around, and he was self-consciously grateful for that realization.

This young man realizes that he has always acted on the basis of tacit responses to these questions of value – that has been his participation in the natural attitude. His illness precipitated the explicit posing of these questions. He now has to *decide* these questions of value; he cannot let valuing simply occur in the course of things. Decisions about how to honor value in one's self and in others are not rational-instrumental decisions. They are decisions that one talks oneself around, into, and eventually through, and that talk is the story. I did not say much at that bedside, and an observer might even have asked why I needed to be there. I was, as I have written elsewhere, just listening (Frank, 1998). Yet just listening can be action enough. If storytelling is, like intersubjectivity, somehow primordial and thus inexplicable, the effect of listening is equally inexplicable as a constituent part of storytelling. The issue is not to seek to explain storytelling and listening but to create conditions for them to occur.

The moral occasion of illness is necessarily a dialogical occasion. This young man's issue was not how he alone would rise to the occasion, but what the intersubjective occasion requires of him in response to others, and as the dependency necessitated by illness made him self-conscious of all he needs from them, what the occasion requires of these others. My listening became part of that dialogical process – a process far more complex than any linear metaphor of concentric circles.

Returning to the quotations with which this essay begins, I affirm Broyard's not entirely common sense: storytelling is a "natural reaction" but not only to illness. Storytelling is a natural reaction to the shock after which the natural attitude no longer unifies and renders coherent the world. It is a natural reaction to consciousness having to deal with its fundamental anxiety no longer being bracketed but becoming thematic. The description of Gerry Rogers is also good phenomenology: humans are "natural storytellers." We have to be because the natural attitude is perpetually being shocked, leaving us to rise to the occasion of forming not only "finite provinces" of meaning for ourselves but expanding communities of meaning. These communities begin with those others who share our shock and reach out to those who ought to realize that they gain nothing by their fear of what we are living. In Lorde's (1980, p. 200) bruising and comforting phrase, "Your silence will not protect you." The fragile illusion of the natural attitude is that its evasion of fundamental anxiety can protect it. The moral occasion of illness is to live in full recognition of that anxiety. Telling-illness is the process and practice of that recognition.

University of Calgary
Calgary, Alberta
Canada

NOTES

[1] Toombs (1993) uses phenomenology to describe the discrepant realities of the patient and the physician; the illusion of their relationship is that the objective body of the patient constitutes a common basis of intersubjectivity. The dilemmas of "doctor-patient communication" stem from lack of recognition that each acts from within a different phenomenological reality. The medical stories of Campo (1997) and Verghese (1994, 1998) are notable as depictions of the recognition of the difference between their realities as physicians and the realities of their patients. Verghese (1998) is an especially powerful evocation of the limits of transcending this otherness.

[2] In the terms of C. Wright Mills' (1959) famous sociological distinction, the Run for the Cure provides a compelling aggregation of personal troubles generated by breast cancer—participation in them is a personally moving experience on multiple levels—but these events do not (yet) seek to make breast cancer a public issue.

[3] Simmel's "tragedy of culture" (Levine, 1971) may be the best description of this process through which the original energy invested in social practices becomes ossified into forms that prescribe such continuing practices as routines.

[4] Since Frank, 1995, the literature on this topic has expanded exponentially; of particular note are Eakin, 1999 and Mishler, 1999, both of which contain useful reviews (especially Eakin on research exploring the neurological basis of storytelling) in addition to their own contributions. Scholarship has no monopoly on the idea of story as a vehicle for self; for a particularly insightful example of many "popular" works see Stone, 1996.

[5] The relevant theoretical argument is Taylor (1994, p. 25): "our identity is partly shaped by recognition or its absence, often by the misrecognition of others, and so a person or a group of people can suffer real damage, real distortion, if the people or society around them mirror back to them a confining or demeaning or contemptible picture of themselves. Nonrecognition or misrecognition can inflict harm, can be a form of oppression, imprisoning someone in a false, distorted, and reduced mode of living." As I have suggested with respect to Lorde, Taylor is

proposing that the natural attitude is political, built of selective recognitions and absences of recognition.

BIBLIOGRAPHY

Alsop, S.: 1973, *Stay of Execution: a Sort of Memoir*, J.B. Lippincott Company, New York.
Bauman, Z.: 1992, *Mortality, Immortality, and Other Life Strategies*, Stanford University Press, Stanford, CA.
Becker, E.: 1973, *The Denial of Death*, Free Press, New York.
Becker, G.: 1997, *Disrupted Lives: How People Create Meaning in a Chaotic World*, University of California Press, Berkeley.
Broyard, A.: 1992, *Intoxicated by My Illness*, Clarkson Potter Publishers, New York.
Callahan, D.: 1993, *The Troubled Dream of Life: In Search of a Peaceful Death*, Simon & Schuster, New York.
Campo, R.: 1997, *The Poetry of Healing: A Doctor's Education in Empathy, Identity, and Desire*, Norton, New York.
Charmaz, K.: 1991, *Good Days, Bad Days: The Self in Chronic Illness and Time*, Rutgers University Press, New Brunswick, N.J.
Couser, G.T.: 1997, *Recovering Bodies: Illness, Disability, and Life Writing*, The University of Wisconsin Press, Madison.
Eakin, J.P.: 1999, *How Our Lives Become Stories: Making Selves*, Cornell University Press, Ithaca, N.Y.
Elias, N.: 1985, *The Loneliness of the Dying*, E. Jephcott (trans.), Blackwell, Oxford.
Frank, A.W.: 1998, 'Just listening: Narrative and deep illness', *Families, Systems & Health* 16, 197-212.
Frank, A.W.: 1997a, 'Enacting illness stories: When, what, why', in H.Lindemann Nelson (ed.), *Stories and Their Limits: Narrative Approaches to Bioethics*, Routledge, New York, pp. 31-49.
Frank, A.W.: 1997b, 'Illness as moral occasion: Restoring agency to ill people', *Health* 1, 131-148.
Frank, A.W.: 1995, *The Wounded Storyteller: Body, Illness, and Ethics*, University of Chicago Press, Chicago.
Garfinkel, H.: 1967, *Studies in Ethnomethodology*, Prentice Hall, Englewood Cliffs, N.J.
Hawkins, A.H.: 1993, *Reconstructing Illness: Studies in Pathography*, Purdue University Press, West Lafayette, Indiana.
Hoffmaster, B.: 1992, 'Can ethnography save the life of medical ethics?' *Social Science and Medicine* 35, 1421-31.
Leder, D.: 1990: *The Absent Body*, University of Chicago Press, Chicago.
Levinas, E.: 1998, *Entre nous: Thinking-of-the-Other*, M.B. Smith and B. Harshav (trans.), Columbia University Press, New York.
Levine, D.N. (ed.): 1971, *Georg Simmel: On Individuality and Social Forms*, University of Chicago Press, Chicago.
Lorde, A.: 1980, *The Cancer Journals*, spinsters/aunt lute, San Francisco.
Mairs, N.: 1996, *Waist-High in the World: A Life Among the Nondisabled*, Beacon Press, Boston.
Mattingly, C. and Garro, L.C. (eds.): 2000, *Narrative and the Cultural Construction of Illness and Healing*, University of California Press, Berkeley.
May, W.F.: 1991, *The Patient's Ordeal*, Indiana University Press, Bloomington.

Merleau-Ponty, M.: 1962, *Phenomenology of Perception*, C. Smith (trans.), Routledge and Kegan Paul, New York.
Mills, C.W.: 1959, *The Sociological Imagination*, Oxford University Press, New York.
Mishler, E.: 1999, *Storylines: Craftartists' Narratives of Identity*, Harvard University Press, Cambridge, MA.
Murphy, R.F.: 1987, *The Body Silent*, Henry Holt, New York.
Robillard, A.B.: 1999, *Meaning of a Disability: The Lived Experience of Paralysis*, Temple University Press, Philadelphia.
Schutz, A.: 1971, 'Symbol, reality and society', in M. Natanson (ed. and intro.), *Alfred Schutz: Collected Papers I*, Martinus Nijoff, The Hague, pp. 287-356.
Stone, R.: 1996, *The Healing Art of Storytelling: A Sacred Journey of Personal Discovery*, Hyperion, New York.
Taylor, C.: 1994, 'The politics of recognition', in A. Gutman (ed.), *Multiculturalism*, Princeton University Press, Princeton, N.J., pp. 25-73.
Toombs, S.K.: 1993, *The Meaning of Illness: A Phenomenological Account of the Different Perspectives of Physician and Patient*, Kluwer Academic Publishers, Dordrecht.
Toombs, S.K.: 1995, 'Sufficient unto the day: A life with multiple sclerosis', in S.K. Toombs, D. Barnard, and R. Carson (eds.), *Chronic Illness: From Experience to Policy*, Indiana University Press, Bloomington.
Verghese, A.: 1994, *My Own Country: A Doctor's Story of a Town and Its People in the Age of AIDS*, Simon & Schuster, New York.
Verghese, A.: 1998, *The Tennis Partner: A Doctor's Story of Friendship and Loss*, Harper Collins, New York.
Zaner, R.: 2001, 'Thinking about medicine,' in this volume.
Zaner, R.: 1970, *The Way of Phenomenology: Criticism as a Philosophical Discipline*, Pegasus, New York.
Zaner, R.: 1993, *Troubled Voices: Stories of Ethics and Illness*, Pilgrim Press, Cleveland.
Zaner, R.: 1994, 'Experience and moral life: A phenomenological approach to bioethics', in E.R. DuBose, R. Hamel, and L.J. O'Connell (eds.), *A Matter of Principles?Ferment in U.S. Bioethics*, Trinity Press International, Valley Forge, PA., pp. 211-39.
Zerubavel, E.: 1979, *Patterns of Time in Hospital Life*, University of Chicago Press, Chicago.

S. KAY TOOMBS

REFLECTIONS ON BODILY CHANGE: THE LIVED EXPERIENCE OF DISABILITY[1]

My interest in the phenomenology of illness and disability has grown out of my own experience as a person living with multiple sclerosis, an incurable, progressively disabling disease of the central nervous system. Over the past twenty-eight years (since the age of 30) my physical capacities have altered in a startling number of ways. At one time or another my illness has affected my ability to see, to feel, to move, to hear, to stand up, to sit up, to walk, to control my bowels and my bladder, and to maintain my balance. Some abilities, such as sensing the position of a limb, I have lost abruptly and then slowly regained. Some, such as clear vision in one or the other eye, I have lost and regained numerous times. Other physical capacities have disappeared and never returned. I can, for example, no longer walk because I am unable to lift my legs. This latter change has, however, been gradual. For a number of years, although the muscles in my legs gradually weakened, I was able to get around "on my own two feet" using first a cane, then crutches, and finally a walker for support. Several years ago I was forced to give up the walker and begin full-time use of a wheelchair for mobility.

All these physical changes can, of course, be described in terms of central nervous system dysfunction and explicated with respect to a demyelinating disorder. Indeed, it may even be possible, through the use of sophisticated medical technology, to visualize lesions in the brain to account for specific physical incapacities. Yet, such a mechanistic description (based as it is on a biomedical model of disease) captures little, if anything, of my actual experience of bodily disorder. I do not experience the lesion(s) in my brain. Indeed, I do not even experience my disorder as a matter of abnormal reflexes. Rather, my illness is the impossibility of taking a walk around the block, of climbing the stairs to reach the second floor in my house, or of carrying a cup of coffee from the kitchen to the den.

In this chapter I will suggest that phenomenology provides a powerful means to illuminate the experience of loss of mobility – a bodily dysfunction that is common in neurological and other degenerative diseases. In particular, in rendering explicit the dynamic relation between body and world, the phenomenological notion of lived body provides important insights into the disruption of space and time that are an integral element of physical disability. Furthermore, a phenomenological account of bodily disorder discloses the emotional dimension of physical dysfunction. In providing a window into lived experience, phenomenology gives invaluable information about the everyday world of those who live with disabilities. Such information is of enormous practical significance when devising effective therapies in the clinical setting and in determining how best to address the personal, social and emotional challenges posed by chronic disabling diseases.

In considering the meaning of disability it is helpful to recall the phenomenological notion of lived body (Sartre, 1956; Merleau-Ponty, 1962). As an embodied subject, I do not experience my body primarily as an object among other objects of the world. Rather than being an object for me-as-subject, my body as I live it represents my particular point of view on the world (Merleau-Ponty, 1962, p. 70). I am embodied not in the sense that I have a body – as I have an automobile, a house, or a pet – but in the sense that I exist or live my body (Toombs, 1992). In this respect the lived body is not the objective, physiological body that can be seen by others (or examined by means of various medical technologies) but, rather, the body that is the vehicle for seeing.

Furthermore, the lived body is the basic scheme of orientation, the center of one's system of coordinates. I experience myself as the Here over against which everything else is There. As orientational locus in the world, my body both orients me to the world around by means of my senses and positions the world in accord with my bodily placement and actions (Husserl, 1982, pp. 116-17; Husserl, 1989, pp. 165-66; Schutz, 1962, pp. 222-26).

Additionally, the lived body is the locus of my intentions. I actively engage the world through the medium of my body. I not only find myself within the world, I continually move towards it and organize it in terms of my projects. Objects present themselves as invitations to my body's possible actions. For instance, the piece of chalk presents itself as "an instrument for writing" or "an item to be replaced in the drawer," the desk inside the office is encountered as "a location for working" or, perhaps, as "something to walk around in order to get to the door." The surrounding world is always grasped in terms of a concrete situation. Contained in the action of reaching for the cup is the intention to bring it to one's lips. One reaches for the cup in order to drink from it.

Physical space is thus for my body an oriented space. Points in space do not represent merely objective positions but rather they mark the varying range of my aims and gestures. For example, the narrow passageway through which I must pass represents a "restrictive potentiality" for my body, requiring a modification of my actions. I must perhaps turn sideways in order to make my way through it (Merleau-Ponty, 1962, p. 143). Surrounding space is experienced as functional space – that environment within which I carry out my various projects. From my center outwards the world around me arranges itself in terms of near and far goals.

Motor disorders resulting in loss of mobility engender a profound disruption in the lived body. For instance, such disorders transform the experience of the body as orientational and intentional locus – thus transforming the experience of space. In the normal course of events locomotion opens up space, allowing one freely to change position and move towards objects in the world. Loss of mobility anchors one in the Here, engendering a heightened sense of distance between oneself and surrounding things. A location that was formerly regarded as "near" is now experienced as "far." For example, when I could walk, the distance from my office to the classroom (about thirty yards) was unremarkable – as were the stairs I climbed to reach the third floor of the building. As my mobility decreased the office appeared near to the classroom on the way to the lecture, but far from it on the return journey; the stairs became an obstacle to be avoided, as much as possible, by using the elevator. Today, if I were to be without my wheelchair, the distance from the office to the classroom would appear immense – absolutely beyond my capacity to reach it. And the third floor is unattainable when the elevator malfunctions, leaving me stranded waiting for the repairman. (If an elevator repairman is not readily available, then my inability to climb stairs necessitates rescheduling my classes on the first floor, exchanging rooms with a professor from another department, having to send someone to carry my books and papers down to me from my office.)

Loss of mobility illustrates in a concrete way that the subjective experience of space is intimately related both to one's bodily capacities and to the design of the surrounding world. The answer to the question, "Is it too far to go?" has little to do with the distance that can be measured in feet or yards. For the person with mobility problems the answer depends, in large part, on what is between here and there. Are there obstacles that make it impossible to maneuver with crutches or a cane? Is the terrain suitable for a wheelchair?

Thus, with the loss of various bodily capacities physical space assumes an unusually restrictive character. Sidewalks may be too uneven to walk on, carpets too thick to wheel over, doorways too narrow to navigate with a wheelchair, slopes too steep to climb. When I could walk, a local shopping mall seemed relatively "flat." Now, as I propel myself around it in my wheelchair, it appears positively mountainous! (A cyclist friend gave me an interesting analogy. He said that before he tries a new cycling route he no longer asks a non-cyclist to describe the terrain. Since non-cyclists have usually only navigated the route in cars, they will assure him there are no hills. However, once he rides his bicycle, he discovers long, steep inclines of which the automobile riders were totally unaware.)

The dimensions of high and low also vary according to the position of one's body and the range of possible movements. From a wheelchair the top three shelves in the grocery store are too high to reach since they have been designed for shoppers who are standing up. (Consequently, the wheelchair user must either be accompanied by an upright person or find a store employee to take items off the shelf. Either option is inconvenient and renders one dependent on others for the simplest of tasks.) Similarly, regardless of its dimensions in inches, a curb is as high as a wall if one uses a motorized wheelchair and cannot get off or onto the sidewalk. The floor is infuriatingly distant and impossibly low, if one is unable to bend to retrieve the car keys that have slipped out of one's hand – especially if there is no one around who can pick them up.

With respect to the changed character of physical space, it is important to recognize that those of us who negotiate space in a wheelchair live in a world that is in many respects designed for those who can stand upright. Until recently all of our architecture and every avenue of public access was designed for people with working legs. Hence, people with disabilities (and those who regularly accompany them) necessarily come to view the world through the medium of the limits and possibilities of their own bodies. One is always "sizing up" the environment to see whether it is accommodating for the changed body. For instance, I well remember that my first impression of the Lincoln Memorial was not one of awe at its architectural beauty but rather dismay at the number of steps to be climbed. This bodily perception is, of course, not limited to those with disabilities. (Recall the example of the narrowed passageway that is encountered as a "restrictive potentiality" for the body because of the necessity to make certain adjustments in bodily bearing in order to pass through it.) What is peculiar about this "seeing through the body" in the event of changed bodily function is that it renders explicit one's being as a being-in-the-world. A problem with the body is a problem with the body/environment.

The disruption of this unified body/world system also includes the disruption of the body as intentional locus. Surrounding objects that were formerly used unthinkingly are now encountered as overt problems to the body. (Recall how my experience of the stairs to my office changed as my ability to climb them declined). For the person with a tremor, a bowl of soup is not simply "something to be eaten." It is a concrete problem to be solved. How does one get the liquid on to the spoon and then the spoon to one's lips without spilling the contents? For those with gait disturbances swinging doors are not just "a device to be opened" but "an obstacle to be negotiated with care" from a wheelchair or with crutches. Ordinary things assume a maddeningly resistant quality. For instance, attempting to put panti-hose on immobile legs requires the most elaborate contortions and exceptional patience! As bodily capacities change it is necessary to develop alternative ways of interacting with objects, to formulate rules of action other than those used when one had other abilities. The necessity for continually finding new ways to solve the challenges posed by objects differentiates the experience of someone who has had abilities and then lost them from the experience of a person who has never had those abilities. Indeed, some of the uncertainty experienced in degenerative diseases relates to the fact that one has to learn and relearn how to negotiate the surrounding world on an ongoing basis. The "I can do it again" can never be taken-for-granted.

To give some idea of the changed interaction with the surrounding world that occurs with loss of mobility, let me recount a typical experience. On arriving at a small regional airport to begin a professional trip, I look for a parking space that is wide enough to lower the wheelchair lift so that I can get out of my handicapped equipped van. There is no handicapped parking space that is "van accessible" so I must find a space at the end of a row in order to lower the lift. If there is no space – and I am alone – I must find a person who is willing to park my van for me and drop me off. Once I am out of the van I look for a ramp to get into the terminus on my scooter. There is only one in the whole length of the sidewalk. However, the rental cars awaiting customers are parked in front of the ramp blocking access. This means I must alert someone inside the terminus and wait for the cars to be moved before I can get into the building. Now that I have a cellular phone – as long as I have the appropriate phone number – I can alert someone in the building. Otherwise, I must wait for an upright person to arrive and request them to go inside and get assistance for me.

Once in the terminus I go to the airline check-in counter. In my battery operated wheelchair I am approximately three and a half feet tall and the counter is on a level with my head. All my transactions with the person behind the counter take place at the level of my ear. The person behind the counter must stretch over it in order to take my tickets, and I must crane my neck and shout in order to be heard.

The small commuter plane arrives and there are steps leading into the cabin. Since I cannot walk, I must be carried on board – a process that requires a transfer from my wheelchair to a device that looks rather like a barstool on wheels. To transfer I manually lift my legs off the footrest of my wheelchair and put my feet on the ground. I can then be pulled into a standing position and lowered onto the carry-on device that has been placed adjacent to my wheelchair. Someone then picks up my legs, places them on the footrest and I am strapped in and requested to fold my arms over my chest (in the manner of a corpse). I am then wheeled out to the plane and two people lift the carry-on device, as one does a stretcher or gurney, and carry me up the steps. Once on board, I again transfer to get into my assigned seat. (This whole operation is, of course, performed in full view of all the people in the airport lobby. They also watch as I am lifted into the plane.)

At the first destination, when all other passengers have alighted, I am carried off the plane (necessitating a repeat of the transfer procedure from airplane seat, to carry-on device, and then to wheelchair). We arrive at the terminal building to find that the only entrance is up two flights of stairs. Once again I must transfer so that I can be carried up the stairs into the terminal. (The necessity to repeat this procedure numerous times is not only difficult and frustrating for me but, judging from their words and actions, it is obviously irritating to those who are carrying me.)

In the terminal building I am outside the security checkpoint. However, it is not possible to go through the security gate in a wheelchair. Consequently, I am taken to the side (once again in full view of everyone who happens to be in the area) and my whole body from armpits to fingertips, from the top of my head to the base of my spine, from groin to ankle, from under my chin and over my breasts and abdomen, is "patted down" ("frisked") by an airport employee.

I leave and go to the restroom. Although the door is wide enough, the "handicapped accessible" stall is too short to accommodate the length of my battery-operated wheelchair. I must, therefore sit in it with the door open. (Obviously, if that particular stall is occupied – whether by a person with a disability or by an "able bodied" individual – I must wait until it is vacant since I cannot get through the door of the "normal" sized stalls.)

I leave on the second "leg" of my trip. Although I do not have to negotiate steps to board the larger plane that will take me to my destination I must still be carried to my seat (and off the plane) since the aisle between the seats on an airplane is not wide enough to accommodate a wheelchair. This also means that, in the event of a long trip, I must request an "aisle chair" (a smaller narrower version of the "barstool on wheels") to get to the restroom during the flight. Not only is it extremely undignified to request this assistance but I must wait until it is "convenient" for the personnel servicing the flight to attend to my needs. (In practical terms I have learned that this means I should request assistance about thirty minutes before I think I will need it, and never at a time when the meal/drink carts are out of the galley). I am then wheeled to the restroom in full view of most of the passengers who watch as I transfer from aisle chair to toilet.

After arriving at my destination, de-planing, and getting to the hotel – a trip that necessitates taking a taxi since the airport limousine cannot accommodate a person in a wheelchair – I find that (the Americans With

Disabilities Act notwithstanding) not one of the four hotel restaurants is accessible from inside the building. Each one can only be reached by going up steps from the hotel lobby. Limited access is available from outside the building but this necessitates going two blocks in the pouring rain. Later, I discover that a colleague and I are unable to walk from the hotel to a nearby place of interest since we can go no farther than a block in any direction – there are no curb cuts that would enable me to ride my wheelchair off one sidewalk and on to another.

I should note that these examples represent just the "tip of the iceberg." As my body has further weakened I can no longer travel alone. More often than not, "handicapped accessible" rooms in hotels do not have wheelchair accessible bathrooms, let alone accessible showers or commodes. Consequently, my traveling companion has to be able to manually assist/lift me/hold me up if I am to be able to use the bathroom. It is simply a fact of life that the majority of bathrooms/restrooms in modern buildings (including private houses) are constructed in such a way that the doorways are not wide enough for wheelchairs to pass through them. A routine invitation to a restaurant, a theater, someone's house for dinner, a professional meeting, a friend's apartment, or a shopping excursion requires that I first ascertain if I can get into and out of buildings, if there is adequate parking, if there are ramps and elevators, if I am blocked by stairs (either into the house or inside it), if rooms/hallways/bedrooms are so arranged in private houses that it will be possible to negotiate around furniture in my wheelchair, if I can get in and out of automobiles, if I can transfer from my wheelchair to the toilet. As bodily capacities decrease, the challenges posed in the negotiation of space obviously become more numerous.

These routine experiences illustrate another aspect of the changed experience of space that occurs with loss of mobility. The surrounding world appears (feels) different than it did prior to bodily dysfunction. In particular, the world is experienced as overtly obstructive, surprisingly non-accommodating. Actions are sensed as effortful, where hitherto they had been effortless. On occasion the world threatens even. And often it presents itself as questionable. The "knowing how" of one's engagement in the world is rendered circumspect. The effortful nature of worldly involvement that is characteristic of incapacitating disorders can engender a sense of fatigue that I shall call "existential fatigue." To organize and carry out projects requires not only physical ability but, as importantly, an exercise of will. When ceaseless and ongoing effort is required to perform the simplest of tasks (getting out of bed, dressing, taking a shower, going on a trip), there is a powerful impulse to withdraw, to cease doing what is required. The person with a disability is tempted severely to curtail involvements in the world.

Surrounding space also changes in the sense that objects assume a different meaning. The bookcase outside my bedroom was once intended by my body as "a repository for books," then as "that which is to be grasped for support on the way to the bathroom," and is now apprehended as "an obstacle to get around with my wheelchair." Additionally, space changes in the sense that permanent loss of function requires the reorganization of functional space. I must remove the beautiful, glass topped coffee table from my study if I am to get my wheelchair in there. This re-interpretation and reorganization of space is not without emotional import. Removing the coffee table not only alters the aesthetics of my study – and thus changes the "feeling" of the room (a "feeling" that is very meaningful to me) – but it symbolizes a permanent change in my mode of being-in-the-world. In this regard it is important to recognize the lived body as possibility, potentiality for action in the world. At the pre-reflective level my leg is not an object but, rather, the possibility of walking, running, playing tennis (Sartre, 1956, p. 402). Permanent loss of function represents a modification of the existential possibilities inherent in the lived body.

The lived body manifests one's being-in-the-world not only as orientational and intentional locus but in the sense that distinct bodily patterns (walking, talking, gesturing) express a unique corporeal style, a certain bodily bearing that identifies the lived body as peculiarly me (Merleau-Ponty, 1962, p. 150). Motor disorders transform corporeal style. For instance, in my own experience of neurological disease, I have found it difficult to identify with changed and changing patterns of movement. When I see myself on a home video, I experience a sense of puzzlement. I catch myself wondering not so much whether the body projected on the screen is my body but, rather, if the person in the video is really *me*. However, if I see old pictures of myself when I was walking, or leaning on a cane, I find it hard to remember how it was to be that person, or, even *who* I was when I moved like that. (Interestingly, those closest to the affected person also experience this loss of bodily identity. Recently, on seeing an old photograph that showed me standing up, my husband remarked wonderingly, "Weren't you tall?" – as if he, too, found it hard to identify me with the image of this "other" standing person.)

Moreover, I now perceive "normal" movement to be an extraordinary accomplishment. I catch myself watching students running across the campus, or colleagues taking the stairs two-at-a-time, and I marvel at their effortless ability to do so. Try as I might, I can no longer remember how it was to move like that. It is not simply that I cannot recall the last occasion when I walked upright. It is that I cannot recollect, or re-imagine, the felt bodily sense of "walking."

Merleau-Ponty would argue that my inability to recall or re-imagine "walking" can be understood in terms of bodily intentionality. For instance, he notes that the phenomenon of "phantom limb" is best explained in terms of the body's involvement with the environment. The person who feels the phantom limb does so for as long as the body remains open to the types of actions for which the limb would be the center if it were still operative (Merleau-Ponty, 1962, p. 76.) Thus, my inability to re-imagine "walking" might be understood in terms of a permanent change in bodily intentionality. My limbs are no longer open to the possibility of moving in a certain manner (i.e. in the mode of "walking").

The transformation of corporeal style is particularly profound when it involves the loss of uprightness. The significance of this loss is not confined to technical problems of locomotion. We assign value to upright posture. We applaud the man who can "stand on his own two feet." We praise those who are "upstanding." We look askance at those who are "unstable" or "weak kneed." I am most aware of the diminishing effect of loss of uprightness whenever I attend stand-up gatherings such as receptions. In my wheelchair I am approximately three and a half feet tall and the conversation takes place above my head. When speaking to a standing person, I must look up at them and they down at me. This gives me the ridiculous sense of being a child again surrounded by very tall adults.

Loss of upright posture not only concretely diminishes one's own autonomy (as a person who routinely uses a wheelchair I have no choice but to request assistance in a world designed for upright bodies) but it causes others – those who are still upright – to treat one as dependent. Whenever I am accompanied by an upright person, in my presence strangers invariably address themselves to my companion and refer to me in the third person. "Can SHE transfer from her wheelchair to a seat?" "Would SHE like to sit at this table?" This almost always happens at airports. The person at the security barrier looks directly at me, then turns to my husband and says, "Can SHE walk at all?" We now have a standard reply. My husband says, "No, but SHE can talk!" When I am unaccompanied people often act as if my inability to walk has affected not only my intelligence but also my hearing. When forced to address me directly they articulate their words in an abnormally slow and unusually loud fashion – in the manner that one might use to address a profoundly deaf person who was in the process of learning to lip-read.

With respect to the interaction with others, normal social conventions with regard to the absolute control one exercises over one's body are also disregarded with loss of upright posture. For example, when I am wheeling myself around a local shopping mall, strangers will (without my permission) start pushing my wheelchair, explaining they are "helping" me. On one occasion, in order to speed up the process of getting me from the backseat of a cab out on to the curb, a taxi driver simply picked me up with no warning and no inquiry as to whether I would like him to do so. (While I have no doubt that these actions are taken with the best of intentions, imagine how you would feel if a taxi driver simply picked you up, lifted you out of the cab, and set you down on the curb.)

With respect to the discomfort I feel when people push my wheelchair without my permission, it might be helpful to refer to Merleau-Ponty's (1962, p. 143) insight that, through the performance of habitual tasks, we incorporate objects into bodily space. For example, the woman who routinely wears a hat with a long feather on it intuitively allows for the extension of the feather when she goes through a doorway – just as a 6' 7" man unthinkingly allows for his height. Similarly, when one is proficient at typing, one no longer experiences the keys of the computer as objective locations at which one must aim. To get used to a computer keyboard is to incorporate it into one's bodily space. For the person who routinely uses a wheelchair the device becomes a part of the body. One intuitively allows for the width of the wheels when going through a doorway; one performs the necessary hand/arm movements to move forwards and backwards without thinking about it. With habitual use the wheelchair becomes an extension of one's bodily range. Thus, when a stranger pushes my wheelchair without my permission, it is invading my personal bodily space.

A phenomenological analysis of bodily change discloses the interrelationship between emotions and bodily being. For instance, shame is an integral element of disordered body style, a non-rational response to the change in body. Shame manifests itself in several ways that can be understood in terms of lived body disruption. At one level, to use Sartre's (1956, pp. 445-60) terminology, shame is the experience of being-for-the-Other. One sees one's disordered body style through the eyes of the Other and thus constitutes it in a negative fashion. This is not a culture that celebrates physical difference or dependence. There is great emphasis on health, physical fitness, productivity, sexuality and youth. The person who staggers, uses crutches, a cane, or a wheelchair is far from the ideal (Goffman, 1963; Asch and Fine, 1988; Murphy 1987). Although to feel shame it is not necessary that one actually be observed by another person (Goffman, 1963, pp. 7-8), more often than not the "gaze" of the Other is concrete (recall my experience at airports). It is important to emphasize the felt dimension of interaction with others. I directly experience your responses (facial expressions, gestures, averted eyes, words directed away from me to my companion, irritation). It is not merely a matter of what I think you feel about me. In this regard I continue to be astonished at the insensitivity of those who can walk. As an example: On one occasion I was wheeling myself around a local shopping mall. The place is a favorite "haunt" of "walkers" who go there to exercise. In the space of thirty minutes two complete strangers made the following remarks to me: "Why don't you change places and let me ride in your wheelchair for a while and you walk. My legs are tired!" and "If you want to exercise, you should get out of that wheelchair and walk around like the rest of us." Remarks of this kind are commonplace – so much so that I no longer pay much attention to them.[2]

Feelings of shame inevitably accompany the changes in body that elicit negative responses from others. Every time I have had to adopt a new way of getting around the world, first a cane, then crutches, then walker, and finally a wheelchair, I have experienced feelings of shame. I recall that, when I first used my battery-operated wheelchair, I refused to ride it into the classroom because I thought my students would think less of me if I were unable to stand "on my own two feet." So for one whole semester I rode the wheelchair down the hallway, parked it outside the classroom, and staggered into the room – hoping I would not lose my balance altogether before I reached the podium. This was ridiculous, of course, I should have known better. But, at the time, it was part of the sense of diminishment that accompanied my changed and changing embodiment.

As this example illustrates, shame also accompanies the frustration of intentionality. The "inability to" interact with the world is not only the experience of loss, "I can no longer do," but diminishment, "I must depend on others to do for me" and, further, "I should be able to do for myself." Once again this is a non-rational response to the change in body and perhaps reflects the strong emphasis in our culture on the importance of independence and autonomy, of "being able to look after oneself," of being capable of "running one's own life."

With regard to the interrelation between emotion and bodily being, as Sartre (1956, p. 455) and others have noted, "psychical qualities" are ways in which the body is lived. The "voices of aggressive people are hard, their muscles bunched, their blood pulses more fiercely through the vessels" (Van den Berg, 1955, p. 42). In many types of neurological disorder shame manifests itself as an increase in the severity of symptoms. In the experience of the Other's "gaze" an already existing tremor invariably intensifies, spastic limbs become more rigid, difficulty in controlling movement is more pronounced. (I remember the occasion of receiving my doctoral degree. I could still walk a few steps if supported, and I could climb stairs with assistance. The day before the ceremony, a colleague and I went over to the auditorium to practice climbing the few steps on to the stage where I was to be seated. I negotiated the steps without too much difficulty. However, on the following day – in the concrete experience of about sixteen thousand eyes watching me – I was completely unable to lift my right leg. It was not just that I was conscious of the fact that my body moved in a peculiar fashion. Rather, I was temporarily paralyzed. No amount of willing resulted in a corresponding contraction of the muscles. I did eventually "drag" my leg up the steps but my incapacity was much more pronounced than it had been the previous day when no-one was present in the auditorium.)

The transformation in being-in-the-world that occurs with disability incorporates not only a change in surrounding space and a disruption of corporeal identity, but also a change in temporal experiencing (Toombs, 1992; Toombs, 1990). Just as lived spatiality is characterized by an outward directedness, purposiveness and intention, so time is ordinarily experienced as a gearing towards the future. Normally we act in the present in light of anticipations of what is to come, more or less specific goals relating to future possibilities. With bodily dysfunction this gearing into the future is disrupted in a number of ways.

For instance, temporal experiencing changes in the sense that the sheer physical demands of impaired embodiment ground one in the present moment, requiring a disproportionate attention to the here and now. One is forced to concentrate on the present moment and the present activity rather than focusing on the next moment. Mundane tasks take much longer than they did prior to the change in abilities. For instance, when habitual movements are disrupted, the most ordinary activities such as getting out of bed, rising from a chair, getting in and out of the shower, knotting a tie, undoing a button, demand unusual exertion, intense concentration, and an untoward amount of time. (Think, for example, of the difference between the time and effort required to tie one's shoelaces using one, as opposed to both, hands – especially if one is right handed and only able to use the left hand to perform the task.) In this respect persons with disabilities find themselves "out of synch" with those whose physical capacities have not changed. This temporal disparity is not insignificant in terms of relations with others. "What's taking so long?" others ask impatiently.[3]

In the case of progressive disability time may be disturbed in the sense that the future, rather than the present, assumes overriding significance. The uncertainty associated with diseases such as multiple sclerosis may cause one to project into the future rather than living in the present moment. For instance, a newly diagnosed patient may start living as if already severely incapacitated, or as if the threat is imminent. Indeed, a study carried out at the large multiple sclerosis clinic at the University of Western Ontario showed that receiving a diagnosis of M.S. was, for many patients, equivalent to "moderate disability," regardless of physical status.[4] In this event the actual present is forfeited and transposed into an imagined future.

In progressively degenerative diseases time may also be disrupted in another way. Given the nature of such disorders, the future assumes an inherently problematic quality. In thinking of goals and projects, one can never be sure what one's physical status will be at some future point in time. As was noted earlier, the assumption "I can do it again" can no longer be taken-for-granted by a person whose physical abilities are constantly diminishing. This uncertainty can be extremely debilitating in the sense that it makes it difficult to continue to strive towards future goals. Are such goals unrealistic in light of the probability of decreased and decreasing abilities? Moreover, the future is not only problematic but overtly threatening. The truism, "Things could be worse" takes on a whole new meaning for those living with progressively incapacitating disorders. This change in the relation to the future contributes to personal meanings, in particular one's sense of what is possible in one's life (Toombs, 1995).

In sum, then, phenomenology provides important insights into the lived body disruption that is intrinsic to the experience of loss of mobility. In particular, phenomenology discloses the ways in which this kind of bodily change necessarily alters the experience of surrounding space, disturbs the taken-for-granted awareness of (and interaction with) objects, disrupts corporeal identity, affects relations with others, and changes the experience of time.

Such insights have practical application in the clinical context. Indeed, if therapy is to be successful, it is essential to address the global sense of disorder that permeates the patient's everyday life. Let me give a couple of examples to illustrate this point. Technical solutions, such as the adoption of a cane or crutches to enhance mobility, are less likely to be effective if explicit attention is not given to the affective responses, such as shame, that inevitably accompany such solutions – patients will not use these devices if they are ashamed to be seen doing so. Patients must learn to accept such objects as extensions of bodily space, rather than as symbols of disability, before they can effectively incorporate them into the lived body.

In this regard (and recalling Merleau-Ponty's discussion on incorporating objects into the body) it is important to note that different modes of extending bodily space have varying psychological effects. Before I purchased a lightweight wheelchair I was unable to wheel myself around because a standard model was too heavy for me to operate. Consequently, I had to be pushed. I hated "being in" a wheelchair. It made me feel utterly dependent on others. It was a symbol of limitation. I used it as little as possible (even though that meant sometimes cutting back on social engagements). Then I obtained a lightweight wheelchair I could operate myself. I no longer needed to be pushed. "Using" rather than "being in" a wheelchair is an affirming, rather than a demeaning, experience. This phraseology is not just a matter of semantics. When I manipulate the chair myself, I am in control. I can go where I want to go "under my own steam." Thus, wheeling represents freedom rather than limitation. My wheelchair has become, in effect, my legs – an integral part of my body.

In terms of effective therapy, it is also important to pay explicit attention to such factors as the change in temporal experience occasioned by various physical incapacities. For instance, therapists can assist those with degenerative disorders to address future fears in a realistic fashion, thus enabling them to live more effectively in the present. Fears for the future are almost always concrete: Will I be able to continue working at my job? Will I have enough physical strength to care for my children? Will I embarrass myself in a public setting? Once these fears have been explicitly articulated, then strategies can be developed to deal with them in the present. This allows the person facing the challenges to remain in control of his or her situation. If, however, the change in temporal experiencing is not addressed, a person living with chronic, debilitating disease may feel (and probably does feel) that his or her situation is totally out of control. This makes it difficult, if not impossible, to deal in a positive fashion with the bodily changes that are occurring

A phenomenological account of loss of mobility also has practical implications in the social context. For instance, in detailing the types of lived body disruption that accompany loss of uprightness, phenomenology provides important information for those engaged in activities such as developing ways to re-constitute public space (both physical and social) so that it is accommodating to differing modes of being-in-the-world. Such an account also forcefully reminds us of the overt barriers (including commonplace attitudes and prejudices) that prevent those with disabilities from flourishing in our society.[5]

Baylor University
Waco, Texas
U.S.A.

NOTES

[1] This chapter is a revised version of my article, 'The lived experience of disability', *Human Studies*, **18** (1995): 9-23. Reprinted by kind permission of Kluwer Academic Publishers.

[2] In this respect it has been my experience that, in spite of the passage of the Americans With Disabilities Act, negative attitudes towards persons with disabilities remain unchanged. As an example, on a professional trip to New Orleans, a friend went to check whether a certain well-known restaurant was accessible for a wheelchair. The building was accessible. However, the employee at the restaurant made it very clear that people in wheelchairs were not welcome in the restaurant and insisted we would be more comfortable at another establishment.

[3] For a detailed account of the disruption in communication that occurs with such a change in temporality, see Robillard (1999, pp. 48-63).

[4] This information was given to me by Dr. George C. Ebers, Department of Clinical Neurological Sciences, University of Western Ontario, London, Ontario.

[5] It is vital to recognize that the surrounding world can be restrictive not only in a physical sense but, more importantly, in the sense of restricting existential possibilities. If there was no ramp into the building where I teach, or a curb cut that allows me to get on to the sidewalk in my wheelchair, or an elevator to get to the third floor, or an accessible restroom, I would be unable to teach in the philosophy department. And if I did not have the economic means that allow me to own a van with a wheelchair lift, and a wheelchair, and to build a ramp providing me access into and out of my home, I would be unable to live an independent life. And, obviously, these economic means are – to a large extent – dependent on my access to employment, so it is a vicious circle.

BIBLIOGRAPHY

Asch, A. and M. Fine, (eds.): 1988, *Women With Disabilities*, Temple University Press, Philadelphia.

Goffman, E.: 1963, *Stigma: Notes on the Management of Spoiled Identity*, Prentice Hall, Inc., New Jersey.

Husserl, E.: 1982, *Cartesian Meditations: An Introduction to Phenomenology*, D. Cairns (trans.), Martinus Nijhoff, The Hague.

Husserl, E.: 1989, *Ideas Pertaining to a Pure Phenomenology and to a Phenomenological Philosophy: Studies in the Phenomenology of Constitution*, Rojcewicz and A. Schuwer (trans.) Kluwer Academic Publishers, Dordrecht, Holland.

Merleau-Ponty, M.: 1962, *Phenomenology of Perception*, C. Smith (trans.), Routledge and Kegan Paul, London.

Murphy, R.: 1987, *The Body Silent*, Henry Holt and Company, New York.

Robillard, A.B.: 1999, *Meaning of Disability: The Lived Experience of Paralysis*, Temple University Press, Philadelphia.

Sartre, J.P.: 1956, *Being and Nothingness: A Phenomenological Essay on Ontology*, H. Barnes (trans.), Pocket Books, New York.

Schutz, A.: 1962, 'On multiple realities', in M. Natanson (ed.), *The Problem of Social Reality*, Volume 1, *Alfred Schutz: Collected Papers*, Martinus Nijhoff, The Hague.

Toombs, S.K.: 1995, 'Sufficient unto the day: A life with multiple sclerosis', in S.K. Toombs, D. Barnard and R.A. Carson (eds.), *Chronic Illness: From Experience to Policy*, Indiana University Press, Bloomington.

Toombs, S.K.: 1992, *The Meaning of Illness: A Phenomenological Account of the Different Perspectives of Physician and Patient*, Kluwer Academic Publishers, Dordrecht, Holland.

Toombs, S.K.: 1990, 'The temporality of illness: Four levels of experience', *Theoretical Medicine* 11, 227-41.

Van den Berg, J.H.: 1955, *The Phenomenological Approach to Psychiatry*, Charles C. Thomas, Springfield, Illinois.

IRENA MADJAR

THE LIVED EXPERIENCE OF PAIN IN THE CONTEXT OF CLINICAL PRACTICE

I. INTRODUCTION

Walking down a gleaming corridor of a modern hospital, glancing inside rooms filled with electronic equipment that seems to surround today's patients even after relatively minor treatments, it may be tempting to assume that the agonies that accompanied illness and trauma in the centuries past, are no longer a part of the experience of being ill, of having surgery, or of recovering from trauma. At least when access to modern health services is available, it may be reasonable to expect that we should no longer have to dread the pain that our ancestors would have known in similar circumstances. But pain is more than a scientific problem and modern science and technology can address only a part of the puzzle that experience of pain presents to patients, to clinicians[1] and to people generally.

Pain is said to be one of the most common reasons why people seek medical attention. The quiet, subtle, barely noticeable symptoms of high blood pressure or of slowly developing diabetes can be easily overlooked. Even when noticed, they can be successfully ignored or set aside for some time, while we get on with out busy lives and attend to other, more pressing demands. Perhaps if we do not pay them too much attention we may hope that such symptoms will disappear. But pain, especially sudden and severe pain, such as that of an injury, a renal stone, or a myocardial infarction is different. It forces us to stop; it grips the body; it literally takes our breath away. This is "my body" and "my pain" – not some abstract philosophical idea of pain, or a report of another person's pain. While observers look for some objective, preferably measurable evidence of pain, patients know their pain with direct certainty of lived experience. They know it from inside, and they know it in its complexity.

Clinicians and scientists try to separate pain from anxiety, depression, suffering, and other emotional responses, sometimes going so far as to assign these aspects of human experience to different specialists. Patients, on the other hand, do not experience pain as a pure sensation – for them pain "arrives as a complete package... painful, miserable, disturbing..." (Wall, 1999, p. 149). They often struggle for meaning and for words with which to communicate their experience to others. Yet rather than being assisted to share their experience, they are often expected to reduce their individual concerns to standardized terms and numerical values of assessment tools and survey forms. It is such differences in experience and perspective that create a gulf between patients and clinicians, and make it difficult for them to form therapeutic partnerships needed for true healing to take place.

In arguing for the importance of a hermeneutic interest in the understanding of health and modern medicine, Hans-Georg Gadamer (1996, p. vii) writes that, "The physics of our century has taught us that there are limits to what we can measure." He goes on to say that "...doctors cannot help anyone over a serious difficulty or even a minor affliction when they do no more than simply exercise the routinized skills of their particular disciplines" (Gadamer, 1996, p. 101). What Gadamer's comments suggest is that it is not possible to reduce a personal experience of pain to a single number, a "five" or a "seven" on a ten-point scale – as clinicians so often ask their patients to do – and at the same time to understand what pain is, or what it is to be in pain. Neither is it possible to minister in any meaningful way to a person whose story we do not know. Routine interventions, whether based on personal expertise, clinical guidelines, or the latest meta-analysis of best scientific evidence, can only go so far. For the rest, we need to hear the individual story, and to engage in a human-to-human encounter in which the patient feels listened to and heard, and the clinician is made aware of the difference he or she has made – or failed to make – to the person, not just the disease or symptom.

In adopting a phenomenological gaze, as far as this is possible, I want to give voice to the patients' experience as it unfolds in the context of clinical practice.

II. BEING A PATIENT AND HAVING PAIN

To begin, I want to sketch briefly two personal accounts of pain by two women with a diagnosis of breast cancer. The first is Fanny Burney, an educated and well-travelled novelist, living in Paris when she was found to have breast cancer and underwent a radical mastectomy. It was almost a year later, in a letter to a friend, that she recorded her painful ordeal. The surgery, which she underwent with great reluctance and fear – with a drink of fortified wine as her only anaesthetic – took place in 1810. It was performed in Miss Burney's home by a surgeon accompanied by four colleagues and two medical students. Her face was covered with a piece of white gauze, while the medical men consulted and the surgeon pointed with a knife where he intended to make the incisions. Her attempts to sit up and protest were ignored and the operation commenced. This is how she describes what followed:

> Hopeless, then desperate, and self-given up, I closed once more my eyes, relinquishing all watching, all resistance, all interference, and sadly resolute to be wholly resigned ... this resolution once taken, was firmly adhered to, in defiance of terror that surpasses all description, and the most torturing pain. Yet – when the dreadful steel was plunged into the breast – cutting through veins – arteries – flesh – nerves – I needed no injunctions not to restrain my cries. I began a scream that lasted unintermittingly during the whole time of the incision – and I almost marvel that it rings not in my ears still! so excruciating was the agony. When the wound was made, and the instrument was withdrawn, the pain seemed undiminished, for the air that suddenly rushed into those delicate parts felt like a mass of minute but sharp and forked poniards, that were tearing the edges of the wound, - but when again I felt the instrument – describing the curve – cutting against the grain ... then, indeed, I thought I must have expired, I attempted no more to open my eyes ... [she then gives a vivid description of the sternum and ribs being scraped until the surgeon was satisfied that every bit of cancerous tissue had been removed. The operation lasted no more than 20 minutes.] ...When all was done, and they lifted me up that I might be put to bed, my strength was so totally annihilated, that I was obliged to be carried, and could not even sustain my hands and arms; which hung as if I had been lifeless; while my face, as the Nurse has told me, was utterly colourless... (cited in Porter & Porter, 1988, pp. 109-110).

This vivid and haunting account of what it was like to undergo surgery in the days prior to the use of anaesthesia may seem foreign, distant, and quite irrelevant to the clinicians of today. But is it?

The second woman is "Jane", rendered anonymous by her participation in a research study a decade ago in New Zealand, but no less articulate than Fanny Burney about her own experience. Jane developed a lump in her breast that gradually started to ache, especially at night. Fearing the diagnosis, and what would inevitably follow, she covered the growing ulcer that broke to the surface of her breast with ever larger and thicker dressings. It was only when the smell of the suppurating flesh became impossible to hide from her work colleagues that she consulted her physician and was referred to an oncologist. By this time, with cancer spread across her chest wall, to the lymph nodes, and into the other breast, a mastectomy was no longer seen as an option. Instead, Jane was offered a course of chemotherapy. Over time, the breast healed a little but was still painful and infected, so the case was referred to a surgeon. Following a brief phone call, and while Jane was still lying on the examination couch in the oncology clinic following a course of chemotherapy, a surgical resident and a medical intern appeared. After a brief examination, and almost no communication with Jane, the resident suggested that the area should be debrided, that the procedure could be done with the aid of some local anaesthetic, and that since he had some spare time, the procedure could be carried out immediately.

The pallor of Jane's face and the note of panic in her voice were unmistakable as she tried to protest. She pleaded to have the breast ulcer left to heal on its own, or at least to have the procedure postponed. This is how she later described the experience:

> You steel yourself [for medical treatment], at the same time putting on a very brave face. Only one thing needs to go wrong and straight away it is very hard to handle it all. My first instance was when they wanted me to have that debridement [the surgical resident] said 'straight away'. And of course I asked 'today? today?'. I was in a panic, because I didn't know I was going to have that, and that was a whole new ball game to me ... I felt I'd had enough that day ... I didn't expect that happening ... I didn't think I could have coped with it that day, because I'd had enough with the chemo. [The prospect of having the procedure under local anaesthetic was her main concern.] You always think that you may feel it if you are awake, and I thought to myself, 'if it is done under local then obviously they will put some injections somewhere around the breast and ... that would be quite painful (Madjar, 1998, p. 92).

In the end, Jane's surgery was postponed until the next day. In spite of a sleepless night and anxious expectation of pain she thought she would have to endure, Jane appeared composed and accepting of what was to come. She almost fainted with relief, however, when shortly before being taken to the operating room, she was told that the procedure would be carried out under general anaesthesia.[2] To an observer, Jane's experience would not seem to have been as painful, nor as traumatic as that of Fanny Burney, yet despite a separation of almost two centuries and all that that entails, these two stories have much in common. They point to the vulnerability and fear that pain brings, to the wounding nature of pain, and to the distancing that so often exists between those who know pain as a personal, embodied experience, and the clinicians who are charged with relieving others' pain but who so often find themselves failing in this task. These aspects of pain are some of the key phenomenological themes explored in the rest of this chapter.

III. WRITING ABOUT PAIN

What I have attempted to write is not the phenomenology of pain – an almost impossible task – but an essay that examines patients' experiences of pain in order to say something about its existential nature, and why this kind of understanding is important for those engaged in the healing professions. In doing so, I point to the paradoxical nature of clinical work that presents clinicians as caring professionals whose aim is to relieve pain and suffering, yet requires that they also inflict pain in the normal course of their everyday work. Too often, hospitals and clinics are places where pain goes unrecognized or underestimated, rather than being alleviated; its presence or severity doubted or dismissed altogether (Madjar, 1998).

Pain is a particularly apt topic for phenomenological description. It's defiance of scientific attempts to objectify and externalise it, means that the only way we can know the pain of another is through their communication of their subjective experience. A phenomenological gaze facilitates understanding of the experience, as well as how people in different situations make sense of and communicate it. Inevitably, my writing is reflective of who and what I am. In this context, the most relevant is the fact that for more than half of my life I have been a nurse. I therefore write as a person both limited and enabled by my clinical experiences and the research I have conducted with people in pain (Madjar, 1985; 1998; Madjar & Higgins, 1996).

The project of presenting pain through the lived experience of others is a difficult and a paradoxical one. It begins with a personal and a largely private experience of another; experience that one cannot enter or share directly. Yet, it goes on to assume a capacity of the person in pain to communicate something of the essential nature of that experience, and a capacity of an interested other to grasp that which is being communicated. By transforming such understanding into language that captures the essential qualities of the phenomenon, the lived experience may be re-presented in a way that creates shared meanings and possibilities for social connection (Reinharz, 1983). Such epistemological transformations can help to bridge the experiential gap that divides patients from those who minister to them, and provide a basis for sensitive and compassionate care that can and often does create a healing environment even in situations of extreme pain and suffering.

IV. PAIN AS LIVED EXPERIENCE

Scientific developments and their clinical applications over the past 150 or so years have led to what Caton (1985, p. 493) has termed "the secularisation of pain". The Aristotelian idea of pain (and pleasure) as "the passions of the soul", and the equally ancient notion of pain as some form of divine punishment (Livingston, 1998, p. 147), have been superseded by more scientific explanations of pain; at least in the medical domain. Today's health care professionals see pain as a natural phenomenon, best understood in terms of the underlying physiology and pathology and best treated by medical, most often pharmacological therapies. However impressive and essential such explanations are to clinicians in the diagnosis and management of pain, at the level of human experience they leave many questions unanswered. In the words of John Banja (1999), "patients don't experience delta fibers, nonmyelinated C fibers, or the periaqueductal gray area ... . Rather, they experience pain, anger, fear, depression, weakness, disability".

A personal experience of pain, particularly when that pain is severe or continues over a long period of time, is a mystery. It calls for more than a scientific explanation of what is going on at the cellular and molecular level of our bodies. Personal experience calls for meaning, for answers to questions of "why me?", "why does it have to hurt so much?", "how do I handle the pain and the fear that comes with it?", and for some people "how do I live a life in pain?" and even, "is life in pain worth living?" Such questions defy purely scientific explanations of pain, diagnostic categories, and symptom scales. They hint at suffering and call for a dialogue between patients and their healers that recognizes the patients' need for support, for understanding, and for therapeutic accompaniment, while they search for meaning of their pain as well as relief from it.

A. The Embodied Painfulness of Pain

The essential quality of all pain is its embodied painfulness. This is at once self-evident and yet, for some reason, obscure. It is thus possible for pain to be simultaneously "something that cannot be denied and something that cannot be confirmed". In other words, as Scarry (1985, p. 13) goes on to say, "to have pain is to have *certainty*; to hear about pain is to have *doubt*". This is the challenge that pain poses to health professionals who cannot know their patients' pain directly, but must believe the words and overcome their natural inclination toward doubt. For those who suffer pain, however, there can be no doubt. Pain, says Addis (1986), imposes its presence and calls for attention, not only because of its intensity, but also, as Merleau-Ponty (1962) too has commented, by it's spatiality. For the person experiencing it, pain is not something out there; it occupies a space within the very fibres of one's body and makes it inseparable from the essential self.

As I have argued elsewhere (Madjar, 1998, 179), the hurt and painfulness of pain is "more than nociceptive activity within the anatomical body. Rather, it challenges the familiar, taken-for-granted body of everyday life, entering it and creating a 'pain-infested space' within it (Merleau-Ponty, 1962)." The metaphorical language of pain reflects this embodied painfulness. When patients describe their pain as crushing, burning, searing, excruciating, or revolting, they are trying to communicate the painfulness of their pain, but they are also trying to do more than that. Their words call to others to feel something of what they feel, and for a pathic response that acknowledges their pain and reaches out in sympathy and solicitude.

B. The Wounding Nature of Pain

Pain, as Scarry (1985) so eloquently reminds us, is also something essentially wounding. In the context of illness and hospitalization, the tactile encounters between patients and their healers often expose patients' vulnerability and carry the possibility of wounding. "The bitter kiss of needles", encountered in the course of hospital care (Clendinnen, 2000, p. 23), speaks of the transformation of the clinical encounter into something technical and cold, yet still human and laden with other possibilities. (A kiss is meant to bring pleasure, comfort and closeness. It is meant to say that one is loved, desired, cherished. A bitter kiss is a kiss of betrayal, of separation, of pain.) The wounding is physical – hence the common metaphors of crushing, or burning, or stinging – but it is also existential, exposing the person as a fragile, vulnerable being whose integrity and even existence may be threatened by the pain.

Pain inflicted in the course of treatment is more obviously wounding, since it is the actions of another that create the pain and the instruments of treatment that are often experienced as weapons (Madjar, 1998). Such wounding can transform a place of safety and comfort into a place where vigilance and anticipation of pain predominate. The comfort and safety of home is replaced by a much more fragile sense of safety in a hospital. Often it is only the hospital bed, and sometimes not even that, that defines the boundaries within which the patient feels safe. Spatiality of pain thus relates to both the internal location of the pain within the body, and the character that experiences of pain give to spaces such as clinics or hospital treatment rooms where pain is often inflicted but not always acknowledged. The possibility of wounding and pain create spaces in which vigilance replaces ease, and anticipation of the next episode of pain creates tension and dis-ease.

C. Being With Those in Pain

Pain is one of the most private, isolating experiences, yet it also has an essential element of relationality, of being with others. The embodied immediacy of being in pain is mediated by cultural imperatives that dictate how we should behave when in pain, what language we should use to communicate it to others, and what expectations we can have of them.

Pain often has to be endured in the presence of others thus shifting the focus from the subjective experience to the socially mediated expression of pain. But it goes further than that. Serious illness and its treatment require that people become patients; that they hand their bodies over to others to probe, palpate, investigate and dissect in order to diagnose and treat. It is in the handing over of their bodies to others that patients enter into a social contract that allows their healers to hurt and to wound them. To lie on a trolley being wheeled into the operating room, knowing that one is about to undergo a mastectomy, for example, is to hand one's body over to others to cut and mutilate, to wound and to cause it pain. Subsequent interventions – dressing changes, chemotherapy, follow-up monitoring – repeat the demand for this complicity with others. The inborn inclinations to resist such assaults on the body, and to avoid pain, have to be overcome in order for the work of diagnosis and treatment to go on. In this way, patients are seldom passive recipients of care.

Enduring pain, and doing so in culturally acceptable ways, often requires that patients "restrain their bodies and their voices" (Madjar, 1998). The natural inclination may be to cry, to curse, or to scream when in pain, but it is remarkable how seldom patients in hospitals in countries such as Australia or New Zealand (and no doubt many others) do. By restraining their bodies and their voices patients in pain make possible the work of the health care providers. But in doing so patients also exercise some degree of control over their situation, some choice over how they behave, and some sense of dignity and humanity. In the process, they constitute their experience as well as being constituted by it (Benner & Wrubel, 1989), and they give meaning to the pain they endure.

V. GIVING MEANING TO PAIN

A noted Australian historian, Inga Clendinnen, begins her recently published memoir with a recollection of a serious illness that culminated in a liver transplant:

A decade ago, when I was in my early fifties, I fell ill. 'Fall' is the appropriate word; it is almost as alarming and quite as precipitous as falling in love. It is even more like falling down Alice's rabbit hole into a world which might resemble this solid one, but which operates on quite different principles. Pain, death and loneliness are domestic presences there... (Clendinnen, 2000, p. 1).

Given the nature of her illness and the events that ensued, she says remarkably little about pain:

People come and go; you learn their comings and goings. Your flesh learns to flinch from metal and the bitter kiss of needles. There is no other flesh to comfort it. Carnal contact with your many handlers only parodies intimacy. Too often it also means pain (Clendinnen, 200, p. 23).

There may be many reasons for the restraint evident in Clendinnen's words, but the overall tone is one of inevitability. Pain, including pain inflicted in the course of therapy, is what happens when one falls down a "rabbit hole" and enters the world of illness and hospitalization.

A. Pain as Inevitable Consequence of Illness

Patients accept much of the pain they experience, whether arising out of some underlying pathology or clinically inflicted in the process of treatment, as something natural, expected and inevitable. It is as though, once ill, people revert to some kind of elemental logic that pain is what one expects when the taken-for-granted normality of life is lost and illness invades the body. Empirical evidence of what actually happens does little to suggest that patients' fears and expectations are unfounded.

Studies conducted in some of the best resourced clinical settings continue to report high levels of pain in patients whose pain can and should be relieved. A recent study from Sweden, for example, showed that during the first 24 hours after elective surgery 43 per cent of the patients (n=200) experienced moderate to severe pain at rest and, during the same period, 88 per cent of them experienced moderate to severe breakthrough pain, particularly on movement (Svensson, Sjöström & Haljamäe, 2000).

Along similar lines, a prospective longitudinal study of pain in 95 AIDS patients in Denmark found that the majority received less than optimal analgesia, even in the last weeks of life. Yet, only nine percent of the patients expressed dissatisfaction with their pain management, while of those who reported that they were satisfied with how their pain was managed 79 per cent also reported "constant pain" (Frich & Borgbjerg, 2000, p. 344). What is it in the patients' experience that stops so many from challenging this situation? This is a puzzle that is not easily solved.

Part of the answer may lie in the way clinicians behave and the way they help patients to constitute their experiences of illness and pain. This is particularly so when staff are themselves the source of pain, as happens in settings such as burn care and oncology units (Madjar, 1998). Having to inflict pain in the course of therapy intended to benefit the patient – changing burn dressings, inserting intravenous needles or cannulae, or mobilising an already painful part of the body – is often seen as inevitable by both staff and patients. But herein lies the problem, for when nurses whose job it is to perform such procedures perceive the pain they inflict as inevitable, they (as observed in a burns care unit (Madjar, 1998, p. 134)) "also perceived themselves as unable to change the situation for the better. Rather than seeing themselves as the ones with the knowledge, resources, and power to relieve and manage pain, nurses often felt helpless and immobilized in the face of pain they had inflicted." The burden of enduring and of controlling the expression of pain, what we call "coping", thus falls on the patient. And it is the patient who is left to make sense of the pain and of the suffering it so often brings.

B. Pain as Failure

Pain is not a singular phenomenon. It comes in many shapes and sizes. And so do the patients' responses to it. Research literature on pain acknowledges individual variability in sensitivity to, and tolerance of pain, as well as responsiveness to specific therapies. Yet, as Mogil (1999) admits, "clinicians and scientists detest variability" for it makes prediction in relation to individual patients unreliable and frustrating. Variability between individuals, and even in the same individuals over time, serves to confound the science that focuses on the group mean. Quite out of the blue, so to speak, individual patients report no pain when they are expected to have pain, complain of lack of relief after a dose of morphine that is apparently effective for everyone else, or "fail" to get over the acute episode of pain and develop pains that are difficult to explain.

This constant 'defying difference' between diagnosis or prognosis on the one hand and contingency and concreteness on the other, is what makes each person, each patient, uniquely who he or she is – which is never the same as the diagnostic portrait that the expert constructs. The individual human being always falls to a certain extent 'outside' of the dossier, the diagnosis, the description, the prognosis (van Manen, 1999, p. 34).

Pain, more than many other ailments, calls for recognition of difference as a basic premise of effective practice, yet so often, individual variability is construed as some kind of failure.

There are many ways in which pain is experienced as failure. It signals a breakdown, a failure of normal function, and unreliability. An academic who suffers frequent and often debilitating migraine headaches works hard at keeping her predicament hidden from her students and colleagues. It is as though there is something weak or shameful about having migraines, a failure to cope with the demands and stresses of the job that other able bodied people apparently manage to do. Worse still, letting others know is to risk suspicion that the headaches are not "real" but a handy excuse for a failure to conform to the corporate image of a committed academic.

Sometimes words are all that is needed to communicate failure. We say that a patient has "failed to respond" to a particular intervention; not that the intervention did not suit the particular patient, or that of all the available interventions we had chosen one that was not effective in this case. People with chronic pain whose pain persists, despite their own and their healers' best attempts to alleviate it, often feel that they have failed, as patients and as people. When labels such as "the failed back surgery syndrome" are applied as though they had diagnostic rather than merely descriptive value, they contribute to the sense that persistent pain signifies personal failure, of the patient, the surgeon, or both.

C. Pain as Punishment

Like Inga Clendinnen, "falling down a rabbit hole" where things are unpredictable and strange, patients are often confused and bewildered by pain. Pain makes no sense. It may be inevitable, it certainly has to be endured, but it cannot be understood as merely a sensory experience. There has to be a reason for it and it has to have come from somewhere. "Pain forces the question of its meaning, and especially of its cause, insofar as cause is an important part of its meaning. In those instances in which pain is intense and intractable and in which its causes are obscure [although not only then], its demand for interpretation is most naked, manifested in the sufferer asking, 'Why?'." (Bakan, 1968, p. 58). Pain is indicative of a break, a discontinuity of usual being, an interference. It is a visitation, an other. Patients struggle for words, often resorting to metaphors to capture something of the essence of their experience. The metaphors are, and are meant to be, discomforting to the listeners. Soderberg and Norberg (1995), in their study of Swedish people with fibromyalgia, for example, found that patients spoke of their pain as aggressive, deforming, torture-like experience. Such aggression comes from something or someone alien and hostile and it places the sufferer in a position of subjugation and dependence.

Like illness or trauma that lead to it, pain may be seen as fair or unfair, as deserved or undeserved, but often as punishment or a trial to be endured. Even patients who profess no particular religious beliefs often look to a god, to laws of karma, or to some more nebulous idea of cosmic justice, to explain a sense of having been singled out for their experience.

Reflecting on her husband's illness and death, Lyn Payne records the inexplicability of his and of her own experience:

It is March 1999 and he is about to embark on a new chemotherapy regime that will ravish his body but will not affect his cancer. By June he will find the secret malevolence within him has travelled to his spine. He will start to experience pain in his back and his liver and he will grow mad on morphine they give him – so mad his oncologist will order a scan of his brain. By July he will experience internal bleeding, his body invaded by toxins. On the 19th day of the month he will softly die. Throughout this horror ... I wash his clothes, shop for organic food, cook special meals he will not eat, carry wood and light fires to warm him, arrange palliative home-care, wait for doctors. I accompany him to his chemotherapy treatments ... in the last week, he withdraws from me, no longer wants my care ... He does not understand what is happening to him. He thinks he is being punished. "What have I done to deserve this?" he whispers. "I should have been a better person" (Payne, 2001, p. 67).

This is what the world of illness and pain is about for some patients – the separation from the world of the healthy and the utter helplessness in the face of what being ill and in pain entails. It means being dependent on others' knowledge, resources, ministrations and explanations. And because pain is unpleasant, isolating, it hurts and it threatens to overwhelm, it cannot be an inconsequential 'nothing'. It must mean something, however unwelcome or unpalatable the message it contains. The response may be anger and rage at the injustice of undeserved punishment or the unfairness of it all, a searching for the supposed misdeeds, or a submission to one's apparent fate.

The language of modern medicine is often insufficient to account for such experiences, even though patients need to speak of them and have them recognized and legitimized (Kleinman, 1988; Schweizer, 1995). This was the experience of a young woman I shall call Geraldine, trying to come to terms with severe burns, her disfigured body, and the daily pain involved in her treatment:

I don't think it was fair what happened. I don't think I deserved that...I've never ever thought that this would happen, never, never, never ever! Never!... It is usually the physical pain that sets my emotions going and... usually they'll make me look at what's happened, and when I look at what's happened I think about what's happened, and then, when I think about what's happened, I cry. So it's all interrelated... This is another body that I don't know. Not me. I don't deserve it. It's not me. I don't deserve it.... (Madjar, 1998, p 105).

Her experience is of painful punishment; of violence done to her body – her only means of being in the world – that has become alien and uncomfortable and something that she wishes to, but knows she cannot, escape. Her rage is deep, even as her words are soft. Each painful treatment takes her back to the original pain and the sense of unfairness of a life marred forever by what she is going through. She cannot separate the bodily pain from the anguish, even though she is constantly asked to do that. The physiological explanations for her pain are totally inadequate; they provide no answers that her existential crisis demands.

It is puzzling, and perhaps even irritating to modern health professionals, that the ancient idea of pain as punishment should persist into the current age of secularism, science and technology. Yet it does persist, even if patients often feel unable to fully express their sense of what it means to them, and may not be given a space within which to discuss it as an integral part of their experience.

VI. OPENING UP THE DIALOGUE

The vocabulary of medical science is for the most part "gnostic", or rather "diagnostic" (van Manen, 1999), trying to capture in language, and with as much objectivity and precision as possible, the baffling problem that lies hidden beneath the human skin. Pain assessment tools thus expect patients to assign numbers to their pain, and to apply standardized verbal descriptors to an experience that, amoeba-like, changes shape and seeps out through gaps created by individual need and circumstance.

The diagnostic language of pain is one sided. It serves the needs of clinicians who need the patients to transform their experiences into language they (the clinicians) can understand. Thus it is important to know whether the chest pain is "burning" or "crushing", since the first may point to indigestion, the second to a heart attack. Patients, on the other hand, need more than a diagnosis, vital though it is. As van Manen (1999) points out, cognitively dominated "gnostic" thought and practice need to be complemented by words and interactions that are "pathic" in nature; that relate to the person and provide insights into the personal experience. Gnostic touch that probes for answers and tries to arrive at a diagnosis needs to be complemented by pathic touch that communicates concern, understanding, and support. The idea of pain as punishment or as failure has little diagnostic value, but we should not ignore or dismiss such understandings of pain as irrelevant. They call for pathic relation, for reassurance and support that can help the patient bear the burden of pain and rediscover his or her own sense of wholeness.

VII. CONCLUSION

To understand pain we need to understand the person in pain and a phenomenological gaze can help us to do that. The key is our attentiveness to the lived experience of the person in pain, and our willingness, individually and as members of health care teams, to work as much *with* as *on* our patients. The cognitive and the technical work of pain diagnosis and treatment needs to go hand in hand with the supportive, and the affirming acts that make possible for the patient's voice to be heard and to be valued.

IRENA MADJAR

University of Newcastle
Callaghan, New South Wales
Australia

NOTES

[1] The term is used in a generic sense to refer to physicians, nurses, physical and other therapists – all professionals that in some way attend to patients and their experiences.
[2] Over the following two years Jane suffered other episodes of fearful anticipation and of actual pain, before dying from the cancer that spread to her lungs, bones, and finally, her brain.

BIBLIOGRAPHY

Addis, L.: 1986, 'Pain and other secondary mental entities', *Philosophy and Phenomenological Research* **47**, 59-73.
Bakan, D.: 1968, *Disease, Pain, and Sacrifice: Towards a Psychology of Suffering*, University of Chicago Press, Chicago.
Banja, J.: 1999, 'The art of healing and the art of medicine: A discourse on medical epistemology', Paper presented at the 20th Annual Scientific Meeting of the Australian Pain Society, Freemantle, WA, April 1999.
Benner, P. and Wrubel, J.: 1989, *The Primacy of Caring: Stress and Coping in Health and Illness*, Addison-Wesley, Menlo Park, Ca.
Caton, D.: 1985, 'The secularisation of pain', *Anesthesiology* **62**, 493-501.
Clendinnen, I.: 2000, *Tiger's Eye: A Memoir*, Text Publishing, Melbourne.
Frich, L.M. and Borgbjerg, F.M.: 2000, 'Pain and pain treatment in AIDS patients: A longitudinal study', *Journal of Pain and Symptom Management* **19**, 339-347.
Gadamer, H.G.: 1996, *The Enigma of Health*, Stanford University Press, Stanford, Ca.
Kleinman, A.: 1988, *The Illness Narratives: Suffering, Healing and the Human Condition*, Basic Books, New York.
Livingston, W.K.: 1998, *Pain and Suffering*, IASP Press, Seattle.
Madjar, I.: 1985, 'Pain and the surgical patient: A cross-cultural perspective', *The Australian Journal of Advanced Nursing* **2**, 29-33.
Madjar, I. and Higgins, J.: 1996, 'The adequacy of pain relief measures in elderly nursing home residents', Paper presented at the 8th World Congress on Pain, Vancouver, Canada (22 August).
Madjar, I.: 1998, *Giving Comfort and Inflicting Pain*, Qual Institute Press, University of Alberta, Edmonton.
Merleau-Ponty, M.: 1962, *Phenomenology of Perception*, (C. Smith trans.), Routledge & Kegan Paul, New York.
Mogil, J.S.: 1999, 'Individual differences in pain: Moving beyond the "universal" rat', *American Pain Society Bulletin* **9** (5), accessed electronically at http://www.ampainsoc.org/pub/bulletin/sep99/research.htm
Payne, L.: 2001, 'Charles –29.9.36– 19.7.99', *Illness: A Journal of Personal Experience* **1** (1), 67-72.
Porter, R. and Porter, D.: 1988, *In Sickness and in Health*, Fourth Estate, London.
Reinharz, S.: 'Phenomenology as a dynamic process', *Phenomenology + Pedagogy* **1**(1), 77-79.
Scarry, E.: 1985, *The Body in Pain*, Oxford University Press, New York.

Soderberg, S. and Norberg, A.: 1995, 'Metaphorical pain language among fibromyalgia patients', *Scandinavian Journal of Caring Sciences* **9**, 55-59.

Svensson, I., Sjöström, B. and Haljamäe, H.: 2000, 'Assessment of pain experiences after elective surgery', *Journal of Pain and Symptom Management* **20**, 193-201.

Schweizer, H.: 1995, 'To give suffering a language', *Literature and Medicine* **14**(2), 210-221.

Van Manen, M.: 1999, 'The pathic nature of inquiry and nursing', in I. Madjar and J.A. Walton (eds.), *Nursing and the Experience of Illness: Phenomenology in Practice*, Allen & Unwin, Sydney, and Routledge, London, pp. 17-35.

Wall, P.: 1999, Pain: *The Science of Suffering*, Weidenfeld & Nicolson, London.

JO ANN WALTON

THE LIVED EXPERIENCE OF MENTAL ILLNESS

I. INTRODUCTION

Mental illnesses come in many different shapes and forms. The term 'mental illness' may be used to describe something relatively short-lived, a problem manageable enough that a family doctor might be the only health professional called on for help. Or it might denote a major illness, or one of lasting duration, for which several specialist services may be needed.

It seems significant that such a wide range of illness types should be covered with one umbrella term: *mental* illness, while we have no similar term for the vast array of physical illnesses. Pretty obviously, the idea stems from the apparent location of mental illness *in the mind*.

As I hope to show in this essay, that location of the problem, and its assumed similarity with other 'mental' illness states is, in itself, problematic if our interest is in experience as it is lived. People experience illness in their whole being. I will return to this idea later. For the moment, in spite of my reservations about the words, I will stay with the idea of mental illness as a term that can be understood by both reader and writer.

My intention in this essay is to discuss some of the elements that are central to the mental illness experience, and to point to studies in which particular illness experiences have been examined through a phenomenological lens. In doing so I am assuming that there are some common themes that transcend illness type, diagnosis, prognosis, duration. I hope to be able to show that the phenomenological approach has much to offer in our understanding of what people experience when mentally ill, and that phenomenology also offers insights into what it means to live as someone 'with a mental illness.' The two are not the same thing, and the distinction will be drawn out later in this essay. In undertaking this exercise I hope also to show how phenomenology, as research, offers the kinds of insights which can also be gained from literature, and what the effect of the phenomenological approach to practice may be for the clinician him or herself.

Phenomenological inquiry, as Fisher (1978, p. 175) succinctly puts it "seeks to answer two intrinsically interrelated questions: what is the phenomenon that is experienced and lived, and how does it show itself?" What then is mental illness and how is it experienced and lived? Lived experience is multifaceted, as Allen (1990, p. 160) details below.

> The lived experience has as its moments the single perceptions, recollections and anticipations, fearing, dreading, and willing, etc., which together form the unbroken unity of a human life. The experience flows on constantly. It encompasses my own stream of experience, to which I have immediate access, as well as those experiences of other individuals and of the social community at large, experiences to which I have access indirectly.

Allen's (1990) inclusion of indirect experience, of the experience of society at large, is particularly helpful in relation to mental illness. A great deal of what it means to be mentally ill is bound up with the social context in which illness occurs, and the reactions of others to the mentally ill – indeed toward mental illness in itself.

There are assumptions about mental illness deeply rooted in our society. Barham and Hayward (1991, p. 2) refer to the work of Roy Porter, a medical historian who has written extensively on the social history of madness and psychiatric treatment, in their claim that "As social historians of madness have shown, we have inherited from the last century a deep disposition to see madness as essentially Other." Such an inheritance can be found not only in historical accounts but also in cultural media such as literature, art, newspapers, film and television. Recognition of their state of perceived difference is also to be found in the writings of the mentally ill themselves, who "complain that 'alienness' is a false identity thrust upon them, or indeed a non-identity, a sense of being rendered a non-person" (Porter, 1987, p. 25). Porter suggests that this has readily become an excuse as to why the mentally ill should not be heard.

While it is tempting to conclude that health professionals do not share this perception, in his work on the representations of disease Gilman (1988, p. 48) emphasizes the tendency for professionals to incorporate understandings held by the wider society. "However much clinicians (not to mention the lay public) believe themselves to be free of such gross internal representations of difference, they are present, and they alter the relationship with the patient or client." In Gilman's analysis it is actually our own fear of collapse and dissolution that we in the Western tradition project onto the world in order to have power over it – as is shown by our images of illness. "Then it is not we who totter on the brink of collapse, but rather the Other. And it is an-Other who has already shown his or her vulnerability by having collapsed" (Gilman, 1988, p. 1).

The legacy of understandings of mental illness that have come down to us over generations influence the ways in which we *are* towards those who suffer from mental illnesses. Such understandings also underlie assumptions about what mental illness is and how it should be dealt with. Even today the reactions of society to those considered mentally ill remain colored by commonplace beliefs, superstitions and prejudice.

Fortunately things are changing, albeit slowly, and the process is helped by those who are courageous enough to describe their experiences. Examples include Kay Redfield Jamison's (1995) *An Unquiet Mind* in which the author, herself a psychiatrist, gives an account of her life with manic-depressive illness. Another is the beautifully packaged *A Gift of Stories* (Liebrich, 1999) recently published in my own country, New Zealand. Of course there are many more.

Because mental illnesses are stigmatizing and (usually) invisible, people who suffer from mental illness have many decisions to make about whether to disclose their illness (and/or a resulting disability) or whether it is better to attempt to 'pass,' to let the illness remain undeclared. Such choices are not made once in a lifetime, but in fact over and over again, as situations, health status and relationships change. The situation is different for the person with a chronic illness than it is for someone who is ill only occasionally. And yet this distinction is in itself a marker of public attitudes. Hope for change is bound to a growing recognition of the person behind the illness (as also identified by, for example, Barham and Hayward, 1991; Davidson, 1992).

One powerful means by which the person, rather than the illness, can be recognized, is through narrative accounts or stories. As Liebrich (1999, p. 5) says:

Stories are one of the world's oldest and surest methods of teaching ideas. They help us *real*ise ideas; they make ideas real because they ask us to use our intellect and feelings at the same time, and sometimes they also speak to our spirit. ...[Stories]...give us an opportunity to understand something – to stand *under* knowledge – and the chance to gain insight – to see from *within*.

Some literature about illness, or about practice, while concerned with particular times, places and circumstances, offers a truth that transcends those particulars, and can be read on several levels – "literal, symbolic, metaphorical, or allegorical" (McLellan, 1996, p. 1014). Reading such works often requires self-evaluation and reflection on the part of the practitioner (McLellan, 1996).

Phenomenology offers something additional. Because phenomenological studies are concerned with the life world of actual people who have undergone a specific experience, they are able to illuminate our understanding of that experience as it occurs in the real world. Further, good phenomenological studies point back to our own preconceptions, our pre-understanding, our conscious or unconscious theorizing. They do this because their aim is to show the phenomenon up in a way not previously seen. Thus they stimulate the reader to reflect on what is different about this way of looking at the phenomenon, and how this changed understanding will impact on his or her actions.

Burch (1989) suggests that the practical benefit of phenomenology is to be found in what the method "does to us." A phenomenological study is useful not simply because of what it "finds," or the written account which is presented as a result, but because it aims to change our understanding. It is fundamental to a phenomenological analysis that we (researcher, clinician, reader) endeavor to put aside our prior theoretical understandings, returning instead to the things themselves in order to gain a renewed understanding of something that we thought we already knew.

In the mental health field theoretical understandings may have a powerful effect in limiting the way we look at clients' experiences. Even the words we use to define illness can work in this way. Comparing the word 'melancholia' with the more recent 'depression,' William Styron (1990, p. 56) calls the latter "a true wimp of a word for such a major illness". "Nonetheless" he writes "for seventy-five years the word has slithered innocuously through the language like a slug, leaving little trace of its intrinsic malevolence and preventing, by its very insipidity, a general awareness of the horrible intensity of the disease when out of control."

Sadly, in spite of our good intentions, as health professionals we sometimes do not listen to what it is the patient would like us to hear (Baron, 1985; Karp, 1996; Reynolds and Scott, 2000). In part this may be because we think we already know, yet we do not always recognize *what* we think we know, or how it could be different. Part of the problem is that without the words and concepts to communicate clearly to each other, clinician and patient inhabit different worlds. In her critical work about women and madness, Astbury (1996, p. 26) suggests that "The absence of a conceptual space in which the concerns and experience of a muted group might be heard has the effect of making the experiences, needs and wishes of the group invisible and incomprehensible." Astbury refers to Freudian theory, which still colors much of contemporary practice, as 'neurotic science'. It is, she asserts, "a science that says infinitely more about the psychology of the observer than the observed, and especially about the observer's often malign fantasies of the other" (Astbury, 1996, p. 2). While Astbury's concerns are primarily about women, we are also returned to the notion of the mentally ill as Other.

A large part of the problem also lies in the difficulty that those with a mental illness may have in communicating their experience. Wolpert (1999, p. 9) notes that "severe depression is virtually indescribable to anyone who has not experienced it". Similarly, Karp (1996, p. 42-3) discusses how people who have never experienced depression don't, as he says, "get it". "Cognition alone of a human condition always falls short of complete understanding" (Karp, 1996, p. 42). As a result people with depression spend much of their time attempting to conceal their real feelings from others, in a process he calls impression-management.

Thus, while a diagnosis is an important part of the clinical picture for a treating clinician, DSM diagnoses do not sum up what we know. In fact while they are helpful in determining what possibilities exist for help, in themselves they contain a shorthand for symptoms that removes us from the patient's experience.

A diagnosis of mental illness does not say anything about a person's capabilities, personality, or future. The vast majority of people who have some kinds of mental illness: get better, can hold down jobs, make good partners, lovers, parents, and are not dangerous, and have a great deal to give the world. In fact, the very act of dealing with a mental illness often gives people extraordinary strength of character (Liebrich, 1999, p. 7).

Liebrich's (1999) emphasis on strengths and recovery echoes several strong themes in current literature about illness and disability, i.e. that people are more than their illnesses, that individuals do recover, and that they have a large part to play in their recovery (eg. Davidson, 1993; Davidson and Strauss, 1992; Lindsey, 1996; Walton 1999, 2000). It is here that the distinction between what it is like to live with a mental illness (that is to be unwell) and what it is like to live with the social stigma of having a mental illness (of living as someone 'with a mental illness') can be drawn.

At this point I will turn to some studies that have examined aspects of the mental illness experience, specifically related to the illnesses schizophrenia, postpartum depression and anorexia nervosa. I have chosen these particular studies because they are familiar to me, because they each have something different to offer in their perspective on the illness experience, and because they have all moved me. One is my own work, the others are studies reported in the literature.

II. THE LIVED EXPERIENCE OF SCHIZOPHRENIA

Amidst the vast literature on schizophrenia are several phenomenological studies that examine aspects of the illness. Some work looks in depth at a particular symptom experience, while other research has examined the experience of schizophrenia more broadly.

The *Stimmung,* an experience often occurring in the early stages of schizophrenia, is the subject of a phenomenological analysis by Sass (1988). The *Stimmung* is a state in which the world becomes strangely different, in which objects seem either to be fragmented, or particularly but intangibly different, and in which they lose their previous everyday significance. While this experience has tended to be regarded as one radically separate from the understanding of clinicians and others who have not experienced it, Sass has managed to capture the ineffable quality, the strangeness and the specialness of the experience. To help the reader understand he uses examples from early modernist art as a parallel. In doing so he is able to demonstrate powerfully that such symptoms are, in fact, not incomprehensible at all. In this way he shows that such symptom experiences are open to empathic understanding, rather than in need of theoretical analysis, psychoanalytic or otherwise.

In his study Sass (1988) suggests that psychotic symptoms such as delusions and ideas of reference can be understood as stemming from the existential foundation of the *Stimmung*. In this state patients experience what Sass calls a kind of 'anti-epiphany' in which 'there is often a contradictory sense of meaning*ful*ness and meaning*less*ness' (Sass, 1998, p. 224). In effect delusions or ideas of reference may be a way of making meaning of the events that surround the *Stimmung* experience. Sass (1992) has extended his ideas about the parallels between art and the mental illness experience in his book *Madness and Modernism*.

Delusions are also the subject of a phenomenological study by Davidson (1993) in which he relates and then examines the story of Nancy, a 37-year-old woman who found magical powers in a spool of navy blue thread. Nancy's story is placed in context in terms of her life, her aspirations and the interactions she has with her mother. In doing this Davidson demonstrates that Nancy's experience is not only understandable, but in fact useful in mediating her recovery from illness.

Davidson (1993, p. 200) suggests that delusions can be viewed as 'stories that people with schizophrenia tell about their lives'. As such they are both accessible to clinicians, and may be understood as regulating the amount of change to which the person has to adapt at any one time in the process of recovery. The individual's experience can thus be seen to encompass his or her own meaningful explanations for events, rather than these being categorized and judged as 'false' beliefs.

My own phenomenological study involved interviews with ten adults about what it is like to live with schizophrenia. Parts of my work, particularly discussion of the experience of symptoms and the relational aspects of the illness and its sequelae have been published elsewhere (Walton, 1999; 2000). In this essay I will address something of the temporal aspects of the illness experience: the way the study participants looked both backward on their illness and forward toward the future. Here some kinds of delusional ideas are re-conceptualized as belonging to the category of hopes and dreams.

The participants in this study were ten adults, living in the community in New Zealand. Each of them was taking medication for their illness at the time I spoke with them, and all were considered well at the time.

The illness has had an effect on all aspects of the participants' lives (Walton, 1999). Thus the notion that schizophrenia is a *mental* illness belies the experience of those who suffer it. Not only does medication have physical effects on the body, but the illness was felt to have bodily effects also. In addition, the uncanny experience of having thoughts, beliefs and convictions altered by medication gave people a lived sense of the connection between mind and brain. In effect they were confronted by the mind-body connection, not briefly as is the case of someone who takes a hallucinogenic drug for instance, but deeply, seriously, profoundly (Walton, 1995; 1999, 2000). Along with other aspects of the illness experience this has had a significant effect on the way in which the participants' now conduct their lives.

Each of the participants in the study had had the experience of at least one episode of acute schizophrenic illness and each continued to take medication of some kind. This was part of their experience of Being-in-the-world and affected their Being-with-others. If they were to integrate the diagnosis of schizophrenia into their sense of self, then the participants needed to recognize and accept the experiences that accompanied illness and treatment. However, the concern the participants demonstrated in terms of pursuing a "normal life" also required that they not dwell only with their illness but also look forward and endeavor to make something else of their lives.

There is a fine line to be drawn here. While participants can and do look forward to the future, they can never completely forget the experiences they have had while ill. Indeed they must not; their efforts to maintain their health actually rest on knowledge of their illness. Similarly, coming to terms with lost opportunities in life, and with restrictions and limitations, is done in relation to the facts about what has happened in the past.

The terms 'recalling' and 'letting go' are not synonymous with remembering and forgetting. Rather, they are used to describe the processes by which the participants brought their experiences of the past to mind in the present, and then, as it were, put them aside in order to look ahead toward the future. Such a process requires an understanding and an honest appraisal of what has happened, of the participants' own past.

We project ourselves into an anticipated future as the ultimate aim of our endeavors. But this is not the only temporal dimension that is at work in our projection, because our projection is not a free choice of the future. According to Heidegger, we cannot make any such projections without an existing understanding of the world and ourselves in it, an understanding determined by the past we have been and still are. Therefore, not only do we carry our past with us, as one carries weighty memories, but we always already understand ourselves and our projects in terms of the past and out of the past. Finally, in all our enterprises, whatever they may be, we are tied to the present, because we are in and with the world that absorbs us and ties us down to our everyday endeavours (Frede, 1993, p. 64).

The effect of past experience and of the way that the participants found themselves in-the-world was sometimes very difficult. One young man, Chris, was philosophical about his illness and accepting of his diagnosis, yet as he explained, there are times when life looks very dismal to him.

> It's hard being paranoid. It's not hard being high but it's hard being paranoid. And it's hard being schizophrenic because you are not part of society any more, you haven't got a job, you're not at university any more, you've got nothing left. Other people who I would think of as having a hard life are people who appear to be very brutish and whose only enjoyment appears to be drinking and who work like dogs to... what appears to me to be like they work like dogs all their lives, but they wouldn't feel like that, they'd feel on top of the world I think. But you don't feel like that when you are a schizophrenic. There's no thing of, oh OK you can feel on top of the world when you are on a high, but when you are not on a high ... you are at the bottom. And as much as schizophrenia can give great imaginative things, but [sic] it can also take all your imagination away and leave you with ... with just rock hard reality. No comforting job, no comforting relationships, no comforting anything. It is out of control, it is something that just happens. You get to the stage where you're thirty ... you feel like you've accomplished nothing, you're just walking down the street and that's all you are doing ... you're not thinking, you're not planning, you're not looking forward to anything, you're just walking down the bloody street. It gets a bit tiresome after a while.

But not all the consequences of having a schizophrenic illness are negative. Several participants were able to find positive aspects in their experience, and to identify strengths or opportunities that had arisen for them as a result of their life situation. This was especially true of those who had a creative bent, and who were able to capitalize on the process of recalling the past, incorporating their recall into their creative projects.

On the other hand some participants reported specific, frightening or unpleasant aspects of illness that they felt it best to let go. A 21-year-old woman, Lucy, kept a diary during some of the time she was most unwell. She describes how and why she decided to destroy this evidence, which was a reminder of her illness:

> I had to record the next sighting of whatever I saw [hallucinations]. Because I had everything in exact detail. And in March this year I just got so scared and I thought oh shit, I can't keep this, 'cos it was so painful looking back. And it was like I was reliving it. And I thought if I'm ever going to break free from it I'll have to burn them I'm afraid.

Reconciliation of the fact of having been ill, of some residual difficulties, and of her now altered future is summed up in the statement made by another woman, Judith.

It's wrecked so many things for me. Because I really want to succeed in doing something. I know I can do that, but I don't know how to go about that. I don't know what steps to take. It's like the sun setting in front of me, and I'm going to go for it, but I know that there's going to be a lot of set backs on the way.

Understanding and accepting the nature of their illness, and recognizing that care is needed in managing their lives from here on has enabled most of the participants to integrate their past into their present. They are able to recall their past, to recognize and come to terms with that experience and to let it go. They avoid dwelling with their illness or in the past, yet continue on with life in light of the understanding they have from these experiences, which are part of who they are in-the-world. It is not only the participants themselves who must let go; in their experience they also find that family and health professionals must learn to accept them as no longer unwell, and to let go of the "patient" or "sick" ways in which they were previously known.

According to Heidegger's description of human existence, to know what we *are* is at the same time to know what we *can and should be* if we are to achieve coherence and unity in our lives (Guignon, 1993, p. 231). In integrating their experiences of illness and treatment, their Being-in-the-world and Being-with-others, participants also looked forward to the future; they were concerned in their involvement in the world in a future-oriented way. Thus the participants maintained a hopeful state-of-mind and held up for themselves goals and dreams to which they felt they could aspire.

In *Being and Time* Heidegger (1927/1962) is clear that hope is not simply a forward-looking expectation. Rather, hope is intensely bound to past experiences. Hope, he suggests, is related to our "having-been," to past hurts, disappointments and pain. As the opposite of "depressing misgivings" hope is possible only when the past is acknowledged and taken on board. Hope is related to what we wish and look forward to for ourselves. In order to look forward in this way we must acknowledge what has already been and accept our thrownness, the way we are in-the-world. It is only from this acceptance that we can move forward to what we hope for.

The different situations of the participants meant that different aspects of life called for hope. Those participants who were in stable relationships, for instance, had no need of hope in regard to their locating a partner in the way that several of the unattached people did. Those with children all held hopes for them in the future, even if such hopes were very broad and unspecific.

Most hopes expressed by the participants related to living a "normal" life. Those who lived in supervised accommodation spoke of a desire to live with "ordinary" people, or to own a house of their own. One couple lived in a small council flat in a supervised block and hoped that they might eventually be able to rent on the open market and thus escape the welfare system in that regard.

Achievement in a field that interested them was the subject of the hopes and dreams of several participants. Here, as in the dreams about home and family life, my feeling was that some of the hopes of the participants could seem grandiose and unlikely to be fulfilled. However, whether or not the dream is attainable is not really important in terms of its function as a counter to a sad and difficult past and present, as Heidegger (1927/1962) suggests hoping to be.

Jack, a sixty four year old man who had first been diagnosed with schizophrenia some forty years ago, placed a great deal of trust in science and education. Not having been in a position to gain much formal education in his youth, he now dreamed of leaving money for an Institute of Higher Learning, which would include social, medical and scientific research. The Institute would be named after him, and would carry on "for all eternity." Given Jack's concern for the environment, politics and science, his dream might be seen to cover both his desire to do good in the world, and to make possible some recognition of his worth and personal contribution.

Chris, mentioned earlier, became ill while he was a university student and subsequently had to withdraw from his studies. He, too, holds an ambitious dream of recognition, this time for his intelligence and problem solving ability and his personal moral stance:

I think that I might be as important as Einstein because when I was at university the first time I came up with a legal argument in a Privy Council case that had never been thought of before by anyone. And then I did another assignment which I think was pretty much on the same level but I never went back to collect it because I had achieved everything with my third assignment and I didn't want anyone to say that I got less than an A. I just couldn't handle that any more, having some idiot tell me what I'd got; I thought I'd reached a stage ... I had reached a stage where I was not prepared to listen to other people's evaluation of my work. And I like to think that I did something there that was taken notice of and of course I wasn't interested in any reward for it or anything. I like to think that it became important, that it was important. And now I like to daydream that it was as important as the theory of relativity or more important so now I'm walking around everywhere thinking I'm as good as Einstein. And in some ways I think I'm better. He had a son who was in a mental hospital and he did something to his first wife which wasn't very nice, divorced her or something, and never went to see the son. I don't know, something like that. But he wasn't perfectly good, this is from his ... one of his other children, they were saying he wasn't God you know. And that people have this image of him as being totally incorruptible and everything when he was as corrupt as most people. And I'm not.

Chris laughed several times while telling me this, demonstrating the amusement he felt at holding such dreams, which he acknowledged to be far-fetched. Yet the solace they provide in the context of his past life can also be poignantly felt.

Pure and unambiguous definitions that would differentiate between hopes, dreams, wishes, faith and beliefs would be neither possible nor useful in a study such as this, which concentrated on the life stories of the participants. It would be artificial to divide the participants' aspirations into those that are "realistic" and those that are "delusional" or "grandiose." While such divisions may form a useful part of an interview designed to gather information for the purposes of diagnosis and treatment, any such categorization in the life stories of people who are to all intents and purposes well (as were the study participants) would, of necessity, lose sight of their imaginative, creative and meaningful nature. There is also the potential of abusing the honest communication in which the participants shared some of their most unusual thoughts. Most people probably nurture dreams or fantasies of fame or wealth in some form, yet they are not in a position to have such ideas diagnosed or judged as to their feasibility. By whatever names the thoughts and dreams of the participants are known, they serve a clear purpose of directing the participants into the future, a future which can be seen in a positive light whatever their past experiences might have been.

III. THE LIVED EXPERIENCE OF POSTPARTUM DEPRESSION

In contrast to the ostensibly 'well' participants in the study reported above, participants in two studies into the experience of postpartum depression (Beck, 1992, 1996) are shown to have held little in the way of hope for their future. Beck's earlier paper (1992) is a powerful description of seven women's experience of postpartum depression. In it Beck demonstrates the use of Colaizzi's method of data analysis, and presents eleven themes to describe the experience of postpartum depression.

As Beck summarizes the themes in the journal abstract (1992, p.166):

> Postpartum depression was a living nightmare filled with uncontrollable anxiety attacks, consuming guilt, and obsessive thinking. Mothers contemplated not only harming themselves but also their infants. The mothers were enveloped in loneliness and the quality of their lives was further compromised by a lack of emotions and all previous interests. Fear that their lives would never return to normal was all-encompassing.

Clinicians reading this summary will recognize symptoms that indicate a diagnosis of postpartum depression, although as Beck points out the themes do not correlate well with the generalized depression scales often used in practice settings.

But there is much more here. In her choice of words, and in illustrations from the data, Beck conveys a great deal of the urgency, distress and tension of the participants. We are left not just with a list of symptoms, but with a pervasive feeling of the women's isolation and despair.

Further, by identifying the women's need to be mothered themselves, Beck (1992) suggests a return to a kind of nursing intervention that has been out of favor for some considerable time. The idea that nurses might themselves step in and act for women beleaguered by their illness, or that they might encourage other support people to act in this way, is not a fashionable one in an era where independence is valued over interdependence. In this way Beck throws a new light on our professional actions.

In a later study Beck (1996) extends her previous work. Through interviews with twelve women she demonstrates the impact of postpartum depression on mothers' interactions with their infants and older children. Here the sense of overwhelming responsibility felt by the mothers, and their own lived inability to act toward their children as they themselves felt and knew they should, leads to some important conclusions about effective support services and the kind of interventions that might help foster improved maternal-infant interactions.

IV. THE LIVED EXPERIENCE OF THE RELENTLESS DRIVE TO BE EVER THINNER

In presenting the findings of a phenomenological study about the relentless drive to be ever thinner, Santopinto (1989) sheds the theoretical lens from the outset, describing the experiences of two women who had lived the experience, but using no familiar diagnostic terms in doing so. This researcher employs Giorgi's modification of the phenomenological method to analyze the data, and reflects on the fit of the findings with a complex contemporary nursing theory, that of Parse (1981). A general structural description is given, and the results are compared with the writings of both Binswanger (1958) and Boss (1963). Some useful suggestions are made for changes to practice as a result of the understanding generated by the study.

The generalized description is too long to be quoted here in full, but the first section gives a taste of the whole. Santopinto (1989, p. 34) says,

> The relentless drive to be ever thinner is a lived agony emerging in mutual interrelationship with others. One becomes aware of being seen in different ways by others who understand the experience only superficially. Buoyant with the reaffirmation of a new self, one intensifies the struggle to be ever thinner in a leap beyond one's images of the past. Steadfastly confirming this way of being-with-world, one lives a pattern of skirmish and encounter with close others.

In this description we are transported past the common everyday understandings we hold about the relentless drive to be ever thinner, and past the theoretical understandings we have of anorexia nervosa. We are thereby helped to gain an intuitive grasp of the experience, both in its general form, and as it might be lived by an individual client.

V. CONCLUSION

I began this essay with the suggestion that I am uncertain about the commonalities inherent in mental illness. Apart from the social stigma that such illnesses bring, I am of the opinion that we would be safer to try to generalize about the experience of illness *in itself* than to conclude that mental illnesses are in some way set apart from physical ones. I have come to this conclusion largely as a result of my explorations into the phenomenological view. As I see it, the trajectory of illness and its impact on the person in terms of acuity or chronicity have more common effects than whether the illness is thought to be one situated in the mind, the brain or in some other part of the body. There is scope here for considerable research in clarifying our understanding and better illuminating the human experience. Perhaps there *are* aspects of the 'mental' illness experience that sets such problems apart from the experience of 'physical' illness.

As for the other possibilities for future research in this field, they seem to me to be almost endless. As my brief reference to delusions has shown, there is merit in being able to consider a variety of perspectives, drawn from studies by people with different ways of understanding the phenomenon. In this way the shadows of our understanding may be filled in and rounded out. Similarly there are numerous possibilities for understanding different illness experiences, and for comparing these with each other.

Since virtually no illness experience is isolated to an individual but is rather a social occurrence, there is also scope for examining the experience of those who care for people in times of mental illness. I have not had the scope to explore such literature here, but would point interested readers to studies by Chesla (1994); Jeon and Madjar (1998); Meighan, Davis, Thomas and Droppleman (1999); and Tuck, Dumont, Evans and Shupe (1997) as examples of phenomenological work that explores the family caregiver's experience.

There have been several times in this essay when I have identified contradictions, different ways of approaching things, new ways of seeing. Along with many others I believe that it is only through attempting to use *all* such avenues that we are truly able to say we are doing our best in terms of the evidence we have for practice (see also Bischo, 1998; Coles, 1995;

Greenhalgh, 1999; Sweeney, MacAuley, and Gray, 1998). The phenomenological approach cannot stand alone, but excellent practice cannot do without it.

The more one understands the struggles that a fellow human being undergoes when ill, the more one recognizes the sick person as just that – a *fellow* human being. Phenomenological research points to our essential humanness – a humanness that is evoked on reading such work. In provoking thoughtfulness in practitioners, a phenomenological approach also has the potential to have a significant impact on the experience of those who come to us for help.

Auckland University of Technology
New Zealand

BIBLIOGRAPHY

Allen, J.: 1990, 'The role of imagination in phenomenological psychology', *Review of Existential Psychology and Psychiatry,* **20**, 159-167.
Astbury, J.: 1996, *Crazy for You. The Making of Women's Madness,* Oxford University Press, Melbourne.
Barham, P., and Hayward, R.: 1991, *From the Mental Patient to the Person,* Routledge, London.
Baron, R.J.: 1985. 'An introduction to medical phenomenology: I can't hear you while I'm listening', *Annals of Internal Medicine* **103**, 606-611.
Beck, C.T.: 1992, 'The lived experience of postpartum depression: A phenomenological study', *Nursing Research* **41(3)**, 166-70.
Beck, C.T.: 1996, 'Postpartum depressed mothers' experiences interacting with their children', *Nursing Research* **45(2)**, 98-104.
Binswanger, L.: 1958, 'The case of Ellen West: An anthropological-clinical study', in R. May, E. Angel, and H. Ellenberger (eds.), *Existence: A New Dimension in Psychiatry and Psychology,* Basic Books, New York.
Bischo, D.: 1998, 'The art of nursing: The client-nurse relationship as a therapeutic tool', *Nursing Case Management* **3(4)**, 148-150.
Boss, M.: 1963, *Psychoanalysis and Daseinsanalysis,* Basic Books, New York.
Burch, R.: 1989, 'On phenomenology and its practices', *Phenomenology + Pedagogy* **7**, 187-217.
Chesla, C.A.: 1994, 'Parents' caring practices with schizophrenic offspring', in P. Benner (ed.) *Interpretive Phenomenology,* Sage, Thousand Oaks, pp. 167-184.
Coles, R.: 1995: *The Mind's Fate. A Psychiatrist Looks at His Profession* (2nd ed.), Little, Brown and Company, Boston.
Davidson, L.: 1992, 'Developing an empirical-phenomenological approach to schizophrenia research', *Journal of Phenomenological Psychology* **23(1)**, 3-15.
Davidson, L.: 1993, 'Story telling and schizophrenia: Using narrative structure in phenomenological research', *The Humanistic Psychologist* **21**, 200-20.
Davidson, L. and Strauss, J.S.: 1992, 'Sense of self in recovery from severe mental illness', *British Journal of Medical Psychology* **65**, 131-45.

Fisher, W.F.: 1978 'An empirical-phenomenological investigation of being-anxious: An example of the meanings of being-emotional', in R.S.Valle and M. King (eds.), *Existential-Phenomenological Alternatives for Psychology*, Oxford University Press, New York, pp. 166-81.
Frede, D.: 1993, 'The question of being: Heidegger's project', in C. Guignon (ed.), *The Cambridge Companion to Heidegger*, Cambridge University Press, Cambridge, pp. 42-69.
Greenhalgh, T.: 1999, 'Narrative based medicine: Narrative based medicine in an evidence based world', *British Medical Journal* **318**, 323-25.
Gilman, S.L.: 1988, *Disease and Representation*, Cornell University Press, Ithaca, NY.
Guignon, C.B. (ed.): 1993, *The Cambridge Companion to Heidegger*, Cambridge University Press, Cambridge.
Heidegger, M.: 1927/1962, *Being and Time* (J. Macquarrie and E. Robinson, Trans.), Basil Blackwell Ltd, Oxford.
Jamison, K.R.: 1995, *An Unquiet Mind*, Random House, Toronto.
Jeon, Y. and Madjar, I.: 1998, 'Caring for a family member with chronic mental illness', *Qualitative Health Research* **8**, 694-706.
Karp, D.A.: 1996, *Speaking of Sadness. Depression, Disconnection, and the Meanings of Illness*, Oxford University Press, Oxford.
Liebrich, J.: 1999, *A Gift of Stories*, University of Otago Press/Mental Health Commission, Dunedin, New Zealand.
Lindsey, E.: 1996, 'Health within illness: experiences of chronically ill/disabled people', *Journal of Advanced Nursing* **24**, 465-72.
McLellan, M.F.: 1996, 'Literature and medicine: Some major works', *The Lancet* **348**, 1014-16.
Meighan, M., Davis, M.W., Thomas, S.P., and Droppleman, P.G.: 1999, 'Living with postpartum depression: The father's experience', *MCN, The American Journal of Maternal/Child Nursing* **24**, 202-208.
Parse, R.R.: 1981, *Man-Living-Health: A Theory of Nursing*, Wiley, New York.
Porter, R.: 1987, *A Social History of Madness*, E.P. Dutton, New York.
Reynolds, W.J. and Scott, B.: 2000, 'Do nurses and other professional helpers normally display much empathy?' *Journal of Advanced Nursing* **31**, 226-34.
Santopinto, M.D.A.: 1989, 'The relentless drive to be ever thinner: A study using the phenomenological method', *Nursing Science Quarterly* **2**, 29-36.
Sass, L.A.: 1988, 'The land of unreality: On the phenomenology of the schizophrenic break', *New Ideas in Psychology*, **6**, 223-42.
Sass, L.A.: 1992, *Madness and Modernism: Insanity in the Light of Modern Art*, Basic Books, New York.
Styron, W.: 1990, 'Darkness Visible' in S. Dunn, B. Morrison and M. Roberts (eds.), 1996, *Mind Readings. Writers' Journeys Through Mental States*, Minerva, London.
Sweeney, K.G., MacAuley, D., and Gray, D.P.: 1998, 'Personal significance: The third dimension', *The Lancet* **351**, 134-36.
Tuck, I., Dumont, P., Evans,G., and Shupe, J.: 1997, 'The experience of caring for an adult child with schizophrenia', *Archives of Psychiatric Nursing* **11**, 118-25.
Walton, J.A.: 1995, *Schizophrenia: A Way of Being-in-the-World*, Unpublished PhD thesis, Massey University, Palmerston North, New Zealand.
Walton, J.A.: 1999, 'On living with schizophrenia', in I. Madjar and J.A. Walton (eds.) *Nursing and the Experience of Illness: Phenomenology in Practice*, Allen and Unwin, St Leonards, NSW and Routledge, London, pp. 98-122.
Walton, J.A.: 2000, 'Schizophrenia and the life-world of others', *Canadian Journal of Nursing Research* **32**, 69-84.
Wolpert, L.: 1999, *Malignant Sadness. The Anatomy of Depression*, Faber and Faber, London.

SECTION FOUR

CLINICAL PRACTICE

CARL EDVARD RUDEBECK

GRASPING THE EXISTENTIAL ANATOMY: THE ROLE OF BODILY EMPATHY IN CLINICAL COMMUNICATION

I. INTRODUCTION

The aim of this contribution is to give some clarity to "body experience": the ways it unfolds in the experiences, presentations, and reception of presentations of symptoms in the clinic. The idea behind this venture is that the apparently self-evident notion that symptoms are about body experiences results in "body experience" attracting less attention than called for by its actual importance. Since symptom presentations are usually thought to be the unequivocal expressions of the situation in the body, the possible pathology lying behind a certain symptom tends to become more intriguing to medicine than the experience itself. In the consultation the doctor tends to *observe* the symptom presentation and judge whether it falls within preconceived disease categories, rather than share it. This attitude incurs two immediate risks. First, the doctor's apprehension of what the patient presents may be shallow, leading to misunderstandings of various extent and importance even from a diagnostic point of view. Second, the patient may feel the doctor lacks interest in her particular situation, causing her to withdraw, which, in turn, may threaten the outcome of the consultation. On a broader scale, the scant attention paid to body experience limits the competence of doctors. If we establish a way of "thinking" body experience that is in consonance with its real character in the life of the individual, doctors may have a frame of reference that will make it easier for them to become attuned to, and learn from, symptom presentations. They will also have a more explicit reference for the experience of their own bodies, thus increasing their self-awareness and enabling them more fully to grasp the experiences of their patients. At the level of the medical profession, the "cases" of clinical discourse may be transformed from the anonymity of diseases to the particular experience of the individual, but still with the focus on the body. Lastly, the gap between practice and the reflection on practice may be narrowed. It is especially important for a practice-based discipline like family medicine/general practice to make practice itself the core of professional reflection.

II. THE SYMPTOM PRESENTATION

Doctors never meet symptoms independently of the persons having them. In fact, what doctors judge are symptom presentations (Rudebeck, 1992, p.72). This may seem self-evident. Nevertheless, the distinction between symptom presentation and symptom is crucial in understanding practice in its own right. Saying "symptom presentation" is an unequivocal reference to practice, and thus, to the factual starting point of the clinical interaction. Furthermore, the symptom presentation refers to a unique individual and a unique situation, just as every instant in real life is unique. In the symptom presentation we may find out why apparently similar symptoms evoke very different responses. For instance, pain in the upper right abdomen is a very broad categorisation of a huge number of individual presentations. The symptom presentation is an act of communication and, as such, it is a whole in which form and content are inseparable. A crucial aspect of the doctor's acquired competence is the ability to distinguish the symptom – the very experience – within the presentation.

III. SYMPTOMS AND BODILY EXPERIENCE

When examining the meaning of symptom presentations, it is vital to answer the question, "What kind of subjectivity is body experience?" Phenomenological analysis (Merleau-Ponty, 1962; Sartre, 1956; Toombs, 1987, 1992; Leder, 1990) points to the decisive gap between the body as a lived reality – lived body – and the body as object for medical judgment and intervention. The lived body is the body as non-thematised. In this perspective, I do not have a body, but I am my body. The relation with my body is "existential" rather than "objective." I am the possibilities and intentionalities of my body projecting itself into the world of time, space and particulars. The lived body is "my particular viewpoint on the world" and "it is always with me" (Merleau-Ponty, 1962, p.90). However, the meaning of the body, as body, is not explicitly attended to. It is "surpassed" (Sartre, 1956, p. 309) or, to use Leder's (1990, pp. 1-3) language, the body is, in fact, "absent." Illness is experienced as a disruption of the lived body (Toombs, 1992, pp. 62-70) – a disruption that includes a shift in consciousness to an explicit awareness of the body. The symptom experience draws attention to the body through discomfort and/or disability and/or worry to such a degree as to motivate a medical consultation.

Through what Leder (1990, p.78) calls the "hermeneutic moment," the person seeks for explanations and strategies to deal with the situation. The threshold for judging the changed state as a symptom, rather than a variation within the borders of normality, varies among individuals due to personal and contextual features (Morell, 1978). This variation is not the main issue here, but it conforms to the fact that experience is subjective. However, the person's cognitive approach to the changed existential predicament in which she becomes aware of her body as *body*, implies a certain degree of objectification, or reflection. In phenomenological writings this objectification/reflection is considered to be one aspect of a dichotomy in which the lived body is the other. Thus, what the patient tells her doctor about is an abstraction from the real experience. There is thought to be a definitive gap between the experience as lived, and the experience as it is narrated. However, this "gap" may not be so obvious in everyday life. When somebody tells me about her terrible fatigue, or headache, I usually presume that she is talking about the actual experience. Perhaps a paradox is lurking here.

IV. THE EXISTENTIAL ANATOMY: THE STRUCTURE OF THE SELF-AS-BODY

Experiences from clinical practice, and from other areas as well, suggest there is a meaning structure of the body in lived experience. This structure is not a uniform level of body awareness, but represents varying degrees of conceptualization of body experience within varying situations and contexts. If we name this overall thematization of the lived experience of the body *self-as-body*, we may accordingly name its structure *existential anatomy* (Rudebeck, 1992, pp. 45-46; Rudebeck, 2000).

As existential anatomy, the body projects itself into the world; its meaning is the meaning of life. But still, it is a bodily meaning of life. Accordingly, the heartbeat is life, the legs are walking and independence, the hand is grasping or showing the way, the eyes are seeing or weeping, the sex organs are male or female, and so on. From the particulars of those experiences of embodied life, bodily objectification takes place when the person tries to come to grips with experiences that cause ailment or fear. Maybe "intellectualization" captures what this process is about more precisely, an expansion of the cognitive aspect of the totality of experience, where the other aspects are perception and affect.

The anatomy of the physical body, based on the findings in dissections, has its very definite outer border in the skin. Inwardly the parts of the body are demarcated from each other through borders with exotic names such as margo, fascia and fossa. The self-as-body also has its borders, but they are more like horizons, gradual transitions, than sharp limits.

According to Husserl, in order for a phenomenon to appear as a discrete entity in the world, it has to be built up through a series of intentional and perceptive acts which link up with each other in a harmonious way (Karlsson, 1993, p. 62). Every new intentional act (milliseconds) tries and confirms what has emerged through the preceding acts. Every figure-background form is in that sense also a form in time. An external object appears only in relation to the embodied subject who projects the world (Merleau-Ponty, 1962) and to other entities of the external world. The subject, in turn, also appears to itself in the continuum of intentional acts in relation, on the one hand, to the external world, and on the other, to the physical reality of the own body, which is the "body-as-nature" (Rudebeck, 1992, pp.40-47). The existential anatomy unfolds as an aspect of the subject between those two meaning-producing horizons. Rather than stating that the body is either "absent" or "objectified," I would suggest that the importance of the body experience varies within the totality of intentionalities of the living subject. However, it makes sense to distinguish situations in terms of their dominating intentionality.

In the *outward* intentionality, corresponding to "the ecstatic body" in Leder's (1990, pp. 11-36) terminology, the subject addresses the physical and social world with the own-body more or less subsumed. The body is tacitly taken for granted. When, for instance, one states that something is big, or heavy, the own-body is a tacit reference. Even when actually lifting, the impression of weight supersedes one's actual effort as long as one's strength is considered normal, and the lifting is not a matter of competition.

As Svenaeus (1998, pp. 181-82, 2001) points out, in extending the analysis of Heidegger, the role of the body in outward intentionality is tool-like. The meaning of the body is hidden in the preoccupation with the world until, for some reason, intentionality shifts to become reflexive or inward, and thus, the body detaches itself from the background. But very little has to change to make that happen. The existential anatomy is like a hidden pattern in the web of the experience of the world. It is hidden only because there are other patterns that are more appealing.

When the body comes to the fore in the *reflexive* intentionality, its meanings are the existential anatomy. The *self-as-body*, is intended where it ambivalently, and concomitantly, relates to the outer world and to its inner horizon of nature. The seed of the reflexive position is the awareness of one's existence, which is like an ever-returning unveiling of oneself in the midst of life. As long as one lives, existence is permanent, but the recognition of its "unbearable lightness" is not. Therefore, the intentionality tends to move from life itself to the content in life – in this case the simultaneous and mutually defining relations of the self-as-body to the world, its being-in-the-world (Heidegger, 1962, pp.78-90), and to the body-as-nature. At both interfaces the self-as-body is giving and attaining meaning. In the reflexive intentionality, the outer horizon is not just something the subject tries to stretch beyond in grasping the world. Rather it is the point in consciousness where the self-as-body becomes prominent in the interaction with the world. This may be a result of a deliberate mental act, or happen accidentally as when one slips on a spot of butter on the kitchen floor. The outer horizon may also be brought into concrete experience through a simple practice of body awareness: When standing without shoes on the floor, one can concentrate on the existential anatomy of the soles (self-as-body) against the floor (the world), or, the reverse, of the floor against the soles.

The constitution of meaning at the inner horizon of the reflexive intentionality is similarly reciprocal. The continuous flow of perception and proprioception gets its overall meaning through the subject's being-in-the-world. A well-working organism amounts to a sense of normality, capability and security. Homelikeness is the term Svenaeus (1999, pp. 151-57, 2001) suggests to describe this existential situation. However, the inner horizon is also where the self-as-body detaches itself as the numb fact of the body-as-nature. The self is continuously born as "more-than-nature" from the background of biology.

In the *inward* intentionality, the body-as-nature is actively addressed. It has links to Leder's (1990, pp. 36-38) concept "the recessive body" but, whereas Leder here refers to the interior of the body in a topographic sense, the inward intentionality refers to the very facticity of the body (Heidegger, 1962, p.174). This facticity is disclosed when the overriding meaning of "normality" of the self-as-body cracks. If one really tries to understand one's body-as-nature as an aspect of being, one will get lost in a journey of consciousness that ends asymptotically in the microcosm of the own-body. "Other-than-me" is the only answer available (Toombs 1987, 1992, p. 75). Yet, this is an endeavor no-one can resist, especially when the body-as-nature presents itself through the implacability of malfunction or breakdown. Meaninglessness is hard to endure, and so is the fear that very often accompanies it. Intellectualization – the hermeneutic moment (Leder, 1990, p.78) – therefore tends to become the attitude in face of the inward horizon. Of course, plain curiosity or fascination may also be valid motives for an objectifying scrutiny into one's body-as-nature. When actually approached as experience – for instance through mindfulness meditation, or simply accepted as an irrefutable fact – meaninglessness may offer an entrance into spiritual life (Varela, Thompson, Rosch, 1993, pp.253-54).

To sum up, the existential anatomy can be characterised through the following propositions:
➢ It is a defined field of meaning – intentionalities – in human experience.

> The field extends between, and gets its meanings from, on the one hand, the horizon of the external world, and, on the other, the horizon of the body-as-nature.
> The core meaning is "my body", the self-as-body.

V. THE *GESTALT*

The existential anatomy is woven into the totality of human experience and expression, but it is explicit only in the mode of self-as-body. As long as the body does not signal from within due to derangement, it is not the impulse flow itself that decides whether the body is in the foreground or the background of experience. Rather, it is the overriding meaning of a certain situation, the *Gestalt* that decides the focus of attention (Perls, Hefferline and Goodman, 1973, pp. 13-21). The *Gestalt* includes the experiencing person as well as that which is experienced. The very implication of the concept of the *Gestalt* is that intentionality is never focused on a single phenomenon. There is always a background, never nothingness, of the figure in the figure-background form, which is the prototype of experience. Also, there is always movement, the potential shift of intentionality, which means that aspects of the background may step into the fore. These transferrals are glides rather than jumps between definite categories, something that is very obvious in everyday experience. At the start of a cold there may be a hardly noticeable irritation behind the nose. But once this irritation becomes an "object" within consciousness, and parallel to the emergence of a more definite state of infection, the symptoms from the upper respiratory tract will gradually become more prominent in consciousness.

Thus, it depends on their role in the *Gestalt* whether the individual actively perceives her posture and state, and whether she perceives them to be normal or not. Whether she is seriously frightened, or running up the stairs, her quickly bumping heart is a tie of meaning between herself and the event or action in question. If struck by an equal increase of pulse rate when lying in bed and expecting to fall asleep, the same person may find the possible meaning of that event quite threatening, thinking something may be wrong with her heart. The sequence is here from a normal, floating attention, through a sudden awareness of the existential anatomy of the heart – my heart, my life – to a cognitive approach of more or less elaborate character. The original experience, however, is usually still there at the basis of most symptom presentations. If the doctor knows his existential anatomy, he can follow the patient back to the origin of the experience, where it may be more easily understood. As stated earlier, since the reflexive intentionality is a disclosure of the self-as-body, as it emerges between the outward and the inward horizons respectively, it brings about the meanings of those two horizons: I walk with my legs, I see with my eyes, I feel in my chest, etc. The emphasis is on the self, but the self cut off from its relations is nothing. Therefore, the meanings of the existential anatomy (beside the bare meaning of "self") cannot be grasped other than in the implicit forms, the hidden patterns, of the outward and inward intentionality. But here we may trace them in the three basic and interrelated dimensions of experience: perception, emotion, and cognition.

A. Perception

Merleau-Ponty (1962) makes it very clear that, to the human being, perception is meaning. In experience there are no such things as pure sense data. That which is perceived is perceived within the total form of the *Gestalt*. In playing the piano, the virtuoso's whole intentional being is expressed in the contact with the music itself. He hardly senses keys. His body is ecstatic, a human instrument treating a material one. However, in extremely demanding technical passages, he probably becomes aware of his hands and fingers touching the keys as the bodily aspect of playing.

If, as an act of sympathy or love, an intimate puts her hand on my shoulder, I sense this intention and feel good accordingly. On the other hand, if I am walking alone on a dark street late at night and feel the same touch on my shoulder from behind, I perceive this as a serious threat. The neuronal, sensory excitation is originally the same in the two situations, but the experiences – the meanings – are very different.[1]

In this respect, the sense organs are "ecstatic." They tend to diffuse as the hidden dimension of the experience in question. But the act of perception also has its bodily and proximal aspect. This becomes obvious in situations that, for some reason, stand out as remarkable within the normal flow of perception. Examples are rubbing one's eyes when doubting what is within the perspective, or sniffing when trying to recognize a smell, or turning the head and cupping the hand behind the ear when really wanting to hear. From the point of view of the subject, the sense organs amount to the *acts* and *faculties* of perception themselves. The existential anatomy of the eyes is seeing, of the ears hearing, and so on. This has important implications for experiencing and presenting of symptoms. For example, even minor affections of the outer eye may imply the risk of impaired vision to many patients. Within the existential anatomy there is no decisive distinction between the outer and the inner eyes. Likewise, in cases of occluding earwax, it is very difficult to distinguish the local discomfort in the ear from the existential predicament of being shut off from the world like a captive within one's own senses.

B. Emotion

The emotions have a distinct anatomy that enhances what is experienced and expressed. When I am sad, I feel a lump in my throat. When I am happy, I may feel like singing, which is an outward stance of the whole bodily being. In deep grief I really feel a weight in my chest, sometimes even a sharp pain. The tension of grief in the chest may be freed in the deep sigh of relief. In anger withheld, I clench my fist in my pocket, while in the opposite mood I may almost embrace the whole world in a very open bodily existential posture. The emotion is proximally a bodily state, and distally it affords to the certain situation its overall meaning – interesting, frightening, joyful and so on. Once one becomes aware of the existential anatomy of a certain emotion, this may increase body awareness as well as emotional self-awareness. In the former, the body is waking up. Sets of impulses from within the body converge into distinct experiences, which may become recognized as such when they return, and when described by others. In the latter, one may recognize that a certain bodily state bears an emotional tension, and once this has occurred such recognition may be immediate in new situations.

In the clinic common symptoms may be recognized as "frozen" emotions. One very obvious example is the "globus" phenomenon, which is the permanent and disturbing feeling of a lump in the throat. Certain forms of chest pain, and of epigastric pain, are probably others. The perpetuating *Gestalt* of anger and/or fright may involve muscular dysfunction all through the body. The bodily existential perspective differs from the mainstream psychosomatic perspective in that the bodily state or symptom is held to be an aspect of the emotion rather than an effect of it. The symptom is itself intentional (Bullington, 1999).

Tomkins (1992, pp. 64-96) holds an array of the basic affects to be inborn. He divides them into positive (interest, happiness, astonishment) and negative (sorrow, fear, anger, shame, disgust and dismay). From early on in life all these can be recognized in the facial expressions. The face is the "emotional organ," par excellence. It is in itself expressive, both from inside the subject – you may feel your face as part of the bodily aspect of a certain emotion – and from outside, to others. This becomes obvious to doctor and patient when examining the functions of the facial nerve. Insofar as there is no severe damage, making faces for a neurological examination is much more awkward than lifting one's leg in a lying position to test gross strength.

In fact, as a bodily region, the face expresses the idea of the "absence" of the body with ease. The face is very much the person herself, both in the mirror and in the gaze of others. A smile, or a look of sorrow, does not need any interpretation unlike, say, a feeling such as discomfort in the chest. But we may also all recognize the way habitual emotional expressions sculpture the anatomy of the face. Some persons have sorrow as their "basic look," others a constant politeness and interest in their face. Once recognized as constant, such "looks" become difficult to interpret. When petrified as body, the expression loses its meaning in the immediate sense. Probably though, in a *Gestalt* where the time dimension is extended into years, the habitual look will bear a very important meaning. In the clinical context, jaw problems and tension headaches may be manifestations of that meaning.

According to Tomkins (1992) the physiology of the affect (emotion) – such as blushing or becoming pale, increase in pulse rate, or raising of the hairs of the skin – enhances the affect and, thus, becomes an aspect of the affect itself (its proximal aspect). He also points at distinct links between the affects and the voice and the breath. In all, we see an emerging theoretical basis for what is evident in experience; namely that the affects (the emotions) are an important dimension of the lived experience of the body.

C. Cognition - Language and the Existential Anatomy

The observations and conclusions of Jean Piaget (1976, pp. 54-61) suggest a model for cognitive development in the child. Starting from a point of "primary knowing," a meaning relation "afforded by sensory-motor experience" as a "direct response to the world" (Toombs, 1992, p.54) and from a set of reflexive, habitual movements, the infant embodies the experiences of space and time. From there, and through the processes of generalization/differentiation and assimilation/accommodation, and through memory, the symbolising capacity develops parallel to an ever-refined interaction with the world. Crucial in symbolising is language, which, according to Piaget, has its obvious roots in pre-conceptual experience. Mark Johnson (1988, pp.18-100) points to the fundamental role of originally embodied "image schemata," and their metaphorical projections across domains of experience, in the constitution of meaning. Even in our most abstract arguments, body experiences ground linguistic meaning. "What are often thought of as abstract meanings and inferential patterns actually do depend on schemata derived from our bodily experience and problem-solving" (Johnson 1988, p. xx). Logic may be seen as abstraction from relations of objects and forces in space-time. Examples of such "image schemata" are: "container," "balance," "centre-periphery," and "restraint-removal" (Johnson, 1988, p.126). Obviously here Johnson is close to the thought of Merleau-Ponty (1962, p.101) who discusses the relation between intelligible and objective space, and states that the meanings of prepositions like "on," "under," or "beside" emanate from the implications of an inhabited space where the "I" always moves around as its center. If I am not oriented, nothing else could be oriented either. The target of Johnson's critique is the objectivistic notion of meaning, according to which propositional cognition, through literal concepts, is a direct and valid representation of reality.

Language is embodied, but like – and even more than – the lived body, it extends far outside the limited "here and now" of the situated body. The force of language is abstraction. That force is, indeed, very strong, so strong that Plato inferred that the ideas were the "true" reality. But, irrespective of the degree of abstraction, the bodily anchorage of language is never totally lost. In the context of the existential anatomy it makes no pragmatic sense, however, to try to trace the bodily, proximal aspect of just any word. Rather, the very idea of the concept is to capture the variety of experiences that have the lived (which, of course, logically is the only) experience of the body as their common denominator. But having said that, we also have to recognize that the borders between the language of the existential anatomy and language in general are, in fact, rather broad areas of ambiguity, a soil for metaphors and poetry. Language expresses the ambiguity of existence itself, the "in between" so eloquently formulated by Merleau-Ponty (1962, p.430): "The world is inseparable from the subject, but from a subject which is nothing but a project of the world, and the subject is inseparable from the world, but from a world which the subject itself projects."

D. Body Language

Body language is a special case in which bodily being and signification merge in a direct and observable way. It displays the human presence and experience that is rendered invisible in abstract language. Body language is not a language of its own. Rather, it enhances, clarifies and calibrates interhuman communication in all its aspects. Bodily posture and movement become relational and social through the features of the specific context in which they are displayed. Through the warning finger, the raised hand, the nodding head, I disclose myself in the world of meanings. Although the outward stance of the whole body is prominent, it is foremost the very finger, hand, and so forth, that are transformed into distinct meanings. Therefore, the latent, expressive faculty of the distinct parts of the body is clearly a dimension of their existential anatomy. For instance, the rigid face of the person who suffers from Parkinson's disease incurs a serious handicap of communication, even though the pronunciation of words may be satisfactory.

E. The Language of Body Experiences

The words referring to body experience, per se, form the backbone of verbal symptom presentations. In the balance between the inward and outward horizons, they tend to approach the inward one. Nevertheless they also relate to the outer horizon, and then mainly as aspects of withdrawing from it. Homelikeness is under challenge. Because they are inherently meaningful as lived experiences, their "I-form" dominates over objectification.

Fatigue, ache, itching, numbness, weakness, dizziness are all names of distinct body experiences. However, like other words, they are abstract as long as they are not brought into the realm of individual life. Clinical algorithms refer to the generalized symptom experience of a class of persons having a certain disease; yet, the doctor in the real life situation has to interpret an individual message. Between itching, per se, and the itching experienced by an individual patient, are potential levels of intersubjectivity from the merely semantic to a shared understanding in a deeper sense. What is peculiar to the words referring to body experience, however, is that, even in their abstract sense, their only meaning is existential anatomy. It is impossible to refer to a state of itching, or fatigue, without referring to one's own, internal reality. Of course, the solipsistic argument is that human experience is always internal, and that intersubjectivity is a more or less random phenomenon. However, the qualitative and decisive difference between internal and external references is hard to deny. Although it may be difficult to give words to a certain bodily sensation that lies between the major semantic markers – for instance, a persistent headache, which is neither intense nor negligible, and which moves around the scalp with no obvious focus of pain – words may take shape in the process of dialogue. A prerequisite is, then, that the partner in the dialogue – for instance, a doctor – does not ask for the non-existing "objective" experience. If the endeavor to verbalize a body experience turns into intellectualization, the experience itself will vanish. So, if one really wants to hear the truth of the body, one has to encourage accounts of subjectivity.

In situations of bodily extremes, metaphors become the most effective messages. Accordingly, when tired, my legs are "heavy like lead" while, on another occasion I may "feel like bouncing about." "My stomach is empty," or "I am completely full," describe in an almost literal sense the situations of hunger and satisfaction. When struggling with an intellectual challenge, "I am thinking so hard I can feel the wheels turning," whereas in situations of intellectual fatigue, "my head is totally empty."

To sum up:
- The existential anatomy is the proximal, bodily aspect of perception, emotion and language.
- The existential anatomy of the sense organs amount to the acts and faculties of perceiving.
- Emotions have their distinct existential anatomy which is a combination of localized bodily experiences, facial expression, and vegetative reactions.
- Language as a whole has its resonance in the existential anatomy, but the language of the existential anatomy limits its references to body experience.
- Body language displays the role of the body in abstract language but it is not a language of its own, nor the language of body experience.

VI. THE EXISTENTIAL ANATOMY IN THE CLINICAL ENCOUNTER

The aim of the paper so far has been to suggest the existential anatomy as a way of "thinking" body experience. The endeavor has been to discern the pattern of the existential anatomy in the totality of experience comprised of perception, emotion, and cognition, here defined as language.[2] The existential anatomy takes its overriding meaning from the fact that it is the proximal, bodily aspect of experience – the self-as-body. The existential anatomy holds that pre-reflective body experience is an obvious and variable dimension of self-experience, in contrast to the lived body, in which the body is taken to be either "absent" (Leder, 1990), or an anonymous background – "with me" (Merleau-Ponty, 1962). I shall now bring the framework of the existential anatomy more systematically into the clinical reality starting with an analysis of the intentionalities of symptoms, and from there moving into the clinical encounter itself.

A. The Extroverted Symptom

The primary intentionality of symptoms, irrespective of their causes, is reflexive. The self-as-body stands out in the *Gestalt*. The natural and self-evident becomes questioned. But, as stated earlier, the self-as-body tends to turn into the meanings of the inward and/or outward intentionalities. Bodily experiences related to strain, or conflict, may turn into symptoms when the obvious tie of meaning between them and the external situation is cut off, or when discomfort or disability is perceived to be intolerable. Imagine a woman who works in an electronic industry with a constant and static load on her right shoulder. Normally, after every shift, she has an experience of fatigue, sometimes combined with pain in her shoulder and upper arm, and normally this has vanished when she wakes up in the morning. Her judgments and actions may follow two alternative routes. On the one hand, she may regard the signals from her shoulder as quite normal. Her body tells her she is tired and that this degree of strain is just about what she can take. It may well be that her habitual bodily posture, and a tendency for general muscular tension in response to the strains of life, makes her susceptible to the kind of load to which she is exposed. All the same, the meaning of the experience is that of tiredness which, in turn, gives a certain meaning to her work. Intentionality is mainly outwards, towards the horizon of the world, although – depending on the magnitude of the sensations – her body-as-nature may also attract her conscious attention as a fact of the inner horizon.

On the other hand, she may have little tolerance for pain, and a low threshold for seeking medical advice. Those traits drive her to see her doctor at this stage. Her symptom could then be looked upon as more or less synonymous with the fatigue from a static load on her right arm at work. The main intentionality of the symptom is "extroverted" (Rudebeck, 1992, p.45). In immediate fashion it is context related and conditional. The confusing aspect here is when, in seeking explanations, the woman exclusively addresses her inner horizon, the body-as-nature. The doctor often experiences this clinical situation as unclear, insofar as its logic is not apparent to him. Patients with "vague complaints" are a common complaint of doctors (Malterud, 1987; Johansson, et al., 1996).

B. The Introverted Symptom

The intentionality of the experience underlying the presentation of a primary bodily symptom is inwards, in contrast to the intentionalities of the outer physical and social world. The direction of the symptom presentation, per se, is, of course, outwards. It is an effort to make the doctor understand. The inwardness of disease is something most people have experienced. A bad cold with a high temperature is enough to loosen one's connections with the world, something that Svenaeus (1999, pp. 137-38, 2001) describes expressively as a certain form of "unhomelikeness." In really serious disease this is, of course, even more obvious. When a human being is fighting to survive, existence and biology are very close. This same inwardness is a feature of most presentations of symptoms of predominantly organic disease with which the patient is not previously acquainted, and irrespective of the severity of the condition in the medical sense. The meanings of the symptoms are the body-as-nature, which means that they lack the familiar logic, or normality, of the existential anatomy. It is, therefore, difficult to name and describe what is experienced, and the signals of meaninglessness and threat may also be received from beyond the inner horizon. In the existential sense, this is even so when signs of disease are visible on the body surface. For example, the facticity of a rash or a wound is salient (Heidegger, 1962, p.174). This means that the own-body is the intimate reality into which the patient is thrown without control – a reality that is disclosed in situations of bodily change. While the clearly extroverted symptom with its intentionality is an aspect of the communication with the doctor, the doctor may sense the introverted symptom hiding within the depth of the patient. This is especially so when the patient is frightened.

C. The Ambiguous Symptom

Imagine, then, that the pain of the woman worker described in connection with the extroverted symptom gradually becomes constant and more intense, spreading to her neck, and resulting in difficulty lifting her arm. She no longer copes with full working hours. In this situation the meaning of the discomfort in her shoulder still has much to do with her workload. In one sense, her workload appears even more demanding. On the other hand, her "body-as-nature" is now more prominent. She cannot actively diminish her pain by resting, or relaxing movements. Her symptoms have reached a state of relative *physical autonomy*. The inward intentionality has become an important constituent of her symptom, which can now be characterised as a symptom of *ambiguous* intentionality. There is no explanation model of obvious preference. Since the physical signs are at hand, most doctors would not call her complaints vague. The ambiguity will be disclosed when remedies limited to the physical body do not work other than temporarily.

D. The Two Steps of Diagnosis

Clinical diagnosis is a two-step procedure. The first step is grasping the patient's experience. We may call it the *prediagnostic* step. In the second step the experience is transformed into a *communicated experience,* which is then the factual object for clinical judgment. The richer, and more differentiated the doctor's perception of the symptom presentation (as well as of the original experience within the presentation), the easier it is for the doctor to reach a valid judgment. This judgment may imply putting the communicated experience within the framework of biomedicine or within other systems of formalized knowledge. The judgment may also be a simple recognition of the experience as valid in itself, and a doctor-patient agreement to refrain from any specific actions. The context of meaning of the first diagnostic step is the existential anatomy. To be able to build a bridge of meaning between himself and the patient, the doctor needs here to offer his own bodily subjectivity. The route of communication is bodily empathic. We shall now look at the prerequisites for this form of understanding, which is also a production of clinical knowledge (Rudebeck 1992, p.29, Malterud, 1995).

E. Bodily Empathy Linking Understanding and Diagnosis

Doctor and patient share their existential anatomies in crucial ways. Toombs (1992, pp.98-102) describes how the doctor's own experiences of the limitations, uncanniness, and facticity of the normal body provide important insights into "illness-as-lived," although the sickness experience itself "has a greater existential import" in demonstrating the "ambiguity" of the body which is at the same time 'me' and 'not me'. Engel (1985, p.364) describes how "the life-worlds of both patient and physician provide the starting point for mutual understanding." The fact that the doctor's response is "predicated on a personal life-world and lived body" permits an almost instant understanding that what the patient is striving to communicate concerns not only some aberration in the lived body but also a plea for help. Cassell (1985, pp.46-47) suggests that the physician uses personal experience of lived body to gain an accurate understanding of the patient's experience. He also describes how, in attending to the complaints of patients, the doctor can expand his own knowledge, when he is not previously acquainted with a particular bodily experience. Thus, it is not necessary for the doctor to suffer from all symptoms himself to establish satisfactory communication with the patient. The route of empathy is also available for the communication of bodily experiences. Most often, empathy is referred to as the intersubjective understanding of emotions. To specify the role of empathy in the context of predominantly bodily symptoms, the term *bodily empathy* may be applied (Rudebeck, 1992, pp.79-81). This brings empathy back to the broad connotation of *Einfühlung* implied by Husserl (1970, pp.258, 358) as the mediation of inner experience in general. With such a broad comprehension of empathy, the intersubjective understanding of emotions would accordingly be called *emotional empathy*. [3]

F. The Dimensions of Bodily Empathy: General Bodily Empathy

The origin of bodily empathy is the evident fact that humans share many of their bodily, existential conditions. Bodiness as such, i.e. existing in a physical and social world within and through one's body, and thus being vulnerable and finite, being even "other-than-me," is the very basis for understanding the body experience of another human being. Connected to this are the self-functions of the existential anatomy made up of the senses, movements, and the inner organs. We see, hear, sense, walk, make gestures, grab, freeze, sweat, have intercourse, and empty bladder and bowel as bodily selves. These constituents of bodily being fuse, or stand out in an individual fashion, dependent on the situational aspects of the continuous flow of the existential anatomy. Through sharing the existential anatomy with his patients, the doctor can approach the patient's experience in a meaningful way without having had that same experience himself. In communication, existential anatomy is more of a grammar of body experience than a means of direct recognition of experience. Although intersubjective, it has a certain objective, generalizable power. When we write "I" we may refer to the "I-ness" of all individuals, or to a particular subject. There is the grammar of "I," and the unique experience of an "I." This contingency applies also to the existential anatomy, and makes it possible for doctor and patient to create a shared version of the patient's symptom experience; *the symptom as communicated.*

These basic paths of intersubjective bodily communication we may call *general bodily empathy*. They are valid for everyone and, therefore, in themselves they have no professional implication. However, professional understanding would be totally impossible without general bodily empathy. In addition, the doctor's bodily self-knowledge is expanded in the daily practice of encountering and interpreting symptom presentations. Thus, his general bodily empathy may be refined considerably. A condition here, of course, is that the doctor is open to his own existential anatomy.

G. Specific Bodily Empathy

General bodily empathy makes it possible for the doctor to integrate the symptom presentations of patients with his own existential anatomy. The vast number of presented experiences become internalized and add to his readiness to understand as a "collective bodily self." Memories of symptom presentations are not merely cognitive, since they are grasped through bodily empathy. When consulting with patients, the doctor can activate the inner representations of presented bodily experiences in an intuitive and rapid fashion, offering the patient a degree of understanding that far exceeds that of a layman. A prerequisite for reaching this degree of individual understanding is the doctor's receptivity to the symptom presentations as expressions of the existential anatomy. If he is receptive only to the expressions of disease, empathy will provide only a minor contribution to the diagnostic process.

H. Biological Bodily Empathy

The biology of bodily empathy expresses the linkage of the biomedical knowledge about anatomy of the functions of bodily organs with the doctor's own existential anatomy. This knowledge is not only knowledge about "the other" but about *me* as well. Because of his education the doctor may refer own bodily experiences throughout life to anatomical and physiological entities in a way not accessible to laymen. The bodily self-knowledge of a doctor is, or could be, a very special dialectic of the existential anatomy and the biomedical perspective.

The biological dimension is also important in a direct way in the empathic understanding of symptom presentations. Biomedical knowledge not only serves as a tool in its formal sense. When appropriate, it also helps to meet the patient's feeling of desolation within a threatening and discomforting body, both in the case of disease proper and in the case of functional disturbances. When one understands the pathophysiological changes, it becomes easier to understand also the experiences caused by these changes. Engel (1985) has also commented on this empathic asset linked to biomedical training.

I. The Contents of the Symptom Presentation

Even on the general level the symptom presentation has a complex intentional pattern. First, there are the main intentionalities of the symptom – inward, outward, and ambiguous. Second, we have the intentionalities of the reflection on the symptom, "the hermeneutic moment," which does not necessarily follow those of the symptom itself. Last, we have the intention of the communicative act, which may be anything from "Tell me I am well, doctor," to "Tell me what's wrong, doctor." The intensity of the experience, and the energy of the presentation, adds to the general pattern. On the skeleton comprised of the gross existential anatomy, more precise aspects of lived, and reflected, bodily experience complete the message to become a form.

It follows naturally that the language of the symptom presentation is very much a mixture of spoken language and body language. In organic disease, the body-as-nature may become its own "body language" through the changes of the bodily exterior, a decrease or loss of function, and the signs of lessened vitality. When signs are obvious, presentation and sign are more or less synonymous. Also, when there is considerable coherence of form and content (i.e. when the underlying bodily experience, the reflection on it, the details of its presentation, and the request within the presentation reinforce one another to become a comprehensive whole) understanding may be obtained without too much empathetic effort. The obvious sign and the complete coherence of the symptom presentation are the paradigmatic situations of biomedicine where its knowledge works straightforwardly. Therefore, when we look at bodily empathy as an acquired skill, it should, firstly, be expected to add to diagnosis in situations of ambiguity. However, it should always be relevant for understanding the predicament of the patient.

In the process of dialogue and examination moments of mutual attentiveness and discovery emerge. These moments may correspond to moments of coherence in symptom presentation, facilitated by the bodily empathic "being-there" of the doctor. The "flash" – the mutual and sudden understanding of important aspects of the patient's problem (Balint and Norell, 1973, pp. 33-44) – may refer to communication of this kind. Affective resonance, described in relation to the empathy of emotions (Basch, 1983), which is thought to include the neuro-humoral state of the particular affect, seems also to have relevant connections with bodily empathy. However, the idea of resonance has its drawbacks. If the notion of bodily empathy evokes the expectation that the doctor should experience more or less the same as the patient, this might lead to a depreciation of the degree of validity of *the symptom as communicated* that can realistically be achieved. The best would then turn into the enemy of the good. This is a situation no clinician can afford, since the problems of his patients have to be dealt with, although they are never completely understood. Thus, it is clear that understanding has to be based on the ethical stance towards the patient.

Carl Edvard Rudebeck
Västervik
Sweden

NOTES

[1] A possible, neuro-anatomical basis of this relativity of perception is illustrated by human vision (Varela, Thompson and Rosch, 1993, pp.93-96). Of the total impulse flow reaching the lateral geniculate nucleus, a region in the thalamus in the brain where the optic nerve connects and from where impulses are led further to the visual cortex, only 20 percent stem from the retina. 80 percent come from other parts of the brain. In addition, more fibres lead from the visual cortex back to the lateral geniculate nucleus, than to it. Thus, a structure is laid, which allows for infinite reactions and for the emergencies of new configurations. The authors state that "The behaviour of the whole system resembles a cocktail party conversation much more than a chain of command." In relative terms, it seems that seeing, and the other modes of perception, are

predominantly inner processes, which find their meanings through mental activities such as memory, association, selection, defense and creation.

[2]Throughout this text experience is referred to when describing the existential anatomy. Why not, then, use the term "experiential anatomy"? The reason behind the choice made is that the underlying meaning of experience is existence rather than the other way around. However, to the individual human being existence is experience, irrespective of whether this is recognized as fact only, or if it implies the person's naked recognition of the human condition.

[3]If the reader would regard this broadening and differentiation of the empathy concept to enhance dualism, our answer is that language itself implies a thematization of experience, which, however, does not contradict the wholeness of reality in itself. As illustrated by the existential anatomy of emotions, body experience is an important dimension of emotions, but still, when we talk of emotions, we usually look for their meanings – what they tell about our relation to the world. Likewise, we may state that every conscious bodily state has its emotional tone, or aspect, although this is not usually our focus of attention when we talk about body experiences. In fact, any act of consciousness has its perceptive, affective, and cognitive elements. The context decides which of the three will name the act.

BIBLIOGRAPHY

Balint, E. and Norell, J. (eds.): 1973, *Six Minutes for the Patient: Interactions in General Practice Consultation*, Tavistock Publications, London.
Basch, M.F.: 1983, 'Empathic understanding: A review of the concept and some theoretical considerations', *The Journal of the American Psychoanalytic Association* **31**, 101-26.
Bullington, J.: 1999. *The Mysterious Life of the Body: A New Look at Psychosomatics*, Tema, University of Linköping, Linköping, Sweden.
Cassell, E.J.: 1985, *Clinical Technique*, Vol. 2, *Talking With Patients*, The MIT Press, Cambridge, Massachusetts.
Courtenay, M.: 1996, 'A new look at the nature of the doctor/patient relationship', *The Journal of the Balint Society* **24**,19-20.
Engel, G.: 1985, 'Commentary on Schwarz and Wiggins: Science, humanism, and the nature of medical practice', *Perspectives in Biology and Medicine* **28**, 362-65.
Heidegger, M.: 1962, *Being and Time*, Basil Blackwell, Oxford.
Husserl, E.: 1970, *The Crisis of European Sciences and Transcendental Phenomenology*, Northwestern University Press, Evanston, Illinois.
Johansson, E.E., Hamberg, K., Lindgren, G. and Westman, G.: 1996, "I've been crying my way' - Qualitative analysis of a group of female patients' consultation experiences', *Family Practice* **13**, 498-503.
Johnson, M.: 1988, *The Mind in the Body - The Bodily Basis of Meaning, Imagination, and Reason*, The University of Chicago Press, London.
Karlsson, G.: 1995, *Psychological Qualitative Research from a Phenomenological Perspective*, Almqvist & Wiksell International, Stockholm, Sweden.
Leder, D.: 1990, *The Absent Body*, Chicago University Press, Chicago.
Malterud, K.: 1995, 'The legitimacy of clinical knowledge: Towards a medical epistemology embracing the art of medicine', *Theoretical Medicine* **16**,183-98.
Malterud, K.: 1987, 'Illness and disease in female patients. I. Pitfalls and inadequacies of primary health care classification systems - A theoretical review', *Scandinavian Journal of Primary Health care* **5**, 205-209.
Merleau-Ponty, M.: 1962, *Phenomenology of Perception*, Routledge & Kegan Paul, London.

Morell, D.: 1978, 'The epidemiological imperative for primary care', *Annals of the New York Academy of Science* **310**, 2-10.
Piaget, J.: 1976, *The Child and Reality - Problems of Genetic Psychology*, Penguin Books, Harmondsworth.
Perls F., Hefferline R.F. and Goodman P.: 1973 (1951), *Gestalt Therapy. Excitement and Growth in the Human Personality*, Penguin Books, Harmondsworth.
Rudebeck, C.E.: 2000, 'The doctor, the patient, and the body', *Scandinavian Journal of Primary Health Care* **18**, 4-8.
Rudebeck, C.E.: 1992, 'General practice and the dialogue of clinical practice - On symptoms, symptom presentations, and bodily empathy', *Scandinavian Journal of Primary Health Care*, **Supplement 1**, 3-87
Sartre, J.P.: 1956, *Being and Nothingness: An Essay on Phenomenological Ontology*, Routledge, London.
Svenaeus, F.: 2001, 'The phenomenology of health and illness' in this volume.
Svenaeus, F.: 1999, *The Hermeneutics of Medicine and the Phenomenology of Health: Steps Towards a Philosophy of Health Practice*, Linköping Studies in Art and Sciences, Department of Health and Society, Linköping, Sweden.
Tomkins, S.S.: 1992, *Exploring Affect: The Selected Writings of Silvan S.Tomkins*, Cambridge University Press, Cambridge.
Toombs, S.K.: 1992, *The Meaning of Illness: A Phenomenological Account of the Different Perspectives of Physician and Patient*, Kluwer Academic Publishers, Dordrecht, The Netherlands.
Toombs, S.K.: 1987, 'The meaning of illness: A phenomenological approach to the patient-physician relationship', *The Journal of Medicine and Philosophy* **12**, 219-40.
Uexküll, Th. and Wesiack, W.: 1988, *Theorie der Humanmedizin: Grundlagen Ärztlichen Denkens und Handels*. Urban & Schwarzenberg, München.
Varela, F.J., Thompson, E. and Rosch, E.: 1993, *The Embodied Mind: Cognitive Science and Human Experience*, MIT Press, London.

PAUL KOMESAROFF

THE MANY FACES OF THE CLINIC: A LEVINASIAN VIEW

O trees of life, when will your winter come? We're never single-minded, unperplexed, like migratory birds. Outstript and late, we suddenly thrust into the wind, and fall into unfeeling ponds. We comprehend flowering and fading simultaneously. And somewhere lions still roam, all unaware, in being magnificent, of any weakness – Rainer Maria Rilke, Fourth *Duino elegy*[1]

People go to doctors for many reasons: for the diagnosis and treatment of illnesses, to make sense of unusual experiences in their bodies that they suspect may be related to illness, for reassurance that they do not have a serious condition, for advice about how to avoid such conditions, for help in dealing with life crises like unemployment, bereavement, marriage breakdown, or for assistance with the passage through natural processes like pregnancy and menopause. Often, several such reasons occur together.

Take for example, a patient, Rebecca, whom I have known for about ten years. Now a woman in her late fifties, she was born in Poland during the Second World War where she suffered painful experiences that deeply affected the later course of her life. She came to Australia as a young child, married young, had two children and later separated from her husband. She worked as a teacher in country schools for many years and became quite a talented artist. In her early fifties she experienced menopausal symptoms, which caused a degree of discomfort and anxiety that surprised her. Flushing occurred many times during the day and night, together with aches and pains, headaches, bloating, breast tenderness. She felt, she said, that her body was disintegrating, that it was out of control, possessed by a devil. She did not like going to doctors and was normally even less inclined to take medications. However, on this occasion she needed help. She needed assistance in keeping her body together and making her life bearable again.

Traditional accounts of medicine do not recognize the multifaceted intricacy and complexity of the clinical process. Since at least the mid-eighteenth century, medicine has been viewed as a singular discursive system, built around an epistemological theory of the body as an anatomical and later physiological and biochemical structure, unified by the methods and concepts of science. Even in more sophisticated recent accounts, which acknowledge the cultural contingency of the assumptions underlying modern medicine, medicine is considered to constitute a unitary discourse with a fixed concept of truth embedded in it. Ethical reflections on medicine have largely left these assumptions unquestioned. Finding the epistemological system to be limited to facticity, with no endogenous source of morality, they sought to append an ethical system, separated from the cultural sources that generated it, and subject to exogenous imperatives. In the appended morality, doctors and patients are invested with the capacity to choose between alternatives, to exercise their freedom, to protect and realize their human rights. Clinical medicine, however, does not limit itself to a well-defined, rigorously delimited discourse of truth or morality. Rather, the encounter between patient and doctor – or more generally, patient and carer – is highly heterogeneous and contingent on contextual features. The encounter is highly volatile and ambivalent, as any encounter between two embodied subjects must be. The clinical dialogue evokes and creates memories, causes and alleviates pain, disassembles and rejoins, dissembles and enjoins promises.

* * * * *

The patient's story never comes out as a single, coherent account. It is fluid, unpredictable and fecund. As Rebecca tells the story of her illness she moves seamlessly between different kinds of language. There is the language of scientific medicine, of symptoms and causes – menopause, hormonal changes, hot flashes and arthralgias. There are experiential descriptions of their impact – the sense of decay, the palpable experience of mortality. There are also psychological reflections, philosophical meditations, sexual conflicts and challenges, the crystallisation of past experiences.

This heterogeneity is typical of clinical discourse. The language of science, which is directed toward mastery over mute objects, brute things, that do not reveal themselves in words, that do not comment on themselves, is only one among many in the clinic. Just as important are those discourses that seek to accomplish the task of establishing, transmitting and interpreting the meanings of the illness and of the scientific facts. In the clinic, and no doubt more widely, there is no unitary, singular or sacrosanct language. Rebecca can relay facts carefully and precisely, she can speak passionately and poetically of her feelings, or she can imagine and invent. Language is present only as something stratified and heteroglot (Bakhtin, 1992, p. 332). This is not unlike the appearance of language in the novel, which also incorporates a diversity of speech types and voices. A single sentence can contain scientific jargon, colloquial language, language specific to age or ethnic groups, sometimes conveying social prejudice or political outlook (Bakhtin, 1992, p. 262). It can convey precise scientific formulations or vague psychological speculations or deep philosophical reflections. All these linguistic and discursive modalities are specific points of view on the world, ways of conceptualising it in words, each characterised by its own objects, meanings and values. As such they may be juxtaposed to one another, mutually supplement one another, contradict one another and be interrelated dialogically (Bakhtin, 1992, pp. 270-72).

Rebecca's attendance in the clinic, like her story and her language, is erratic and unpredictable. She sometimes disappears for months at a time, then returns or telephones, demanding the opportunity to talk and to be listened to. She cries, argues, shouts, or just sits. When she talks, she mixes up facts and dates, and there are large gaps in her memory. She speaks about her current feelings, her relationships, her art. Most of all, she talks about her early life. As a small child, she spent three years in Auschwitz, where she experienced unutterable horrors, and both her parents died. After the liberation she came to Australia under the care of foster parents, whose relationship to her was never made clear. As it turned out, they were as scarred as she was, and she was subjected to terrifying ordeals in the new country, where it was never assumed that the family was safe from a repetition of the horrors of Nazism. Rebecca recounts experiences suggestive of much cruelty from her foster parents, though to be fair, the mutual pain was so profound that it is no doubt impossible to distinguish factual occurrences from fears and fantasies.

Rebecca's marriage was not a success and her relationships with her children were always contaminated by her wartime memories. She tried to establish a normal life and went to university and trained as an art teacher. This seemed to work for a time and she was able to maintain a job as a high school teacher for about ten years. She was evidently committed to her work and, it appears, well liked by her students. However, the inextinguishable memories of her early childhood repeatedly recurred and, once again, spoiled everything. She developed a range of symptoms – severe chest pain, arthritis, rashes, imagined cancers – which rendered her bedridden for long periods. She drifted between relationships, jobs, cities. She became a talented and successful artist, winning much praise and many prizes for her tortured and evocative depictions of the pathos and depravity of the human condition; however, she gave all the winnings away to charities, leaving herself in a constant state of poverty and privation.

* * * * *

The variability of linguistic usage in the clinical dialogue is not a purely technical matter: it is literally the means by which the patient reveals her "self." Classically, the self was conceived as a unified subject, in which being is homogeneous and indivisible. However, this conception of the self was just another manifestation of the universalizing project in both epistemology and ethics of Western thought, which culminated in the European Enlightenment. The subject of classical philosophy was regarded as separated from the world, a solitary mind or ego from which truth appears to emanate. Modernity declared an identity between Being and Knowledge. "The labour of thought wins out over the otherness of things"(Levinas, 1989a, p. 78).

Of course, this view of the subject was never uncontested, and it was pointed out again and again that there is a fundamental omission. As is demonstrated repeatedly in the daily experience of the clinic, the subject is not a unified entity but is radically pluralistic. It does not arise independently of social life or interpersonal experience. Subjectivity does not derive from a pre-formed cogito, or from language or society; it cannot be encompassed as merely an openness upon the world. Rather, it is generated out of the experience of otherness or, more precisely – to use the term introduced by Emmanuel Levinas – the "alterity" between self and other.

The subject is strictly not a thing but a relation – between self and other, between subjects in formation, between "intercorporeal" objects (Merleau-Ponty, 1968, pp. 140-42). Similarly, truth is not an interrogation of sameness but an apprehension of difference, an attempt to grasp what is other. Reality itself is complex and heterogeneous, internally divided, a mixture of otherness and sameness. Our sense of exteriority, our inherent relationship to the world and other selves, is irreducible. There is "a duality that concerns the very existing of the subject. Existing itself becomes double" (Levinas, 1989c, p. 53).

The jumbled voices Rebecca articulates are authentic presentations of different facets of her self. She has occupied a variety of roles: as a victim of rape and torture, as a mother, an artist, a teacher. Like the language of the prose writer, her speech presents not only her different roles and experiences but also different strategies for engaging the world and other people:

[C]ertain aspects of language directly...express...the semantic and expressive intentions of the author, others refract these intention; the [author] does not meld completely with any of these words, but rather accents each of them in a particular way – humorously, ironically, parodically and so forth...; and there are, finally, those words that are completely denied any authorial intentions: the author does not express himself in them...rather, he exhibits them as a unique speech-thing, they function for him as something completely reified (Bakhtin, 1992, p. 299).

It is, therefore, not just the patient's language that is multi-faceted and pluralistic: it is, in a sense, the patient herself, as she is formed and presented in the clinical dialogue. Her sense of physical decay, her overwhelming sense of isolation and loneliness, even in the midst of large crowds, her fear of dying but lack of commitment to living, her inability to open herself up emotionally or sexually to another person – all these are not just symptoms and signs of a deeper, more essential, diagnosis, they are not mere epiphenomena: they are Rebecca's world of meaning and value; they are what constitutes Rebecca herself.

Her physical "symptoms" too – the pains, the clouding of vision, the unsteadiness of gait and the falls, the conviction that she is suffering from an incurable cancer – are not merely incidental. They too are fundamentally constitutive of her world and fix not just its physical co-ordinates but also its values. Her body, like her sense of self, is the trajectory she traces out in her dealings with the world. Often, it is assumed by Rebecca's doctors that the illnesses she describes are either psychological in origin or purely imaginary. However, this is a mistake. Her pain, her symptoms, her lack of confidence in her body, are certainly real. The hidden pathologies that so torment her body are the outcomes of the unhealed wounds sustained at Auschwitz, of the "great death" within her – to use Rilke's (1995, p. 90) words – that continues to "[hang] green, devoid of sweetness, like a fruit inside [her] that never ripens."

The body constitutes an axiological system: it is the site of meanings and values on the epistemological, ethical and aesthetic planes.[2] The values attached to the body are themselves subject to social processes and have undergone profound transformations throughout history. Since the Enlightenment, the body has been conceived as a self-contained entity wherein it "degenerates at the end into an organism as the sum total of the needs of natural man" (Bakhtin, 1990, p. 58). Illness involves a process of dissecting and reintegrating meanings of bodily experience. An important moment of the dynamic of the clinic is this process of revision and critique. The clinical space provides a setting within which this fundamental process can take place (Komesaroff, forthcoming).

* * * * *

Just as the patient utilizes a range of discourses and speaks in multiple voices, so does the doctor. There is the language of science, of diagnosis and treatment; there are social and psychological discourses; there is interpretation, provocation, supportive dialogue and persuasion. But of fundamental and irreducible significance is the fact that the doctor is there at all, as a partner in dialogue. For not only do doctors and patients speak with many different voices: they speak to each other. In the clinical dialogue the patient lays open her world, facing the mirror of alterity. In the controlled space of the clinic, she addresses herself to another person. She reveals herself for another, through another, and with the help of another. Only in this way, by a relationship toward another consciousness, "by that which takes place on the boundary between one's own and someone else's consciousness, on the threshold," does she becomes conscious of herself (Bakhtin, 1990, p. 252).

Dialogue is not an abstract linguistic exchange occurring between two pre-formed entities. The phenomenon of selfhood is constituted through the operation of a dense and conflicting network of discourses and signifying practices that are themselves bound up with the intricate phenomenology of the self-other relation. As Bakhtin (1984, p. 252) put it, "dialogue... is not the threshold to action, it is the action itself. In dialogue a person not only shows himself outwardly, but he becomes for the first time that which he is...not only for others but for himself as well. To be means to communicate dialogically. When dialogue ends, everything ends."

In responding to the call of the other I – Rebecca's doctor – divest myself of intentionality and rational calculation. This is because in the crucible of alterity – and this applies equally to both partners in the dialogue – the self "has no name, no situation, no status. It has a presence afraid of presence, afraid of the insistence of the identical ego, stripped of all qualities" (Levinas, 1989a, p. 31). In its nakedness, "in its mortality, the face before me summons me, calls for me, begs for me, as if the invisible death that must be faced by the Other, pure otherness, separated, in some way, from any whole, were my business" (Levinas, 1989a, p. 83). In the pre-reflective proximity to another person I myself am stripped bare; my "ego (is stripped) of its pride and the dominating imperialism characteristic of it" (Levinas, 1989b, p. 100).

The starting point for meaning, value and truth is the face-to-face confrontation with another (Levinas, 1981, pp. 18-19). Subjectivity does not derive from a pre-formed cogito, or from language or society; it cannot be encompassed as merely an openness upon the world. It is the complex and pluralistic ethical bond to alterity that is prior, that is over and beyond Being and its truth. "The prophetic word responds to the epiphany of the face...

(T)he epiphany of the face...attests the presence of the third party, the whole of humanity in the eyes that look at me" (Levinas, 1981, p. 213).

It is impossible not to be drawn into the dark poignancy of Rebecca's world. In her cries – and in her art – she calls for witness to her suffering. All her life she has been a hostage to what has been called the age of atrocity. By becoming her witness one becomes her hostage, one's body aches while standing before her, bearing the guilt we all share for the monstrosities so many have endured.

Before the neighbour I am summoned and do not just appear...(T)he stony core of my substance is dislodged...(I)n calling upon me as someone accused who cannot reject the accusation, it obliges me as someone unreplaceable and unique, someone chosen...I cannot evade the face of the other, naked and without resources...I am pledged to (her) without being able to take back my pledge (Levinas, 1987, p. 167).

To witness suffering is a complex thing. It is impossible not to be drawn into it, even if, as a doctor, one has seen many people immersed in suffering. It is often said of doctors that they become hard and calloused as a defensive measure, to protect themselves. It seems unlikely that it is ever possible to become fully hardened to another's pain, to escape – as Hannah Arendt (1977, p. 106) put it – "the animal pity by which all normal men are affected in the presence of physical suffering." The experience of coming into contact with suffering is described by Alphonso Lingis:

We do not simply see the pallid surfaces, the contorted hands and fingers; we feel a depth of pain... There is contagion of misery. For one does not view the pain behind the surfaces of his skin; one feels it troubling one's look, one feels it up against oneself. The sense of sharing the pain of another, the sense of the barriers of identity, individuality, and solitude breaking down hold us. There is anonymity but also communion in suffering. One suffers as anyone suffers, as all that lives suffers...(Lingis, 2000).

This does not mean that the clinical dialogue should be understood fundamentally as an expression of empathy, arising out of a relationship of reciprocity. A doctor does not provide support by expressing empathy or by establishing friendship: indeed, such relationships actually undermine the possibility of precipitating critical reflection within the many different kinds of discourse inhabiting the clinic.

The premise of anyone going to a doctor is that someone will listen to his or her story. The first act of the clinical encounter is the primordial first person singular utterance, "Here I am", which is heard and, like an outstretched arm, responded to by another.[3] The declaration is not an abstract act of self-positing. It presupposes alterity and the existence of the other. Thus the face to face confrontation is not a relationship as ordinarily conceived: it is asymmetrical and non-reciprocal. The otherness of the other is irreducible and unfathomable. It is not reciprocal because humans are not interchangeable, or able to be substituted for each other (Levinas, 1969, p. 298), but it does entail an act of self exposure and vulnerability.

> To be means to communicate. Absolute death (non-being) is the state of being unheard, unrecognized, unremembered. To be means to be for another, and through the other, for oneself. A person has no internal sovereign territory, he is wholly and always on the boundary; looking inside himself, he looks into the eyes of another or with the eyes of another (Bakhtin, 1984, p. 287).

The clinical space is the space of face-to-face confrontation, of the subtle and irreversible play of alterity. It is also, of course, the space of language: after all, the face to face founds language (Levinas, 1969, p. 207). But language is not just a vehicle for transmission of meaning, an untroubled promoter of communication, for it is the conditions for communication that are at stake. "Saying is communication, to be sure, but as a condition for all communication, as exposure" (Levinas, 1981, p. 48). Nor is it merely a medium for the play of ideas: it is the incarnate movement of the ideas themselves. "Meaning is the face of the Other, and all recourse to words takes place already within the primordial face to face of language" (Levinas, 1969, p. 206).

* * * * *

Both doctor and patient are in their own ways witnesses to suffering. There are two meanings of the word "witness," which, as Agamben (1999, p. 17) has pointed out, correspond to the two Latin words *superstes* and *testis*. The former sense refers to "the survivor," the bearer of an experience, the one who has lived through it from beginning to end, and has acquired knowledge about it from the inside, so to speak. It is in this sense that Rebecca – the patient – is a witness. The latter refers to the person who, in a trial or lawsuit between two parties, is in a position of a third party. This is the position in which I – the doctor – stand: as the one who observes, who sees from the outside, who observes and records the suffering.

The doctor hears and records the patient's experiences. He constitutes an active memory of them (Agamden, 1999, pp. 143-44), an "archive" of events, of experiences, of suffering. Thereby, he becomes a kind of timekeeper, a custodian of the past and one who defines the possibilities that might be realized in the future. His job, or part of it, is to assist the witness –

the *superstes* – to integrate his or her experiences into a larger whole, and to pass them back to the accumulated knowledge and wisdom of the community. As an archivist, in this Foucauldian sense, therefore, the doctor does not just accumulate facts or propositions, symptoms and signs, dates, times, laboratory results. He assists the patient, often slowly, painstakingly, to construct a "history," to define the conditions of the system of statements that constitutes the sum of his or her experience.

The archive that is constructed through the clinical dialogue is not an undifferentiated mass of statements that "unifies everything that has been said in the great, confused murmur of a discourse"; rather, it is a dynamic, reflective "system that governs the appearance of statements as unique events," which "differentiates discourses in their multiple existence and specifies them in their own duration" (Foucault, 1972, p. 129). That is, the clinical archive recognizes the multiple voices of the clinic and seeks to provide them with a topographic schema within which they can preserve their distinctive character, but nevertheless engage each other in a kind of internal dialogue. Thus the medical diagnosis, like the archaeological one,

[D]oes not establish the fact of our identity by the play of distinctions. It establishes that we are different, that our reason is the difference of discourse, our history the difference of times, our selves the difference of masks. That difference, far from being the forgotten and recovered origin, is this dispersion that we are and make (Foucault, 1972, p. 131).

The act of witnessing and of creating the archive is a process of interpreting the traces, the inscriptions, laid down within Rebecca's body both of the horrors of the twentieth century and of her more mundane experiences (Derrida, 1996, p. 26), of unjumbling her multiple voices, and of interpreting them, or at least seeking a place for them within a dynamic whole.

* * * * *

Over time, Rebecca's symptoms have increased and became more debilitating, and she has come to believe that she harbors a malignant tumor. A visit to the doctor has become a fearful experience, even if she also finds it affirming and comforting. She rarely keeps appointments, but when she does she will talk about her symptoms and her presumed illnesses at great length and with great intensity. In spite of this, she refuses to undergo tests, in the fear that the diagnosis would be proved right – or perhaps because it would be proved wrong.

The pain and other symptoms are undoubtedly both physical and psychic. Rebecca's appeals to her doctors for help are in a sense calls to heal the unhealable, the deep scars opened up in our civilization by the abominations of the century just ended. But her pain is more than a metaphor; it is real, incarnate, physical suffering. Her pain reflects the anguish that has characterized almost every moment of her life, together with an underlying physical process, the nature of which at this time remains undetermined. Her symptoms cannot be understood either as merely psychic phenomena or as the effects of a biological process: they are both together; one gives meaning to, and provides the framework for, the other.

For Rebecca, every confrontation with a doctor is almost as challenging as the illness itself. In it all her fears and hopes are crystallized, to a point that has become almost unbearable. On the one hand, she sees the prospect of a relationship of caring with another person who would assume responsibility for her welfare and in whom she could place her trust, a relationship that has evaded her throughout her life, a relationship that could perhaps, in a limited way, make up for the chasm opened up when her parents were murdered before her eyes. On the other hand, the very same relationship carries with it the possibility of the pronouncement of her own death sentence, the confirmation of her terminal diagnosis. She understands that the moment of trust is also a potential moment of truth, and betrayal. Thus the clinical encounter contains a promise, a challenge, and a threat, both of which she well recognizes.

Rebecca's ambivalent relationship to the medical profession draws attention to another aspect to the confrontation between the partners in the clinical relationship that plays a critical role in its dynamics: its potentially disruptive and menacing quality. Illness is experienced as a perturbing force, often dark and threatening. The exposure of the experience of illness in the clinic always therefore has a potentially disturbing, disruptive aspect to it. This is one of sources of the peculiar power of clinical medicine: by addressing the patient as a singular being, by engaging her in her specificity, it can gain access to the most intimate recesses of that person's being. It can challenge meanings, undermine assumptions and provoke fundamental reassessments and revaluations.

The confrontation between the partners in the clinical encounter is irreducible; the moral exigency to which alterity gives rise is unsatisfiable. The other person cannot be thematized, objectified, or totally defined. He or she is ambiguous, irreducible, infinite – infinitely transcendent, infinitely foreign (Levinas, 1969, p. 194). The face is present in its refusal to be contained. This is the obverse side of the ethical relationship which subtends discourse: it is threatening, challenging, "it puts the I in question" (Levinas, 1969, p. 195). The clinical relationship is not just one of support and affirmation, but also of subversion and confrontation. Its processes involve not just the integration of experience, but disintegration, critical questioning, anatomizing.

The institutions of totality, including law, science and ethics, can only judge the individual as a universal "I," as the instantiation of a universal rule. Others as marginal – as face-to-face interlocutors – do not, however, fit as particular instantiations of universals. Totality always speaks in the third person, singularity in the first person: it always has a name. When one comes face to face with another person in his or her singularity, one is exposed to his or her powers, and is vulnerable to his or her strength; but at the same time, one resists that power by calling it into question (Levinas, 1969, p. 199). Through this mutual resistance and vulnerability each party is challenged in a manner that is both threatening and affirming; each is called upon, "to take the bread out of (his) own mouth," as Levinas (1969, p. 75) puts it, "to nourish the hunger of another with (his) own fasting."

Usually, this obduracy, this unsettling character of the clinical interaction is not a deficiency. Rather, it is one of its greatest strengths. By calling into question the equilibrium formerly considered unshakeable – indeed, not even recognized as an equilibrium – by challenging the economy of the body, the possibility of new experiences and new meanings is evoked. The clinic is often the site for creative change. Illness, even supposedly trivial illness, can release powerful processes of meaning creation or rearrangement, of insight and reflection.

* * * * *

Alterity, the irreducible contact with otherness, is the engine of the clinical discourse. Despite its fundamental role, its boundaries remain indistinct. This is because the relationship between self and other cannot be formalized or subsumed under universal principles, since this would undermine its spontaneous, local character. Alterity simply is: it is a relation of pure immediacy. The mutual commitment of two people is a presupposition of all dialogue. In a sense, therefore, the other person is always already present within my self.

My relationship with another person is at once a relationship through language and through the body. Conversely, my experience of my own body presupposes my relationships with other people. For my body is made of the same flesh as the world, as Merleau-Ponty (1968, p. 248) said – and this flesh of my body is shared by the world and by other people. There is, therefore, no brute otherness, and no brute world. There is "only an elaborated world; there is no intermundane space, there is only a signification 'world'"(Merleau-Ponty, 1968, p. 48).

As a doctor, I do not grasp the patient so as to dominate her; rather, I respond to her as if to a summons that cannot be ignored. "In the exposure to wounds and outrages, in the feeling proper to responsibility, the oneself is provoked as irreplaceable, as devoted to the others, without being able to resign, and thus as incarnated in order to offer itself, to suffer and to give" (Levinas, 1989b, p. 95). The relationship with alterity is finding oneself under a bond, commanded, contested, having to answer to another for what one does and for what one is. It is finding oneself addressed, appealed to, "having to answer for the wants of another and supply for his distress":

To acknowledge the imperative force of another is to put oneself in his place, to answer to his need, "to give to the other the bread from one's own mouth." On the other hand, (t)o put oneself in the place of another is also to answer for his deeds and misdeeds, to bear the burden of his persecution, to endure and answer for it... (Lingis, 1981, p. xxii)

Like all other relationships, the clinical relationship is immanently ethical. However, as this account shows, this ethical character cannot be explicated in terms of a formal prescriptive sets of norms or principles. The microethical contents are irreducibly dependent on the specific details of an individual's personal history and the present context of the clinical dialogue, as the latter manifests itself corporeally and through language. The moral bond is irresistible and irrevocable, but it is also intractable and limitless.

* * * * *

There is no *dénouement* to Rebecca's story – either happy or unhappy. Years have gone by and, despite hard won insights, her pain persists. Despite the small victories, she still hurtles from one crisis to another, each as desperate as the last. Her drawings and her paintings continue to reflect the horror with which her life started, and will no doubt end.

She agreed to undergo testing for cancer, and came to accept that she was not, after all, about to die. She has been able to establish a rich and nurturing relationship with a new partner, though one not without ambivalence. She still hankers after the lost community and family of her parents, and her parents' parents. But in this, perhaps, she is not different from many other members of her generation.

Rebecca's story illustrates some fundamental points about clinical medicine. It shows that there is no single, unitary discourse characteristic of the clinic, but that rather the latter is the site of a multiplicity of conflicting and overlapping discourses. It emphasizes that this heteroglossia is not just a peculiarity of language, but that it reflects the deep structure of selfhood as it is formed and presented in the clinical encounter. It reveals that this moment of marginality that produces both uncertainty and anxiety is in some ways the real source of the clinic's inventiveness and potency. In other words, it is the peculiar structure of alterity in the clinic that is responsible for the paradoxical fact that medicine embraces both caring and confrontation. It emphasizes that the clinical encounter is inherently a moral relationship, but that the conditions of morality here are very different from those presupposed in classical ethical theory. Finally, as in Rilke's poem, "never single-minded, unperplexed,like migratory birds," Rebecca and her story remain infinite, open to new meanings and new interpretations, always unfathomable, always unfinished.

Baker Medical Research Institute
Melbourne
Australia

NOTES

[1] Rilke, R.M. *Duino elegies*. (New York, WW Norton and Co., 1939), p. 41.
[2] Bakhtin provides an account of the history of the axiological assessment of the body in his early essay "Author and hero in aesthetic activity". (Bakhtin M, *Art and answerability: Early philosophical essays* (Austin, University of Texas Press, 1990), pp. 254-256.
[3] Cf. Rilke, R.M. *Duino elegies*, p.65: "Like an outstretched arm is my call. And its clutching, upwardly open hand is always before you as open for warding and warning, aloft there, Inapprehensible."

BIBLIOGRAPHY

Agamben G.: 1999, *Remnants of Auschwitz. The Witness and the Archive*, Zone Books, New York.
Arendt H.: 1977, *Eichmann in Jerusalem: A Report on the Banality of Evil*, Penguin, Harmondsworth, U.K.
Bakhtin, M.M.: 1992, 'Discourse in the novel', in M. Holquist (ed.), *The Dialogic Imagination: Four essays*, University of Texas Press, Austin, Texas, pp. 259-422.

Bakhtin M.M.: 1990, *Art and Answerability: Early Philosophical Essays*, University of Texas Press, Austin, Texas.
Bakhtin, M.M.: 1984, *Problems of Dostoevsky's Poetics*, Manchester University Press, Manchester, England.
Derrida J.: 1996, *Archive Fever*, University of Chicago Press, Chicago.
Foucault M.: 1972, *The Archaeology of Knowledge*, (trans. A.M. Sheridan Smith), Tavistock Publications, London.
Komesaroff P.A.: Forthcoming, 'Sexuality and ethics in the medical encounter', in P.A. Komesaroff, P. Rothfield, and J. Wiltshire (eds.), *Sexuality and medicine: Bodies, Practices, Knowledges*.
Levinas E.: 1989a, 'Ethics as first philosophy', in S. Hand (ed.), *The Levinas Reader*, Blackwell, Oxford, pp. 75-87.
Levinas E.: 1989b, 'Substitution', in S. Hand (ed.), *The Levinas Reader*, Blackwell, Oxford, pp. 88-126.
Levinas E.: 1989c, 'Time and the other', in S. Hand (ed.), *The Levinas Reader*, Blackwell, Oxford, pp. 37-58.
Levinas E.: 1987, *Collected Philosophical Essays*, Martinus Nijhoff, Dordrecht.
Levinas E.:1969, *Totality and Infinity*, Alphonso Lingis (trans.), Dusquesne University Press, Pittsburgh.
Levinas, E.: 1981, *Otherwise Than Being or Beyond Essence*, A. Lingis (trans.), Martinus Nijhoff, The Hague.
Lingis A.: 2000, *Dangerous Emotions*, University of California Press, Berkeley, California.
Lingis A.: 1981, 'Translator's introduction', in *Otherwise Than Being or Beyond Essence*, Martinus Nijhoff, The Hague.
Merleau-Ponty M.: 1968, *The Visible and the Invisible*, (trans. Alphons Lingis), Northwestern University Press, Evanston, Illinois.
Rilke R.M.: 1995, *The Book of Hours*, Salzburg University Press, Salzburg, Austria.
Rilke R.M.: 1939, *Duino Elegies*. W.W. Norton and Co., New York.

IAN R. McWHINNEY

FOCUSING ON LIVED EXPERIENCE: THE EVOLUTION OF CLINICAL METHOD IN WESTERN MEDICINE

I. INTRODUCTION

The clinical method practiced by physicians is always the practical expression of a theory of medicine – a theory which embraces such concepts as the nature of health and disease, the relation of mind and body, the meaning of diagnosis, the role of the physician, and the conduct of the patient-physician relationship. In recent times, medicine has not paid much attention to philosophy. When our efforts have been crowned with such great successes as they have in the past century, why be concerned if someone questions our assumptions? Indeed, we often behave as if they are not assumptions, but simply the way things are. Crookshank (1926) marks the end of the nineteenth century as the time when medicine and philosophy became completely dissociated. Physicians began to see themselves as practitioners of a science solidly based on observed facts, without a need for inquiry into how the facts are obtained and, indeed, what a fact is (Fleck, 1979). We believe ourselves to be at last freed from metaphysics, while at the same time maintaining a belief in the theory of knowledge known as physical realism.

Most notably, modern medicine has failed to make any serious contribution to an understanding of the effect of the mind on illness (Hoffmeyer, 1996, p.71). Acceptance of Descartes' separation of *res extensa* from *res cogitans* enabled biology and medicine to make great progress on the assumption that the human body is a machine. But the problems left unsolved by Descartes have been gnawing away at the conceptual foundations of science and medicine, questions such as: what is the relationship between the observer and the world of phenomena, and how can a non-material mind act on a material substance?

Modern medicine and its institutions are strictly divided along dualistic mind/body lines. A fault line separates internal medicine, surgery, pediatrics and geriatrics from psychiatry, child psychiatry, and psychogeriatrics. On one side of the line, physicians attend in their clinical method to the body, but not to the emotions; on the other, they attend to the thoughts and emotions, but do not examine the body. Each side has its own textbooks, taxonomies, and institutions. One could hardly wish for a better example of how ideas can shape social structures. The terms we use betray our dualistic mindset, though our form of dualism differs from that of Descartes in one important respect. Contrary to modern assumptions, Descartes maintained that most aspects of affective states are primarily somatic (Brown, 1989). Meanwhile, the evidence accumulates that all illness is a function of the human organism as a whole, at all levels, from the molecular to the cognitive and affective, and that the organism's environment, both social and physical, is an important influence on illness and healing. E.A. Burtt (1924) concluded his classic study, *The Metaphysical Foundations of Modern Physical Science*, by saying that "An adequate cosmology will only begin to be written when an adequate philosophy of mind has appeared." Nearly a century later, we have still not achieved this goal.

II. THE EVOLUTION OF CLINICAL METHOD

The clinical method of Western medicine has its roots in Greek antiquity. Until the 5th century B.C., Greek medicine was a combination of the empirical and the magical. It was empirical in the sense that treatments found to be beneficial were repeated; it was magical in the manipulation of powerful superhuman forces by special rituals. After the 5th century B.C. a new method emerged: technê iatrikê, the art of healing. The physician was expected to act "with some degree of scientifically exact knowledge of what was being done and why" (Entralgo, 1969).

The Hippocratic physician needed three kinds of knowledge: what the illness is; what the remedy is; and why the remedy acts in this disease and not in others. The ideal relation between doctor and patient was one of friendship (philia), which consisted of desiring and procuring the friend's good. For the doctor, friendship was a combination of love of man (philanthropia) and love of the art of healing (philotechnia). Greek medicine, however, was not a monolith. Two schools of thought, the Coan or Hippocratic, and the Cnidian, practiced different clinical methods based on different concepts of disease. The Coans saw the essential unity of all disease, with various presentations depending on personal and environmental factors; the Cnidians were concerned with the diversity of diseases and the distinction between them. For the Coans, the purpose of diagnosis was descriptive – an assay of the patient's state. The classical account is given in the *First Epidemics* of Hippocrates, in which he says that he framed his judgments by paying attention to:

What was common to every and particular to each case; to the patient, the prescriber, and the prescriptions: to the epidemic constitution generally and in its local mood: to the habits of life and occupation of each patient; to his speech, conduct, silences, thoughts, sleep, wakefulness,

THE EVOLUTION OF CLINICAL METHOD 333

and dreams – their content and incidence; to his pickings and scratchings, tears, stools, urine, spit and vomit; to earlier and later forms of illness in the same prevalence; to critical or fatal determinations; to sweat, chill, rigour, hiccup, sneezing, breathing, belching; to passage of wind, silently or with noise; to bleedings; and to piles.

For the Cnidians, the purpose of diagnosis was to classify the patient's illness in accordance with a taxonomy of diseases. To them, diseases had a reality independent of the patient. The Coans, on the other hand, did not separate the disease from the person, or the person from the environment. The Coans employed individual description, the Cnidians abstraction and generalization. These differences were also reflected in attitudes to therapy. The Cnidians employed the specific remedy deemed efficacious for the named disease. The Coans treated each patient individually and symptomatically, attempting to assist nature in restoring the functional unity of the organism. The Cnidians prescribed remedies, the Coans a regimen (Crookshank, 1926).

These two schools of thought, named by Crookshank the natural (Coan) and the conventional or academic (Cnidian) have run through Western medicine since antiquity, sometimes one being dominant, and sometimes the other. In philosophy and the theory of knowledge, the two schools have had their parallel in the nominalists, who denied the reality of entities such as species and diseases, regarding them rather as convenient mental constructs, and the realists who believed in the existence of these entities.

Each school of thought has its strengths and weaknesses. Natural diagnosticians do not separate disease from the patient and the patient from his environment. For them, causation is complex, multifactorial, and varies with the individuality of each patient. Being skeptical of disease labels, they may lack a taxonomic language and teach by example rather than by the word. On the other hand, their distrust of classifications may lead to a lack of analytical rigour and to clinical indecision. Conventional diagnosticians have the advantage of a taxonomic language and, in certain categories of illness, gain predictive and therapeutic power from their precise classification. On the other hand, their single-minded pursuit of a name for the illness, and for a single cause, may leave them powerless when faced with an illness which is not so easily explained. In all ages, the greatest physicians have tended to take something from each school.

Although Hippocratic physicians paid attention to habits, behavior, and dreams, Greek medicine after the 5th century B.C., both Coan and Cnidian, rejected psychological explanations of disease and sought for explanations in nature (physis), which they endeavored to understand scientifically. Disease was explained according to nature alone, not according to the personality of the patient (Entralgo, 1956). At the same time, faith healing survived in the form of the ancient healing practices of divination, incubation, and catharsis, even though rejected by the body orientation of the Hippocratic tradition.

Galen (131-201 A.D.), the greatest physician of the post-Hippocratic era, broke away to some extent from Coan medicine, becoming identified with the Cnidians. In Renaissance Europe, his name was attached to the conventional diagnosticians who were locked in controversy with the Hippocratists about clinical method, whether it should be natural diagnosis in terms of experience, or conventional diagnosis in terms of reasoning (Crookshank, 1926). Galen was a pioneering dissector and experimental physiologist and neurologist, but his anatomy was based on animal dissection and his extrapolations to human anatomy were erroneous. He was a theorist, but his theories were *a priori* speculations without any basis in the clinical observation taught by the Hippocratic tradition.

With Galen, Hellenic naturalism reached its culmination. Moral irregularity came to be regarded as an aspect of nature, and therefore within the purview of the physician. But moral irregularity was not an explanation of disease: on the contrary, a disorder of nature was the cause of moral irregularity (Entralgo, 1956, p.64). The pathological effects of mental processes, said Galen, came from their localization in the body. "Galen's pathology was not able to become biographical or personal" (Entralgo, 1956, p.65). It was the philosophers, not the physicians, who developed a "therapy of the word" for problems of living. Nussbaum (1995) suggests that the Hellenistic philosophers anticipated psychoanalysis.

The Greeks believed that a physician should cultivate a capacity for friendship in great things and in small. For the Greeks, it was necessary to be a good person in order to be a good physician. Since even good people are fallible, goodness requires constant vigilance for those egoistic and self-serving motives that can damage or destroy a friendship. There were, however, limits to a Greek physician's beneficence. They had a strong sense of the power of nature; its healing power and also its destructive power: the necessity (anenke) before which we must all bow; the certainty of suffering and death. It was not expected, therefore, that the Greek physician would attend the dying.

With the Christian era, a new ethos came into being. Jesus taught that there were no natural limits to human beneficence. The physician's beneficence was due to the vulnerable and the dying. Care (caritas) was due the neighbor and the stranger as well as to the friend. Even if they were not themselves religious, Hippocratic and Galenic physicians approached the body with awe and respect, since for them it was the visible form of the divine nature. In the patient's body, the doctor encountered the transcendent. The body was "tremendum" (causing one to tremble) and "fascinans" (fascinating when it revealed the wisdom and beauty of nature) (Entralgo quoted by Drane, 1995 p.122). To the Hippocratic and Galenic physician, medicine was a sacred calling.

Early Christian theologians drew a clear distinction between sin and disease, but recognized a parallel with clinical medicine in the healing of souls. At the same time, they acknowledged a secondary link between sin and disease. Excessive appetites and uncontrolled anger obviously had their repercussions on the body, and disease could lead to moral failings in the patient. Medicine, however, clung to its somatic interpretation of human pathology.

There are profound differences in the way the body is experienced today and was experienced by physicians and patients up to the 17th century. Until this time, there was no sense of the body as a space containing organs or bodily functions and of diseases as located in the organs. To heal itself, the body "had to flow: pus, blood and sweat had to be drawn out" (Duden, 1991, p.17). The boundaries between mental and physical afflictions were blurred (Duden, 1991, p.12). To medieval man, the interior of the body was a microcosm of the outside world. There were correspondences between the inner and the outer, and people experienced themselves as part of the cosmos. Given this experience of the body, it is easy to understand why an anatomy of the human body was impossible before the Renaissance, and why medieval physicians did not examine the body. For all the theoretical distinction between mind and body, reinforced by Descartes' analysis, until the 19th century Western physicians continued to practice in the belief that the emotions could be causal in bodily illness, and that mental illness could benefit from bodily therapies (Rather, 1965). It was not until the nineteenth century that textbooks of general medicine began to dispense with the emotions as causes of bodily illness (Rather, 1965). An interest in the emotions, however, did not alter the somatic interpretation of illness, which remained a constant of Western medicine until modern times. Nor did the personal relationships with patients, however important to physicians, lead to any concept of psychosomatic pathology or of the patients' inner life as an explanatory context for illness.

III. THE SCIENTIFIC REVOLUTION AND THE MODERN PARADIGM OF KNOWLEDGE

The early European science which developed in the 17th century – A.N. Whitehead's "century of genius" – was not without its precedents in the Middle Ages, which were "an epoch of orderly thought, rationalist through and through" (Whitehead, 1975). But it was a rationalism based on *a priori* assumptions, not on facts. The willingness to challenge the assumptions by reference to observed facts ushered in the new worldview of science and medicine. In *The Advancement of Learning*, Bacon urged the revival of the Hippocratic method of recording case descriptions, with their course toward recovery or death; and the study of the pathological changes in organs – the "footsteps of disease" – with a comparison between these and the manifestations of illness during life (Faber, 1923, pp. 5-6).

In the intellectual climate of the 1600s arose the first modern physician to use systematic bedside observations, Thomas Sydenham. Sydenham described the symptoms and course of disease, setting aside all speculative hypotheses based on unsupported theories. He classified diseases into categories – a novel idea at the time – believing that they could be classified by description in the same way as botanical specimens. Finally, he sought a remedy for each "species" of disease, exemplified by the newly introduced Peruvian bark (quinine). His great innovation, however, was to correlate his disease categories with their course and outcome, thus giving them predictive value. His method bore fruit in the distinction, for the first time, of syndromes such as acute gout and chorea. Sydenham was a close friend of John Locke, who took a great interest in Sydenham's observations, sometimes accompanying him on visits to patients, an interesting example of the connection between medicine and the dominant ideas of the time.

After Sydenham, the work of classifying diseases was taken up by others, notably Sauvages of Montpellier, a physician and botanist who sought to group diseases into classes, orders, and genera in the same way that biologists classified plants and animals. Sauvages was a strong influence on Carl von Linné, the Swedish physician and botanist responsible for the Linnaean system of botanical classification – another instance of the connection between medicine and the ideas of the Enlightenment. The classifications of diseases by Sydenham's successors were however, of little practical value, since they were not correlated with the course and outcome of disease and represented only random combinations of symptoms with no basis in the natural order.

Sydenham died in 1689, and for the next 100 years no system for classifying diseases proved to be of lasting value. The next great step, and the one that laid the foundations for the modern clinical method, was taken by the French clinician-pathologists in the years after the French Revolution. The method was described by Laennec, the greatest genius of the French school:

THE EVOLUTION OF CLINICAL METHOD 337

The constant goal of my studies and research has been the solution of the following three problems:
1. To describe disease in the cadaver according to the altered states of the organs.
2. To recognize in the living body definite physical signs, as much as possible independent of the symptoms...
3. To fight the disease by means which experience has shown to be effective: ... to place, through the process of diagnosis, internal organic lesions on the same basis as surgical disease. (Faber, 1923, p.35, Translation by Jean Martin).

For the first time, clinicians examined their patients by using new instruments such as the Laennec stethoscope. They then linked together two sets of data: signs and symptoms from the clinical inquiry, and the descriptive data of morbid anatomy. At last, medicine had a classification system based on the natural order of things: the correlation between symptoms, signs, and the appearance of the organs and tissues after death. The system proved to have great predictive value and was further vindicated when Pasteur and Koch showed that some of these entities had specific causal agents. The clinical method based on this system developed gradually during the 19th century, until by the 1870s it had taken the form familiar to us today. The new method (I will refer to this as the modern clinical method) was a major swing from the natural to the conventional method of diagnosis. The objective, which had before been to diagnose a patient, was now to diagnose a disease. Disease was viewed as having an anatomical location in the body and an existence of its own, conceptually separable from the patient. Mental and physical diseases were separated and the manuals of clinical method, which made their appearance in the early 20th century, gave "excellent schemes for the physical examination of the patient while strangely ignoring almost completely, the psychical (sic)" (Crookshank, 1926).

With its predictive and inferential power, the new clinical method was highly successful. Indeed, the application of new technologies to medicine depended on it. It had other strengths: it gave the clinician a clear injunction to identify the patient's disease or rule out organic pathology; it broke down a complex process into a series of easily remembered steps; and it provided canons of verification: the pathologist was able to tell the clinician whether he or she was right or wrong. Although beneficence remained the overriding purpose of medical practice, pathology was still viewed as somatic, as it had been since the days of Hippocratic medicine.

So successful was the method that its weaknesses became apparent only much later, as its abstractions became further and further removed from the experience of the patient. True to its origins in the age of reason, the method was analytical and impersonal. Feelings, and the life experience of the patient did not figure in the process. The meaning of the illness was established on one level only – that of physical pathology. The focus was on diagnosis, with much less attention to the detailed care of the patient. In keeping also with its Cartesian origins, it divided mental from physical disorder.

The central idea on which the modern clinical method was based came into being at a time when Enlightenment ideas had become the dominant worldview of the West. Reason was enthroned, but it was a reason defined as formal logic, divorced from human experience. The fruits borne by these ideas in our own time include this clinical method and all of the benefits and problems of modern medicine. With the Enlightenment came the modern paradigm of knowledge, which has dominated the medical schools and has been a strong influence on the modern clinical method. Knowledge that matters is impersonal, public, productive, and empirically verifiable. Knowledge that is personal, experiential, or tacit is much less trusted and valued. After the 17th century, knowledge became decontextualized, reflecting:

A historical shift from practical philosophy, whose issues arose out of clinical medicine, juridical procedure, and moral case analysis... to a theoretical conception of philosophy.... Thus, from 1630 on, the focus of philosophical inquiries has ignored the particular, concrete, timely, and local details of every day human affairs: instead it has shifted to a higher, stratospheric place on which nature and ethics conform to abstract, timeless, general, and universal theories (Toulmin, 1992, pp. 34-35).

IV. THE REFORM OF CLINICAL METHOD

It is not surprising that criticism of the modern clinical method from within the profession has come mainly from fields of medicine that most experience the ambiguities of abstraction and the importance of the patient's life story, notably general practice and psychiatry. In the 1950s Michael Balint, a medical psychoanalyst, began to work with a group of general practitioners, exploring difficult cases and the doctors' affective responses to them. Balint distinguished between "overall" diagnosis and traditional diagnosis; he emphasized the importance of listening and of the personal change required in the doctor; and he introduced new terms such as: "patient-centered medicine," "the patient's offers" and the doctor's "responses," the doctor's belief in his "apostolic function," and the "drug doctor" – the powerful influence for good or harm of the doctor-patient relationship. The idea that physicians should attend to their own emotional development as well as the emotions of the patient was revolutionary in its day. In other ways, Balint's method conformed to the dualistic approach of the period. The method was intended to apply only to certain patients with "neurotic illnesses," not to those with straightforward clinical problems.

In the 1970s Engel, an internist and psychiatrist with a psychoanalytic orientation, used systems theory as a model for integrating biologic, psychologic, and social data in the clinical process. Engel's critique of the modern clinical method focused on the unscientific nature of the physician's judgments on interpersonal and social aspects of patients' lives, based on "tradition, custom, prescribed rules, compassion, intuition, common sense, and sometimes highly personal self-reference."

Any successor to the modern clinical method must propose another method with the same strengths: a theoretical foundation, a clear set of injunctions about what the clinician must do, and canons of verification by which it may be judged. Lain Entralgo (1956) attributes the failure of Western medicine to integrate the patient's inner life with the disease to the lack, among other things, of a method – "a technique [for] laying bare, to clinical investigation and to subsequent pathological consideration the inner life of the patient ... an exploration method – the dialogue with the patient...." Balint and Engel both provided a theory, but were less clear about what the clinician must do and how the process is to be validated. Although Engel emphasized that the verification must be scientific, validation of both models was bound to depend on qualitative methods that were barely accepted as scientific. A model is an abstraction; a method is its practical application; and medicine had to wait longer for the transition to occur.

Since the work of Balint, general practitioners have continued to develop the concept of patient-centeredness, notably Stevens (1974), Tait (1979), Byrne and Long (1976), Wright and MacAdam (1979), Stephens (1982), Levenstein (1984), and McWhinney (1986). Byrne and Long developed a method for categorizing consultations as either doctor-centered or patient-centered and found that many practitioners had developed a static style of consulting that tended to be doctor-centered.

Critiques of modern medical theory have come from the philosopher Laurence Foss (Foss and Rothenberg, 1987; Foss, 2001), and from medical anthropologists Fabrega (1974) and Kleinman (Kleinman, Eisenberg, and Good 1978). Foss proposes replacing the biomedical model with an infomedical model, thus allowing medical science to incorporate the compelling evidence for the influence of information on the function of the human organism in health and disease, including intentional changes by individuals in their own bodily processes. Fabrega emphasizes the conceptual distinction between illness (the human experience), and disease (the mental construct), and the importance of attending to both. Another important critique has come from the narratives of illness published by patients.

V. ILLNESS NARRATIVES

In the past three decades there has been a remarkable increase in he number of books and articles describing personal experiences of illness. These writings, by patients themselves or by their relatives, are often bitterly critical of clinicians and, by implication, of the modern clinical method. Hawkins (1993) sees this literature as a possible reaction to a medicine "so dominated by a biophysical understanding of illness that its experiential aspects are virtually ignored." Two themes recur in these stories: "the tendency in contemporary medical practice to focus primarily not on the needs of the individual who is sick but on the nomothetic condition we call the disease, and the sense that our medical technology has advanced beyond our capacity to use it wisely" (Hawkins, 1993).

Some illness narratives are written by patients who have a professional perspective as physicians, philosophers, sociologists, or poets. Sacks (1984) viewed his experience of a body image disorder from the perspective of an existential neurologist and medical theorist. Stetten (1981) found that his fellow physicians were interested in his vision, but not in his blindness. Toombs (1992), a phenomenologist who has multiple sclerosis, noted that physicians' attention is directed to their patients' bodies rather than to their problems of living. The patient feels "reduced to a malfunctioning biological organism." Toombs writes: "... no physician has ever inquired of me what it is *like* to live with multiple sclerosis or to experience one of the disabilities that have accrued ... no neurologist has ever asked me if I am afraid, or ... even whether I am concerned about the future." Writing of his experience with testicular cancer, Frank (1992), a sociologist, observed that the more critical his illness became, the more his physicians withdrew.

The theme of so many narratives is that the patient's experience is not valued or attended to. Diagnosis is a process of classification and abstraction, exemplified by the clinico-pathological conference in its classic form and illustrated in Figure 1. The three irregular figures represent patients with similar illnesses. Although similar, there are many individual differences between them. The three squares represent what they have in common. In the process of abstraction, the common factors are reified to form a disease entity, while the individual differences are tacitly ignored. Among the differences is the patient's experience of illness. The disease is a digital coding of the patient's experience. Our system of abstractions is a map of the territory of human illness. Abstraction distances us from experience and strips away the affective aspects of the illness. If the abstraction is not balanced by attention to the patient's experience, the result may be the patient's alienation, described in so many illness narratives. In its extreme form, physicians may come to believe that their abstractions are the real world, thus mistaking the map for the territory.

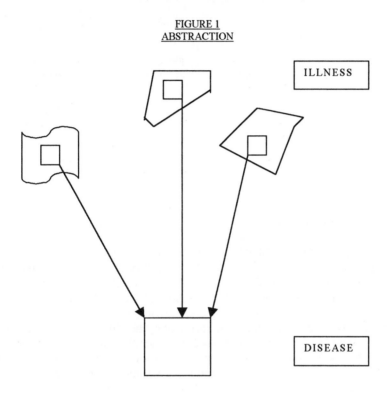

FIGURE 1
ABSTRACTION

Adapted from McWhinney, I.R., *European Journal of General Practice*, Dec. 2000, p.135. Reprinted with permission of the publishers, Mediselec G.V.

The power of generalization increases with each level of abstraction. In medicine, some of our categories are low level abstractions with limited predictive power, such as clusters of clinical observations called syndromes. Categories like irritable bowel syndrome and low back syndrome do not separate us much from the patient's concrete experience. Further degrees of abstraction, illustrated in Table 1, interpret the meaning of the illness in terms of pathology, chemistry, or imaging, taking us further and further from the experience.

TABLE 1:

LEVELS OF ABSTRACTION IN A PATIENT WITH MULTIPLE,
FLUCTUATING, NEUROLOGICAL SYMPTOMS AND SIGNS

Level 1	Patient's sensations and emotions	Preverbal	Illness experience
Level 2	Patient's expressed complaints, feelings interpretations	First order abstraction	"Illness" (doctor's understanding)
Level 3	Doctor's analysis of illness: clinical assessment	Second order abstraction	"Disease" (clinical diagnosis: multiple sclerosis)
Level 4	M.R.I. scan	Third order abstraction	"Disease" (definitive diagnosis M.S.)

From: McWhinney, I.R.: A Textbook of Family Medicine, 2nd ed., p.77. Reprinted with permission of Oxford University Press, N.Y.

The power of the abstraction depends on its having the same structure as the illness it represents – a feature exemplified by the case of multiple sclerosis. However, no abstraction is ever a complete picture of what it represents: it becomes less and less complete as the levels of abstraction and power of generalization increase. Every patient's illness is different in some way. As we increase the levels of abstraction, the differences are ironed out in the interests of increasing our power of generalization. As we increase the levels of abstraction, the affective contribution to our understanding of the original experience becomes less and less. We cannot experience the beauty or the terror of a landscape by reading the map. Of course, one can get passionate about maps. There is a thrill in making a good diagnosis (finding our place on the map), and there can be beauty in a radiograph. But this is not the same as a feeling for the patient's experience of illness.

VI. THE PATIENT-CENTERED CLINICAL METHOD

The method described here uses insights from Balint and Engel, but goes further. It specifies the clinical task in operational terms and establishes criteria for validation. It was formulated as a clinical method by Levenstein (1984) in his own practice, and further by Stewart and her colleagues at The University of Western Ontario (Stewart et al., 1995; Levenstein et al., 1986).

The method consists of six interconnected components: exploring the patient's experience of illness and integrating this with conventional diagnosis; understanding the whole person; finding common ground with the patient; incorporating prevention and health promotion; enhancing the doctor-patient relationship; and being realistic about available resources.

The patient-centered clinical method, like its predecessor, gives the clinician a number of injunctions: "Listen to the patient's story" recognizes the importance of narrative and context. "Explore the patient's experience of the illness" invites the clinician to enter the patient's life-world, recognizing the uniqueness of each individual. "Seek to understand the meaning of the illness for the patient: his ideas, beliefs, and fears" enjoins the doctor to attend to the patient's inner life. "Ascertain the patient's expectations" recognizes the importance of knowing why the patient has come. "Make or exclude a clinical diagnosis:" the new clinical method transcends, but includes, its predecessor. "Seek common ground" enjoins the clinician to develop with the patient a common understanding of the illness – an understandind that reconciles the meaning of the illness in terms of pathology with its meaning for the patient's life-world. Ideally the common understanding will lead to agreement on therapy. "Assess your relationship with the patient and filter your own responses for the intrusion of bias, self interest, and counter transference" recognizes the importance of the relationship and the potential for unexamined negative emotions to destroy or impair it.

The key to the patient-centered clinical method is for the clinician to allow as much as possible to flow from the patient. The crucial qualities and skills required of the physician are openness, empathy, attentive listening, and responsiveness to those verbal and nonverbal cues by which patients express themselves. It is important not to think of the method as a strictly defined process, with sequential stages and standardized procedures and interviewing styles. If the flow of the consultation is to follow the cues given by the patient, its course will depend on how and when these cues are given, and will vary from patient to patient. The process does not exclude the interrogatory style of the predecessor method when this is appropriate. The consultation will typically weave to and fro between illness and disease, concrete experience and nosological abstraction, territory and map (Figure 2).

FIGURE 2
THE PATIENT-CENTERED CLINICAL METHOD

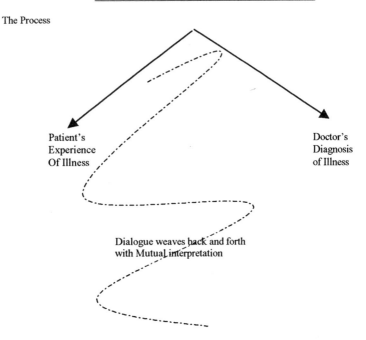

Freely adapted from Stewart, M., Brown, J.B., Weston, W.W., McWhinney, I.R., *et al.*, *Patient-Centered Medicine: Transforming the Clinical Method*, p. 37. Reprinted by permission of Sage Publications Inc.

VII. QUESTIONS ABOUT THE PATIENT-CENTERED CLINICAL METHOD

Four questions are commonly asked about the patient-centered clinical method. First, is it always necessary to use the method? Suppose the problem is very straightforward: an injury, for example, or an uncomplicated infectious disease. The answer is that we do not know unless we ask. Patients have fears and fantasies even about common and minor problems. In emergencies, of course, the medical priority must take precedence. But when these needs have been met, no patient is in greater need of being listened to than the one with sudden and severe acute illness or trauma.

Second, what if there is a conflict between the patient's expectations and the medical assessment? Suppose, for example, that a patient wishes to manage her diabetic ketoacidosis without admission to the hospital. The physician must then try to reconcile the two conflicting views. The more the physician can understand about the reasons for the patient's position, the more chance there will be of a satisfactory conclusion. The reluctance to go into the hospital, for example, may be due to a feeling of responsibility for a child or elderly parent. In some cases, there will be an irreconcilable conflict, as in a demand for a narcotic drug, and the physician will have to refuse to meet the patient's expectations. In the more usual situation, doctor and patient have different interpretations of the illness or conflicting notions about its management, as when the patient, believing his or her pain indicates an organic disease, cannot accept the doctor's view that this is not the case. The doctor's contribution to reconciling conflicting views is threefold. First, the doctor can acknowledge the validity of the patient's experience and take his or her interpretation seriously, even if he cannot accept it. Second, he can be aware of the danger that his own prejudice, rigidity, dogmatism, or faulty logic may be the reason for the difference. Third, he can make sure that the patient has all the information we can provide. Conversely, some humility may be called for, as when a very well informed patient knows more than the doctor does about his or her condition.

One of the most frequent reasons for disagreement is the so-called somatoform disorders, defined as bodily symptoms of emotional states, incorrectly ascribed to organic pathology. The physician's task here is to bring the patient to a different understanding of his illness without damaging the doctor-patient relationship. Seeking common ground takes place in the context of compassion for the patient's suffering, but compassion should not be "idiot compassion" (Wilber, 1999, p. 99), as in the uncritical support of self-destructive behavior. There are times when confrontation is therapeutic, as long as it is done in the patient's interest.

Third, is there not a risk of invading the patient's privacy? Suppose the patient does not want to, or is not ready to, reveal her secrets? If privacy is invaded, then the method has been misunderstood. The essence of it is that the doctor responds to cues given by the patient, allows and encourages expression but does not force it. If cues are not given, feelings are explored with general questions that invite openness. If the patient does not wish to respond, the matter is not pursued. At least the doctor has indicated that such matters are admissible.

Fourth, what about the time problem? How can the doctor afford the time to listen to the patient? It is difficult to answer this, because little research has been done on the relation between consultation time, clinical method, and effectiveness. From work done so far, we can say tentatively that patient-centered consultations take a little longer, but not much longer, than doctor-centered ones. Beckman and Frankel (1984) found that when patients are uninterrupted, their opening statements lasted only two and a half minutes on the average. Stewart *et al.* (1995) reported that nine minutes was the critical duration for patient-centered consultations in general practice. What we do not know is how much time is saved in the long run by an early and accurate identification of the patient's problems.

VIII. VALIDATION

The ultimate validation of a diagnosis in the modern clinical method is the pathologist's report. In the clinico-pathological conferences modeled by *The New England Journal of Medicine*, a clinician is presented with a case report and develops a differential diagnosis, which is then confirmed or otherwise by a pathologist. Other forms of validation include the response to treatment of the illness.

Validation of the patient-centered clinical method includes the above, but places them in the context of a validation of the patient-centeredness of the process. The ultimate validation is the patient's report that his or her feelings and concerns have been acknowledged and responded to, and that there has been a search for common ground. In the normal course of practice, validation comes from the course of the illness, the success of therapy and the openness of the doctor-patient relationship.

A number of studies have examined the relationship between patient-centered scores of audiotaped consultations and health outcomes. Twenty-two studies relevant to patient-centered communication and health outcomes have been identified up to 1998. Seven out of eight studies focusing on exploring the illness experience and diagnosing the illness obtained significant positive findings, the important aspects being physicians asking questions about the patient's experience of illness, showing support for the patient, and allowing patients to express themselves. Thirteen out of sixteen studies that examined the seeking of common ground found significant positive findings, the important factors being the sharing of information, mutual decision-making and agreement between doctor and patient (Stewart, 1995; Stewart and Brown, 2000).

A major study (Stewart *et al.*, 2000) examined the relationship between health outcomes and patient-centered scores by observers and by patients. Patients' perceptions – especially of finding common ground – were directly associated with positive outcomes. Observer scores were correlated with the patients' perceptions, but not directly with the health outcomes, thus supporting the view that patients are the most perceptive judges of patient-centeredness.

IX. LEARNING THE PATIENT-CENTERED CLINICAL METHOD

To assist learning, the process may be broken down into a number of rules, tasks, and stages. But no list of components can include all the tacit knowledge that can only be acquired by experiencing and "dwelling in" the process. It is difficult to be aware of the components (subsidiary awareness) and of the whole process (focal awareness) at the same time (Polanyi, 1962). If we think of the rules while trying for the first time to ride a bicycle, we probably fall off. The tension between these two levels of awareness, and the need to alternate between them is difficult at first, but when the doctor finally "dwells in" the process, it becomes second nature and there is no need to attend to rules and stages, unless there is need to review the process after the fact.

The learning of techniques and skills of communication is necessary, but can never be enough to become a practitioner of the method. Practicing the method is a different way of being a physician. As Balint observed many years ago, it requires a change in the physician's personality. Once this change has taken place, everything the physician does will be different. The change is a transformation rather than an addition to the physician's knowledge.

X. HEALTH ENHANCEMENT AND DISEASE PREVENTION

Some determinants of health are beyond the control of individuals. Others are strongly related to the individual's beliefs, values, and motivation. The physician as teacher can educate patients about ways they can enhance their health and reduce their risk of disease. Whether or not patients will make changes will depend on many factors, including other aspects of their lives and the reality of threats to their health. Knowledge of a patient gained by patient-centered practice can enable the physician to sense the point at which he or she is ready to act: to make dietary changes, to take more exercise, to stop smoking, to take their diabetes more seriously, or to change a self-destructive way of life.

Physicians' responsibility for health promotion and disease prevention varies with their field. Those in primary care disciplines, with long-term doctor-patient relationships, have the broadest responsibility and the greatest opportunity for prevention. But there is no field so limited that a patient-centered approach cannot provide such opportunities.

XI. ASSESSING AND DEVELOPING THE RELATIONSHIP

What can be achieved in the consultation is very dependent on the quality of the relationship – what Balint called the "mutual investment company" between doctor and patient: its openness and honesty; the level of mutual commitment and trust; its intimacy; and the patient's confidence in the doctor's judgement. On this assessment will depend the timing of interventions such as reassurance or confrontation.

A patient-centered clinician will attend to the patient's needs, both great and small. When time is short, it is tempting to miss out the small courtesies, the little things that are so important in the cultivation of friendship (Drane, 1995, p.90). The more serious the patient's illness, the more important it is to attend to the small comforts which can add so much to a patient's peace of mind and body, and it is in the most desperate situations that it is easiest for the clinician to withdraw from the patient while concentrating on the technological aspects of care. The more the common courtesies of the relationship can become habits, the more secure the virtue of friendliness will be, and the less vulnerable to the stresses and difficulties of medical practice. "Individual acts, chosen and repeated – have an effect on the kind of person one becomes" (Drane, 1995 p.138).

For the clinician, enhancing the relationship involves attending to his or her emotions, especially those unacknowledged narcissistic emotions that can be so destructive in a relationship: helplessness in the face of incurable disease, fear of failure, anger at patients who annoy us, eagerness for our professional enhancement (McWhinney, 1996). Even qualities we regard as virtues may have destructive consequences. We have to be careful that our practice does not become dominated by an ideology of health which drives our responses to patients, irrespective of their individual needs. We also need to be aware of the possibilities for countertransference in any doctor-patient relationship.

XII. CONCLUSION

What we now call *the* patient-centered clinical method is the most recent version of the historic struggle to reconcile two often competing notions of the nature of disease and the role of the physician. The last century has seen the increasing dominance of abstraction and the devaluation of experience. The patient-centered method can be viewed as a move to bring medical practice and teaching back to the center, to reconcile clinical with existential medicine (Sacks, 1982). It may seem paradoxical that the modern clinical method does not have a name. It is simply the way clinical medicine has been taught in medical schools in modern times. Giving the successor method a name has its dangers, notably that of conveying different meanings to different people. In this transition period however, it does seem necessary to have a name for the new method. But when the transition is complete, perhaps it can simply become "clinical method."

The new method should not only restore the Hippocratic ideal of friendship between doctor and patient, but also make possible a medicine which can see illness as an expression of a person with a moral nature, an inner life, and a unique life story: a medicine that can heal by a therapy of the word and a therapy of the body.

Centre for Studies in Family Medicine
The University of Western Ontario
London, Ontario
Canada

BIBLIOGRAPHY

Balint, M.: 1964, *The Doctor, His Patient and the Illness*, Pitman Medical Publishing, London.
Beckman, H.B. and R.M. Frankel: 1984, 'The effect of physician behavior on the collection of data', *Annals of Internal Medicine* **101**, 692.
Brown, T.M.: 1989, 'Cartesian dualism and psychosomatics', *Psychosomatics* **30**, 322-31.
Burtt, E.A.: 1924, *The Metaphysical Foundations of Modern Physical Science*, 2nd ed., Routledge and Kegan Paul, London.
Crookshank, F.G.: 1926, 'The theory of diagnosis', *Lancet* **2**, 939.
Drane, J.F.: 1995, *Becoming a Good Doctor: The Place of Virtue and Character in Medical Ethics*, Sheed and Ward, Kansas City.
Duden, B.: 1991, *The Woman Beneath the Skin: A Doctor's Patients in Eighteenth Century Germany*, Harvard University Press, Cambridge, MA.
Entralgo, P.L.: 1969, *Doctor and Patient*, McGraw-Hill, New York.
Entralgo, P.L.: 1961, *The Therapy of the Word in Classical Antiquity*, Yale University Press, New Haven.
Entralgo, P.L.: 1956, *Mind and Body*, PJ Kennedy, New York.
Faber, K.: 1923, *Nosography in Modern Internal Medicine*, Paul B. Hoeber, New York.
Fabrega, H.: 1974, *Disease and Social Behaviour*, MIT Press, Cambridge.
Fleck, L.: 1979, *Genesis and Development of a Scientific Fact*, University of Chicago Press, Chicago.

Foss, L.: Forthcoming, *Biological Medicine Under A Microscope*, SUNY University Press, New York.
Foss, L. and K. Rothenberg: 1987, *The Second Medical Revolution: From Biomedicine to Infomedicine*, Shambhala, Boston and London.
Frank, A.: 1992, *At the Will of the Body: Reflections on Illness*, Houghton Mifflin, Boston, MA.
Hawkins, A.H.: 1993, *Reconstructing illness: Studies in Pathography*, Purdue University Press, West Lafayette, Indiana.
Hoffmeyer, J.: 1996, *Signs of Meaning in the Universe*, Indiana University Press, Indianapolis.
Kleinman, A., Eisenberg, J., Good, B.: 1978, 'Culture, illness and care: Clinical lessons from anthropologic and cross-cultural research', *Annals of Internal medicine* **88**, 251.
Levenstein, J.H.: 1984, 'The patient-centered general practice consultation', *South Africa Family Practice* **59**, 13-19.
Levenstein, J.H., McCracken, E.C., McWhinney, et al.: 1986, 'Patient-centered clinical method: A model for the doctor-patient interaction in family medicine', *Family Practice* **5**, 276-282.
McWhinney, I.R.: 1996, 'The importance of being different', *British Journal of General Practice*, **46**, 433.
McWhinney, I.R.: 1986, 'Are we on the brink of a major transformation of clinical method?' *Canadian Medical Association Journal* **135**, 873.
Nussbaum, M.C.: 1995, *The Therapy of Desire: Theory and Practice in Hellenistic Ethics*, Princeton University Press, Princeton.
Polanyi, M.: 1962. *Personal Knowledge: Towards a Post-Critical Philosophy*, University of Chicago Press, Chicago.
Rather, L.J.: 1965, *Mind and Body in Eighteenth Century Medicine*, University of California Press, Berkeley, CA.
Sacks, O: 1984, *A Leg to Stand On*, Gerald Duckworth, London.
Sacks, O.: 1982, *Awakenings*, Pan Books, London.
Stetten, D., Jr.: 1981, 'Coping with blindness', *New England Journal of Medicine* **305**, 458.
Stewart, M., Brown, J.B., Donner, A., McWhinney, I.R., et al.: 2000, 'The impact of patient-centered care on outcomes', *The Journal of Family Practice* **49**, 796.
Stewart, M.: 1995, 'Effective physician-patient communication and health outcomes: A review', *Canadian Medical Association Journal* **152**, 1423-1433.
Stewart, M., Brown, J.B., Weston, W.W., McWhinney, I.R., et al.: 1995. *Patient-Centered Medicine: Transforming the Clinical Method*, Sage, Newbury Park, CA.
Toombs, S.K.: 1992, *The Meaning of Illness: A Phenomenological Account of the Different Perspectives of Physician and Patient*, Kluwer Academic Publishers, Dordrecht, The Netherlands.
Toulmin, S.: 1992. *Cosmopolis: The Hidden Agenda of Modernity*, University of Chicago Press, Chicago.
Whitehead, A.N.: 1975, *Science and the Modern World*, Collins, Fontana, San Francisco.
Wilber, K.: 1999, *One Taste*, Shambhala, Boston.

PATRICIA BENNER

THE PHENOMENON OF CARE

Illness is, in the last analysis, not the established result which scientific medicine declares as illness but, rather, the experience of the person suffering it...The distinction between health and illness is a pragmatic one and the only person who has access to it is the actual person who is feeling ill, the person who can no longer cope with all the demands of life and so decides to go to the doctor (Gadamer, 1993, p.55, 162).

I. INTRODUCTION

Distinctions between health, illness, and disease have gained increased acknowledgement within medicine (Kleinman, 1988). Health remains an enigma but entails equilibrium within the body, and also within the person's lifeworld so that the person experiences a sense of well-being and wholeness (Gadamer, 1993; Benner and Wrubel, 1989). Doctors are trained to intervene in and/or cure disease, freeing the person to return to his or her usual life. The disciplinary strategies physicians' use for objectifying the causative agents and appropriate interventions are accomplished by what Foucault (1963/1973) has called the clinical gaze. The clinical gaze was ushered in by the introduction of dissection into medical science. The physician uses detached concern (Halpern, 2001) to observe and validate objective signs that can be verified as pathology at the tissue, cellular or organ levels. This is a powerful focusing strategy of de-worlding the disease. The lived body and world are passed over in order to focus on the genomic, biochemical and physiological substrates of the physical-chemical body. Levin (1999) traces the clinical gaze to Descartes' view from the window, and notes that it epitomizes the transformation of the modern subject as "the one who looks" rather than one who is "looked upon":

Since Descartes adopts the observations of a spectator as model for a philosophical understanding of self-awareness, it should not be surprising that he would take the spectator's observation as paradigmatic of our relationship with others ...
 Descartes stands at the window, silently looking out. Even though he is prepared to recognize the speech of the other as an irrefutable evidence of the other's humanity, he makes no attempt to go outside, to meet the "men" he seems to see, to talk with them. The philosopher *prefers* the distance of vision, even when this distance means uncertainty – even when it means dehumanization (Levin, 1999, pp. 41, 47).

The metaphor of the window works well for the objectification required for doing surgery or working at the microscope. However, it is less well suited for doctoring. Being a good physician requires that the physician has some understanding of the person who falls ill and some understanding of the disturbances to the person's functioning in the lifeworld as a result of the illness. Left out of a strictly Cartesian picture of medicine is the embodied person who suffers, requires caregiving or self-care, and who recovers and must re-integrate herself in her lifeworld (Cassell, 1991). The skillfulness of the doctor's diagnostic and intervention strategies depend upon the physician's relationship with the patient in at least three crucial ways: (1) The relationship and the mood or emotional climate of the physician-patient encounter determines what aspects of the patient's ailments and suffering will be disclosed; (2) Understanding the contributions to the health, illness or recovery that a particular person's lifeworld makes, requires knowing the patient as a person; (3) The physician's caring practices and rhetorical skills determine how and what information is disclosed to the patient about diagnosis and treatment, and how these may or may not help with the re-integration of the person back into her lifeworld (Benner and Wrubel, 1989; Cassell, 1991; Gadamer, 1993; van Manen, 1999). The physician must learn to be critically reflective about what a detached concern or clinical gaze discloses, what possibilities it creates, and what it excludes within patients and within the physician's own emotional engagement and attunement in the world. When the clinical gaze spreads out into all areas of the physician's life, richness and connectedness are lost. A consistent style of detached observation prevents adequate engagement for living out one's concerns. Being in relationship with others conditions the possibilities for experiential learning that open one's world and foster character and skill development (Benner, Tanner, and Chesla, 1996; Dunne, 1997).

This chapter explores: (1) Ontological care structures, defined as those most basic ontological structures of connection and concern that suspend persons in a human world complete with relationships, identity, recognition and concerns; (2) Ontic caregiving practices (specific styles of caregiving practices) that nurture recovery and promote health; (3) How good medical practice requires caring practices that take into account ontological care and the lifeworld of the patient, as well as the disciplinary strategies of Cartesian medicine. Like suffering, ontological care and caregiving practices are typically covered over in the clinical gaze that dissects and analyzes disease processes (Cassell, 1991).

II. ONTOLOGICAL CARE

Care has ontological privilege. Patients never seek health care unless a health worry (or symptoms) intrudes upon their network of concerns or actions. Illness and disease do not exist in a one-to-one relationship. One can be cured of one's disease and still experience illness or, alternatively, disease can be silent with no illness experience (Frank, 2001). Illness and recovery refer to the experience of losing and regaining a sense of wholeness and integration in one's world. Once symptoms arise, by way of intrusion or disruption of daily concerns and activities, the disease itself is – to some extent – altered and shaped by the person's illness experience. Oliver Sacks (1985) states:

> I feel myself a naturalist and a physician both; and that I am equally interested in diseases and people; perhaps, too, that I am equally, if inadequately, a theorist and continually see both in the human condition, not least in that quintessential human condition of sickness. Animals get diseases, but only man falls radically into sickness.

The lived body opens one to particular lifeworlds and makes it necessary and possible to care. Human lifeworlds are structured by care. Care is ontological in that it structures being human – what and how something matters, and what can be encountered (noticed) and known. Whatever autonomy we gain and effect in our lives depends on the structures of care that have shaped us and open us to the world. Worlds and relationships are entered through affectivity (Benner and Wrubel, 1989; Agich, 1995). Care and caregiving are often treated as if they were merely sentiment or attitudes, void of cognition, skill, and particularized relationships. Yet care structures the human lifeworld of concerns and possibility. Specific caring practices (ontic regions or domains of knowledge related to caregiving practices), such as those developed in family care and professional caregiving, contain within them knowledge and skill about everyday human needs for recognition, nurturance, shelter, food, hygiene, protection, and so forth. In order to be experienced as a caring practice, caregiving must attend to what matters (ontological structures of care) in the particular lifeworld of the person, infant, or community (Benner and Wrubel, 1989; Benner, 2000).

Self-care (or neglect) and the quality of caregiving practices of others shape the illnesses, recoveries or rehabilitation that can be experienced or appropriated. Since care shapes the human lifeworlds in which we dwell, obviously the human worlds of care in which we are born must be supportive enough for human survival. For example, in the extremes of the technical worlds of Neonatal ICU's, parents, nurses and physicians struggle with the quality of environments created for premature neonates. It is never a question of whether an infant can make it on her own. Astute discernment (phronesis) is required for three major inter-related goals: The first goal is to meet and assess the particular infant. Discerning the infant's maturity and capacities is crucial since maturation rates vary among infants and maturity is more than just a product of size and intrauterine time. Based on this discernment, the second goal is to place the infant in a particularized technical/human environment that will support the infant, while minimizing technological hazards. The third goal is to foster the social, human birthing of the infant. For example, managing discomfort and pain in these technical environments requires judiciously introducing and teaching human comfort and solace in concert with the infant's ability to tolerate these. Overstimulation can be dangerous. However, social birthing is arrested if the infant does not learn to respond to human touch, comfort and voice. Introducing these depends on the infant's embodied capacities and readiness. Physiological demands exist in concert with the demand that body/social/world relations are adequate for the baby's well being. Thus, the premature infant's survival and flourishing depend on technical support structured by human care, as is the case for every newborn infant.

In the extreme case of the premature neonate, it is easy to grasp the human necessity to create a human world structured by care and specific caring practices. As human beings mature, the awareness of requisite supportive structures of care of self, others, and world recede into the background. Nevertheless, throughout our lives, human capacities and action are constituted and structured by human worlds of care. Radical freedom and radically independent separateness are illusions (Taylor, 1985; MacIntyre, 1999).

Human beings dwell in human worlds constituted by care, relying on others and the human lifeworlds that they both constitute and are, in turn, constituted by. The knower and the known are intertwined. The intertwining of our embodied self and world is so pervasive that it lies in the taken-for-granted background. Thus, we fail to see – and, therefore, forget – the concerns that daily suspend us in the webs of care that make up our lifeworld. Without networks of care or concern we would rattle around capriciously in a vast, random universe, lacking the structures in which to ground our actions and choices.

Losses such as the death of a loved one, or one's own illness/disease/injury can disrupt (if not shatter) one's taken-for-granted world. Therefore, recovery comes not only from curing the body, or repairing the injury or loss, but also in restoring the integration and functioning of the embodied self in his or her particular world.

Both the ontological forms of care that constitute persons and their lifeworlds, and particular ontic forms of local caregiving practices provided in the home, school, neonatal ICU, long term care settings, and so forth, are inter-related. In order to be experienced as nurturing and supportive, ontic caregiving practices must be attuned to the person's ontological concerns or structures of care. By ontic forms of caring practices, I mean caring practices that are located in specific lifeworlds and specific cultural traditions of caring. For example, in the Enlightenment tradition – with its emphasis on the individual – caring practices have developed to support formation of self-sufficient selves who aim for independence and who tend to be forgetful of their dependence on others (Taylor, 1985; MacIntyre, 2000). However this forgetfulness is easily interrupted by breakdowns in embodied capacities, or through the loss of familiar lifeworlds that suspend even the most self-sufficient person in everyday structures of interdependence and care. When familiar lifeworlds or major concerns and projects are lost through the death of a close other, or through accident, injury or disease, the person may experience a sense of *falling* or feeling anxiously lost (Dreyfus, 1991; Heidegger, 1926/1962).

Modernist medicine has separated the social from the physical, and has focused upon diseases that can be treated or cured with biomedical interventions. This project of separation of the social from the medical – an initial condition for the secular development of medicine – has been sustained, in part, by the many successes of biomedical cures, and by our mythic dreams of transcending our human conditions of finitude, embodiment and dependencies. But, paradoxically, the more successful biomedicine becomes in treating disease, while ignoring the personal and social aspects of health and illness, the more the shadow world of everyday care (ontological structures and specific traditions and styles of caregiving practices) manifest themselves. For example, medically treating chronic illnesses of the elderly (or of neonates) and excluding attention to environment, nutrition, exercise and social well-being is medically unsustainable. Medicine can only accomplish its work if these world-sustaining care structures are reasonably intact. While this is patently obvious, healthcare policy and the economic and institutional structures for biomedicine are geared to provide expensive treatments with little attention given to caregiving structures that sustain and/or prevent the need for medical treatment. For example, extreme economic pressures prohibit physicians from admitting patients to hospitals or from providing home care for merely "social" needs. Yet, ignoring the need for nutrition, rest, shelter, and supportive care for the chronically ill or frail may cause death or severe illness. Medical interventions tend to escalate in the absence of everyday care, compounding the problems that are systematically ignored by strictly medical interventions (Foucault, 1963/1973; Benner, 1994b).

Ontological care refers to the ways that all human existence daily depends on embodied dwelling in particular human lifeworlds. Ontological structures of care shape everyday comportment and understandings, allowing a person to gear into a particular lifeworld. Ontological care refers to the range of connectedness and mattering that human beings experience in particular lifeworlds. It is crucial not to read into *ontological care or concern* a particular style or tradition of caregiving, as in unconditional regard (agape), liberating transformative care, or even consumerist simulacra for care (as in the "used car" salesperson who cares). Particular forms of connecting and mattering are as diverse as particular lifeworlds. Typically dwelling in a particular lifeworld is accomplished in non-reflective ways because we are always born into and *thrown* into human worlds that are already in progress. Ontological care is primary in the ways that perception and embodied existence are primary (Benner and Wrubel, 1989; Merleau-Ponty, 1962; Casey, 1997) and the knowledge and skill embedded in particular ontic forms of caring are primary and essential for any curing or health maintenance. Ontological structures of care demonstrate that human beings are the kind of beings that allow other beings to show up and to be. Structures of ontological care create clearings or diclosedness (Dreyfus, 1991; Wrathal, 2000; Heidegger, 1926/1962).

In lifeworlds that stress competitive individualism and independence, interpersonal concerns are structured around being separate, competent and detached. Or at the extremes of social breakdown, one's predominant connectedness to the world may be in anger, distrust and avoidance. For example, in war torn areas, old traditions of enmity and violence may create engagement with the "other" who has become "wholly other" (Levinas, 1998/1991; Bauman, 1993) in forms of systematic dehumanization. Even though recognition practices deteriorate into anger, enmity and devaluation, interactions and relationship still matter to people, i.e. form structures of *care* or *concern* that set up meaningful worlds where some things stand out as more or less salient. Therefore: *Because caring sets up what matters to a person, it also sets up what counts as stressful and what options are available for coping. Care sets up embodied intentionality that allows the person to get around in the world. Caring creates possibility. This is the first way in which is caring is primary* (ontological).

Care and caring are world constituting. *This enabling condition of connection and concern is another way in which caring is primary.* Embodiment sets up situated freedom, locating the person in a particular place and time and enabling the living out of particular concerns. Without structures of care, one would not know how to get around in the world. One can imagine being lost in a large universe of infinite possibilities or of nothingness. Either way the individual is radically alone and detached. Yet no individual can sustain such empty disconnected existence. Caring (i.e., having things matter) allows one to be available in the situation, to notice what is at stake or what needs to be done. Emotional connection to the situation creates the possibility of acting in the situation, along with the possibilities of practical rationality (Damasio, 1999; Benner 2001). Emotional connection allows for recognition based upon perceptions born of emotional and embodied responses to the situation. We notice and attend to what is salient to us. Mattering is emotionally valenced, a form of embodied intentionality that may or may not include deliberate or conceptual intentionality.

Finally, caring is primary because it sets up the possibility of giving help and receiving help. Caring sets up the possibility for acting in the world. Hannah Arendt pointed out that we have come so far in our elaboration of the self that we have lost our understanding of the public world and the place of action in the world. Action that brings something new into one's world, or that opens the horizons of a diminished world, can be bold or simple. For example, Joan, a critically ill patient who had been in the Intensive Care Unit for two months, had become so institutionalized that her world had shrunk down to the narrow center of her bed. She was awaiting triple heart valve surgery to replace heart valves damaged by rheumatic fever in her youth. But, first, her congestive heart failure and liver failure had to be corrected. Tethered to the bed by a respirator and IV's, Joan, her family, nurses, and doctors managed to create a safe and familiar haven. The nurses thought Joan needed a rehearsal for going to, and returning from, surgery because almost any journey seemed impossible to her. The nurses, therefore, arranged a festive, light-hearted journey to look at the city lights from a newly built vantage point in the hospital. Though it took a number of personnel to accomplish this journey, the effects on Joan's spirit were remarkable. The trip expanded her horizons, and increased her imagined ability to leave, and return to, the unit intact – a feat unimaginable during the worst of Joan's illness (Benner, 2000). This is a dramatic example of how an acute illness can change one's sense of self in the world and how recovery entails recovery of self and world. Prior to Joan's acute illness, she had adapted to chronic heart failure. Recovery after the valve replacement surgery entailed living in a different world – for instance, one where only one pillow was needed at night – and a regained ability to participate in social activities. As this example clearly shows, the lived boundaries of "I can" and "I can't" change with illness, recovery and relapse (Kesselring, 1990). To be human is to dwell in a particular world of embodied capacities, concerns and relationships. Without this situatedness in a lifeworld, one loses the ability to feel at home in the world.

III. ONTIC CAREGIVING PRACTICES

Caring practices are necessarily related to local specific lifeworlds and to specific ontological structures of care. Understanding ontological care as world and person-disclosing enables us to reflect on specific caring practices that may support, or undermine, the person's self and world-defining concerns (ontological care). The final section of this chapter examines caring practices of families, friends, and helping professionals. Caregiving practices manifest specific styles and content. Care is experienced as supportive, only if the one caring takes into consideration the local specific lifeworld of the one cared for. For example, for individuals who seek autonomy in ways that deny their interdependence and dependencies, caring practices will have to be supportive in ways that empower the person's quest for control and independence. When the world of the independent individual is shattered by an incapacitating illness, empowerment as a caring practice may help the person sustain hope and a sense of independence in those spheres where personal control is still possible. Or alternatively, illness may become an occasion for the 'independent self' experientially to come to terms with necessary dependence and inter-dependence. Everyday embodied existence requires that we learn the limits of personal control without feeling completely helpless. Sooner or later all human beings fall ill and face their concrete embodied limits to independence. Re-visiting, instead of covering over, care as the ontological basis for being human and flourishing provides a non-normalizing and non-pathologizing understanding of the relationships between ontological structures of care, self-care, caregiving and health.

Western societies in general, and medicine in particular, have come so far in constructing the possibility of *objectifying or deworlding* our perspectival stance that we ignore our necessary situatedness and pass over our social, sentient, emplaced bodies (our common humanity) to address the de-worlded physiological body (Casey, 1997). Aristotle was the first philosopher to warn about the forgetfulness of what all human beings hold in common: (1) We are embodied and thus vulnerable, requiring the care of others for survival; (2) we are historical and finite; (3) being historical we are thrown and constituted by local, particular human worlds that are subject to opening and closing down; (4) we hold in common, as human beings, the difference and otherness created by our situatedness in particular worlds (Levinas, 1989). In addition to being rational animals in the Western tradition, we are also dependent rational animals (MacIntyre, 1999; Benner and Wrubel, 1989; Dreyfus, 1991).

In this time of pluralism, we are prone to mis-read "sameness" into the word "common." However, holding some things in common does not necessarily create sameness. Holding some things in common makes it possible to encounter the other, who may be both different and similar in some respects. For example, even though my local world and human experiences may be distinct from others, it is still possible to meet another with *some* understanding of his plight based upon the following aspects of our common humanity:

A. *Situatedness*

Though the details and content of one's situatedness may differ, the human experience of being situated is held in common. The resistance and distance of difference make understanding possible through dialogue and imagination based upon being embodied and dwelling in particular places and circumstances (Casey, 1997). Human or natural catastrophes reveal the commonalities we hold in being situated in local worlds. For example, after the Loma Prieta Earthquake in San Francisco, many people experienced a common situatedness in terms of fear of aftershocks, altered transportation, and very different situatedness regarding specific losses. But San Francisco Bay Area residents held in common the shared events and history of the earthquake.

B. *Thrownness*

We are always experiencing the future in some relationship to our present and past. At birth we are thrown into particular child rearing practices and lifeworlds that shape the kinds of selves we can become. This is part of our facticity or givenness. We can gain new possibilities by entering, and learning to dwell in, new lifeworlds (e.g. cross-cultural exchanges, historical changes, changes in our natural habitats). However, we are thrown into new lifeworlds from the lifeworlds of our past.

C. *Finitude*

An essential fact of my existence is that I expect to die. In common with all other human beings, I hold biological death as a future event. But biological death is not the only finitude that I experience. Throughout my existence, I confront the passing (or closing down) of previous embodied and social capacities, lifeworld concerns and possibilities. As the mother of a particular infant, my concerns and particular caring practices in relation to caregiving must change as the infant grows to child and adulthood. I hold in common with all other human beings my biological finitude and the finitude of social sustenance.

D. Embodiment

Social, sentient, skilled embodiment constitutes commonalities, as well as differences, with other human beings. We are always perceiving and experiencing from the perspective of being embodied. As Samuel Todes (2001) points out, our embodied perceptions and experiential trajectories ground our understanding. Moving from one situation to another is an embodied move that leaves bodily traces. Our embodied capacities give us the sense of "I can" (or "I cannot") do something or be someone. The world I inhabit is experienced differently according to whether I am prone or walking, standing or running. Perception is an embodied capacity that changes with experience over time. Learning to see distinctions is a skill governed by lifeworld concerns and physical environment. For example, the field of threats, harm and possibilities that I experience daily is expressed in my skillful habitual body. As a San Francisco Bay Area resident, I dread traffic, have a healthy respect and intermittent fear of earthquakes, and count on mostly pleasant weather. I can extend the sense (and range) of my embodiment through the use of a ski pole, a cane, a prosthesis, or an instrument. I can predict what levels of exertion will make me short of breath or fatigued. I know firsthand the vulnerabilities and possibilities of embodiment, as in the experience or fear of pain. My access to the world is embodied and emotionally imbued. This is illustrated in Jean Johnson's (1973) research. Johnson demonstrated that rehearsal of actual perceptual sensations – such as the noise, heat, and vibration of cast cutting – is more effective preparation, in terms of diminishing anxiety during a procedure of removing a cast, than are verbal descriptions alone.

Moral emotions, such as shame, change my countenance and posture. The experiences of fatigue, weakness, clumsiness, or grace are known as others and I embody them. Disagreement and difference are only possible if there is some embodied common ground – at least some commensurability. Experiential learning leaves its traces in the skillful body. For example, I know what it is like to experience pain in different measures, since I am embodied. I know the pain of childbirth. But I can only imagine the pain of a severe burn or a kidney stone. Furthermore, my memory of a past pain is dim compared to someone else's experience in the moment. Nevertheless I can extrapolate and imagine and, therefore, empathize with another's pain. I may even experience compathy, a bodily response that mirrors or responds physically to another's pain (Morse and Mitcham, 1997). In different qualities and measures, I may know what it is like to be mis-recognized and marginalized, though I must not imagine that I can experientially know another's particular human experience of exclusion and racism in his or her particular local worlds. This is what I understand Levinas (1989; 1998) to mean when he states that we meet the other as *other* (distinct in their particular lived experience and world) but not *wholly other*, that is excluded from the possibility of meeting, learning from, and even understanding aspects of the other's existence.

Understanding our common humanity requires that we acknowledge our dependence on both ontological and ontic structures of care (or concern) in local worlds. Our heroic myths in Western societies allow us to imagine what it would be like to have a radically independent, disembodied freedom. We seek freedom from local specific worlds and workplaces through the disembodied engagement made possible through the Internet, telephones and written communication. However spanning spatial embodied distances through whatever means still depends on practical knowledge of local embodied worlds, our sense of place (Casey, 1997). We live as human beings, dependent upon fellow human beings and the social structures of recognition, dialogue and care. As members and participants we are constituted by our concerns and relationships – our lifeworlds – for better and worse. We are never truly independent and, only in our worst mental illnesses, are we radically separate and completely autonomous. The phenomenon of care captures many aspects of human engagement in a lifeworld. Care or caring refers to all the ways a person may be engaged in a lifeworld. It is a phenomenological paradox that caring both constitutes, and is constituted by, a person's lifeworld. What one does not care about, or care for, cannot easily come into perceptual awareness or attentiveness. Thus, caring shapes a world, and allows other beings to be noticed (or to disclose themselves) in particular ways.

IV. HOW ONTOLOGICAL CARE AND CAREGIVING PRACTICES DISCLOSE THE POSSIBILITIES OF FEELING AND ACTING AUTONOMOUSLY

We sometimes imagine that caregiving is menu-driven by the care needs of the one cared for, yet ontological structures of care and ontic caregiving practices actually set up the conditions of what it is possible to ask for, and receive, from those doing caregiving work. We develop cultural practices and scripts about what it is reasonable to ask in the way of caregiving, and have socially generated notions about the attendant burdens of our dependencies. Even though the human face of the other calls me to care (Levinas, 1998), I imagine what I may or may not expect from others as a result of my social class, as well as normative structures about dependencies and responsibilities. Therefore, with insight, those with disabilities correctly point out the slippery slope of "the right to die" becoming the obligation to die, once caregiving burdens grow too heavy. The one who witnesses illness does not have the same experience or imagination as the one who is ill. It is easy for those looking on to imagine that a life with obvious dependencies and disabilities is not worth living, while the one suffering may find great pleasure in the situated possibilities available to him or her.

In the U.S. commodified healthcare environment, healthcare and caregiving practices contain many "entitlement discourses" about who is entitled to care and how much. Thus, a person diagnosed with a catastrophic or chronic illness may confront a set of unwarranted prejudices that assign ultimate responsibility to the afflicted individual. Moral dimensions of shame and responsibility are built into our ways of understanding the mind as the locus of control and moral agency, and the emotions and body subject to executive control by the mind (Benner and Wrubel, 1989; Benner, Janson, Ferketich and Becker, 1994). The following interview excerpt illustrates the burdens of self-blame or recrimination that many cancer patients face:

> I've always taken good care of myself and my body, especially health wise. You know, I've been a vegetarian for over 10 years, but a careful one. And I've been very health conscious – no drugs, no cigarette smoke, no coffee, caffeine, soda, anything....Cancer is something I'm just starting to identify with because that's a difficult word, because I guess it's associated with like toxins, and I just never did any of those carcinogenic things.... I'm kind of angry at the stuff that I read where people say that it was your fault that you got it. You know, I wonder, and I've thought about it, but you know could it be that because I was under stress or unhappy or depressed or had a difficult year, and it could be, but I would rather think that it is not. Because I look around at my friends, and they're all having struggles and they're all having concerns about their careers, and they're under stress from school, and I just don't think that it was my fault. So that's how I'm choosing to look at it right now. (Benner, 1991, p.75)

This person expresses the pain caused by assuming moral responsibility and blame for areas beyond her control. She resists the blame even as she struggles with it. Such moral attribution is an alienated view that suggests that the person possesses and controls feelings as if they were raw material or resources to be managed (Benner and Wrubel, 1989; Benner, 2000). This young woman experientially learns that she has to find a way back to experiencing her feelings as connected to her life, rather than causal agents of disease that she has to suppress:

> I started thinking that because I was depressed that I *really* got caught up in being depressed...but now I'm starting to pull back a little bit and realize that I don't have to find out all the answers, that everybody else also thinks about those things. Just because I'm sick, doesn't

mean I'm the only one that gets depressed or the only one that thinks about those things. That took the focus off of me, that I can be like that sometimes and not be like that, just as other people are in their wellness (Benner, 1991, p.76).

Denying or managing her feelings, in ways that generate secondary emotions of responsibility and blame for the feelings, creates more estrangement and alienation from the ordinary flow of emotional life. In a similar vein Arthur Frank (1995) contrasts "chaos narratives" and quest narratives. Though the culture provides ample storylines about heroic or transcendent quests as a result of illness or disability, such storylines are sustainable only intermittently in even the most heroic lives. Quest storylines spawn sensational newspaper coverage of a heroic overcoming of disability that few people – well or ill – can achieve or sustain. The distance between the living, and telling, of transcendent moments can be great. And, as exemplified above, denying the experience of sadness and chaos created by illness creates alienation and distress.

Caregivers may seek to assign moral responsibility for the disease to the other as a way of defending against the threat of the illness themselves (Benner and Wrubel, 1989; Benner, 1991). The young woman quoted above had to defend herself from her mother's untested assumptions about her responsibility for illness:

My mother feels that it is somehow my fault, so that's kind of strange because it is very important for me not to feel that way. She doesn't feel that I am to blame necessarily, but she is of a different belief than I am right now. And it is hard because I did kind of come from that belief, that somehow, if you got sick, you have to take some sort of responsibility for that...[One of the difficult things this week] has been when I was arguing with my mother about the two different ways that we saw why I am sick. (Benner, 1991, p.76)

Locating responsibility within the person can give an illusion of control, or an illusion that the world is "just and completely interpretable" (Rubin and Peplau, 1975). Assigning personal responsibility to the ill person can create a false sense of immunity to the one making such an assignment. But it unjustly creates feelings of alienation and guilt on the part of the one who is ill. "Stress management" strategies require too much when they require persons to step outside their own history, when the task at hand is rather to come to terms with one's lived history in the process of forging of new possibilities. Creating a healing environment and decreasing negative feelings of alienation and despair are worthy ends in themselves and, coincidentally, more attainable when they are not taken up instrumentally to effect a cure.

V. CREATING A DIALOGUE BETWEEN CARTESIAN MEDICINE, LIFEWORLDS AND CARE

In its Cartesian tradition medicine gains its power by passing over human lifeworlds and care. In practicing scientific medicine, the doctor looks for objective signs, using subjective reports of the patient's symptoms and lived experience only as a means to point to the objectified, mechanical body (Foucault, 1973). In many instances patients look to physicians for this objectified stance. Rather than seeking causes in their lifeworlds, people often prefer that physicians find "external" causal agents for disease – such as viruses or bacteria or mechanical/structural problems that can be fixed with minimal effort or personal responsibility. Yet the phenomena of ontological and ontic care provide an ever-present shadow world for medicine. Medicine effects its cures only if it is supported by reasonably well-functioning lifeworlds with caring practices that sustain the person in the daily activities of living, nourishment, and health promotion. When persons fall ill, their lifeworlds sustain them through care, supporting their dependencies. The familiar lifeworld also entreats them to recover and take up their familiar concerns again. Sickness, pain, and disability infringe on a person's self and lifeworld, diminishing (if not collapsing) horizons of possibility. If the "medical" gaze is projected onto lifeworld concerns and human identity, then caring practices are disrupted and medical discourse becomes totalizing, effectively shutting out the healing and human possibilities inherent in particular lifeworlds. Medicine works not by creating health, but by intervening in disturbances in health and functioning. Once the disturbance is cleared (or identified), the doctor must assist in freeing up the patient to become re-integrated into their lifeworld at whatever level possible (Gadamer, 1993).

"Activities of daily living" is a code phrase often used in the helping professions to point to all that keeps the embodied person emplaced in a world. When measured, these activities are usually reduced to functional abilities to get dressed, bathe and feed oneself. Even these minimal abilities depend on a particular lifeworld and a skilled habitual body. Sacks (1985) points out that when psychological traits and cognitive abilities are evaluated in standardized, decontextualized, assessment approaches, the examiner is unaware of the performance capacities persons exhibit when skillfully comporting themselves in their familiar worlds, solicited by concerns, opportunities and threats.

From a strictly Cartesian understanding of symptoms, the mind receives and interprets impressions and sensations from the body. Consequently physicians may view symptoms as subjective interpretations of the body's real disease. The mind is considered a less reliable reporter of symptoms than those that can be documented objectively with medical instrumentation and measurement. From a perspective of care, this way of interpreting symptoms is problematic. The patient's credibility as a witness is called into question by the Cartesian quest to determine whether the symptoms "really" reside in the body and not in the mind. From a care perspective any symptom must be heard, and attended to, in its own right and not just as evidence for an accurate diagnosis. The patient's narrative understanding of the illness can be solicited, in order better to understand the patient's lived experience of the illness and his or her concern related to treatment and daily functioning. For example, Nurse Linda Sawyer encountered a young woman whose symptoms of brittle diabetes were considered to be thoroughly explained by the pathophysiology of diabetes. This young woman had extensive skin breakdown in her feet due to pressure when walking. Thus, she was confined to a wheelchair. She experienced episodes of high and low insulin almost daily causing her to be seen frequently in the Emergency Department. All of this was *explainable* in terms of her extremely brittle diabetes – the steep accelerations and declines of her blood sugar curve. She was mentally handicapped and, thus, people had all but stopped working with her on learning to manage her diabetes. One might say that she had been "explained" out of existence. What was necessary was that she be understood back into existence in her lifeworld and situated possibilities. Linda Sawyer found that, together, the patient and her mother were capable of closer monitoring of blood sugars. The patient's pressure sores were due to poorly fitting shoes. This problem was corrected by having the orthopedic department take impressions of her feet for special shoes, and by retraining her to walk with canes to alleviate some of the pressure. Painstakingly, many aspects of her daily care were monitored and slowly became understandable, and therefore modifiable. Her Emergency Department use was dramatically reduced. Her level of diabetic control increased her ability to resume her social activities. Since symptoms are integral to the person's lived experience of the illness, they are also integrally related to the meaning the illness has for the person. Listening with care to the lifeworld implications of symptoms is equally as important as using symptoms to detect "true" pathology and diagnosis. Just because a disease is explainable in biomedical terms does not necessarily mean that it is understandable, or livable, from within the person's lifeworld and lived body.

Both explanation and understanding are required for adequate treatment of the embodied person dwelling in a particular lifeworld. The social, sentient body that dwells in familiar worlds and relationships – and seeks out growth, novelty, intimacy, distance or enclosure – is not adequately accounted for by descriptions of the body as a collection of physiological systems, cells and genes that can be cured or altered. Merleau-Ponty (1962; Benner and Wrubel, 1989) pointed to the habitual, skilled body that dwells in lifeworlds, common meanings, practices and language as a collection of *middle terms* (in between the physiological body and the abstract mind). These middle terms get ignored when we think of the mind as pure intentional will, knowledge, and beliefs, and the body as a collection of biochemical and mechanical physiological systems. Dividing mind and body up in disparate, oppositional ways ignores the manner in which our sentient bodies exist in meanings, skills, relationships, placing us in pre-formed worlds, and *constituting* newly forming worlds as we move out into new experiences. For example, the person who has lost a leg must learn to get around in the world with various supports and functions that accommodate the missing limb. He or she must dwell in a world that is no longer available in the same way, while learning to dwell in a world with different affordances, possibilities and constraints. Negative phantom limb pain can be debilitating and baffling when imagined, painful, gangrenous toes still recall painful, embodied sensations that demand relief. However, the helpful phantom limb allows the person to recall the feeling of lower leg and foot and enables the person existentially to fill out, and inhabit, a prosthetic limb as a part of the inhabited body. The Cartesian paradox is somehow to sever the errant nerves and memory of the painful toes while not inhibiting the person's ability to remember how it was to get about with two limbs. Quieting the existential memories of the habitual body that can fill out the prosthetic limb would lead to further disability. Here the shadow world of the habitual, skillful and sentient embodied person reasserts itself, and the physician must find a way to bridge the Cartesian gap, while not losing the Cartesian gains that created the possibility of surgically removing the life threatening painful limb, or severing the remaining inflamed or hyper-reactive nerves.

Allopathic medicine gains its power by applying the objective sciences of patho-physiology, biochemistry, genomic science, and so forth, to the objectified physiological body. Each of these scientific disciplines pass over the middle terms of the person's lifeworld and the social, sentient embodied existence, in order to treat the physical-bio-chemical aspects of diseases and injuries. The powerful results achieved by placing caring practices and healing arts in the background can appear to be the condition of possibility for curative sciences. However, the efficacy of these sciences in effecting cures breaks down, when taken-for-granted caring practices (and healing arts) grounded in particular lifeworlds, break down. Perhaps this has never been more obvious than when hospital emergency rooms attend only to physical ills and ignore the gaping social wounds of homelessness, in heeding their economic mandate to "treat them and street them" (Malone, 1998).

Cure of a disease or medically controlling a chronic illness are not the same as recovery from an illness and regaining a livable embodied relationship to world and others. The objectifying gaze of biomedicine discloses disease and cures in ways that allow for cures and surgery that the patient may not be able effect or imagine. Partialling out the mechanical body and effecting cures on that body can alleviate much suffering. But in addition to curing, the physician also forms a relationship with the patient that may in itself be healing. The art of doctoring requires relating to the patient as a person, and not just a disease. Paying lip service to this maxim is not enough. Physicians and nurses need to be taught how to form respectful dialogues and alliances with the person's concerns, skills and lifeworld. We begin by making the patient's care structures visible through dialogue and listening to the patient's lived experience with the illness.

The goals of a phenomenology of illness, care, and recovery, are not to achieve a single paradigm medicine, so that increasingly medicine incorporates more of the patient's lifeworld in understanding pathologies and cures. Rather the goal is to retain the scientific advances of modern medicine, while enlarging the dialogue and formally acknowledging, in policy and practice, the primacy of care in effecting cures. Such a dialogue facilitates the patient/family's ability to draw on their particular lifeworlds of care and recovery for health promotion, illness management, rehabilitation and recovery. New healthcare policies that support, or provide, caregiving structures are needed – rather than policies that continue inappropriate expenditures focused on highly technical interventions in the absence of essential structures of care in the lifeworld.

Currently healthcare policies do not adequately attend to informal caregiving structures to sustain health. Much of the downsizing in acute care settings has shifted the burden of caregiving of even critically ill patients to families (Benner, Hooper-Kyriakidis, and Stannard, 1999). With an aging population, and increasing levels of chronic illnesses, everyday management of care and illness symptoms determines much of medical utilization.

Acknowledging at the public societal levels, both our indebtedness to others and our inescapable dependencies and interdependencies, would move us to improve public caregiving to address our common humanity (MacIntyre, 1999). Each stage of human existence requires rich networks of care from birthing to dying. Acknowledging necessary inter-dependence and dependencies could change public understandings and public policies about what it is to be a person, and what "independence" can mean within the limits of one's particular embodiment and lifeworld.

University of California
San Francisco, California
U.S.A.

BIBLIOGRAPHY

Agich, G.J.: 1995, 'Chronic illness and freedom', in S.K. Toombs, D. Barnard, and R.A. Carson (eds.), *Chronic Illness: From Experience to Policy*, Indiana University Press, Bloomington, Indiana.

Bauman, Z.: 1993, *Postmodern Ethics*, Blackwell, Oxford.

Benner, P.: 2001, 'The roles of embodiment, emotion and lifeworld for rationality and agency in nursing practice', *Nursing Philosophy* 1, 5-19.

Benner, P.: 2000, 'The quest for control and the possibilities of care', in *Heidegger, Coping and Cognitive Science: Essays in Honor of Hubert L. Dreyfus*, Vol. 2, MIT Press, Cambridge, Massachusetts.

Benner, P., Hooper-Kyriakidis, P., and D. Stannard: 1999, *Clinical Wisdom and Interventions in Critical Care, A Thinking-in-action Approach*, Saunders, Philadelphia.

Benner, P., Tanner, C.A. and C.A. Chesla: 1996, *Expertise in Nursing Practice, Caring, Clinical Judgment and Ethics*, Springer, New York.

Benner, P., Janson-Bjerklie, S., Ferketich, S., et al.: 1994, 'Moral dimensions of living with a chronic illness, autonomy, responsibility and the limits of control', in P. Benner (ed.), *Interpretive Phenomenology: Embodiment, Caring and Ethics*, Sage, Thousand Oaks, California, pp. 225-54.

Benner, P.: 1994b, 'The role of articulation in understanding practice and experience as sources of knowledge in clinical nursing', in J. Tully and D.M. Weinstock (eds.), *Philosophy in a Time of Pluralism: Perspectives on the Philosophy of Charles Taylor*, Cambridge University Press, Cambridge.

Benner, P.: 1991,'Stress and coping with cancer', in *Cancer Nursing: A Comprehensive Textbook*, W.B. Saunders Company, Philadelphia, pp. 74-81.

Benner, P. and. Wrubel, J.: 1989, *The Primacy of Caring: Stress and Coping in Health and Illness*, Addison-Wesley, Menlo Park, California.

Berkman, L. and Syme, S.L.: 1979, 'Social networks, host resistance, and mortality: A nine-year follow-up study of Alameda County residents', *American Journal of Epidemiology* 94,105.

Casey, E.S.: 1997, *The Fate of Place: A Philosophical History*, University of California Press, Berkeley.

Cassell, E.J.: 1991, *The Nature of Suffering and the Goals of Medicine*, Oxford University Press, Oxford.

Damasio, A.: 1999, *The Feeling of What Happens, Body and Emotion in the Making of Consciousness*, Harcourt Brace Company, New York.

Dunne, J.: *Back to the Rough Ground: Practical Judgment and the Lure of Technique*, University of Notre Dame Press, Notre Dame, Indiana.

Dreyfus, H.L.: 1991, *Being-in-the-world, A Commentary on Heidegger's Being and Time, Division 1*, MIT Press, Cambridge, Mass.

Foucault, M.: 1963/1973, *The Birth of the Clinic: An Archeology of Medical Perception*, A.M. Sheridan Smith (trans.), Vintage, New York.

Frank, A.: 2001, 'Can we research suffering?' *Qualitative Health Research* 11, 353-62.

Frank, A.W.: 1995, *The Wounded Storyteller: Body, Illness, and Ethics*, University of Chicago Press, Chicago.

Gadamer, H.J.: 1993, *The Enigma of Health*, Stanford University Press, Stanford, California.

Halpern, J.: 2001, *From Detached Concern to Emotional Reasoning*, Oxford University Press, Oxford.
Heidegger, M.: 1926/1962, *Being and Time*, J. Macquarrie and E. Robinson (trans.), Harper and Row, New York.
Johnson, J.: 1973, 'Effects of accurate expectations about sensations on the sensory and distress components of pain', *Journal of Personal and Social Psychology* **27**, 261-75
Kesselring, A.: 1990, 'The experienced body, when taken-for-grantedness falters: A phenomenological study of living with breast cancer', doctoral dissertation, University of California, San Francisco.
Kleinman, A.: 1988. *The Illness Narratives: Suffering, Healing and the Human Condition*, Basic Books, New York.
Levin, D.M.: 1999, *The Philosopher's Gaze: Modernity in the Shadows of Enlightenment*, University of California Press, Berkeley, California.
Levinas, E.: 1989/1991, 'Time and the Other', in Sean Hand (ed.), *The Levinas Reader*, Basil Blackwell, Oxford, pp. 37-58.
Levinas, E.: 1998, *Thinking-of-the-other Entre Nous*, M.B. Smith and B. Harshav (trans.), Columbia University Press, New York.
MacIntyre, A.: 1999, *Dependent Rational Animals: Why Human Beings Need the Virtues*, Open Court, Chicago.
Malone, R.: 1998, 'Whither the almshouse? Overutilization and the role of the emergency department', *Journal of Health Politics, Policy and Law* **2**, 795-832.
Merleau-Ponty, M.: 1962, *Phenomenology of Perception*, C. Smith (Trans.), Humanities Press, New York.
Morse, J.M. and Mitcham, C.: 1997, 'Compathy: The contagion of physical distress', *Journal of Advanced Nursing* **26**, 649-57.
Sacks, O.: 1985, *The Man Who Mistook His Wife For a Hat and Other Clinical Tales*, Simon Schuster, New York.
Rubin, Z., and Peplau, L.A.: 1975, 'Who believes in a just world?' Journal of Social Issues **31**, 65-88.
Taylor, C.: 1989, *Sources of the Self: The Making of Modern Identity*, Harvard University Press, Cambridge, Mass.
Taylor, C.: 1985, 'Theories of meaning', in *Human Agency and Language, Philosophical Papers*, Vol. 1. Cambridge University Press, Cambridge, pp. 248-92.
Todes, S.: 2001, *Body and World*, M.I.T. Press, Cambridge, Massachusetts.
Van Manen, M.: 1999, 'The pathic nature of inquiry and nursing', in I. Madjar and J.A. Walton (eds), *Nursing and the Experience of Illness, Phenomenology in Practice*, Routledge, New York.
Wrathal, M.: 2000, 'Background practices, capacities, and Heideggerian disclosure', in *Heidegger, Coping and Cognitive Sciences Essays in Honor of Hubert L. Dreyfus*, MIT Press, Cambridge, Massachusetts.

ERIC J. CASSELL

THE PHENOMENON OF SUFFERING AND ITS RELATIONSHIP TO PAIN

I. INTRODUCTION

Pain and other symptoms such as dyspnea (shortness of breath) have a physical basis, but suffering is personal. The path, stretching from the physiological response evoked by a physical stimulus to the disintegration of the person experienced as suffering, is important to understand not only for itself but also as an exemplar of mind-body interactions. As is well known, the domains of body and mind have been kept separate in the intellectual life of Western culture from its earliest history to the present. This is despite the fact that, in stage after stage in the development of Western medicine from the Hippocratic tradition to the present, there has always been an understanding that the mental life – called the moral life at some periods – has an impact on the body and its afflictions (Entralgo, 1956). It is difficult to make a statement about the relationship of these two realms that does not sound dualistic. This essay, to the contrary, is based on the belief that we are of a piece – anything that happens to one part affects the whole, what affects the whole affects every part. All the parts are interdependent and not one functions completely separate from the rest. For example, although the functions of the lung are unique to that organ, their results influence the whole organism: respiration, in all its complexity, flows through the whole. Similarly, reason and the passions (classically considered functions of the mind) both require and act through the body, they flow through the whole. If you think this a strange idea, ask yourself how there could be complete separation of any part from the whole, so that what happened in the part had no effect on the whole?

A second categorical statement is also important: the body participates in everything. And, so does the mind and virtually everything else that is part of being human – everything emotional (feelings), social, purposeful, reasoning. All these things affect the body. What do you do that is social that doesn't involve a physical element? Aren't feelings felt physically? There are no emotions that just exist in your head. Purposeful behavior requires the body to participate and so does reasoning – the brain is part of the body. The reverse of these statements is also true. Everything physical is also social, emotional, involves thought, and is purposeful. Categories like physical, social, and psychological have many uses, but such divisions get in the way of understanding pain and its progression to suffering. The purpose of these statements is not to enter the long controversy about what the mind is, or even what the living body is, but to deny that there exists a dualism in the conventional senses of that word and to provide a basis for the place of meaning in the production of pain, suffering, and other distress.

One of the difficulties in understanding is the persistence of the idea that a symptom as experienced – pain or dyspnea – is the unmediated result of the pathophysiology (the physiological chain of events leading to the abnormal state). In the case of pain this means the persistence of the model of acute pain, and the idea that pain is secondary to tissue damage – exemplified by what everyone experiences as a result of a splinter, a burn, or twisted ankle. No doubt there is usually a direct relationship to tissue damage in acute pain and that, where there is tissue damage, pain often follows. But fractures, acute localized inflammation, and rapidly expanding lesions (such as abscesses or tumors) are present in only a fraction of patients who experience pain or suffering.

Another part of the problem of understanding is that we all (physicians and non-physicians) love mechanism and it pulls us along – as though we really knew more than we actually do about patients with pain in chronic or terminal illness from what has been published about acute pain physiology.

Another thing that impedes understanding is the ease with which it is possible to control much pain (particularly in a hospital or hospice) and the efficacy and safety – at least in the short run – of many of the pain relieving treatments. While pain is often relatively easy to control (at least in good part) in most patients in pain, physicians tend to move pain that is not so easily relieved into other categories by emphasizing, for example, the psychological aspects of the patient's problem. This is, in fact, the final obstacle to understanding – seeing things as either physical or psychological. This seems particularly troublesome when people act as though there are only two possible kinds of mechanisms – physiological or emotional – involved in pain or other symptoms. If you cannot explain some feature of pain (or a patient in pain) physiologically, then it must be psychological.

II. PHYSIOLOGY

There is no question that there has been a recent explosion in knowledge about the physiology of pain. For example, a recent National Library of

Medicine Medline search on the keywords "physiology of pain" brought up more than 4000 references from 1996 to 2000. Searching from 1966 to 1970 retrieved 136 references. The mechanisms revealed are so interesting that it is difficult to avoid returning again and again to this research to explain this or that manifestation in an individual patient, or as a basis for the choice of treatment for a patient's pain. On the other hand, as the complexity of the phenomenon called pain appears to double, and then redouble, as a result of new discoveries about it's physiology, a relatively simple conclusion becomes possible. What has been clearly revealed by newer research (something already apparent in a simpler fashion years ago) is that there is a not a one-to-one correspondence between the stimulus that elicits pain and the pain itself. In fact, it is evident that there exist many pathways by which the person's experience of pain can be enhanced, diminished, and changed in character or distribution.

If there are so many different physiological ways to alter the nature of the pain as felt, what determines the person's actual experience? It cannot be the pain stimulus itself – whether there is evident tissue damage or not – because that does not match the complexity of pain physiology. It cannot be person's body alone – although physical phenomena such as fever or other discomforts may change the person's experience of the original pain – it must be something about the nature of the person in pain. I will return to this point shortly.

III. EMOTIONAL

It is commonly believed that when a physical basis for pain cannot be demonstrated – no lesion sufficient to explain the pain can be found, or the pain does not conform to expectations – then the pain is emotional (or psychogenic) in origin. Systematic understandings of psychological contributions to illness grew from the beginning of the twentieth century when Freud and others demonstrated the domain of the unconscious in the mental life – in this view, primarily a place of libidinous or other primitive forces, repressed conflicts, needs, wishes, and desires. This material was convincingly shown to find expression in thoughts, dreams, language, and behaviors, even patterns of life whose sources in the unconscious were unrecognized by the person. It was demonstrated that pain or other symptoms – sometimes whole illnesses – could be the expression of repressed unconscious ideation. The idea of psychosomatic illness spread, particularly in the United States, so that by the 1950s it was a common belief. George Groddeck, a practicing clinician, gave many graphic descriptions of such cases, as did the psychoanalysts themselves (Groddeck, 1977). The illness or symptoms were considered to represent a symbolic expression of the otherwise unconscious conflicts – ideas that were too painful to be expressed in words found expression in the body. Put another way, rather than the unconscious expressing the meaning of the repressed ideation in words, the symptoms or illness became the expression of that meaning. Unfortunately, as time has passed, illness caused solely by psychological difficulties has come wrongly to be seen too often as not real illness; real illness (from this perspective) is something that can be found in the body associated with tissue pathology. In this view psychogenic pain is imaginary pain; the patient feels pain that does not exist. It is true that there are rare patients who report pain, but nothing else in their behavior or activities supports their claim. Their activities are unchanged, their use of pain relieving medications is not increased, they do not utilize medical services, and their interactions with family and others are unchanged. In other words, all the behavioral changes that usually occur when people are in pain are absent. These are not patients in palliative care programs, or patients with serious disease (whether or not lesions to cause the pain they report can be demonstrated). Similarly, problems of malingering or using pain as part of drug-seeking behavior are uncommon and should not distract from the central problems that get in the way of understanding pain and suffering.

Using hypnosis, it is not difficult to have patients re-experience pain that had occurred in the past, e.g. pain at the site of an old fracture, the chest pain of a previous heart attack, and so forth. Clearly, this is not pain associated with current tissue damage nor does it have its origin in current emotional states. It appears to be a recall of bodily sensation and suggests that such sensation is independently stored in memory apart from emotion. Such bodily sensations disappear when the person emerges from the trance even when sensations were reported as severe while in the trance state. The same phenomenon can be shown with other symptoms such as nausea or dyspnea. Here, also, the symptoms experienced in the trance are not a result of a primarily noxious or disease-related stimulus. In the trance, the recalled pain or other symptom may, however, be accompanied by emotions similar to those experienced at the time of the original insult.

There is no question, on the other hand, that the emotions of fear, panic, or even anger may aggravate pain – as may suffering – making it difficult to control. Increasing opioid or analgesic dose for pain that is unrelieved because of concurrent emotional distress, anguish, or suffering leads to persistence of the pain or opioid toxicity, escalating drug costs, diminished patient quality of life, increased family distress, and greater demands on caregivers. This situation is not rare. Indeed, it is a common problem in patients such as those on palliative care services, often accounting for doses of opioid analgesics that are sometimes astonishingly high but with inadequate pain relief.

IV. MEANING

I believe it is the phenomenon of meaning that brings together what is known about pain – the physiological mechanisms and the role of emotion – with the experience of the person in pain. It is difficult to discuss the subject of meaning because there have been so many disputes over the ages about the meaning of meaning. On the other hand, there can be no understanding of pain and suffering without considering the place of meaning in these feelings. Indeed, despite the fact that non-psychiatric symptoms are body sensations, it is difficult to separate a body sensation from the meaning that has been assigned to it. Meanings, as I am using the term, are the ideas or notions to which a thing refers. They are the concepts a person has of words, objects, events, or relationships that both explain them in terms of the person's other ideas, beliefs, or concepts, and relate them to the concepts of the communities of which the person is a part. Meaning is not simply an aspect of cognition – the logical processes of thought – and not solely represented by words or language. Because of the disputes on the topic and the fact that (at least in academic philosophy) the subject of meaning has been largely appropriated by analytic and ordinary language philosophers, it is not widely understood or accepted that meanings are not only mentalistic things. Meanings do include dictionary type, denotative, designative, rational information – what words refer to or what things signify; for example, "chest pain, hemoptysis, and dyspnea signify pulmonary embolus," or, "the frown on the doctor's face signifies bad news," and also, "chest pain signifies coronary heart disease." Meanings also include the connotative element of importance, which carries the value elements of meaning. For example, the attitude the person has toward the thing. Importance is always personal, although many persons may share the same values related to meaning. This aspect of meaning is exemplified in statements like; "the clot in my lungs could have killed me," or "cancer means pain and suffering," and also "If my test is negative I'm going to celebrate." In addition to the primarily cognitive element of significance and the value dimension of importance present in all meaning, concepts also contain a content of emotion – happiness, sadness, fear, anger, and all the other variants of emotion.

Common experience may tell us that virtually everything we confront from the seeming injustice of a child's pain, to chewing gum, to a copy of Hegel's *Phenomenology of Spirit,* to our own unfairly inflicted wound, evokes emotion of varying degree and type. We all know this, but tend to forget that these emotions and even emotional states must then be part of the meaning of the thing or event. To say that the specific concept that is a particular meaning for each of us contains emotions merely rephrases the former sentences. It is difficult to conceive that *all* meanings include emotion until it is pointed out that emotion includes three distinct phases. The feeling of emotion that is the immediate flash of feeling – i.e. the almost instantaneous, and as quickly subsiding, sensation of joy, pleasure, humor, sadness, irritation, anger, and so forth – is an aspect of emotion. So also is the next step, the emotional state, in which the person is occupied by the state of feeling – he or she is angry, happy, sad, filled with uncertainty, joyous, feeling neutral, apathetic, and so on. (Of less immediate concern to meaning is the final phase of emotion which occurs when the person *acts* angrily, apathetic, sad, happy, or whatever.) Transitory feelings or states of feelings, then, that are associated with things like words, objects, events, or relationships are part of the meaning of those things – part of the personal intension of concepts.

All meanings involve the body and all meanings have a physical dimension. On the one hand this statement must be correct as a result of the fact that all concepts include emotion. Emotions – emotional feelings – are always felt physically, from apathy to joy, and anger to fear. Bodily sensation is an ineradicable aspect of emotion so it must be an ineluctable part of meaning. Think of the knot in the stomach that comes with fear, or the heart that leaps with joy, or swells with pride, or the blush of shame. Using hypnosis it can be demonstrated that physical sensation is a dimension of all meanings (Cassell, 1985). There are patients for whom nausea is part of the meaning of chemotherapy, so much so for a few that they develop nausea (and sometimes even vomit) on entering the chemotherapy suite. In *The Body in the Mind,* Mark Johnson (1989) shows how physical experiences common to us all – for example, balance, containment, locomotion along a path, and compulsion – are basic to human understanding of common words and utterances. Johnson and Lakoff (1999) carry this observation further.

Bodily experience is intrinsic to understanding the meaning of common events. For example, the word chair denotes an object designed to sit on. It has attributes that are part of its meaning – despite the fact that there are so many different kinds of chairs – so that it can be recognized visually. Similarly chairs have attributes in bodily sensation that make it possible to recognize them when sitting down in total darkness. Sit down, feel the seat against buttocks and thighs – sensation immediate. Just as immediate is the assignment of meaning "my desk chair." Lacking the appropriate sensory information that makes up the concept, the identification will be denied. Try it and you will see how complex is even such a simple set of sensations as sitting in a chair. (And how distressing it must be to persons with altered sensory apparatus when they do not feel the expected sensation.) Variations on those sensations would alter the *meaning* of the chair – what kind of a chair it is. Like those fancy descriptions of wine – "rich fruit, lots of character, supple tannins, overtones of chocolate" – with each word assigning a different simile or metaphorical meaning label to part of a complex sensation, and the complex sensation being part of the meaning of the particular wine and wine in general. It becomes clear that separating sensation from meaning is difficult and that meaning *must* play a part in symptoms.

Or put another way, symptoms which are bodily sensations (except psychiatric symptoms) must have meanings associated with them. This is also the case because meaning is attached to all events, and symptoms (even if they come out of the blue,) are events. Because of this, symptoms are not simple brute facts of nature; the pre-existing meanings of the persons in whom they occur actively influence the symptoms. To see this clearly it is necessary to understand that meanings are not fixed or static, they are dynamic – in two senses of the word. First, the content of the conception that makes up meaning includes ideas of where things come from and what they become. To say of a chest pain that it means coronary heart disease is to invoke ideas the person has of where coronary heart disease comes from and what happens (e.g. heart attacks). Thus meanings contain predictions and predictions lead to the foretaste of experience. Think, for example, of how your foot feels when the next step going down is not in its expected place – like walking down a stationary escalator. The failed expectation of sensation produces its own sensation. The second way in which meanings are dynamic is that learning modifies them. Whenever something is experienced it confirms or modifies meanings. Therefore, meanings and their resultant predictions are not only a result of sensation but they modify the ongoing sensation (from moments to months), so that it is reinforced, amplified, other sensations are recruited, or it is diminished.

V. DYSPNEA

I would like to clarify these ideas by starting with the relatively simple symptom dyspnea (shortness of breath) because pain is so complex, as is its relationship to suffering. Dyspnea is the subjective experience of effort with breathing. It is exercise limiting in some persons with cardiac or respiratory disease and it is the manner by which they experience their illness. There are several different mechanisms that may lead to the feeling. With pulmonary edema (wet lungs) or pulmonary embolus (blood clot in the lung) dyspnea is associated with a greater effort to breathe in the presence of decreased physiologic drive to breathe (reduced ventilatory drive). Dyspnea may be present when there is a sensation of choking, actual hindrance to breathing as in obstruction of the airways, or something actually occupying lung tissue (e.g. pneumonia). Decreased neuromuscular power, e.g. muscular dystrophies, ankylosing spondylitis, or weak muscles secondary to illness or aging, may also produce dyspnea. Dyspnea may also occur as part of anxiety – here the person may be unaware of the anxiety but is aware of the dyspnea so that the dyspnea *is* the feeling of anxiety. Exercise is the most common cause of shortness of breath – the sensory aspects of exercise and their determinants (dyspnea) have not been appreciated until recently because of a reluctance to accept subjective data (Jones and Killian, 2000). Using exercise as an example, it is easy to demonstrate that the degree of dyspnea is anticipated in relation to the exercise performed – given the load carried and the degree of exertion, a person expects a certain amount of shortness of breath. Similarly, within a short time, increased respiratory drive, hindrance to breathing, or decreased power to breathe become open to prediction by persons who experience these as part of their disease. As a result the dyspnea – the experience of a necessary increase in respiratory effort resulting from an increased demand for respiration – may begin in anticipation of the increased demand. The dyspnea is the result of a prediction (a meaningful prediction), as a result of experience, and occurring below awareness that dysfunction will occur. Thus, although it is common to speak of dyspnea as though it arises solely from the physiology or pathophysiology of respiration, an element of the person – meanings and experience – may enter it. This warrants the conclusion that, except in the simplest acute situations, meanings, predictions, and interpretations modify the symptom expression of the pathology. Put another way, except in acute settings, such symptoms are, in part, expressions of the person and are not merely depersonalized diseases or the body speaking. Modification of symptoms also occurs at a conscious level. People change their behavior to avoid the symptom, initiate other countervailing activities – medications, visits to physicians, change in occupation, exercise programs – all with varying success which also becomes part of the meanings.

VI. PAIN

For many of us, pain continues to be predominantly represented by the pains of acute injury – needle sticks or broken bones. From childhood we grow up believing that pain is a sign of tissue injury, that hurt always means harm. But the pain that looms large in chronic disease, in serious diseases such as cancer or rheumatoid arthritis, in patients with terminal illness in palliative care programs or hospices, does not fit such a simplified understanding.

Just as with dyspnea, personalization of meaning, interpretation, prediction, avoidance, and countervailing behavior, are also part of pain. The pathophysiology of pain is so complex, however, with mechanisms that enhance, diminish, alter the character, or change the distribution of the pain, that the impact of personalization on pain is much greater than it is on other symptoms. In addition, pain and the modalities to treat it (principally opioids) have many well-established meanings derived from personal experience, family experience, and the society and culture. Therefore, the symptom of pain is virtually always an interpretation of both the source and its meaning. Or to go further, the pain, as experienced, is itself an interpretation. For example, a person experiences chest pain of the kind he previously associated with angina pectoris. On this occasion, the pain does not subside but increases in severity and duration. As the pain continues to escalate the patient becomes frightened that it represents a heart attack. The usual statement might be that the fear aggravated the pain. But this is not the case. The escalating pain and associated fear are *the interpretation* of an impending heart attack (true or not). The verbalization that a heart attack may be occurring is another secondary and separate expression of the same interpretation of the same information.

At first, this thought may be difficult to accept. We are so used to keeping the domains of mind and body apart that we automatically apply such thought to body and person. But persons are more than minds and more than bodies and not just the two joined like horse and cart. It is a consequence of the methods of modern medical science that the body and its mechanisms can be kept firmly apart from any other human aspect. A necessary feature of the statistical methods that permit generalizable knowledge is that the individual variations present in research results disappear from view, leaving the false impression of uniform physiological mechanisms. In thinking of pain (and other symptoms of illness), it is important to see the fabric of the person as seamless with all its threads interacting.

Part of the pain experience arises from the fact that when there is pain in a body part from whatever source that, in itself, sets in motion responses that change the pain experience and the reaction of the patient and others to the pain. For example, fear may lead to actions that aggravate the pain. Or the person may dissociate from the pain and thus effectively hide it. Body parts may change in response to pain – guarding, muscular tightness, postural change are examples – so that the pain is experienced differently. On the other hand, in the musculoskeletal system pain may be a local phenomenon interpreted wrongly as a statement of injury, inducing counterproductive behavior that aggravates the pain. For example, when an ankle is sprained and not quickly mobilized, motion at the ankle will cause pain. The pain is not the direct result of injury, which (in this example) subsided days earlier, but results from muscle tightness at the joint so that stretching the muscle causes pain. Generally the person responds by avoiding motion, which aggravates the problem. "My ankle lied to me," said one patient. The shoulder that is immobilized after injury is another example of pain arising from motion of the shoulder that does not represent injury but rather an attempt to keep the shoulder from moving. The intimation of purpose in this example is intended. Pain may also occur as a part of, or as the sole expression of, an emotional state. For example, generalized pain ("I hurt all over") may be an expression of panic or of anger. The patient usually (perhaps never) recognizes the panic or the anger, so that the pain is the expression of the panic or the anger. Such pain, which is almost always associated with muscular tension or tightness, does not usually improve with analgesics, even when given in large doses. Howard

Spiro (1977) said, "Chronic pain is called 'wordless' by some commentators, but I take it to be rather a cry in a foreign language. Such pain is a sign, an icon opening a file full of sorrow. To treat chronic pain with analgesics is, sometimes, to make mistakes in translation."

An example of a patient whose pain appeared to express fear is a man who had an expanding metastatic lesion of the femur (from cancer of the lung) that was unresponsive to treatment. Increasing doses of opioid analgesics had minimal effect on his severe pain. It was only when he was reassured that his leg was not going to be amputated – a fear that had not been previously expressed – that his pain became controllable. He also had a psychotic belief that there was a rat inside his leg that also had to be addressed for full pain control to be achieved. In another patient the pain seemed to have intentionality as it called increasing attention to its source, subsiding only after the source was treated: A woman had a painful metastatic lesion in the ilium of pelvis. The diagnosis was not made and the pain was not treated. The pain then gradually spread over the hip area and still was not acknowledged. After it spread up into the neck, the pain was investigated and the original hip lesion was discovered and treated. When analgesics were started the pain again became local. The widespread pain seemed to be the call for help.

These examples raise the suspicion of anthropomorphism, as though the body had intensions and could act like a person. Yet, experience in the clinical setting with pain that is not acute suggests the body does act as an expressive aspect of the person – an aspect that is not part of consciousness. In reflecting on these examples, pain would appear to be doubly subjective. Pain is subjective in the sense of experienced by the subject – the person. Pain is also subjective in the sense of expressive of the subject – the person.

VII. PERSONALIZATION OF DISEASE

For years we have used the distinction between illness and the disease, saying that disease is the actual pathology and pathophysiology while the illness is what the patient actually experiences. We knew that the illness was different in different persons because the nature of the patient had an influence on the illness. These differences in the illness occurred, in part, as a result of psychological and emotional reactions. Further, cultural differences have an impact on how a disease is experienced. The illness/disease distinction appears less useful as understanding builds about the personalization of the disease itself. (Its continued utility lies in understanding that illness represents the impact of sickness on aspects of being human – such as the need for control, participation in the social world, and the importance of understanding events – that occur regardless of the nature of the disease.) It now appears that the person also has an impact on the actual disease process. As an example, a patient who develops life-threatening disease will soon (one way or another) know all the things that can happen. "All the things that can happen" may be the medical facts of the case and the various prognoses, or his or her phantasies, or the knowledge of friends, or folk-tales. In contemporary life, the Internet and the Web provide endless (often conflicting) details. The information obtained in this fashion virtually always foretells doom, in one way or another. Although on the surface patients may assure themselves that they will do well, below the surface dread builds due to the fact that, normally, people who know the worst expect the worst to happen. Moreover, often when people expect bad things to happen, they may act in a way that tends to make them happen. This is the so-called self-fulfilling prophecy. In response to all this information and the behaviors it elicits, patients may begin to personalize the disease – not just the illness. By virtue of their behaviors – for example, the doctors they see, medications they take, changes in life pattern from sleep to food – they change the expression of the pathology and the behavior of the disease, as a result of the person they are. In these examples, meanings assigned by the patient to the events (and influenced by all the information) change the pathology. Another source of meaning arises from the (Freudian) unconscious where repressed material may be awakened by events. This then participates in the effects of meaning (but below awareness). As a brief example, a woman who had been abused severely in infancy developed cancer of the breast. Without awareness she dealt with the disease and its treatment as another episode of abuse – and her physicians as abusers.

This Case is Illustrative

A 53-year old, self-employed, divorced, middle class man developed lumbar back pain. After some weeks he went to a doctor and was ultimately found to have metastatic carcinoma of the prostate in the lumbar spine. There was no evidence of other metastatic disease and he was otherwise healthy. Over the ensuing weeks the pain worsened and it became almost impossible for him to walk. The pain did not respond to the usual treatment although there was no evidence of continued extension of disease. He continued working. He was followed by the pain service at a major cancer center, but they were unable to control his pain. He refused radiation therapy because he was "too busy," but he did take his medication. Finally, he was admitted to the hospital. Extensive work-up again revealed no other disease, and no further extension of his bony disease. He was started on radiation therapy but stayed in the hospital because of inability to walk and continued uncontrolled pain. Opioid analgesics were given parenterally in increasing doses but they did not control his pain. Vomiting with the opioids was a problem. Because he said that no one would be able to control his pain unless they controlled his anxiety, a psychiatrist was consulted who added anxiolytics and anti-depressants to his medication without effect. He was then started on Rispirdal (an antipsychotic agent), again without improvement. Two weeks after admission, he was still unable to walk because of pain, but he believed that the radiation therapy had helped him greatly. By then he was almost emaciated, complained of great weakness, and had virtually no quadriceps or hamstring muscle mass. He appeared chronically ill. Why was this man so sick and suffering when his disease appeared to be arrested or, at least, only slowly progressive?

There are two possible views of the case. The usual view is as he is today – on high doses of opioids, sleepy, unable to walk because of pain and weakness, cries easily, is depressed, and is not eating. Arrangements for discharge from the hospital have been delayed because of his degree of sickness. He is considered to be dying. In considering the chain of events, some might talk about how cancer of the prostate is an aggressive disease in young men. Others would blame emotional factors for his state suggesting that he is depressed and "not trying." He might be considered an overly anxious, dependent man who likes his illness. Others might point out that at this time his state results from multiple drug toxicities. Take your choice. There would be the usual complaint about his doctors; for example, oncologists and surgeons don't pay enough attention to emotional factors. Once he has reached this state of sickness, I don't think it matters much what is done – even in terms of the emotional factors. This cascade of troubles is difficult to reverse. It would require intensive nursing and rehabilitative services, psychological support, as well as total enteral or parenteral nutrition, before he again began to be functional. He will not get these services in New York because everybody will assume that he is a dying patient. In a good hospice he will ultimately die a better death, but the basic problem will not be solved. He is suffering and his suffering would have to be directly addressed, which would probably take place in the hospice.

A different view of the case is possible. To focus only on his awful state in the last few weeks with its multiple determinants loses sight of the most important fact. This trajectory was determined from the moment the diagnosis of metastatic cancer was made without any attempt to intervene in the meanings the patient assigned to events. With adequate treatment he should have achieved slowing of the progression of his disease, pain relief, and maintained his ambulatory functional state. The contrary happened. Things occurred the way they did because of the combination of the nature of the disease process and the meaning to the patient of the disease and its treatment. Generally speaking, when patients like this are first seen everybody concentrates on the tumor, but it is difficult to change what is going to happen to the disease (although he probably could have been convinced to have radiation therapy earlier). Only changing what the diagnosis and treatment mean to the patient can change this patient's course. What I want to show is that addressing problems of meaning (and simultaneously hope) can alter the hard medical facts of the case. For example, the patient said that no one could control his pain unless they relieved his anxiety. What did he mean? The palliative care fellow who had been consulted about pain control thought he meant the medical state characterized by nervousness, agitation, and excessive worry, and requested a psychiatric consultation. The psychiatrist agreed and gave the patient anti-anxiety medications without effect. Neither the psychiatrist nor the palliative care fellow asked the patient what he meant by "anxiety." When he was asked (at a later date) it became clear that he meant the usual non-medical definition of anxiety: uneasiness, concern, and troubled mind about some uncertain event. What would be the troubling event for this educated man, with access to all forms of information about his disease (metastatic prostate cancer) in the year 2000? It was his impending death. Not surprisingly, he had taken his continuing pain to be an indicator of the activity of his cancer, and the continuing activity of his disease to be a sign of impending death. Whatever meaning one might assign to impending death is the meaning he assigns to his pain – not surprisingly. The patient got all his information about prostate cancer from the Internet and the Web where there is a plethora of conflicting information, most of it focused on the chances of survival. He discovered that young men have the worst prognosis, that survival is shortest when metastatic disease is present at diagnosis in young men, and that when metastatic disease is present, cure is impossible.

The patient's belief about the meaning of the pain was incorrect. Continued pain does not always correlate with the degree of disease or the rapidity of progression. On the basis of the facts, as given, he was not soon to die. Further, as we have seen, the pain is not a simple direct result of the prostate cancer. The pain, however, integrates the patient's ideas about the pain, which worsen his experience of pain, which reinforces assignment of meaning, which further alters the pain and ultimately makes it intractable to opioid analgesics. In this patient the continued pain reinforces the original incorrect belief about impending death which sets off a chain of events – drug toxicities, malnutrition, bed rest and inactivity, muscle wasting, weakness, depression – which ultimately confirm his belief that he is dying. What will be listed as the cause of death, carcinoma of the prostate or pathology of meaning? It is common to consider anxiety like his as aggravating the situation – the mind influencing the body. That is not sufficient. Anxiety, it is true, may be expressed as worry, fear, or in "nervous" behaviors. Anxiety is also expressed as changes in the experience of pain and other symptoms, as well as in the disease itself. All of these varied expressions are fundamentally the result of the meaning assigned to events.

Summary thus far of the response to pain: The pain stimulus elicits a sensory response that is perceived and becomes part of the person's experience. Meaning is assigned and meaning begins to condition the response through: (1) bodily response such as local guarding, muscle tightness, and so forth; (2) general changes in posture and activity; (3) emotions such as fear, dread, panic, anger; (4) countervailing behaviors such as avoidance, taking medication, going to physicians, and so forth; (5) predictive statements about what will happen if the pain continues which lead to further countervailing behaviors; (6) activation of the mechanisms that enhance, inhibit, change pain character, or distribution – which, in turn, reinforces or changes meaning with further responses to the meaning. The sum of all these constitutes the personalization of the pain experience. Throughout this process, which unfolds over time, the person can to some degree continue to function as a social entity and maintain central purpose.

VIII. THE PROGRESSION TO SUFFERING

As the pain continues or worsens and the experience of the illness, of which the pain is part, is fed by external events – such as unhappy interactions with physicians, medical care or institutions – the injury to the patient begins to broaden. With continued inflow of information (including no information) and increasing knowledge (which may or may not be accurate or valid), reflection and rumination may begin to change how the events have been interpreted and alter the predictive element of the experience. Thus, events become more threatening and the nature of the threat becomes more important than the pain or other symptoms – sometimes more important than the disease or life experience that introduced the threat. Threat always involves dire predictive statements such as, "My cancer means that I will have a horrible death," or "Sympathetic dystrophy! The pain will never go away, no one will ever be able to help me. I'm never going to have a life, I'll never be normal," "When my daughter was killed, the future ended for us, there will never again be peace or happiness in our house," "No matter how hard I try, I can never feel like other people anymore." What is threatened, or injured, is the integrity or intactness of the person as a person. The injury to the person and the distress of suffering become more important and loom larger than the actual physical symptom, disease or life situation that produced the suffering. Physicians and other caregivers, family members, and even the suffering person, tend to focus on the event or process that initiated the suffering, as though this external event is the problem. But that is no longer the case. With the onset of suffering, the locus of injury has become internal to the person.

The integrity of a person is a phenomenon that involves all the dimensions of the whole individual in his or her particularity. Persons are of a piece, as noted above, what happens to one part happens to the whole. Nonetheless, the injury to the person that initiates the suffering – no matter what the external traumatic event – may occur in relation to any aspect of the person (Cassell 1982, 1991). For example, a person with a rigid personality may be unable to adapt to the life changes demanded by an illness, and suffering starts with the failure of adaptation. Or trusting individuals may have trust badly violated by the behavior of physicians and then find themselves in situations where trust in caregivers is necessary, but no longer possible. The nature of the illness (or the impairments produced by it) may make it appear impossible for the person to live in public without shame, thus rupturing previous relationships with friends or even family. In some situations, the illness or life events may make continued connections to loved ones problematic or impossible, thus setting in motion grief and bereavement. Individual roles – mother, doctor, executive, athlete, and life-of-the-party – may be impossible to sustain, thus destroying self-respect and the person's face to the world. Similarly work, which may have loomed large in the person's life before the process or event that started the suffering, might now be impossible, taking away something that gave meaning to the person's life. Hopes, dreams, and aspirations for the future might seem to disappear, stranding the person in the unsatisfactory present. Unconscious conflicts and repressed trauma are often re-activated by the pain or the illness so that, for example, the person abused in childhood begins to see the pain and illness as a repetition of the abuse, re-inhabiting, as a consequence, the role of marginalized victim. Self-conflict, a constant feature of suffering, emerges between, for example, the needs of the person and the needs of the body, or the demands of the illness and the requirements for normal social life. Self-esteem, the desire to be approved of, the desire to be considered superior, and the desire to be like those one admires, have been recognized as motives for human behavior since the 18th century. In serious chronic illness the sick person still has these desires but illness may prevent their realization, which leads to self-conflict. The chronically person attempts to meet the standards of the everyday world but cannot. These standards appear to be external (the demands of society) but they are contained within verbal categories – mother, teacher, doctor, success, sick, patient, etc. – and as a consequence they lie within the person. The patient who, despite considerable illness, attempts to continue a social or work existence that cannot be maintained lives a life of conflict. So do persons with disabling impairments who try to be "like everyone else" – when ideas about "everyone else" are part of internal concepts again generating self-conflict. Another source of the self-conflict always present in suffering is the battle of the self with the body. The patient may behave as though the body is the enemy, an untrustworthy other, and a potential source of humiliation because of bowel or bladder problems. Self-conflict is even found in acute illness. When pain or serious symptoms arise, the patient fights them. But the struggle magnifies the suffering. If only one did not care or could let the pain simply take over. But the desire to live is too strong to give up the resistance. One wants to live, but not to live the only life that seems to be offered.

Central purpose begins to disappear (Bakan, 1971) as purpose shifts to the sick part. Just as attention goes to the sick part, so does purpose. The focus becomes entirely directed on the self. All purpose becomes directed at the relief of pain, sickness and suffering. The more total and compelling the injury, the more complete the re-direction of purpose. In the pursuit of our purposes, large and small, we need others and we must remain engaged in the social world. In various degrees the suffering person withdraws from the social world and the intercourse of daily life. As suffering progresses, the person withdraws even from spirit – the universal concrete we-ness of everyday life. If we are attentive, when we are with the suffering, we are aware of their absence.

It is apparent why loneliness is a central feature of suffering. The person is lonely because suffering is individual – and no one else can truly know why another person is suffering. (We can know *that* someone is suffering, but exactly *why* – what aspect of the person is disrupted – is almost always closed to others.) Loneliness is worsened by the withdrawal from the social world. Suffering persons may perceive their withdrawal from others and attempt to resist, but usually they cannot, thus worsening the situation. When the source of suffering is long-standing or chronic, suffering becomes a way of life and may appear to an onlooker like depression. Sometimes one sees this written in the sad eyes of children with severe impairments who have been essentially hidden from active life. Sometimes the lonely suffering person believes that it is others who have withdrawn or who have ceased to help. Unfortunately this is often true because healthy people often do avoid people who suffer, or else deal with their difficulties in a superficial or brief fashion. When acute or chronic pain is the initiating source of suffering, all these elements of suffering, and the changes in the person, act on the nociceptive process and the pain becomes increasingly intractable and unresponsive to increasing doses of analgesics. The higher doses lead to increased side effects and further withdrawal of the suffering person aggravated by delirium and confusion. The pain persists, sedation increases, life becomes more miserable and is entirely given over to the illness and its effects. The demands on caregivers increase but, unfortunately, in the absence of recognizing that the problem is now the relief of suffering and not merely pain, their attempts to help seem futile and they frequently withdraw.

This chain of events underscores the fact that, even if the pain or other physical or social processes are controlled, if suffering is not addressed it may continue. Unless the suffering itself becomes the focus of treatment, the pain often remains intractable, whether it is physical, emotional or social. In fact, in established suffering, these distinctions fade in importance – suffering is an affliction of the whole person. On the other hand, it is an error to believe that all suffering is equally severe or long lasting. People may suffer briefly or even slightly. The rapid and complete disappearance of pain or other elements of sickness, a supportive family, and responsive caregivers may all contribute to healing the lesion of suffering – the person is restored to wholeness. (The memory of the suffering never goes away, however.) This is the goal of healing and emphasizes why persons may be cured but not healed – their suffering may continue long after the originating experience. How much chronic suffering is present in a population is not known, but there are groups in the modern world subject to war, widespread death and other losses, injury, disruption of families, and the abandonment of children, where it might be suspected that suffering virtually never ceases. The diagnosis of depression is common, but one wonders how much suffering is masked by, or accompanies, depression.

The phenomenon of suffering cannot be separated from the all to common failure to recognize or treat it. There has been increasing discussion of suffering in the world of medicine, hospice, and palliative care. Why, then, is it not more frequently diagnosed or treated? An important reason may be the fact that suffering, like pain, is subjective and that it cannot be measured. The canons of science that have held medicine in thrall for much of the twentieth century do not have a place for the phenomenon of suffering. Understanding suffering requires an appreciation of the whole patient, while the remainder of medicine is atomistic and reductive. One does not know suffering's presence by independently verifiable facts. In addition the tools by which it is diagnosed lie within the subjectivity of caregivers. Unfortunately, physicians and others caregivers have increasingly distanced themselves from their own subjectivity – feelings, intuition, and even the input of their senses – that would be necessary to detect suffering in their patients. They have done this, in part, because increased reliance on sophisticated contemporary diagnostic and therapeutic technology has diminished whatever skills they had previously. But, perhaps more importantly, many physicians have actively disavowed their subjectivity because of the responsibility to patients that it entails and the relationship between doctors and patients that provides its reliability – a relationship that is also, too often, given short shrift.

In sum, then, suffering is personal and individual, and is marked by self-conflict, and loneliness. While it may arise from physical illness and symptoms, once it starts, suffering itself is the central distress. Suffering can come from any physical, social, or emotional process or event that leads to a loss of integrity of the person. Although removing the person from the inciting cause, or treating the disease and its symptoms, may effectively relieve many of the person's problems, frequently suffering continues unless it is specifically recognized and addressed. In the past it seemed to be true that the personal nature of illness and suffering could be contrasted to the impersonal, completely physical nature of disease and pathophysiology. But this is now seen to be false. Evidence of personalization – alteration of physical processes of symptoms, pathophysiology, and disease due to the unique individual person in whom the processes occur – can be found in many diseases and symptoms of disease. The involvement of persons in their own illness and manifestations of disease can be traced from the earliest symptoms to the ultimate personal injury, suffering. It follows that there can be no true appreciation of sickness and its treatment without an understanding of persons.

Weill Medical College
Cornell University, New York
U.S.A.

BIBLIOGRAPHY

Bakan D.: 1971, *Disease, Pain and Sacrifice: Toward a Psychology of Suffering*. Beacon Press, Chicago.
Cassell, E.J.: 1991, *The Nature of Suffering and the Goals of Medicine*, Oxford University Press, New York.
Cassell, E.J.: 1985, *Talking with Patients*. Volume 1. *The Theory of Doctor-Patient Communication*, MIT Press, Cambridge, Mass.
Cassell, E.J.: 1982, 'The nature of suffering and the goals of medicine', *NewEngland Journal of Medicine* **306,** *639-45.*
Entralgo, P.L.: 1956, *Mind and Body*. P. J. Kennedy and Sons, New York.
Groddeck, G.: 1977, *The Meaning of Illness*. International Universities Press, Inc., New York.
Johnson, M.: 1989, *The Body in the Mind: The Bodily Basis of Meaning,Imagination, and Reason*, University of Chicago Press, Chicago.
Johnson, M. and G. Lakoff.: 1999, *Philosophy in the Flesh: The Embodied Mind and Its Challenges to Western Thought*, Basic Books, New York.
Jones, N.K. and Killian, K.J.: 2000, 'Exercise limitation in health and disease', *New England Journal of Medicine* **343,** 632-41.
Spiro, H.: 1997, 'Clinical reflections on the placebo phenomenon', in A. Harrington (ed.), *The Placebo Effect: An Interdisciplinary Exploration*, Harvard University Press, Cambridge, Massachusetts.

SECTION FIVE

MEDICAL ETHICS

MARK J. BLITON

IMAGINING A FETUS: INSIGHTS FROM TALKING WITH PREGNANT WOMEN ABOUT THEIR DECISIONS TO UNDERGO OPEN-UTERINE FETAL SURGERY

I. INTRODUCTION

For more than ten years now I have been engaged in clinical ethics consultation and, from the start, I have maintained an abiding concern to think through a method inherent to these activities. Two recommendations from the work of Edmund Husserl continue centrally to inform my pursuits. The first involves Husserl's (1960, p. 7) suggestion that philosophers, "each for himself and in himself," should at first "put out of action all the convictions we have been accepting up to now." This suggestion also includes an invitation to "immerse ourselves" in the "scientific striving and doing" (Husserl, 1960, p. 9). For my part, this attitude of "immersing" pertains to the world of clinical medicine, specifically the moral experiences and difficulties that arise in that context. Crucial to this activity is adopting a rigorous reflective attitude, even while becoming immersed in the activities and practices that occur in clinical settings (Bliton and Finder, 1999, pp. 72-75).

The second recommendation from Husserl evolves from the complexity of moral experience in clinical settings. I am thinking particularly of Husserl's (1982, pp. 159-60) remarks that "it is necessary to exercise one's phantasy abundantly," with the idea to "fertilize" it with "the offerings of history, in even more abundant measure from those of art, and especially from poetry." Moral experience cannot be severed from the constituent cultural, social, and historical elements of those environments in which clinical experience and decision-making takes place. Not only are there apparent differences among personal beliefs, professional opinions, and responsibilities, whose meanings may not be reducible to some common theme, but the meanings of physiological factors, let alone moral ones – birth and death, suffering and responsibility – are embedded in complex webs of cultural and social relationships (Zaner, 1988, pp. 29-44). Thus, even for a single perspective within a clinical domain, it is difficult to depict the extent to which possible fluctuations of meaning occur in each unique situation.

To employ free-phantasy variation provides access into otherwise hidden and densely complicated perspectives. In order to discover the possible relevance of moral insights about the rarely stable world of clinical medicine, a particular experience may be considered, and then its variations reflectively imagined, as a means to identify that which remains invariable (Husserl, 1969, p. 248). In this way, a variety of examples may be worked out which clarify important features of what actually occurs in clinical experience and environments. In addition, such examples may "be used as guides for observation, much as one might use a previous observer's description of a landscape as an aid in distinguishing its features while all the time it lies before one's eyes" (Cairns, 1973, p. 227).

Writing more in the form of a documentary report, not unlike what was suggested by Husserl with his image of the philosopher as an explorer of hitherto new or only partially explored territory (Zaner, 1970, p. 33), I bring this approach to bear on one of the most morally intriguing surgical procedures in medicine, open-uterine fetal surgery. Clearly, responsible commentary regarding the ethics associated with open-uterine surgery needs to incorporate insights gained from reflective focus on the ethical features embedded in the actual dynamics of clinical relationships, e.g., the complex layering of values, persons, and decisions involved. Such inquiry must always be guided by careful attention to the moral relationships and understanding of the individuals actually making the decisions and living with their choices. Accordingly, the substantive claim I make here is that the basis for moral discernment in clinical settings must be directed by these specific elements in the clinical context. As a means to identify more accurately the moral significance of the issues and experiences pertinent to a couple's consideration of maternal/fetal surgery, inquiry into the specific circumstances that prompt the appeal to the "ethics" of maternal/fetal surgery must also be clinical, not merely theoretical, or regulative.

To this end I will discuss the moral significance of an array of issues and experiences pertinent to a pregnant woman's (and her partner's) consideration of maternal/fetal surgery. My inquiry into these matters has been prompted by the complexity of the moral considerations identified by Richard Zaner and myself in intensive conversations with the more than 150 pregnant women (and partners) who visited our institution seeking either more information about open-uterine fetal surgery to repair spina bifida or to have the surgery itself.

As an example: many of the women we encounter exhibit a quite enthusiastic, if not anxious, endorsement of the research, the surgeons, and the technologies the surgeons promote and put to use. They avidly want "to do everything we can," regularly noting that their soon-to-be-born children "deserve a chance at a better quality of life," and also, "I couldn't live with myself if I didn't at least try!" Exploring this context of meaning provides important clues as to why these pregnant women agree to undergo, many times eagerly, an innovative, complex, and risky surgery despite a social backdrop of ongoing concerns and controversies about whether, and how, women are coerced into having prenatal diagnosis.[1] Indeed, an important part of the clinical ethics involvement has been to encourage awareness and explicit discussion, on the part of both couples and health care professionals alike, of the possibly coercive factors that influence the women's decisions.

From the outset very serious issues were, of course, quite evident, since no benefits from the surgery could be asserted at the time the original investigational protocol was initiated. The idea was to conduct the discussions by the ethics consultants in a manner that would help each couple, and family, anticipate, identify, and understand the many personal, medical, social and ethical issues they would inevitably experience, whether or not they finally decided to have the surgery. It was likewise essential to ensure that each couple was encouraged to discuss their thoughts, values, and religious convictions, so as to help them identify the implications of this innovative procedure for their own nexus of beliefs.

II. OVERVIEW OF THEMES

Following a brief description of the context and method for these ethics conversations, this essay is organized around a prominent set of recurrent moral themes and beliefs expressed by many of these pregnant women and partners. In particular, their belief that their fetus was already considered a member of their families held special importance for deliberations about their situation.

Identified are a number of themes discovered in the moral deliberations about open-uterine fetal surgery. The first theme concerns the ways that the experience of pregnancy is influenced by prenatal ultrasound imaging and, in particular, the way this sort of visualization is implicated as the means through which 'the fetus' acquires the characteristics of 'being real' in the culturally-made world of medicine. Second, are the uncertainties encountered when fetal anomalies are detected. This second theme will be discussed in terms of the couple's understanding of the fetus as an individual child, with whom they have bonded and named, thus experiencing the felt moral imperatives that accompany the sense of 'parenthood,' and in terms of a fetus as a patient (now correlated with the technological imperative 'to do' something in the world of medicine). A third and obviously complex theme arises in the ways that the threat of disability influences a couple's decision 'to do something' in the face of intractable uncertainty about the extent of possible disability for their child. Discussion of vulnerability constitutes a fourth theme: in particular, the role of religious language and faith in the moral deliberations of these couples. We found that often beliefs regarding 'benefits' from the investigational surgery were directly related to their religious beliefs.

III. THE CONTEXT OF INQUIRY

Since the 1980's physicians have appreciated that the medical and social ramifications of fetal surgeries can be quite complicated. In contrast, their account of the ethical factors has been curiously straightforward, and even a bit naive, possibly as an attempt to mute the complex cultural outrage and fascination that accompanies what some people viscerally regard as the transgression of the pregnant woman's body and uterus. Because the balance of potential risks and benefits of open-uterine surgery must be evaluated in the context of considerable uncertainty, such surgery was initially considered ethically defensible only for those fetal anomalies likely to result in death despite the best available postnatal treatment (Jennings et al, 1992). Decision-making was focused mainly on the pregnant woman and the potential risks to her health and well-being. As regards the fetus, the primary effort was to identify those fetuses with well-diagnosed lethal conditions, thus most likely to benefit from a proposed experimental procedure. In addition, because mere survival is not always perceived as a benefit, guidelines for such procedures were also designed to increase the likelihood that surgical intervention would contribute to an improved quality of life for the infant, and to avoid producing severely compromised survivors (Harrison et al, 1982; Fletcher and Jonsen, 1991).

Surgery for non-lethal fetal malformations like spina bifida raises more subtle and complex ethical questions. Fetuses with non-lethal fetal abnormalities will likely survive whether treated *in utero* or not, although almost all will certainly have some type or degree of disability. With the availability of this surgical option (Bruner, Tulipan et al, 1999; Sutton et al, 1999), decision-makers (pregnant women, their partners, and care providers) are confronted with the challenge of evaluating the potential for improved function for these children over against not only the risks of morbidity from the surgery for both mother and fetus, but also fetal demise during or soon after surgery. This matter is claimed to be at the heart of a paradigm shift (Bruner, Richards et al, 1999, p. 158) in moral and medical issues that has been brought about by this innovative form of open-uterine surgery. The idea is that by extending open fetal surgery to treatment of non-lethal fetal anomalies, the fetus is accorded a more robust status as a *bona fide* patient for whom there is a separate set of risks and benefits distinguishable from those of the mother (Fleischman et al, 1998).

Thus, when experimental open uterine surgery for repair of fetal myelomeningocele, "the most complex, treatable, congenital anomaly consistent with life" (Bunch et al, 1972), was initially proposed at Vanderbilt University Medical Center, the physicians primarily responsible for its innovation recognized that the complex moral and ethical issues indicated a need for much earlier and more extensive involvement of clinical ethics personnel than heretofore had been required for maternal/fetal surgeries. At these physicians' invitation, Richard Zaner and I have been involved with the investigational surgical protocol from the start in April 1997. We helped design the protocol for obtaining informed consent and then began to assist in other activities – primarily to ensure, as far as possible, that each pregnant woman and partner acquires serious understanding about all aspects of the procedure, including the risks and theoretical prospects of benefit, as well as the difficulties to be faced in the future (whether or not they proceed to have the surgery). Because of the experimental nature of the initial surgical protocol, the ethics involvement was designed not merely to review and discuss risks, but more positively to help each couple reach a commensurate level of understanding of the protocol's investigational character, including risks to the fetus, to the woman, and to future pregnancies, as well as various other factors such as disruptions to ongoing family life, effects on current children, etc. (Bliton and Zaner, forthcoming).

In this regard (and because I draw upon my involvement with all the couples who travel to our Medical Center seeking the surgery) I need to emphasize that my experience with these women and their partners – and in some cases other family members – occurs as part of another innovative component of our protocol beyond the surgical techniques, and the fact that we are operating on fetuses. I am referring to a required three-day schedule of counseling and conversations in which each couple participates prior to having the surgery (Bliton et al, forthcoming). This three-day process was designed as a way to elicit ethical concerns on the part of both clinicians and the couples. Each couple meets with my colleague and me on two separate occasions, once at the beginning of the process and then during a "wrap up" session at the end of the second day. They also consult with the neurosurgeon who does the actual fetal repair, the perinatalogist who opens the uterus and then closes after the repair, and the Chief of our Neonatal Intensive Care Unit nurseries, who takes them on a tour of the nursery itself. The process also includes financial counseling, ultrasonography, consultation with anesthesiology, as well as meetings with a perinatal social worker and obstetrics nurses. Following the initial two days, an additional day is set aside for couples to spend on their own, as a way to provide additional time to consider their choices, before proceeding with the surgery. The counseling is thus thorough and time intensive.

IV. WHAT'S "PHENOMENOLOGY" GOT TO DO WITH "ETHICS"?

In order to identify more accurately the moral significance of the issues and experiences pertinent to a couple's consideration of maternal/fetal surgery, it was important to practice close attention to the actual circumstances of their understanding: how what were perceived as the 'issues' had come about, and in what ways they understood the circumstances and factors specific to fetal surgery. In this regard, the method of inquiry reported here is predicated on the idea that clinical ethics consultation must be specifically attentive to the finely textured, subtle context of multiple interactions among physicians, patients, spouses, and others who may be variously involved in such consultations (Zaner, 1993, pp. 24-28). This method is commensurate with the conviction on the part of myself and my colleagues that the basis for any relevant 'ethical analysis' must always be reflectively focused on the moral relationships and understanding of the individuals actually making the decisions and living with their choices. In other words, inquiry into the specific circumstances that prompt the appeal to the 'ethics' of fetal surgery must also be clinical, not merely social, or political, or regulative, in order to gain understanding adequate to the situations that generate the need for ethical guidelines or policies in the first place.

In a similar vein, Susan Sherwin's (1992, p.108-9) discussion of ethics holds for fetal surgery, particularly where she says "fetal development" needs to be "examined in the context in which it occurs, within women's bodies, rather than in the isolation of imagined abstraction." Not only do "fetuses develop in specific pregnancies that occur in the lives of particular women," but "their very existence is relationally defined, reflecting their development within particular womens' bodies" (Sherwin, 1992, pp.108-9). That relationship is what gives these women both the occasion and reason to be concerned about their fetuses. Thus, 'clinical ethics consultation' in this context needs to be defined and determined by the actual circumstances encountered—a proposal consonant with what Alfred Schutz (1967, 1973) advocated as a principle of method for any serious and rigorous approach to human situations, and as articulated by Richard Zaner (1988).

In order to recognize more accurately the moral significance of these matters, such recognition must be grounded in, and derived from, the "things" to be studied. I think that, in practice, such a method of inquiry becomes an instance of what Alfred Schutz (1967b, p. 175) termed an "analysis of the mundane sphere." The idea is that "in the mundane sphere of everyday life I conceive myself as well as the other as a center of activity, each of us living among things to be handled, instruments to be used, situations to be accepted or changed" (Schutz, 1967c, p. 201). As Husserl (1969, pp.278-79) noted, this kind of analysis is no simple matter. When talking to others one needs to be "continually asking what can actually be 'seen' and given faithful expression – accordingly it is to judge by the same method that a cautiously shrewd person follows in practical life wherever it is seriously important for him to 'find out how matters actually are'."

Thus, careful listening constitutes a principle of method of the first order. Since it is not possible to know, beyond common typicalities, which moral issues are actually presented by any specific situation without knowing the actual circumstances in which they occur, these must be learned. A core part of this learning is accomplished in clinical situations by attentive *listening* (Zaner, 1996, pp. 266-69). By that I mean a rigorous listening conducted in the face of a vulnerability characteristic, not simply of the pregnant women and couples, but of their families, and of the care providers. The full phenomenological explication of this sort of listening has yet to be fulfilled. The approach I am documenting here attempts to outline the significance of practicing this sort of attentive listening, while also indicating what may be discovered in the process.

Reflexive Engagement in Moral Experience

In order to help others faced with the necessity of making agonizing decisions, it is imperative that the ethics consultant be – and be understood as – an actual, if only partial and temporary, participant in the situation. That is, not only must s/he reflectively participate in the actual circumstance, but also initiate discussion of the questions and uncertainties at hand. In other words (even though the gravity and meaning of the issues cannot be the same for both of us) how might I possibly find out what is important to you, or anyone else, in such a context unless I am willing to share temporarily the questions and uncertainties you face? Moreover, I need to pursue the same sort of issues for myself in a kind of imaginative variation (Spiegelberg, 1981, pp.172-73) while helping you consider the aspects and ramifications of your situation.

V. FIRST THEME: THE INTIMACY OF PREGNANCY IN AN ERA OF ULTRASONOGRAPHY

Each of the couples was intensely concentrated and committed to doing the best they could for their still-to-be-born-child. Having been thrown into the loss of the imagined child of the pregnancy, many reported that they "felt like they were on an emotional roller-coaster" from the shock and grieving evoked by the fetal diagnosis. Faced with the tenuous uncertainties of having and parenting a child with spina bifida, they presented a focused and total involvement. Accordingly, inherent to my efforts in this essay is a challenge to find a language that conveys the emerging and vivid moral significance of the human form – not as an epistemological matter so much as a recognition of what pregnant women and their partners have presented during conversations in which they are considering maternal/fetal surgery. By "vivid moral significance of the human form," I mean the profoundly evocative potential, and sense, of the human body – not only of a pregnant woman, but also the recent disclosure of the human form of a fetus *in utero*.

Given that evocative and dual aspect of pregnancy, it is necessary to point out two considerations. First, the pregnant human body presents a specific, if not primordial, instance of what Maxine Sheets-Johnstone (1994, p. 63) refers to when she calls "the body," that more "ancient field of possibilities." The pregnant body depicts dramatic and expressive phenomena that arouse controversial views and interpretations about its significance far too extensive to be addressed in this essay. Second, among the aspects of fetal surgery that I find so intriguing are the ways visualization and visibility are implicated as a partial means through which 'the fetus' acquires characteristics of 'being real' in the culturally-made world of medicine. These factors demonstrate both the wider social influences associated with the recent refinement in the technologies of fetal imaging, as well as the more intimate and subtle influences exerted on the uniqueness and intimacy of relationships experienced during pregnancy.

Ultrasound provides a visual window into the pregnant woman's uterus, where many times the ultrasound 'pictures' become blended into a woman's experience of the pregnancy (Sandelkowski, 1994). These images evidently evoke deeply ingrained emotive, valuational, and intellectual responses from the woman (frequently from her mate as well) which shape and guide the sorts of choices and actions that follow (Rapp, 1997). For many women it is the first time they relate visually to the living fetus inside – and intermittently, co-relate that imagery with their own intensely intimate bodily sensings of the fetus growing in them.

Her attention riveted on the video screen, the pregnant woman can join her physician in real time viewing of the 'inside' of her belly. There is even more to this multiple sensory experience. The woman can feel her fetus moving from inside. With her hand on her belly, she can also feel fetal movements while she is viewing the images. What becomes available right then is the dense evocation of both pregnant embodiment and growing alterity of the fetus. Thanks to the simultaneous feeling and 'seeing' of her fetus, the woman comes upon the shimmering astonishment of something at once 'brand new' and yet deeply familiar: what she felt inwardly in her body, now has an almost-face, hands, fingers; and a part of this amazement is the coming upon – 'seeing' – what has issued from your coitus, floating there, right there on the screen. Like the feelings of the woman's belly, this is an experience of wonder and strangeness in the recognition: *It's a 'he' or 'she,' yet is also 'me' and 'him' (or 'her' and 'me').*

Although such experiences can be truly exciting and very reassuring for most parents (Gregg, 1995, pp. 88-93), consider what happens when one is confronted with the shock, confusion, loss, and dread that occur when the ultrasound results in a diagnosis of a fetal anomaly (Manchester et al, 1988; Filly, 1999). The pause in the usual banter of the ultrasonographer, a silence followed by the remark and bustle of "I'll go and find the Doctor," makes anxious tensions prominent – in particular, a kind of fear typically evoked by the prospect of finding a fetal defect; the threat, desperately guarded against, that such visceral wonder and strangeness could turn at any moment into the dread of finding out that the prospects for a child-to-be include that child having a disability. That dread is often repressed or unspoken. This "unspokenness" constitutes another side, the tension, of the emotionally charged expectation of 'seeing' the 'baby' for the first time. It also stimulates another social aspect of ultrasound: a welcoming embrace for ultrasound imaging, with hopes for assurance against the prospects of disability (Petchesky, 1987, p. 75; Taylor, 1998).

Conferring Significance On A Fetus

At issue then are the processes of multiple and complex perceptions that manifest such constructed ideas, beliefs, about the sonographically revealed image of 'the fetus,' that are subsequently accepted on the belief that these have the same ontological status as the naturally given world (Scarry, 1986, p. 125.) Given that obstetric manuals and textbooks regularly include statements to the effect that the visualization of the fetus accentuates parental identification with, and recognition of, 'the baby,' the fetus has already given way to 'patient,' with all that haunts the history and practice of obstetrics and the typical obstetrician.[4] These complexities are especially evident in the ways in which a woman's pregnancy is now increasingly framed by the enhancements in genetic screening and manipulations, as well as by fetal visualization and diagnosis. Conceiving such enhancements as part of "the many layered process involved in the synthesis of this fetus, the invention of fetal norms and needs," Barbara Duden (1993, p. 4, 52) cautions that "increasingly, the public image of the fetus shapes the emotional and bodily perception of the pregnant woman."

The main point is simply stated, albeit controversial. Resulting from numerous interventions into the stages of human life prior to birth, "the 'biological facts' of conception, pregnancy and fetal life are not only powerful as authoritative forms of *knowledge*," but are powerful *symbolically* as well, as these are socially interpreted to connote the implication of 'agency' to the fetus (Franklin, 1991, p. 192). Even while simultaneously reinforcing the multiple forms of intimate interrelations between pregnant woman and fetus, these interventions intensify the belief that a fetal identity is present and separable from the pregnant woman.

Fetal surgery certainly represents a most provocative exemplar whereby the ultrasound image of the fetus has now come to serve as the symbolic prototype of the fetal patient, "making visible through images what will ultimately be revealed with a scalpel" (Casper, 1998, p. 86). Proponents of fetal surgery claim it to be a logical, medical extension of ultrasound imaging, because the "development of invasive fetal therapy has grown directly from advances in prenatal ultrasonography" (Shaaban, et al, 1999, p. 62). Of the current battery of prenatal tests, ultrasound proves most significant because the focus of ultrasound *use* is on identifying skeletal and other structural anomalies, which are considered more manageable by surgical treatment, rather than genetic defects and syndromes (Casper, 1998, p. 84; Madsen, et al, 1998).

As a result, ultrasound imaging and diagnosis of fetal defects both represent and generate the moral context for maternal/fetal surgery and its rationale. Simply using ultrasound sets up the possibility for an increase in the choices and the opportunities *to do something,* thus creating a need to make more decisions – all of which serves to enhance (and tacitly affirm) a kind of technological morality long developing in societies such as ours, by reinforcing "doing" as a positive moral value. It was precisely under the emotionally charged auspices of such a technological morality, with its implied imperatives to "do whatever we can," that we encountered the couples seeking fetal surgery to repair spina bifida.

VI. SECOND THEME: MORAL TRANSFORMATION AFTER FINDING OUT, VIA ULTRASOUND, A DIAGNOSIS OF SPINA BIFIDA

Couples reported that the diagnosis of spina bifida was accompanied by an immediate and shocking initiation of the differentiation that occurs between a pregnant mother and the growing otherness of the fetus: the fetus with a spinal defect is immediately different from the 'imagined child' of the pregnancy. Not only does the fetus now become a 'baby'; it becomes a 'different baby.' Furthermore, the impending difference is of a potential disability created by a diagnostic, and hence medical, reality – a fact that must be reckoned with, if and when termination of the pregnancy is declined. Thus, while these couples routinely expressed feeling grief (in the words of one woman the "diagnosis was catastrophic to me and my husband's dream of a 'normal' child") they in turn have an intense motivation and desire "*To Do Something*" about the impending disability. "Doing," moreover, most often carried the meaning Schutz (1967a, p. 87) delineated as an actual "gearing into the outer world," including bodily movements that contribute not only to a *decision*, but also a *performance* under the pressure of action and time in an actual situation.

"Social Birth" And The Fetus As Patient

As virtually every couple told us, often quite directly, what we were wont to call 'fetus' was in fact understood by them as an already distinct individual. For them, 'fetus' was already 'baby,' and even more than that, it was "Zane or Helen, my son or daughter." It's not yet being a full 'person' didn't matter morally to them, given their views that their fetus could nonetheless be harmed in bodily, personal, and moral ways. Indeed, if anything, fetal vulnerability, its *innocence,* was taken by them as indicative of quite serious moral content. Hence, each of them was already oriented toward their developing fetus as having moral status – a fact that we could neither ignore nor gainsay.

As I mentioned earlier, there is the idea in maternal/fetal medicine that by extending open-uterine surgery to treatment of non-lethal fetal anomalies, the fetus is accorded a more robust status as a *bona fide* patient for whom there is a separate set of risks and benefits distinguishable from those of the mother. That certainly appears to be accurate in the sense of the moral status of a fetus as 'a patient.' In other words, the images of fetal body via ultrasound identify it, the fetus, as a site for surgery, thus initiating a transformation from an entity with 'potential' moral status to an actual patient. Even if we want to be more careful and argue that the ultrasound does not confer that status, such bestowal becomes more difficult to dispute when characteristics typically associated with being a 'patient' are conferred to the fetus by the actual performance of the surgery.

However, for virtually all the couples, the fetus was already named by the time they came to talk with us. In part due to the visualization of the fetus by ultrasound, most often that status as 'patient' was preceded by their understanding of the fetus as an *actual* moral individual with a name and moral presence – even if only in the sense of 'the baby that the fetus will become.' Having been identified by ultrasound imaging as a potential surgical patient, a fetus becomes evidently valued now as 'a baby,' conveying all the patent and subtle forms of thereby relevant moral appeals to others, including family, physicians, and the clinical team. Happening most often at that first ultrasound, although sometimes before, being named is a crucial act, for each named fetus thereby exhibits a special moral presence with important analogies to any infant or child, with the result, as Lynn Morgan (1996, p. 59) notes, that "social birth" can, and does, now often precede biological birth.

Furthermore, conversations with these couples elicited a different, complicated sense of bonding, akin to what occurs at birth. And, *that bonding* also seems to be connected to the frequent number of ultrasound exams these women had undergone in order to clarify and specify a diagnosis. Thus, we find that the circumstances for viewing the fetus are changing, especially regarding the cultural, social, and medical expectations about what those images will reveal. Even so, it remains the case that just because "the fetal patient" may be "a historically produced entity imbued with social and cultural meaning does not negate the lived experience of pregnant women who choose surgery" (Casper, 1999, p. 111).

VII. THIRD THEME: KNOWING MORE THAN BARGAINED FOR; DISABILITY AS THREAT

Issues such as those noted above arose because conversations with these couples often provoked notice of highly sensitive matters rarely openly and so intensely discussed, especially with strangers. When fetal anomalies are diagnosed by ultrasound a very trying and potentially devastating experience is shaped for these parents – an experience which is paradoxically both thoroughly medicalized and intensely private (Manchester, 1988, p. 74). Because many of the candidate couples discovered by ultrasound that their fetus had spina bifida before 18 or 19 weeks gestation, these considerations are morally and ethically significant. Couples were avidly in search of ways to ameliorate their grief and suffering over having 'lost' their 'imagined child.' When they struggled through those gut wrenching feelings they encountered their fears and grief about the likelihood of only being 'partial parents'; and this in the face of their feeling just as acutely their own desire to be 'full' parents.

My participation in these conversations has provided me with a rare and compelling confrontation with the genuine agony and potential moral disruption confronted by these couples. Such focused attention on the sources of moral disequilibrium explicitly raises a question about responsibility: for instance, whether it is more harmful or beneficial to articulate what heretofore may have been not only unspoken, but even possibly unacknowledged by others (Bliton and Finder, 1999, p. 80-81). In this regard, moral candor requires that we not be distracted from the fact that the surgery to repair fetal spina bifida taps into an almost embarrassed uneasiness, "a deep vein of social discomfort directed toward birth defects, physical disabilities, and deformities of any kind" (Casper, 1998, p. 14).

There was often a lack of understanding and inexperience with disability, rarely articulated, yet deeply embedded in the responses and motivations of these couples. The uncertainty and wariness toward disability was further textured by grief, loss and even guilt, often unfocused and vague. At the same time, they were powerfully motivated by hope and relatively unable to, uninterested in, or possibly just weary of dealing with realistic expectations. As a result, their dispositions were often accompanied by a sort of abstractness, an almost tangible sense of distancing themselves from the reality that the surgery neither cures nor changes the fact that their child will continue to face the many challenges of spina bifida.

What was revealed (but extraordinarily difficult for many of these couples to articulate) are the ways that the grim portrayals of spina bifida they received during and immediately after that initial ultrasound (and most often in the context of possible termination of the pregnancy) continued to permeate and haunt the decisions they faced. Often couples told us that the initial counseling they received from their obstetrician was slanted toward terminating the pregnancy, while their own convictions directed them to continue with the pregnancy. What they remembered was that the initial obstetric consultation portrayed a very grim picture about their future child's prognosis. Their immediate sense of desperation was evoked in response to the visceral challenge of disability. Their fears were about mental deficits, "mental retardation," paralysis, and doubts about their ability to manage the obvious challenges.[2] As Zaner (2000) has noted elsewhere:

Add to this a sense of how dreadful it is that any child should have to live with serious disabilities, and it is but a short step to the other, darker side: how awful it will be to live with a disabled child. It must be appalling to have to face this sense of dread in yourself towards your own child, potent fuel indeed for the pragmatic, moral urgency *to* 'do' anything to change things.

Conversations emboldened by the emphatic declaration, "We'll do anything that might help our baby," may just as likely serve to indicate how overwhelming it seems, to anticipate living in the future with a disabled child.[3]

Having already decided against abortion (most couples shared a sense that they could not "do that" to *their* baby), they appeared reluctant merely to "sit around" waiting for their baby to be born, while at the same time resisting the dimly and abstractly understood reality-to-be: a disabled child. In deciding to continue the pregnancy, however, these women are clearly not obligated to undergo the "elective" fetal surgery, but, because they already understand themselves as parents, they may feel morally compelled to choose the surgery. Discussions about the future disabilities waiting for these children – catheters for bladder, braces for legs, shunts for the brain, wheelchairs, numerous hospitalizations – typically exposed serious lack of understanding and plain inexperience, not merely with spina bifida, but with disabled children in general.

When probed further, this sort of uneasiness and uncertainty was most often articulated in a specific form: the couples reasoned that if they decided against the procedure, they would later undergo severe moral and psychological regret, grief, and guilt from foregoing a procedure that might later prove to be "successful." We were regularly told that this would be far more unbearable than if there were few or no benefits from the surgery, even though they well understood the much more obvious risks from such surgical procedures. One can imagine that this rationale is so overtly prominent because it effectively suppresses the corresponding possibility that the surgery "might fail." "Either way," we were told, "it will be hard, so why not try and at least be able to say that we did something?" "At least we would have tried our best to help." Deciding against surgery rarely struck the same chord of moral approbation. As we were persistently told by most couples, their soon-to-be-born children "deserve a chance at a better quality of life," but also, "I couldn't live with myself if I didn't at least try!"

Choosing In The Face Of Uncertainty

Our conversations were textured by both inherent and dynamic forms of uncertainty. Not infrequently, and due to a variety of factors (skill and experience of sonographer, quality of machine, and so on), couples received different diagnostic information regarding the size, location, and specific type of spina bifida lesion when an ultrasound was done at our facility, as compared to the initial ultrasound scans upon which the diagnoses of spina bifida were made. Often, such differences indicated that the predicted outcome for the fetus was worse than previously understood, thus raising specific ethical issues stemming directly from uncertainties about the extent of actual disability. There was also uncertainty about any benefits that might accrue from the remedy of those deficits. Hence, complex ethical dimensions were not merely associated with, but literally composed of, a variety of uncertain factors. At present we are not able to predict the extent of disability for any particular fetus if fetal surgery is not done, and correspondingly remain quite uncertain about potential benefits that might result for that specific fetus if surgery is done.

How then were we to understand the fact that these pregnant women, even the very first ones in the investigational phase, were so eager to undergo a completely untested surgical procedure that put not only the mother-to-be at risk but, even more, risked serious compromise to her developing fetus? This issue has become more intriguing over time. Even though over 120 procedures have been done since April 1997, there has yet to be sufficient evidence to verify whether any benefits can be unequivocally associated with the surgery (Simpson, 1999). What is genuinely puzzling is that these pregnant women (and partners) – who quite genuinely profess love for their "babies" – were yet willing, even eager, to run the risk of one or the other, or possibly both, of the most dreaded outcomes, brought about by their own deliberate choice and action: either death during or shortly after the surgery, or severe disability from post-surgical problems specifically associated with extreme prematurity (Bliton and Zaner, forthcoming).

While these factors sketch the sort of ethical issues actually confronted by these couples, my point is a further one. In what ways are we to make sense of the puzzling, if not downright paradoxical, combinations of beliefs and actions presented by these couples. From very early on in our intensive conversations with them, we began to recognize that their tendency was to take anticipated, and more especially, hoped-for outcomes as what would *eventually justify* the choice for surgery. At least in this respect, the couples at times demonstrated real difficulty in distinguishing between passionate, even religious, *hopes* and reasonable medical *expectations* as regards only potentially beneficial outcomes.

VIII. FOURTH THEME: THE APPEAL TO SPIRITUAL FAITH IN
PARENTAL DECISIONS ABOUT OPEN-UTERINE SURGERY

Through careful although, given the circumstances, challenging probing, we found the underlying stances taken by many couples highly charged, not only in moral, but also in spiritual terms. Couple after couple expressed vigorous and explicit religious commitments, almost all Christian, infusing spiritual and religious dimensions into their experience of the pregnancy. Often, the power, the mystery, and foreboding they faced were rendered as signs of God's presence, even when considering the otherwise obvious technological implications of open-uterine surgery. What is compelling and, quite possibly, necessary to consider are the ways these couples have talked about these experiences, because many of their spiritual references are to religious, mainly Christian, terms and symbols. These religious themes are vibrantly recounted, and emerge in particular when couples recount receiving the diagnosis of spina bifida and subsequent events.

Many couples insistently saw themselves as embarked on a significant 'journey,' even a kind of 'pilgrimage,' through events, places and a course of time burdened with obstacles to be surmounted. Their recounting of this journey highlighted significant aspects and milestones, conveying not only how, or why, the pregnancy was initiated and understood in the first place, but how the prenatal tests came to be proposed (especially the urgency of the ultrasound to confirm the diagnosis of spina bifida), and the ways those results were discussed and accepted.

Particularly in circumstances where couples testify to a longstanding resolve to follow God's will, their parental initiative and courage also seemed to stem from their religious convictions. Given their straightforward statements that their religious beliefs were themselves considered part of the choice, our conversations openly invited discussion of those religious factors, and other beliefs, that contributed to their choosing to undergo the surgery. It could well be, too – given the fierce difficulties and intensification of vulnerability encountered in talking about one's own parental fears of disability – that the journey they recounted harbors even more burdensome meaning for many of them: a way to resolve earlier guilt, hence it may be a sort of test or path leading to needed salvation, if the faith is kept. To paraphrase the words of more than several couples: *There is a reason God chose us* to have this child with spina bifida, and we must bear that burden gladly.

Many couples told us about this sense that they were on an arduous journey was a factor crucial to their decision to have the surgery. "This has hit us hard and it is difficult not to ask 'why us'?" For some the travail was filled with dark and gloomy places and only a glimmer of light (hope) anywhere. Being on such a journey, it seems, may well carry with it the sense that the prospect of surgery emerges from the gloom as a kind of ray of hope, a light, a way to rescue them from disaster. Others said they were on a journey with key 'landmarks' along the way each read as *"signs from God."* Their experience was interpreted as somehow indicating an intuitive sense of following a *"path laid out by God"* right up to the point of coming to Vanderbilt and undergoing the surgery. Accordingly, most couples believe they face a *"serious test by God,"* a trial whose details are not always obvious, but which is nonetheless a *trial-laden odyssey whose burdens must be endured with courage and faith* – itself a robust Christian theme. We have begun to realize that for these couples a form of Christian belief *relating faith and sacrifice in religious terms* is embedded in the prominent themes of loss, grief, and then renewal of hope.

Those couples who saw themselves as having been guided by God – One who could be petitioned and who could intervene – told us they were willing to have the woman undergo this surgery, even if it involved risking the mother and possibly sacrificing their child. They made it clear that for them it really was not a matter of 'choice,' since they were guided by God. In this intimation of sacrifice is a haunting echo of Hans Jonas' (1974, p. 119) remarks regarding "the ethical dimension" of participation in nontherapeutic research, that it "far exceeds that of the moral law and reaches into the sublime solitude of dedication and ultimate commitment, away from all reckoning and rule – in short, into the sphere of the holy."

Risk And Benefit In The Context Of Spiritual Hope

Even if we allow that some type of spiritual (often denominational) language is the primary language available for many in our society for discussion of matters both grave and deeply personal – if not also their only viable anchor for facing such difficulties – that ought not to obscure the deeper implication about the sort of Christian ethos expressed by these couples: *faith requires action*! Thus, we found ourselves confronted with this question: How is our endorsement of respect for a pregnant woman's choice supposed to fit with the fact that for almost every couple the known risks from the surgery, to both fetus and mother, did little except provide stark contrast to the woman's and her partner's marked conviction that the surgery would ultimately prove beneficial?

Given these narratives of divine guidance – when, for instance, even finding out whether their insurance company might pay for the surgery was considered a clue to divine guidance akin to Gideon's trial with the fleece and the dew – is it enough to say that it is the pregnant woman's, or the couple's, decision alone? It is necessary to take a much more careful look at these beliefs and their *performance*. It seems likely that open-uterine fetal surgery may hold much more profound meanings than typically appreciated, especially for the cohort of couples with beliefs and motivations similar to those I have described.

One feature continues to stand out: for many couples, the basis for their decision was not really the facts, theories, and methods of medical science. The sense of benefit they spoke of was, in a word, religious – a kind of fervent optimism and hope, in which they were willing, even eager, to undergo the substantial risks of surgery, not only as a way to affirm how precious life is, but also to try and improve their child-to-be's quality of life; their faith could only be redeemed by undergoing, or at least risking, that sacrifice.

There is, thus, another fundamental reason to pay careful attention to such themes and beliefs. All of this comes at a historical time and in a society in which the medical enterprise is decidedly more secular than not, and typically has little room for religious imagery and language. The resultant stance of secular neutrality risks indifference to a momentous concern in the lives of these people. That is, skirting the explicit discussions of such themes, their underlying motivations and meanings, leaves the pregnant woman (with developing fetus) and father, vulnerable not only to judgments, enthusiasms, and biases of the surgeons and others on the team, but vulnerable to their own beliefs about disability, the surgery, and its potential outcomes.

IX. THE AMBI/VALENT ALLURE OF TECHNOLOGY

These influences I have noted culminate in a decisive event, open-uterine fetal surgery, which yields the actual appearance of the fetus, made *visible* (via the presentation of its back) as a patient. There has also been widespread distribution of color pictures taken during the spina bifida surgery. One reveals a fetal hand extending through the uterine incision[11] and another the face of an anesthetized fetus,[12] both of which serve to elicit and fuel the imagination as much by what is *not* presented as by what is. Evoked by this whole constellation of technologies, complex events, and meanings there is the invitation for parents to confer on their fetus the moral and social status typically afforded to infants, and it becomes "our son" or "daughter," "John" or "Sue." What my colleagues and I are witnessing – and with others like Duden are striving to recognize and hence develop different, more accurate, more adequate ways to talk about – is another cultural upheaval which, like all that preceded it, only barely announces itself until it, the fetus, emerges a full blown social entity. While we may not be all that clear regarding how its ramifications will unfold in contemporary American culture, we must acknowledge the powerful glare that fetal interventions, and especially open-uterine fetal surgery, casts on such issues as maternal responsibility, abortion rights, and disability rights.

Among other things, then, the emblematic fetal form simultaneously reinforces the cultural aura of medical authority, if by nothing other than the implied technical access of sophisticated imaging technology and the allure of medical innovation. The popular use of fetal images carries with it the implication that not only have medical technologists been able to gain access into the previously hidden world of the fetus (which of itself grants a kind of visual independence to the fetus) but since we can now see what was before unseeable, we are beginning to be able to do what was heretofore undoable: and even unimaginable. Our technology, in other words, creates new possibilities hitherto undreamed of. We are captivated by these technologies, in part because by revealing the "previously hidden" (Duden, 1993, p. 15) the technologies are seen as, and taken to be, a locus of power.

Of course, having said that, the complexity of these matters simply cannot be finessed or avoided, in part because an underlying motivation in the quest for such technological power is what Duden (1993, p. 15) calls "the *libido videndi*, the ravenous urge to extend one's sight, to see more, to see things larger or smaller than the human eye can grasp – to see things which have been previously hidden, or even off limits." Although somewhat apprehensive at the ability to reveal things "previously hidden," given how powerful it is – after all ultrasound, not to mention magnetic resonance imaging, can 'see' in me what even I cannot see in myself – we are nonetheless attracted to it, and even willing to submit ourselves to it: "If it's that powerful, then for sure I want to use it to help me!"

As noted above, the diagnosis of spina bifida contributed to the social meaning of the fetus as a relatively separable individual and, therefore, as already a potential patient. There are additional implications that elicit more complex sets of questions: What if this fetus was an actual 'child'? Would the couples so eagerly accept the same sort of surgical risk to the child's life with no established likelihood of therapeutic benefit? And the accompanying question, marked as it might be by a couples' unspoken wariness of disability: What is permissible in caring for, or treating, a vulnerable 'child,' irrespective of the value of the particular disabilities or qualities that child's traits hold for his or her parents? Is it permissible to risk the life of the fetus, in order possibly to improve the 'quality of life' of the child-the-fetus-will-become? If so, to what extent? In other words, rather than dilemmas, aren't there instead potent uncertainties about how best to act in the face of fetal affliction?

And, in the face of such complexity and uncertainty, just what are – and what will become – the parameters for respecting the rights of pregnant patients and their partners to make private decisions? Furthermore, attitudes and responses associated with expressed religious beliefs deserve more detailed and careful consideration, especially given the couples overt and momentous focus on finding some way to respond to the potential deficits associated with spina bifida.

Vanderbilt University Medical Center
Nashville, Tennessee
U.S.A.

ACKNOWLEDGEMENTS

Both the insights and writing presented in the essay have benefited enormously from collaboration with my mentor, Dick Zaner. From the outset of the spina bifida protocol in 1997, we have discussed almost daily what we were confronted by and learned from the couples seeking the open-uterine fetal repair. This essay pursues one set of issues and questions generated in our jointly authored papers. Particularly in the latter sections of this essay, I borrow from our earlier work, as well as several drafts, under the continuing title of "Discovering What Matters in Fetal Surgery."

NOTES

[1] Among the arguments about prenatal diagnosis, there is considerable concern regarding a "technological imperative" to use prenatal diagnosis. One issue of cultural criticism is that both women and their doctors are strongly influenced by economic, class, and patriarchal forces (Rothenberg and Thomson, 1994). There are, on the other hand, significant arguments about the meaning of a womans' reproductive choice, exploitation of women, as well as the effects of prenatal diagnosis on social attitudes toward people with disabilities (Wertz and Fletcher, 1993). The crucial point appears to be the claim that even with the primary purpose of allowing a woman to control her reproductive decisions, prenatal testing could be, and has been, interpreted as absolving the general society of responsibility to foster social climates conducive to bearing (or not) and raising children (Asch, 1999).

[2] See, 'Bad News and Bedside Manners', *Insights into Spina Bifida* Vol XI, No. 1 (Jan/Feb, 2000), 1, 8.

[3] In mentioning this perspective, I am not endorsing the social implications that stem from it. All of us need to be better prepared to respond to something that is perhaps unpleasant to look for and detail. Just here I need to urge that we are a good bit more candid than usual in acknowledging those rather unpleasant details, because likewise I do not want to be misunderstood as moralizing about or judging the responses of these couples. But, to state it in the bluntest of terms, the choice made by these couples might be more *against* disability, and less *for* fetal surgery.

[4]See Beischer, et al, 1997, pp. 311-14. Consider also the notion that the fetus is thought, medically, not as an inert passenger in pregnancy but, rather, as in command of it: "The fetus, in collaboration with the placenta, (a) ensures the endocrine success of the pregnancy, (b) induces changes in maternal physiology which make her a suitable host, (c) is responsible for solving the immunological problems raised by its intimate contact with its mother, and (d) determines the duration of pregnancy" (Findlay, 1984, p. 96).
[5]Echoing Duden, Micheals and Morgan (1999, p. 2) say that any woman's experience of pregnancy is influenced by the predominant fetal images circulating throughout western culture such that fetuses "have come to occupy a significant place in the private imaginary of women who are or wish to be pregnant, as well as in the public arena of contestation over women's and children's rights to health care, to food and to shelter." Thus, these women are likely to have found themselves already embedded in the grips of, huge, complex cultural forces whose working is the continuous generating and re-generating of "new" forms of human life, even prior to coitus-for-pregnancy (Murphy-Lawless, 1998).
[6]See, Davis, 1999. Under the caption, "For baby Samuel," is the following description: "the hand of the 21-week-old fetus emerges from the uterus to grasp Bruner's finger reflexively during the surgery to treat Samuel's spina bifida."
[7]See the popular magazine *People* from May 8, 2000. Included in the story, 'Second Chance for Baby Ethan,' credited to Richard Jerome and Giovanna Breu, is a picture by Keri Pickett with this caption: "Turning the fetus around to its former position, Bruner got a clear view of Ethan's developing nose and mouth."

BIBLIOGRAPHY

Asch, A.: 1999, 'Prenatal diagnosis and selective abortion: A challenge to practice and policy', *American Journal of Public Health* **89**, 1649-57.
Beischer, N.A., Mackay, E.V. and Colditz, P.B.: 1997, *Obstetrics and the Newborn: An Illustrated Textbook*, 3rd Edition, Saunders, London.
Bliton, M.J. and Finder, S.G.: 1999, 'Strange, but not stranger: The peculiar visage of philosophy in clinical ethics consultation', *Human Studies* **22**, 69-97.
Bliton, M.J. and Zaner, R.M.: Forthcoming, 'Over the cutting edge: Moral complexity and ethics consultation in open-uterine fetal repair of spina bifida', *Journal of Clinical Ethics*.
Bliton, M.J., Bruner, J.P., Zaner, R.M., et al.: (Forthcoming) 'Guidelines for open fetal surgery for treatment of non-lethal malformations', *Fetal Diagnosis and Therapy*.
Bruner, J.P., Richards, W.O., Tulipan, N.B., et al.: 1999, 'Endoscopic coverage of fetal myelomenigocele in utero,' *American Journal of Obstetrics and Gynecology* **180**, 153-58.
Bruner, J.P., Tulipan, N., Paschall R.L., et al.: 1999, 'Fetal Surgery for myelomeningocele and the incidence of shunt-dependent hydrocephalus,' *The Journal of the American Medical Association* **282**, 1819-25.
Bunch, W. H., Cass, A.S., Benson, A.S. et al.: 1972, *Modern Management of Myelomeningocele*, Warren H. Green, St. Louis.
Cairns, D.: 1973, 'An approach to Husserlian phenomenology', in F. Kersten and R.M. Zaner (eds.), *Phenomenology, Continuation and Criticism: Essays in Memory of Dorian Cairns*, Martinus Nijhoff, The Hague, pp. 224-38.
Campbell, N. (ed.): 2000, 'Bad news and bedside manners', *Insights into Spina Bifida* Vol. XI, No. 1 (Jan /Feb), 1,8.
Casper, M.J.: 1998, *The Making of the Unborn Patient: A Social Anatomy of Fetal Surgery*, Rutgers University Press, New Brunswick, New Jersey.
Casper, M.J.: 1999, 'Operation to the rescue: Feminist encounters with fetal surgery', in L.M. Morgan and M. Michaels (ed.,) *Fetal Subjects, Feminist Positions*, University of Pennsylvania Press, Philadelphia, pp. 101-12.

Davis, R.: Sept. 7, 1999, 'Spinal surgery in womb tests faith, technology', *USA TODAY*, p. 8D.
Duden, B.: 1993, *Disembodying Women: Perspectives on Pregnancy and the Unborn*, Harvard University Press, Cambridge, Mass.
Filly, R.A.: 2000, 'Obstetrical sonography: The best way to terrify a pregnant woman', *Journal of Ultrasound in Medicine* **19**, 1-5.
Findlay, A.L.R.: 1984, *Reproduction and the Fetus*, Edward Arnold, London.
Fleischman, A.R., Chervenak, F.A. and McCullough, L.B.: 1998, 'The physician's moral obligation to the pregnant woman, the fetus and the child', *Seminars in Perinatalogy* **22**, 184-88.
Fletcher, J.C. and Jonsen, A.R.: 1991, 'Ethical considerations in fetal treatment', in M.R. Harrison, M.S. Golbus and R.A. Filly (eds), *The Unborn Patient*, 2nd Edition, pp. 12-18.
Franklin, S.: 1991, 'Fetal fascinations: new dimensions to the medical-scientific construction of fetal personhood', in S. Franklin, C. Lury, and J. Stacey (eds.), *Off-Centre: Feminism and Cultural Studies*, Harper Collins, London, pp.190-205.
Gregg, R.: 1995, *Pregnancy in a High-Tech Age: Paradoxes of Choice*, New York University Press, New York and London.
Harrison, M.R., Filly, R.A., and Golbus, M.S., et al: 1982 'Fetal treatment 1982', *New England Journal of Medicine*, **307**, 1651-2.
Husserl, E.: 1982, *Ideas Pertaining to a Pure Phenomenology and to a Phenomenological Philosophy*, F. Kersten (trans.), Martinus Nihoff Publishers, The Hague/Boston/London.
Husserl, E.: 1969, *Formal and Transcendental Logic*, D. Cairns (trans.) Martinus Nijhoff: The Hague.
Husserl, E.: 1960, *Cartesian Meditations: An Introduction to Phenomenology*, D. Cairns (trans.), Martinus Nijhoff, The Hague.
Jennings, R.W., Adzick, N.S. and Harrison, M.R.: 1992, 'Fetal surgery', in M.I. Evans (ed.), *Reproductive Risks and Prenatal Diagnosis*, Appelton & Lange, Norwalk , Connecticut, pp. 311-20.
Jerome, R. and Breu, G.: May 8, 2000, 'Second chance for baby Ethan', *People* **52**, 60-65.
Jonas H.: 1974, *Philosophical Essays: From Ancient Creed to Technological Man*, Chicago University Press, Chicago.
Madsen, J.R., Estroff, J., and Levine, D.: 1998 "Prenatal neurosurgical diagnosis and counseling," *Neurosurgery Clinics of North America* **9**, 49-61.
Manchester, D., Clewell, W., Manco-Johnson, M., et al.: 1988, 'Management of pregnancies with suspected fetal anomalies: Clinical experience and ethical issues', in C. Nimrod and G. Griener (eds.), *Biomedical Ethics and Fetal Therapy*, Wilfrid Laurier University Press, Waterloo, California.
Michaels, M.W. and Morgan, L.M.: 1999, 'Introduction: The fetal imperative', in L.M. Morgan and M.W. Michaels (eds.), *Fetal Subjects, Feminist Positions*, University of Pennsylvania Press, Philadelphia, pp. 1-9.
Morgan, L.M.: 1996, 'Fetal relationality in feminist philosophy: An anthropological critique', *Hypatia* **11**, 47-70.
Murphy-Lawless, J.: 1998, *Reading Birth and Death: A History of Obstetric Thinking*, Cork University Press, Ireland.
Petchesky, R.P.: 1987, 'Foetal images: The power of visual culture in the politics of reproduction', in M. Stanworth (ed.), *Reproductive Technologies: Gender, Motherhood and Medicine*, University of Minnesota Press, Minneapolis, pp. 57-80.
Rapp, R.: 1997, 'Real-time fetus: The role of the sonogram in the age of monitored reproduction', in *Cyborgs and Citadels: Anthropological Interventions in Emerging Sciences and Technologies*, G.L. Downey, and J. Dumit, (eds.), School of American Research Press, Santa Fe, pp. 31-48.
Rothenberg, K.H. and Thomson, E.J. (eds.): 1994, *Women and Prenatal Testing: Facing the Challenges of Genetic Technology*, Ohio State University Press, Columbus.

Sandelowski, M.: 1994, 'Separate, but less equal: Fetal ultrasonography and the transformation of expectant mother/fatherhood', *Gender & Society* **8**, 230-45.
Scarry, E.: 1985, *The Body in Pain: The Making and Unmaking of the World*, Oxford University Press, New York.
Schutz, A.: 1967a, 'Choosing among projects of action', in M. Natanson (ed.), *The Problem of Social Reality*, Vol. 1, *Alfred Schutz: Collected Papers*, Martinus Nijhoff, The Hague, pp. 67-96.
Schutz, A.: 1967b, 'Scheler's theory of intersubjectivity and the general thesis of the alter ego', in M. Natanson (ed.), *The Problem of Social Reality*, Vol. 1, *Alfred Schutz: Collected Papers*, Martinus Nijhoff, The Hague, pp. 150-79.
Schutz, A.: 1967c, 'Sartre's theory of the alter ego', in M. Natanson (ed.), *The Problem of Social Reality*, Vol. 1, *Alfred Schutz: Collected Papers*, Martinus Nijhoff, The Hague, pp.180-203.
Schutz, A. and Luckman, T.:1973, *The Structures of the Life-World*, Vol. 2, R.M. Zaner and D.J. Parent (trans.), Northwestern University Press, Evanston, Illinois.
Shaaban, A.F., Kim, H.B., Milner, R., et al.: 1999, "The role of ultrasound in fetal surgery and invasive fetal procedures," *Seminars in Roentgenology*, **34**, 62-79.
Sheets-Johnstone, M.: 1994, *The Roots of Power: Animate Form and Gendered Bodies*, Open Court, Chicago.
Sherwin, S.: 1992, *No Longer Patient: Feminist Ethics and Health Care*, Temple University Press, Philadelphia.
Simpson, J.L.: 1999, 'Fetal surgery for myelomeningocele: Promise, progress, and problems, *Journal of the American Medical Association*, **282**, 1873-74.
Spiegelberg, H.: 1980, 'Putting ourselves into the place of others: Towards a phenomenology of imaginary self transposal', *Human Studies*, 169-73.
Sutton, L.N., Adzick, N.S., Bilaniuk, L.T. et al.:1999 'Improvement in hindbrain herniation demonstrated by serial fetal magnetic resonance imaging followiing fetal surgery for myelomeningocele', *Journal of the American Medical Association*, **282**, 1826-31.
Taylor, J.S.: 1998, 'Image of contradiction: Obstetrical ultrasound in American culture', in S. Franklin and H. Ragone (eds.), *Reproducing Reproduction: Kinship, Power, and Technological Innovation*, University of Pennsylvania Press, Philadelphia, pp. 15-45.
Wertz, D.C. and Fletcher, J.C.: 1993, 'A critique of some feminist challenges to prenatal diagnosis,' *Journal of Women's Health* **2**, 173-88.
Zaner, R.M.: 2000, 'Discovering what matters in fetal surgery' presented at the *Fetal Surgery and the Moral Presence of the Fetus Conference* at Vanderbilt University on March 11, 2000.
Zaner, R.M.: 1996, 'Listening or telling? Thoughts on responsibility in clinical ethics consultation', *Theoretical Medicine* **17**, 255-77.
Zaner, R.M.: 1993, 'Voices and time: The venture of clinical ethics', *The Journal of Medicine and Philosophy* **18**, 9-31.
Zaner, R.M.: 1988, *Ethics and the Clinical Encounter*, Prentice Hall, Englewood Cliffs, New Jersey.
Zaner, R.M.: 1970, *The Way of Phenomenology: Criticism as a Philosophical Discipline*, Pegasus, New York.

CATRIONA MACKENZIE

ON BODILY AUTONOMY

I. INTRODUCTION

There is a scene in Monty Python's film *Life of Brian* in which, in the midst of a discussion amongst a group of revolutionaries enjoying the spectacles at the colosseum, a man from the Judean Liberation Front expresses a strong desire to have a baby. He is told gently that this is an unrealistic desire, but his failure to respond to reason on this issue finally exasperates those around him until one of them (played by John Cleese) shouts at him "But you can't have a baby. Men can't have babies!" The man from Judea shrieks back "Don't you oppress me!" It is obvious that the Judean has misunderstood the meaning of oppression and this is what makes his accusation, and hence this scene in the movie, so funny. But his confusion would not be amusing in the same way if he were simply wrong about oppression. It is only amusing because he has partially grasped the meaning of the concept. What then is the nature of his confusion? The Judean is correct in realizing that, at least in part, oppression involves a denial of a person's autonomy. He is affronted because he feels that his autonomy with respect to his body, or what I am referring to as bodily autonomy, has somehow been infringed or denied. He is also affronted because he understands that oppression is a function of social relationships of power — hence the accusation. However, what the Judean has failed to grasp is that in the exercise of autonomy there must be some correlation or match between our goals and desires, on the one hand, and, on the other hand, our natural or acquired skills, talents and capacities. Although it may be a source of frustration to me that I am not a virtuoso musician, if I lack the necessary talents my inability to satisfy this desire clearly involves no infringement of my autonomy. My autonomy can be infringed only if I have been prevented in some way from realizing my desire to exercise or develop my musical talents, for example because of lack of opportunity due to parental disapproval, or poverty or discrimination. The Judean's thinking scrambles the relationships between desires, capacities, autonomy and oppression, leading him to conclude that if he desires to bear children but lacks the capacity, then his autonomy has somehow been infringed and this must be the result of someone else's attitude or actions.

It is important to the humor of this episode that it involves a desire that is impossible to satisfy, remembering that the action is supposed to have taken place two thousand years ago. From our perspective, however, the possibility of male pregnancy is not nearly so remote, given recent and very rapid developments in reproductive technology. In the event that male pregnancy does become technologically feasible, there would no longer be any mismatch between the Judean's desire and his technologically enhanced bodily capacities. If this should happen, and if the conversation I have just reported were to take place in such a context, would the Judean's outrage then be justified if he, and all men, were denied the choice of pregnancy? Would this denial involve an infringement of men's bodily autonomy? Or, to raise a question that is even more immediate in the present context, how much moral weight should be given to the desire of a post-menopausal woman to carry a pregnancy to term, now that this is a technological possibility? Does such a woman have a right to access those techniques, a right that is grounded in her right to bodily autonomy?

It is not the aim of this paper to provide answers to these first-order moral questions.[1] Such questions are pertinent, however, to the central concern of this paper, which is to critically investigate the theoretical relationships among choice, bodily capacities, and autonomy. In particular, I want to question a cluster of related conceptions of bodily autonomy that prevail among libertarian and, to some extent, liberal views in the bioethical literature. I call these views maximal choice conceptions of bodily autonomy. They include the view that an agent's bodily autonomy is always enhanced by maximizing the range of bodily options available to her; and the view that any expansion of a person's bodily capacities, or of her instrumental control over her bodily processes, *ipso facto* enhances her bodily autonomy. In the second section of the paper I briefly outline and criticize maximal choice conceptions. In the third section I suggest an alternative view, that articulates the concept of bodily autonomy both in terms of the phenomenological notion of an integrated bodily perspective, and in terms of a normative theory of autonomy.

II. MAXIMAL CHOICE CONCEPTIONS OF BODILY AUTONOMY

Maximal choice conceptions of bodily autonomy derive from a number of overlapping justificatory sources within the liberal tradition. One is the minimal, libertarian conception of autonomy as what Isaiah Berlin (1969) calls negative liberty. The view of liberty embodied in maximal choice conceptions is the view that, as long as a person's choices meet two minimal conditions, she should be free from the undue (paternalistic) interference of others or the state to do with her body as she chooses. These conditions are firstly, that the choice does not cause harm to others[2]; secondly, that the choice is non-coerced and reasonably voluntary, that is, at least to a reasonable degree informed and the result of rational deliberation.[3] The view of choice embodied in these conceptions is the view that, a person is entitled to potentially unlimited scope in the range of bodily options available to her and that the more such options she has open to her the better. This is because expansions of choice are *ipso facto* thought to be expansions of autonomy.

Another justification for maximal choice conceptions is the familiar liberal view that autonomy is best promoted through maximizing choice, since there is no universal consensus about the good, and since there are no uncontentious criteria for making qualitative or normative distinctions among options. A liberal society, therefore, ought to make available to its citizens as many options as possible, to best enable them to live according to their own conception of the good life.[4]

Within the liberal tradition, the right to negative liberty and the absence of a universally embraced conception of the good are generally taken to ground the right to autonomy and to maximal choice. But what grounds the right to *bodily* autonomy in particular? This question has not received much consideration in bioethics, due, in my view, to a general neglect of the notion of embodiment. However, within an influential strand of the liberal tradition originating with Locke, the concept of bodily ownership has been taken to explain and justify a person's specific interest in her body and her rights with respect to it.[5] Maximal choice conceptions sit within this strand of the liberal tradition, assuming the notion of bodily ownership as the ground for bodily autonomy.

A final justification for maximal choice conceptions is the equation of bodily autonomy with control over the body and bodily processes. This underpins the assumption that expanding the available range of bodily options enhances a person's autonomy by extending her instrumental control over her body and body processes, thereby making her less subject to unwanted or harmful contingencies.[6]

Under maximal choice conceptions then, unless it can be shown that making some choices available to some people violates the harm to others condition, *prima facie* there can be no good reason, at least on the grounds of autonomy, for not continually expanding the range of available bodily options.

My view is that maximal choice conceptions are misguided for three connected reasons. First, because it is a mistake to construe bodily autonomy simply as control over the body and body processes. In contrast, I shall argue that bodily integration, rather than control over the body and body processes, is constitutive of bodily autonomy. I shall also argue that the notion of bodily ownership misconstrues the sense in which a person's body belongs to her and therefore provides an inadequate grounding for the notion of bodily autonomy. Second, because it is a mistake to assume that expansions of choice automatically increase a person's autonomy. While some expansions of choice clearly do enhance autonomy, it cannot simply be assumed that every expansion of choice must do so.[7] Third, because maximal choice conceptions of choice provide no normative criteria for assessing, among those options available to an agent that meet the harm to others condition, which choices are autonomy-enhancing and which are autonomy-impairing. Under these conceptions, any choice that is not coerced and voluntary counts as autonomous. Thus, a schoolgirl who desires breast implants for her sixteenth birthday because she knows that breast size is important to success in show business, would count as making an autonomous choice.[8] I shall argue that more stringent normative criteria for autonomy are necessary because autonomous choice involves more than non-coerced and voluntary choice. Autonomous agency involves critical reflection both on the worth or value of one's choices, and the desires guiding them, and on the historical and social processes of formation of one's choices and desires.

III. EMBODYING AUTONOMY: BODILY INTEGRATION AND CRITICAL REFLECTION

When we say that there is a moral requirement to respect the bodily autonomy of others we usually mean two things. First, we mean that we may not physically interfere with another person's body or intrude upon her personal space unless, in some direct or indirect way, she has consented for us to do so. Let us call this the right to non-interference. Second, and connected with this, we usually mean that it is up to that person alone to decide what happens in and to her body, and that no-one else may make such decisions for her. Let us call this the right to bodily self-determination. What underlies the requirement to respect another's bodily autonomy is the understanding that respect for a person's body is an integral part of respect for the person. The intimate connection between our bodies and our selfhood is usually immanent in our self-conceptions and in our interaction with others and the world. However, events that disrupt our everyday taken-for-granted bodily existence often make this connection very salient to us. Such events include illness, injury and trauma. They also include bodily changes, gradual or radical, and voluntary or involuntary. When these are sufficiently severe or disruptive, they can induce a sense of loss of self that can have profound effects on an agent's emotional responses, intimate relationships, and sense of the future.[9]

That there is an intimate connection between our bodies and our selfhood is of course well known to oppressors. Humiliating, degrading and torturing a person's body or controlling what she may do with it, are much-used means for crushing resistance, breaking a person's will or simply controlling her life. Given these and other less extreme violations of a person's bodily autonomy, it is clear why the rights to non-interference and bodily self-determination are regarded as central to the concept of bodily autonomy. I agree that they should be central. The issue that concerns me here is what grounds these rights. Libertarians and many liberals, I have suggested, tend to concur in grounding them in the notion of bodily ownership. In what follows I shall argue that the notion of bodily ownership provides an inappropriate model for understanding the sense in which a person's body belongs to her, and an inadequate basis for grounding the rights to non-interference and bodily self-determination.

My argument begins with a discussion of Paul Ricoeur's (1992) critique of Derek Parfit's (1984) approach to personal identity. The relevance of Ricoeur's critique of Parfit is that it helps to clarify the difference between two senses of belonging: belonging in the sense of what one owns or possesses or has, that is, ownership; and belonging in the sense of who one is, or identity. Following Ricoeur, I shall argue that a person's body belongs to her in this second sense of being constitutive of her identity, rather than in the sense of ownership. Ricoeur's critique of Parfit is also relevant because it draws attention to one of the central tenets of a phenomenological approach to embodiment, namely that human corporeality is the invariant condition of our selfhood. On the basis of Ricoeur's phenomenological approach to bodily belonging, I then develop an account of bodily autonomy as constituted by an integrated bodily perspective.

A. Persons, Bodies and Selves

In *Reasons and Persons* Parfit defends a radical variant of the psychological continuity criterion for personal identity, which he calls reductionism. Parfit's reductionism involves a number of claims the most famous of which is that personal identity does not matter.[10] Personal identity does not matter, Parfit argues, because when we are concerned about our future survival, what concerns us is not that a future person who is numerically identical with us will exist in the future. Rather, it is that some person who is sufficiently psychologically connected to us will exist in the future. Parfit takes psychological connectedness and/or continuity to be necessary and sufficient for survival. More radically, he argues that it is necessary and sufficient no matter what causal mechanisms have brought it about. So even though psychological continuity is usually brought about by the continued existence of a person's brain and body, bodily continuity is not necessary for survival. In other words, as long as I am psychologically continuous with some future person, it should not matter to me whether or not this person has my particular body, nor should it matter if my body no longer exists and is replaced by a replica or a sufficiently similar body.

A corollary of the reductionist thesis is the *impersonality* thesis. This is the thesis that we could provide a complete description of our experiences (mental and physical), and the connections between them, in impersonal terms, that is, without needing to attribute these experiences to their subject. Personal descriptions, which necessarily refer experiences to the first-person perspective of the subject to whom they belong, are simply different modes of presentation of the same phenomenon that do not *add* anything to an impersonal description of the same series of events.

Parfit's target is Cartesianism, or what he calls the 'further fact view', the view that consciousness, the 'I', or the self, is a spiritual substance, separately existing from the brain or body. In his view, since the further fact view is metaphysically untenable, the only alternative is reductionism. Ricoeur (1992) argues that Parfit posits reductionism as the only alternative to the 'further fact view' because his approach to personal identity conflates two distinct kinds of identity: *idem* identity, or sameness, and *ipse* identity, or selfhood. Sameness, or 'what' I am, encompasses both numerical and qualitative identity. Selfhood, or 'who' I am, encompasses the particularity, or sense of 'mineness', of my own body, memories, thoughts and experiences.[11] Parfit's reductionism, Ricoeur argues, runs together these distinct dimensions of identity and then proceeds to erase selfhood altogether, as resting on untenable metaphysics, leaving nothing to the person but a series of connected experiences.[12]

Ricoeur's central thesis is that personal identity consists of a dialectical relation between sameness and selfhood, where selfhood is understood in phenomenological, not metaphysical, terms.[13] Personal identity encompasses what we are – this particular biological organism, with this history and these memories, this body, these thoughts, these particular character traits, all of which change over time, but yet are anchored in sameness by virtue of the temporal connections between past, present and future. It also encompasses who we are – the phenomenological subject to whom what we are belongs. Ricoeur argues further that any criterion of personal identity, that is, any criterion of sameness, including Parfit's, must presuppose selfhood. Parfit's reduction of identity to a series of interconnected mental and physical events and experiences ignores the fact that these experiences, this body and brain, belong to *someone*, for whom they are *her* experiences, *her* body, *her* brain. The relevant sense of 'belong' here is the sense that they are who that person is. From my point of view, or my first-personal perspective, my body is not just one body among others, describable in purely impersonal terms and replaceable by a replica. It is who I am, the condition of possibility of my experience.

Ricoeur points out that Parfit's science-fiction thought-experiments are the means by which he effects the erasure of the phenomenological dimensions of selfhood and embodiment. In Parfit's famous teletransportation thought-experiments, for example, we are asked to imagine a new mode of space travel in which we step into a teletransportation machine that destroys our brain and body here on Earth, but produces a qualitatively identical replica of ourselves, from entirely new matter, on Mars. Parfit argues that even though our replica on Mars would not be numerically identical with us, what really matters to us, namely psychological continuity, would survive teletransportation and therefore that fear of teletransportation is irrational. According to Ricoeur, these thought-experiments erase the phenomenological dimensions of selfhood and embodiment by inviting us to identify the person with the brain, that part of the body of which we have no phenomenological experience, and even just with the contents of the brain.

Ricoeur argues that Parfit's examples thus work by radically dissociating those aspects of ourselves that ordinarily cannot be dissociated or disconnected, such as the complex interconnections between psychological and corporeal continuity, or the sense of one's memories and one's body as belonging to oneself.[14] In other words, the technological fictions deliberately function to put into question the givens of human embodiment, rendering these givens contingent. In this respect they function very differently from literary fictions:

> Literary fictions differ fundamentally from technological fictions in that they remain imaginative variations on an invariant, our corporeal condition experienced as the existential mediation between the self and the world. Characters in plays and novels are humans like us who think, speak, act, and suffer as we do. Insofar as the body as one's own is a dimension of oneself, the imaginative variations *around* the corporeal condition are variations on the self and its selfhood (Ricoeur, 1992, p. 150).

In Parfit's technological fictions, selfhood is erased because the very condition of selfhood, namely our corporeal condition, 'our rootedness to this earth', which anchors us in human life, is suspended. Ricoeur suggests that although in acts of imagination we may think we can extricate ourselves from this condition, while leaving what is essential to ourselves intact, this is an illusion. For it is only by presuming our continuing connectedness to our corporeal condition that we are able to make sense of these fictions. In order to imagine that what matters to us still remains after teletransportation, we have to imagine ourselves as the same acting, suffering, living beings. In other words, we have to presume as necessary what Parfit wants to render contingent.[15]

I want to suggest that Ricoeur's distinction between the two senses of belonging, namely selfhood and ownership, helps explain why the notion of bodily ownership misunderstands the sense in which a person's body belongs to her. A person's body belongs to her because it is constitutive of her selfhood, not because it is something she possesses or owns. The notion of bodily ownership abstracts the person from her body and treats the body as a separable item, like property, over which she exercises certain proprietorial rights. Once the person has been abstracted from her body in this way it then seems to make sense to think of bodily autonomy in terms of instrumental control over the body. However, if the relationship between a person and her body is understood as a relationship of constitution, then the notion of instrumental control over the body provides an inappropriate basis for conceptualising bodily autonomy, including the rights to non-interference and bodily self-determination. Below I develop an alternative conception of bodily autonomy that incorporates Ricoeur's understanding of the body as constitutive of selfhood.

I also want to suggest that Ricoeur's critique of Parfit draws attention to an insight that has not received much attention in bioethics, that embodiment is the condition of our selfhood. To draw attention to the givens of embodiment is not to assume that our corporeal condition is static. But it does alert us to the fact that changes to our corporeal condition have the potential to change who and what we are, and that therefore such changes raise normative questions. Technological developments, especially in the area of bio-medicine, provide us with unprecedented capacity for such change. In developing an ethical response to these developments, we cannot simply assume that they are beneficial to humanity, nor, on the other hand, can we assume that they are detrimental. The normative question must always be asked whether there are good reasons for accepting the consequent changes to human embodiment – and hence, to who and what we are – that may ensue.

B. Bodily Perspective

The notion of a bodily perspective aims to articulate the complex relationships among the biological, social and individual dimensions of embodiment in order to explain the sense in which my body belongs to me and is constitutive of who I am. In what follows, I first develop the notion of a bodily perspective. I then argue that an integrated bodily perspective is necessary, but not sufficient, for bodily autonomy, which involves both bodily integration and normative critical reflection.[16]

Our bodily perspective incorporates both certain facts about our embodiment and the social and individual significance of our bodies for us. These facts encompass the givens of human embodiment, specifically mortality, as well as the particular capacities, characteristics, limitations and susceptibilities of the human body. They also encompass the givens of our individual embodiment – our sex; specific talents and abilities as well as disabilities, conditions or impairments (for example, of vision, hearing, fertility); and particular bodily characteristics and features (metabolic rate, body shape and build, skin and hair coloring, facial features, and so on). Another salient feature of our bodily life is that our bodies are continually undergoing involuntary changes, some slow and almost unnoticeable, such as gradual aging, others fairly rapid or radical, such as puberty, pregnancy, or menopause. Some bodily changes are within the scope of our control. For example, some bodily characteristics may be deliberately shaped and altered by relatively non-invasive methods (for example, diet, physical exercise and training) or by fairly invasive methods (for example, steroid use, surgical sex change). Other changes are due to factors outside our control, such as illness, accident or trauma. Some of these changes, whether deliberate or non-voluntary, may have a minimal effect on our capacities, at least for a certain period. Others may radically alter what we are capable of doing, both at a bodily level and in other aspects of our lives.

But perhaps more significant than the fact of our embodiment is our conscious awareness of embodiment. Not only are we aware of at least some of our bodily experiences, but our bodily awareness is always mediated by our attitudes and emotional and physical responses towards our bodies, and by our psychic representations of them. Our attitudes towards our bodies, or bodily changes, can range from narcissistic to self-loathing, and they can include a gamut of feelings from dissatisfaction, frustration and alienation, even disgust, to feelings of identification, pleasure and pride. Our psychic representations are our imaginary representations of our bodies as a whole, as well as of certain bodily capacities or parts. These representations are susceptible not only to our own feelings and attitudes about our bodies, but also to our perceptions and representations of the ways in which others respond, physically and emotionally, to our bodies. It is these attitudes, feelings and representations, which need not be immediately available to consciousness, that form the central core of what I call a person's 'bodily perspective', which can best be characterised as the meaning or significance of one's body for oneself. This meaning, and hence a person's bodily perspective, is not static but changes during the course of a life, in response to changes to one's body, one's situation and one's relations with others.

The relationships among one's bodily perspective, one's given attributes, capacities and characteristics, and the bodily changes one experiences are somewhat complex.[17] On the one hand, a person's bodily perspective involves a representation of her bodily attributes and capacities, a representation that is susceptible to change in response to the biological processes her body undergoes. For example, it would be expected that a person's bodily perspective would undergo a significant shift during and after pregnancy, or after an illness or accident that has brought about a permanent change to her capacities. On the other hand, the significance of these changes for the person cannot be divorced from the social and cultural meanings attributed to particular processes or capacities. Nor can they be divorced from the psychic significance such processes or capacities might have for that particular individual. For example, the shift in a girl's bodily perspective that arises in response to the onset of menstruation probably has as much to do with the social and cultural significance of menstruation as it has to do with biological processes. The same may be true of the kinds of shifts in perspective that occur during the aging process. The loss of a finger, however, is likely to affect the bodily perspective of a musician in a much more profound way than it would for other people. Similarly, our bodily self-representations are susceptible to the cultural and social meanings associated with particular kinds of bodies. For example, the salience of factors such as race, ethnicity, sexuality and class in particular cultural contexts can mark out certain kinds of bodies – the bodies of Aboriginals, or gays, or Asians – as distinct and attracting powerful affective responses in others which are then internalized in self-representations.[18]

A person's bodily perspective thus marks the point of intersection of biological capacities, attributes and processes, social and cultural representations of these, and an individual's particular history, projects, desires, and relations with others. However, the metaphor of a point of intersection should not mislead us into thinking of ourselves as passive with respect to our bodily perspective. One's bodily perspective is not completely perspicuous to reflection, and it is partially constituted in relation to certain givens, or aspects of one's body and situation over which one may have little control. But the constitution of one's bodily perspective is an active process, and is bound up with the process of constituting oneself as a person. The interplay between activity and passivity, or between agency and determination, in the constitution of the embodied self is readily apparent in the aftermath of a traumatic bodily event or change. On the one hand, traumatic bodily events or changes are so devastating because, in constituting our bodily perspective in response to our bodily situation, we constitute ourselves. Any change to that situation has the potential to unravel our sense of self. On the other hand, it is precisely because persons are embodied agents that, in many cases, people are able to reconstitute their bodily perspective, and their sense of self, in the aftermath of bodily trauma.[19] Thus one's bodily perspective is the expression of one's embodied agency and it is actively and continuously constituted in the exercise of one's agency. Further, our bodily perspective is enmeshed with our self-conception and structures our bodily experience, our relations with others, and our interactions with the world. For example, an unusually tall person has little or no control over his height. However, his bodily perspective, whether he sees himself as an awkward, gangly giant, or a towering presence, will have a significant impact on his bodily experience, his conception of himself, and his bodily interactions with others. In sum, one's bodily perspective is partially constitutive of one's point of view.

The idea that my bodily perspective is the expression of my embodied agency clarifies the sense in which my body is constitutive of who I am. To be embodied is to exercise one's agency, that is, to live one's life and to engage with the world, through one's own particular body. The body, thus understood, is not simply the body qua biological organism; it is the body as lived or experienced and as it is represented and signified by oneself and others. In my view, we can only make sense of the importance attributed to the rights to non-interference and bodily self-determination if these rights are grounded in the idea that the body as lived is the expression of our agency. It is because my body is the medium through which I live my life that I must have a right to determine what happens in and to that body. Similarly, bodily violation is a violation of the self not because it involves violation of something one owns or has, but because it involves a violation of one's embodied agency, or of who one is.

So far I have developed the notion of a bodily perspective in order to explicate the sense in which my body belongs to me because it is constitutive of who I am. I have also suggested that this notion makes better sense of the rights to non-interference and bodily self-determination than the ownership model of bodily belonging. However, the concept of bodily autonomy encompasses more than these rights, for it also refers to the capacity that underpins them. Bodily autonomy involves both bodily integration, and normative critical reflection, or the capacity to develop what I call a reflective, integrated bodily perspective.

C. Bodily Integration and Bodily Autonomy

Contemporary theories of autonomy hold that autonomy involves the capacity for critical reflection on one's motivational structure, and the capacity to change it in response to reflection. This view is underpinned by the intuition that there is a difference between those aspects of an agent's motivational structure that she unreflectively finds herself with, and those aspects that, upon reflection, she in some sense identifies with or regards as her own. Disagreements among different accounts of autonomy arise in explicating what is involved in the process of reflection, in explicating how reflection secures autonomy, and in making sense of the notion of 'one's own'.[20] Nevertheless, a number of theorists concur in regarding volitional, or intrapsychic, integration of the different motivational elements of the self, as necessary, if not sufficient, for autonomy. Some theorists, such as Harry Frankfurt, Gary Watson and Charles Taylor, conceptualize integration in terms of a hierarchical approach to the self, distinguishing between an agent's lower-level motivations and her higher-order identifications.[21] Critical reflection is a matter of assessing one's motivations in the light of one's higher-order identifications or values, and autonomy is secured when this process of reflection results in volitional integration. Other theorists, such as Marilyn Friedman (1987) and Diana Meyers (1989, 2000), who reject a hierarchical approach, nevertheless agree that autonomy involves critical reflection upon, and integration among, the different elements of the self.

The notion of bodily integration I develop here builds on these approaches to autonomy. If autonomy involves the capacity to reflect upon and integrate the different motivational elements of the self, bodily autonomy involves the capacity to achieve an integrated bodily perspective. An integrated bodily perspective is achieved when an agent is able to identify with her bodily perspective, that is, when she regards her bodily perspective as expressive of her embodied agency. This kind of integration involves an integration of the biological, socio-cultural and individual dimensions of embodiment in one's bodily self-representation, such that there are no persistent and unresolved conflicts among these dimensions. Such conflicts, I suggest, signal that the agent is in some sense alienated from herself or from aspects of her embodiment. For example, an anorexic is unable to integrate the different dimensions of her embodiment. Not only is there a radical mismatch between the biological state of her body – bordering on starvation – and her imaginary representation of her body as grossly overweight, but she also feels alienated from her lived body, which does not seem to her to express her agency. This kind of alienation may also arise in cases of serious illness, when a person may feel that her degenerating body is an enemy or at least alien to her, that she has lost a sense of the unity of self and body. Or, it may arise in response to bodily violation, as in the case of sexual assault. In such cases, alienation may be a necessary mechanism of self-protection, enabling the ill person to achieve some measure of distance from the disease and its effects on her life, or enabling the survivor of sexual assault to dissociate herself from the assault as a step towards recovery.[22] Nevertheless, agents in these situations do not have an integrated bodily perspective because there is some degree of disintegration among the different elements of their embodiment, preventing them from regarding their bodies as expressive of their agency.[23] By contrast, the agent whose bodily perspective is integrated experiences no such conflicts. Rather, her experience of the biological dimensions of her embodiment, or of its socio-cultural significance, accords with her sense of herself as an agent. Of course identification and alienation, integration and disintegration, is a matter of degree. Most of us, at some time, experience bodily alienation or a sense that our lived embodiment is not expressive of our agency. Just what degree of alienation or disintegration is consistent with autonomy is an interesting question, but one that I will not attempt to answer here. My view does entail, though, that if a person persistently experiences such alienation or disintegration then her bodily autonomy is seriously compromised.

However, although bodily integration is necessary for autonomy, it is not sufficient. A person may identify with her bodily perspective, and experience integration among the biological, socio-cultural and individual dimensions of her embodiment, but fail to be autonomous because her identifications are simply the product of her socialization. Within the contemporary literature, integration theories of autonomy have been criticized on precisely these grounds, that is, for failing to explain how integration secures autonomy. If critical reflection is understood simply as a process of identifying with one's motivational structure, or one's bodily perspective, what guarantees that an agent's identifications are autonomous? Furthermore, why assume that an agent's identifications and values genuinely express her agency? Thoroughly socialized agents may reflectively endorse their socially acquired identifications, but this reflective endorsement provides no guarantee that they are autonomous. In fact, in oppressive social contexts, such agents may feel more integrated than agents who resist or question their socialization. In such contexts, alienation may be a signal that it is the prevailing social conditions that are impairing agents' capacities for autonomy, rather than any incapacity on the part of agents themselves. For example, in a social context in which women's worth as persons is bound up with breast size, an amply breasted woman whose self-esteem rises whenever a man's gaze fixes on her breasts may feel a greater degree of bodily integration than her flat-chested cousin. But why should this be taken as any indication that she is more autonomous with respect to her embodiment?

Some theorists of autonomy have argued that socialization presents a problem for standard procedural theories of autonomy, such as integration theories, because such theories fail to attend to the historical processes of belief-, desire- and value- formation, focusing solely on the agent's occurrent motivational structure (Dworkin, 1988; Christman 1987, 1991). Procedural theories hold that whether or not an agent's desires and values, and the actions that flow from them, are autonomous is determined solely by whether or not the agent has subjected them to the appropriate process of critical scrutiny. The content of those desires and values, and the nature of the actions the agent performs, in other words, are irrelevant as far as autonomy is concerned. John Christman's (1991) counterfactual theory of autonomy is a historical variant of a procedural theory that attempts to answer the problem of socialization, and to secure the autonomy of an agent's higher-order identifications and values, by building certain historical constraints into the process of critical reflection. According to Christman, an agent is autonomous with respect to her values and higher-order identifications if, upon becoming aware of the historical processes by means of which they were acquired, the agent does not resist these processes, or *would* not resist them, were she to become aware of them.

Christman is correct to draw attention to the importance for autonomy of the historical processes of belief-, desire- and value- formation. A virtue of his counterfactual theory is that it provides a compatibilist explanation of the relationship between autonomy and these historical processes. As long as we do not, or would not, resist the way we came by our desires, values, and higher-order identifications, then they are autonomous, even if we have acquired them as a result of socialization. Nevertheless, as Paul Benson (1991) has argued, the counterfactual theory is insufficient to fully answer the problem of socialization. A thoroughly socialized agent, such as the amply breasted woman who identifies with the prevailing social norms regarding the importance of breast size, would be highly unlikely, upon reflection, to resist the historical processes that led to the formation of her higher-order identifications, were she to become aware of them. She would therefore seem to satisfy the conditions for autonomy under the counterfactual theory.

Benson argues that procedural theories, even those that build historical constraints into the process of critical reflection, have trouble handling the problem of socialization because the constraints they set on agents' competence at critical reflection are insufficiently normative. As a result, procedural theories are unable to explain what distinguishes those forms of socialization that promote the capacity for critical reflection from those oppressive forms of socialization that impair the proper development of the capacity. On Benson's analysis, what is specific about oppressive socialization, as opposed to other forms of socialization, is that it works in part by systematically impairing agents' normative competence. Normative competence involves the capacity to recognise relevant reasons for action and to critically assess one's own reasons with reference to the relevant reasons. It also involves having the ability to incorporate this critical assessment into one's deliberations, and being able to act on the basis of these deliberations. Agents who have internalized oppressive social norms are often resistant to critical reflection, because these norms function to restrict the range of reasons such agents are able to recognise. In so doing, they impair agents' capacities to critically assess these internalized norms by reference to a broader set of reasons and norms. Now except in extreme cases, oppressive socialization does not usually result in global impairment of normative competence. Rather it functions to fragment, or compartmentalize, agents' critical or normative competence, so that agents who are perfectly capable of recognising relevant reasons for action in some domains of their lives are unable to recognise them in other domains. Benson gives as an example the norms governing feminine appearance. When internalized, such norms lead women to equate their self-worth with their physical attractiveness to men and at the same time systematically impair their capacities to criticize these norms.[24]

I concur with Benson that autonomy requires normative critical reflection. Bodily autonomy, I claim, involves achieving an integrated bodily perspective in the light of such reflection. It involves not merely identifying with one's bodily perspective, or regarding one's embodiment as the expression of one's agency, but doing so on the basis of a normative assessment of one's perspective. The anorexic whose body image is out of kilter with reality, the amply breasted woman who identifies with the social norms regarding breast size, the woman who seeks cosmetic surgery in the quest to achieve a perfect body shape or to remain ever youthful, are unlikely to be autonomous with respect to these aspects of their embodiment. In answer to the objection that this normative requirement is too stringent, I would reply that autonomy is not liberty; it is an achievement, not an entitlement.[25] Further, as the above discussion has shown, without such a requirement it is not clear how autonomy can be distinguished from heteronomy.

IV. CONCLUSION: CHOICE, EMBODIMENT, AND AUTONOMY

In equating bodily autonomy with maximizing the range of available bodily options or extending a person's instrumental control over her body and bodily processes, maximal choice conceptions conflate autonomy with liberty. Certainly, some degree of liberty is necessary for autonomy, as is some scope for choice. Further, it is true that there are circumstances in which expanding the range of options available, increasing a person's control over her bodily processes, or extending her bodily capacities, does increase her bodily autonomy. But it does not follow that limitless liberty, or making more and more aspects of our bodily life open to choice and change, automatically enhances autonomy. In fact, in many cases, bodily autonomy is more likely to be enhanced by challenging and altering those oppressive social norms and practices that contribute to the experience of bodily alienation. Coming to terms with the bodily changes associated with aging provides a case in point.[26] In a culture in which youth and beauty are paramount and the elderly are regarded as having outlived their usefulness, or even treated as objects of contempt, then cosmetic surgery may seem a rational and desirable means for maintaining self-esteem. But this solution, even if it is reflectively endorsed, signals a failure to reflect upon and criticize those cultural norms that are the source of the problem. Further, given that such a solution is inevitably short-lived, the attempt to achieve an integrated bodily perspective through such means seems doomed to eventual failure.

Following Ricoeur, I would also argue that the givens of human embodiment – birth, death, sex, reproduction, illness, old age – are the condition of human selfhood. Coming to terms with these givens is part of the exercise of embodied agency, it is what is involved in living a human life. To achieve an autonomous integrated bodily perspective is in part reflectively to negotiate the relationship between one's selfhood and the givens of one's human and individual embodiment. This is not to advocate quietism or to suggest that we should passively acquiesce in pain and suffering. Nor is to deny that, at least for those of us who live in privileged first world societies, our life prospects and possibilities have been greatly enhanced by a reduction in the harmful contingencies to which human life is subject. My point is rather that although we can do much to better human welfare by alleviating the effects or extent of harmful contingencies, the givens of human embodiment are nevertheless part of the condition of human life. Further, we may think we can imagine what human life would be like if all these givens were to become matters of choice – for example, if men could have babies, or if we could pre-select all the traits and characteristics of our offspring, or if cryogenics were actually to succeed. We may also imagine that everything that is valuable about human life will still remain. But this may be just as illusory as imagining that what is essential to me will survive teletransportation.[27]

Finally, because altering the givens of human embodiment does have the potential to change both who we are and what we value, there is no escaping the normative question of whether there are good reasons for bringing about these alterations. The standard liberal response to this insistence that normative issues must be addressed before we simply continue to proliferate choices, is that there is no consensus about the good, and there are no uncontentious criteria for making qualitative or normative distinctions among options. This is why a liberal society ought to make available to its citizens as many options as possible, to best enable them to live according to their own conception of the good life. In response to this rejoinder, I would point out that while the liberal commitment to maximizing choice pretends to normative neutrality, this commitment does in fact implicitly endorse certain values, in particular the values of individualism and negative liberty. This is not the place to dispute these values. My point here is simply that they are not value-neutral, nor are they uncontentious. I would also add that the overriding commitment to maximizing liberty (or choice) at the expense of other values leaves a normative vacuum that threatens to undermine the very basis of the value of choice. Choice is valuable because it enables us to enact or express our values. But if our choices are not guided by our values, if choice itself is the only value, then our choices are arbitrary and wanton. Indeed, one of the predicaments of our culture is that while we are steadily transforming the givens of human embodiment into choices, particularly in the areas of medicine and biotechnology, we have no clear idea what norms, other than the value of choice itself, should guide this process. Bioethics is a response to this normative vacuum, but to the extent that bioethics assumes without question maximal choice conceptions of autonomy, it provides little normative guidance.

Macquarie University
Sydney
Australia

ACKNOWLEDGEMENTS

I am grateful to a number of people who provided me with useful written comments on earlier, and rather different, versions of this paper or who discussed with me some of the ideas developed here. In particular I would like to thank Susan Brison, Sarah Buss, Genevieve Lloyd, Diana Meyers,

Michaelis Michael, Lee Overton, Michael Smith and Natalie Stoljar. I would also like to thank participants in discussions of the paper at the Humanities Research Centre and the Philosophy Department at the Australian National University; the School of Philosophy at the University of New South Wales; and the Philosophy Department, York University, Toronto. For comments on later drafts of the paper I would like to thank Peter Menzies, Kay Toombs and an anonymous reviewer for this collection.

NOTES

[1] Developing careful answers to such first-order moral questions is of course a pressing task for bioethics. However, the focus of this paper is the *concept* of bodily autonomy rather than its application to particular moral issues or technological developments. In particular, I wish to show why a prevalent *conception* of bodily autonomy is problematic, and to argue for an alternative conception. While I hope this alternative conception of bodily autonomy may contribute to an understanding of the moral issues raised by developments in biotechnology, each development, and the moral issues raised by it, is specific and must be dealt with on a case by case basis. Further, and I wish to stress this point, in any such moral assessment, autonomy is only one among a broad range of ethical issues that must be considered.

[2] What counts as harm to others, however, and which practices are considered harmful, is open to considerable dispute. Theorists with libertarian leanings tend to define harm narrowly as direct physical harm, or damage to a person's reputation or property. Those of a more liberal persuasion attempt to define it more broadly to include psychological harms to individuals, or socially harmful consequences, or harm to particular groups within society in the form of systemic inequality or disadvantage.

[3] I follow Feinberg here in distinguishing the non-coercion and voluntariness conditions. See Feinberg (1977) and Feinberg (1986, Chapter 17). In much bioethical literature, however, these two conditions are run together in the notion of informed consent, which is usually taken to be sufficient for autonomy. In the principlist approach of Beauchamp and Childress, for example, the principle of autonomy is understood as: "personal rule of the self that is free from both controlling interferences by others and from personal limitations that prevent meaningful choice, such as inadequate understanding. The principle of respect for autonomy requires respecting those choices made by individuals whose decisions are free from external interference or personal limitations". Beauchamp and Childress (1994, p.121). Feminist bioethicists, among others, have questioned the adequacy of both this conception of informed consent and the equation of autonomy with informed consent. See for example Dodds (2000) and Sherwin (1998).

[4] For an example of such a view, see Charlesworth (1993).

[5] Locke, (1690). See Church (1997) for a detailed discussion and defense of the notion of bodily ownership as the basis for a person's rights with respect to her body.

[6] The equation of bodily autonomy with instrumental control over the body and body processes is particularly prevalent in the literature on reproductive technologies. A good example can be found in Schultz (1990).

[7] I am unable to address this issue at any length in this paper. For a general theoretical critique of the assumption that expansions of choice always enhance autonomy, see Dworkin (1988, especially chapter 5). See also Dan-Cohen (1992). For an interesting critique of this assumption in relation to cosmetic surgery, see Morgan (1991).

[8] I have in mind the recent controversy in Britain over Jenna Franklin, whose parents, both cosmetic surgeons, offered her breast implants for her sixteenth birthday. In another recent case, in Quebec, on the basis of a psychiatrist's statement that breast implants would improve a fifteen year-old girl's self-esteem, it was determined that Medicare, the public health care provider, should pay $5,000 to cover the costs of the implants.

[9] Susan Brison (1997) has written powerfully of this sense of loss of self in the aftermath of sexual assault.
[10] The following discussion refers in particular to Parfit (1984), Chs. 10-13.
[11] See Ricoeur (1992, Fifth Study) for a discussion of the distinction between sameness (*idem* identity) and selfhood (*ipse* identity) in the context of Ricoeur's critique of Parfit's approach to personal identity.
[12] Parfit's metaphysics of personal identity is in part motivated by his utilitarianism. It is our preoccupation with selfhood, he thinks, that explains our resistance to utilitarianism as an ethical outlook. However, once we understand that personal identity does not matter, we will realize that the boundaries between self and others are not so important; what matters is the quality of people's experiences, whether they are our own experiences or someone else's. Ricoeur argues that Parfit's argument from his metaphysical thesis to his ethical thesis revolves around, and discloses an ambiguity, in the notion of belonging or 'mineness'. Parfit regards all concern with what is one's own, with selfhood, or 'mineness', as self-interested, egotistical or proprietorial concern. But, as Ricoeur rightly insists, self-concern is different from self-interest. Further, the question of who one is, in other words, selfhood, is also quite different from the question of what one has or what one owns or what belongs to one, in other words, ownership.
[13] Christine Korsgaard's critique of Parfit, in Korsgaard (1989), raises similar issues to those raised by Ricoeur. Where Ricoeur understands personal identity as a dialectic between sameness and selfhood, Korsgaard understands it in Kantian terms as a dialectic between the theoretical and practical standpoints. Similarly, where Ricoeur understands selfhood as phenomenological, rather than metaphysical, Korsgaard understands the necessity of agency as a practical, not a metaphysical, necessity.
[14] It might be objected here that the interconnections between psychological and corporeal continuity are looser than Ricoeur suggests, since we can and do undergo all sorts of bodily changes without psychological discontinuity. I would make two responses to this objection. First, it is based on a misunderstanding of Ricoeur's argument, which is not that any and every bodily alteration challenges psychological continuity. Such alterations, as I shall suggest below, are themselves one of the givens of human embodiment. Ricoeur's point, rather, is to challenge those intuitions and conclusions about personal identity derived from thought-experiments that are premised on a suspension or denial of the invariant givens of human corporeality. Second, if we take seriously the concept of embodiment, that is, of an embodied consciousness which is a unity, then bodily changes will affect both *idem* identity, or what we are, and *ipse* identity, or our sense of who we are. The extent to which identity is affected by such changes will of course depend upon their nature and severity.
[15] Korsgaard makes a similar point, arguing that whether or not we would be different than we are under a different set of technological possibilities is beside the point: 'The relevant necessity is the necessity of acting and living, and it is untouched by mere technological possibilities' (Korsgaard, 1989, p. 115).
[16] Jennifer Church (Church, 1997) develops an interesting account of bodily ownership in terms of the notion of bodily integration. There are some similarities between Church's account of bodily integration, as the reflective integration of a body's psychological states into a self-conception, and the one I develop below. However, where Church sees bodily integration as central to the notion of bodily ownership, I regard it as central to the notion of bodily autonomy. Further, for the reasons discussed above, I think the phenomenological account of the sense in which a person's body belongs to her provides a much better basis for understanding the importance of bodily integration than the liberal notion of bodily ownership.
[17] My discussion of the notion of a bodily perspective, particularly in this paragraph, is much influenced by Iris Marion Young's phenomenology of embodiment, especially her articles 'Throwing Like a Girl: A Phenomenology of Feminine Body Comportment, Motility, and Spatiality' and 'Pregnant Embodiment: Subjectivity and Alienation', both in Young (1990a).
[18] Iris Marion Young discusses the way in which xenophobia is internalized in bodily self-

representations in Young (1990b, Chapter 5).
[19]Brison (1993) and (1997) provides very moving and philosophically reflective first-personal accounts of the way in which the experience of sexual assault shattered her sense of self. She also describes the process of recovery as a process of putting her self and her life back together – but with a difference. The kind of self-conception and relation to the world embodied in her perspective prior to the assault can never be recovered.
[20]For a detailed discussion of what is at stake in these disagreements, see Mackenzie and Stoljar (2000b).
[21]The seminal paper is Frankfurt (1971). Frankfurt's ideas are developed in the papers in Frankfurt (1988) and Frankfurt (1999). See also Watson (1975); Taylor (1976).
[22]Brison (1997) presents a vivid account of such processes of dissociation.
[23]In her analysis of oppression in *The Second Sex*, Simone de Beauvoir argues that sexual oppression is internalized through processes that alienate women's agency from their embodiment. I discuss this issue in more detail in Mackenzie (1998). Sandra Bartky takes up this aspect of de Beauvoir's analysis of oppression in Bartky (1990).
[24]In Stoljar (2000), Natalie Stoljar argues the case for a normative, rather than purely procedural, approach to autonomy through a detailed consideration of case studies of women who have had multiple abortions through failure to take contraceptive precautions.
[25]Liberty contrasts with coercion and constraint and refers to a person's right to freedom from undue interference in the spheres of action, thought and expression. Autonomy contrasts with heteronomy and involves the capacity for critical reflection on one's desires, values and choices, as well as the capacity to incorporate these critical assessments into one's deliberations and actions. In some circumstances autonomy is promoted by restricting liberty. For example, a recovered alcoholic who decides not to go out to bars after work with her colleagues because she knows that there is a risk she would be tempted to order just one drink (and then another and another), restricts her liberty in order to promote her autonomy. For a discussion of the distinction between liberty and autonomy, see Dworkin (1988).
[26]In conversation Genevieve Lloyd pointed out to me the implications of the difference between the bodily integration and the maximal choice approaches to autonomy for coming to terms with the process of aging. This point warrants more careful discussion than I have been able to give it here.
[27]Bernard Williams (Williams, 1973) considers the case of Elina Makropulos, from an opera by Janacek, who is still alive at the age of 342. His argument seeks to show that, just because death is generally regarded as a misfortune, it does not follow that it is better never to die. The case of Elina Makropulos is a counter- example to the kind of illusory thinking that we engage in when we imagine that what is important or of value in our lives will remain intact despite radical changes to the givens of human life.

BIBLIOGRAPHY

Bartky, S.: 1990, *Femininity and Domination: Studies in the Phenomenology of Oppression*, Routledge, New York.
Benson, P.: 1991, 'Autonomy and oppressive socialization', *Social Theory and Practice*, 17, 385-408.
Beauchamp T.L. and Childress, J.: 1994, *Principles of Biomedical Ethics*, 4th edition, Oxford University Press, New York.
Berlin, I.: 1969, 'Two concepts of liberty', in *Four Essays on Liberty*, Oxford University Press, New York.
De Beauvoir, S.: 1949/1972, *The Second Sex*, H.M. Parshley (ed. and trans.), Penguin, Harmondsworth.

Brison, S.: 1997, 'Outliving oneself: Trauma, memory and personal identity', in D. Meyers (ed.), *Feminists Rethink the Self*, Westview Press, Boulder, pp. 12-39.
Brison, S.: 1993, 'Surviving sexual violence: A philosophical perspective', *Journal of Social Philosophy* **24**, 5-22.
Charlesworth, M.: 1993, *Bioethics in a Liberal Society*, Cambridge University Press, Cambridge.
Christman, J.: 1991, 'Autonomy and personal history', *Canadian Journal of Philosophy*, **21**, 1-24.
Christman, J.: 1987, 'Autonomy: A defense of the split-level self', *Southern Journal of Philosophy*, **25**, 281-93.
Church, J.: 1997, 'Ownership and the body', in D. Meyers, (ed.), *Feminists Rethink the Self*, Westview Press, Boulder, pp. 85-103.
Dan-Cohen, M.: 1992, 'Conceptions of choice and conceptions of autonomy', *Ethics* **102**, 221-43.
Dodds, S.: 2000, 'Choice and control in feminist bioethics', in C. Mackenzie and N. Stoljar (eds.), *Relational Autonomy: Feminist Perspectives on Autonomy, Agency and the Social Self*, Oxford University Press, New York, pp. 213-235.
Dworkin, G.: 1988, *The Theory and Practice of Autonomy*, Cambridge University Press, New York
Feinberg, J.: 1986, *Harm to Self*, Vol 3 of *The Moral Limits of the Criminal Law*, Oxford University Press, New York.
Feinberg, J.: 1977, 'Legal Paternalism', *Canadian Journal of Philosophy* **1**, 106-24.
Frankfurt, H.: 1999, *Necessity, Volition, and Love*, Cambridge University Press, New York.
Frankfurt, H.: 1988, *The Importance of What We Care About*, Cambridge University Press, New York.
Frankfurt, H.: 1971,'Freedom of the will and the concept of a person', *Journal of Philosophy* **68**, 5-20.
Friedman, M.: 1987, 'Autonomy and the split-level self', *Southern Journal of Philosophy*, **25**, 19-35.
Korsgaard, C.: 1989, 'Personal identity and the unity of agency: A Kantian response to Parfit', *Philosophy and Public Affairs* **18**, 101-32.
Locke, J.: 1690, *Two Treatises of Government*, reprinted 1973 J.M. Dent & Sons Ltd., London.
Mackenzie C. and Stoljar, N. (eds.): 2000a, *Relational Autonomy: Feminist Perspectives on Autonomy, Agency and the Social Self*, Oxford University Press, New York.
Mackenzie C. and Stoljar, N.: 2000b, 'Autonomy refigured', in C. Mackenzie and N. Stoljar (eds.), *Relational Autonomy: Feminist Perspectives on Autonomy, Agency and the Social Self*, Oxford University Press, New York, pp. 3-31.
Mackenzie, C.: 1998, 'A certain lack of symmetry: Beauvoir on autonomous agency and women's embodiment', in R. Evans (ed.), *Simone de Beauvoir's The Second Sex: New Interdisciplinary Essays*, Manchester University Press, Manchester.
Meyers. D.: 2000,'Intersectional identity and the authentic self? Opposites attract!' in C. Mackenzie and N. Stoljar (eds.), *Relational Autonomy: Feminist Perspectives on Autonomy, Agency and the Social Self*, Oxford University Press, New York, pp. 151-180.
Meyers, D. (ed.): 1997, *Feminists Rethink the Self*, Westview Press, Boulder.
Meyers, D.: 1989, *Self, Society and Personal Choice*, Columbia University Press, New York.
Morgan, K.: 1991, 'Women and the knife. Cosmetic surgery and the colonization of women's bodies', *Hypatia*, **6**, Fall, 25-53.
Parfit, D.: 1984, *Reasons and Persons*, Clarendon Press, Oxford.
Ricoeur, P.: 1992, *Oneself as Another*, K. Blamey (trans.), Chicago University Press, Chicago.
Schultz, M.: 1990, 'Reproductive technology and intent-based parenthood: An opportunity for gender-neutrality', *Wisconsin Law Review*, 297-398.

Sherwin, S.: 1998, 'A relational approach to autonomy in health-care' in *The Politics of Women's Health: Exploring Agency and Autonomy*, The Feminist Health Care Ethics Research Network, coordinator Susan Sherwin, Temple University Press, Philadelphia.

Stoljar, N.: 2000, 'Autonomy and the feminist intuition' in C. Mackenzie and N. Stoljar (eds.), *Relational Autonomy: Feminist Perspectives on Autonomy, Agency and the Social Self*, Oxford University Press, New York, pp. 94-111.

Taylor, C.: 1976, 'Responsibility for self' in A.O. Rorty (ed.), *The Identities of Persons*, University of California Press, Berkeley, pp.281-300.

Watson, G.: 1975, 'Free agency', *Journal of Philosophy* **72**, 205-20.

Williams, B.: 1973, 'The Makropulos case: reflections on the tedium of immortality', in *Problems of the Self*, Cambridge University Press, Cambridge, pp. 82-100.

Young, I.M.: 1990a, *Throwing Like a Girl and Other Essays in Feminist Philosophy and Social Theory*, Indiana University Press, Bloomington.

Young, I.M.: 1990b, *Justice and Politics of Difference*, Princeton University Press, Princeton.

MICHAEL C. BRANNIGAN

MEDICAL FEEDING:
APPLYING HUSSERL AND MERLEAU-PONTY

I. INTRODUCTION

Although medical ethics' vast literature devotes sparse attention to the Claire Conroy case, it remains a legal landmark. Through the case, the New Jersey Supreme Court in 1985 set a legal precedent in ruling that the artificial provision of nutrition and hydration via nasogastric feeding was considered medical treatment in the technical sense and thereby subject to treatment that an individual patient had the legal right to refuse. Imposing such treatment upon the patient without the patient's consent could constitute a violation of that patient's privacy.

Claire Conroy was born in 1900. In her early 80s, she was admitted to a nursing home in Bloomfield, New Jersey, where she suffered from an organic brain disorder, described as a "diffuse or local impairment of brain tissue, manifested by alteration of orientation, memory, comprehension, judgment" (*In re* Conroy, 1983, p. 304). Prior to this, she had been living with her three sisters in Bellville, all of whom had died by the time she was placed in the nursing home. Her condition grew worse as she experienced further complications including urinary tract infection, necrotic gangrenous ulcers on her foot, and dehydration. Eventually, she was no longer able to communicate. Efforts were made to wean her from medical feeding, yet after repeated hospital stays it was clear that she would need to be permanently fed through a nasogastric tube, and she returned to the nursing home with the tube in place. She was described as "bed-bound in a semi-fetal position, suffering from hypertension, diabetes, and arteriosclerotic heart disease; with extensive bed sores, decubitus ulcers on her left foot, leg, and hip, and her left foot gangrenous to the knee; no bowel or bladder control, with a urinary catheter" (Stryker, 1986, p. 230).

Ms. Conroy's only living blood relative was her nephew and legal guardian, Thomas Wittemore. Upon her return to the nursing home, Wittemore requested removal of her tube on the grounds that the feeding was medically futile and was incurring more harm than benefit upon his aunt. Ms. Conroy's physician, Dr. Kazemi, refused, and Wittemore sought a court order. Opposing testimonies regarding her mental state were given during the court hearing. Dr. Davidoff, who also treated Ms. Conroy, maintained that she was irreparably demented and unable to respond. Davidoff further argued that her medical feeding constituted "extraordinary" treatment and was therefore morally non-obligatory. In contrast, Dr. Kazemi claimed that she could respond to some degree although at times she appeared confused. Kazemi totally opposed withdrawing the medical feeding, insisting that it was both morally and professionally required.

Judge Station ruled in favor of the nephew's request. However, the ruling was immediately held in abeyance pending further appeal, and in the meantime, Claire Conroy died with her nasogastric tube in place. Later, after an appeals court overturned Station's ruling, on January 17, 1985, the New Jersey Supreme Court reversed the Appeals Court decision on the basis of the patient's right to privacy and the right to refuse overly burdensome as well as futile treatment.

This case ignites all sorts of ethical concerns. For instance, how should we classify certain treatment modalities? Medical feeding had been traditionally viewed as "ordinary" treatment and therefore morally required. However, in Ms. Conroy's case could we refer to her medical feeding as "extraordinary" treatment and thereby not morally required? Indeed, does applying this traditional distinction between ordinary and extraordinary, a distinction that is an offspring of Roman Catholic theology, even make sense here? I have argued elsewhere that appealing to this distinction rests upon weak grounds and is indefensible (Brannigan, 2001, pp. 516-22). Another issue concerns patients like Claire Conroy who are incompetent to make their own decision. What sort of criteria should we use for assessing what that incompetent patient would want done? What standards of competency do we apply? These give rise to other questions. What constitutes humane treatment for patients like Claire Conroy? What is the symbolic meaning behind feeding, and should this play any role in ethical decision-making?

A more primordial level of concern governs these issues – it is one that pertains directly to Ms. Conroy and to our assessment of her. That is, how should we view her condition? More importantly, *how should we understand her own experience*? The question is critical. Our understanding of her experience and condition lays the groundwork for how we attempt to evaluate it in light of various options. This primordial level of concern is precisely where a phenomenological analysis enters in, for phenomenology provides an avenue by which we can come somewhat closer to comprehending another's experience, in this case that of Claire Conroy. Phenomenology reminds us that we cannot in essence split apart the illness itself from the patient's own experience and understanding of that illness. Phenomenology requires that we approach the illness within the entire context of the illness. And in considering this entire context, particularly the patient's overall condition, experience, and preferences, we refrain from focusing primarily upon the treatment modalities. As in earlier works, I continue to maintain that phenomenology offers a more humanistic framework for addressing the above sorts of medical ethical issues (Brannigan, 2001, 1992, 1990, 1985). Let us now review some fundamental concepts in phenomenology put forth by two of its major representatives, Edmund Husserl and Maurice Merleau-Ponty. Let us also see how their key ideas apply to medical ethical concerns, particularly the case of medical feeding.

II. EDMUND HUSSERL (1859-1938) - THE NEED FOR BRACKETING

As a caution, we need to guard against over-simplifying the complex field of phenomenology, an area that encompasses some rather sharp differences. The existential phenomenology of Gabriel Marcel contrasts with that of Jean-Paul Sartre, just as the thought of Emmanuel Levinas differs from Martin Heidegger. Nevertheless, phenomenology does share some common ground. Its most standard feature addresses the issue of the meaning behind phenomena, or the appearance of things. As we see from earlier essays in this volume, phenomenology seeks to investigate the status of the appearance of things. (As we shall discuss with respect to medical ethics, phenomenology addresses the meaning behind the appearance of illness.) And we can credit Edmund Husserl for providing the first real systematic exposition of this feature. Husserl made it quite clear that "phenomenology as a rigorous science" requires that we be able temporarily to suspend questions about the "being" of what is to be examined, so that we can more closely address all components of the experience itself and thereby "return to the things themselves." Within the medical setting this requires that we make every effort to understand the experience of the patient. After all, medical ethics sustains the fundamental premise that the center of concern in medicine is, and should always be, the patient. Herein lies the challenge for healthcare professionals. The necessarily intrusive business of medicine requires that we make every effort to understand the patient's experience.

Husserl himself does not specifically address issues in clinical medicine, yet his analysis provides an approach by which we can more soundly comprehend the patient's experience of his or her own illness. Husserl's analysis does this through three keys themes: natural attitude, phenomenological *epochē*, and phenomenological reduction. As a start, Husserl's notion of "natural attitude" refers to our usual tendency to assume that objects, events, and all things in the world are in themselves separate from our consciousness of these objects and events and that they therefore possess their own meaning independently of our consciousness. This natural attitude clings to the idea that our ego or cogito is an "island of consciousness," self-enclosed and separate from the world outside of consciousness. This common tendency to bifurcate world and consciousness is reflected in both empiricist and idealist orientations. Whereas empiricism upholds the independent existence of the object at the expense of the subject, idealism defends the absolute existence of subject at the expense of objective reality.

Husserl argues that the "return to the things themselves" is clearly hindered by this natural attitude with its accompanying empiricism and idealism, two avenues along which a good deal of Western philosophy and science has traveled. Husserl further asserts that access to experience (in the case of medicine, access to the experience of the patient) can only come about by putting aside, or "bracketing" this natural attitude. He refers to this parenthesizing of the natural attitude as the phenomenological *epochē* (a Greek term meaning "suspension"). Moreover, phenomenological reduction reduces objects and events to their pre-reflective appearance, that is, to phenomena prior to our conceptualizing about that phenomena. This reduction is critical since any later conceptualization through, for example, scientific theory, stands in danger of eclipsing the primordial ground of our direct, pre-conceptual experience.

Herein lies Husserl's relevance for medical ethics. His analysis suggests to us that the illness from which the patient suffers cannot in essence be separated from the patient's own experience of that illness. This means that the scientific-technical aspects of any illness, due to an objectification of the illness, need to be bracketed in order to ascertain the meaning of that experience *for the patient*. This is morally congruent since the healthcare professional's fundamental moral commitment is to the patient as a person, or, in Kantian terms, as a moral agent. Whereas the natural attitude in medicine limits this commitment by focusing primarily upon the technical, "medical" features of both the illness and its prospective treatment modalities such as nasogastric feeding, through Husserl's phenomenological *epochē*, we suspend this natural attitude and recognize that our own conceptualization of the patient's experience is, in effect, secondary to the patient's own experience.

This *epochē* is imperative given the increasingly scientific-technological character of Western medicine. Indeed, as healthcare grows more scientific and technological, the *art* of healing succumbs to the *science* of treatment as the prevailing medical ethos. As noted above, the center of concern in medicine must always be the patient. Yet on account of medicine's technicalization, patient-centrality stands in constant jeopardy of remaining ancillary to the technical components surrounding treatment. To illustrate, this technicalization altered the original meanings behind the notions of "ordinary" (morally required) and "extraordinary" (morally optional) treatments (Brannigan, 2001, pp. 516-22). According to their original meanings, their distinction only made sense within the entire context of the patient, that is, his or her world as well as his or her perspective of that world. However, medicine's technicalization engendered mechanistic understandings of the terms so that "ordinary" and "extraordinary" were eventually construed in terms of specific treatment modalities and procedures divorced from context and patient perspective. Treatments were considered to be morally required or optional according to whether they were simple or complex, usual or unusual, natural or artificial, inexpensive or expensive, and thus estranged from any significant concern for the patient's overall condition, perspective, context, and the ratio of burdens and benefits.

Focusing solely upon technique violates the phenomenological view. The phenomenological view demands consideration of the entire context of the patient, including the patient's experience and perspective that remains a vital ground for weighing the proportionality of burdens and benefits. Put simply, the phenomenological perspective requires entering into the world and experience of the patient – a daunting task – but nevertheless one that is possible as long as we are willing to bracket the natural attitude.

Suspending the natural attitude, the necessary construct in Husserl's phenomenology, is thereby critical in clinical ethics. It occurs through the phenomenological *epoché* and reduction. Both *epoché* and reduction enable a proximate understanding of the patient's own experience in such a way that the patient's illness is no longer held up for scrutiny as a mere object. The phenomenological perspective does not segregate the illness itself from the patient's own experience of that illness. And this also applies to treatment modalities which the patient undergoes. Thus, the conditions which lead up to medical feeding for the patient cannot in themselves be separated from the patient's experience of these conditions. Furthermore, and more relevant to ethical assessment, medical feeding itself needs to be evaluated in terms of the patient's experience of such modality. In the phenomenological approach, it is crucial to understand how both the condition and treatments impact upon the world of the patient, all in an effort to understand the patient's experience.

III. MAURICE MERLEAU-PONTY (1908-1961) BODY-SUBJECT AND ITS RADICALIZATION

A dualistic ethos still manages to grip our Western understandings of personal identity by assuming some entity apart from physicality. This ethos is pervasive in much of Western medicine in that it sustains a mechanistic approach to the body on the assumption that bodiliness is an insufficient, even irrelevant, condition for personal identity. An example lies in organ transplant technology's attendant public acceptance of the notion of replacing bodily parts and organs. (Ironically, this is the case in spite of the continuing dearth of willing organ donors. That is, although we appear officially dualistic, are we so on a more intuitive level?[1])

This raises the specter of the plaguing philosophical question of human identity. Although earlier chapters already discuss the issue of self, I need to emphasize that our prejudices regarding personal identity in effect provide an ontological framework for any moral system or construct of values. That is, ethics assumes a particular ontological net of assumptions regarding personal identity and self. How we ought to morally behave rests upon fundamental beliefs regarding *who and what, in essence, we are*. Edmund Pellegrino, David Thomasma, and Richard Zaner, to name a few, have certainly labored in this direction to construct a viable ontological context for medicine. We see this clearly in Pellegrino and Thomasma's *A Philosophical Basis of Medical Practice* (1981) and in Zaner's *The Context of Self* (1981). Both works stress not only the importance of understanding ontological roots, but they plainly endorse the phenomenological approach in medicine. I support their claim as well as Pellegrino and Thomasma's that such an ontological foundation can be discerned in the genetic phenomenology of Maurice Merleau-Ponty.

A. Phenomenology of Perception

Merleau-Ponty generates a "genetic" phenomenology in his effort to examine pre-reflective experience, or what Husserl calls the "lived world." This is evident in his pioneering work, *Phenomenology of Perception* (1962), where he reveals a realm of experience prior to, and distinct from, how that experience is later conceptualized. This work introduces his pivotal idea – "body-subject" – a subjectivity that is deeper than the subjectivity associated with reflection and consciousness. Body-subject is precisely the pre-reflective subjectivity that belongs to the body as such, a pre-reflective subjectivity that primordially encounters the world, and this encounter comprises the "lived world."

His discovery is no doubt radical. It points to a subjectivity intrinsic to the body and distinct from the derived subjectivity of an ego or soul entity. In this way, Merleau-Ponty attacks the long-standing Cartesian dualism that implies a subjective ego separate from an objective body. For Merleau-Ponty, bodiliness, in and of itself, is subjective. The body is its own subject. Merleau-Ponty opposes the "linear causality" that tends to view the body as merely an object among other objects. Instead, he proposes a relationship of "circular causality," or "motivation," between the body and the world. Circular causality assumes an intrinsic dialectic between body and world that is mutually defining and interactive. At the same time, the body's own subjectivity empowers it with a "privileged point" in this dialectic, so that the center of meaning in this "lived world" experience centers around this body-subject. In this way, body-subject possesses its own "intentionality" and inherently bestows meaning to any lived experience. Another way of understanding this is that the body, in a way, exhibits its own autonomy, a self-determinative quality that is fundamental to bodiliness itself and is not derived from any notion of a separate self or ego entity. Indeed, this bodily autonomy becomes the ground for any later more deliberate expressions of autonomy.

Phenomenology of Perception (1962) provides us with a phenomenology of illness and conveys an approach to illness that steers clear of strict empiricism and idealism. Furthermore, the text proposes a unique methodology, that of "existential analysis," as a vitally significant and necessary way to understand illness. In view of the body's subjectivity and its own meaning-giving character, existential analysis thereby examines the patient via the patient's own bodily situatedness in the patient's world. In other words, existential analysis redirects our attention to the patient as a "being-in-the-world." Merleau-Ponty illustrates this existential situatedness in his interpretation of the phenomenon of anosognosia, or the phantom limb:

This phenomenon... is, however, understood in the perspective of being-in-the-world. What it is in us which refuses mutilation and disablement is an *I* [body-subject] committed to a certain physical and inter-human world, who continues to tend towards his world despite handicaps and amputations and who, in this extent, does not recognize them *de jure*. The refusal of the deficiency is only the obverse of our inherence in a world, the implicit negation of what runs counter to the natural momentum which throws us into our tasks, our cares, our situation, our familiar horizons. To have a phantom arm is to remain open to all the actions of which the arm alone is capable; it is to retain the practical field which one enjoyed before mutilation (Merleau-Ponty, 1962, pp. 81-82).

Here, Merleau-Ponty clearly identifies body-subject as a bodiliness that is intrinsically meaning-giving in the most primordial sense. This surely applies in the medical setting in that the patient's bodiliness projects a fundamental meaning, and Merleau-Ponty refers to this bodily projection as "intentional arc." The patient's intentional arc underscores the existential situatedness of the *patient as body*. In other words, we live in the world *through* our bodiliness, engaged in a natural dialectic or "motivation." As he states, "The body is the vehicle of being in the world, and having a body is, for a living creature, to be intervolved in a definite environment, to identify oneself with certain projects and to be continually committed to them" (Merleau-Ponty, 1962, p. 82).

B. *The Visible and the Invisible*

Although *Phenomenology of Perception* (1962) offers an understanding of the body-subject as a basis for an ontology of medicine, its terminology still contains some dualistic implications. Indeed, the notions of bodily intentionality and motivation assign a privileged center to bodiliness and, in so doing, strongly foster a corporeal genesis to consciousness. Yet the idea of body-subject still suggests its polar opposite – world-object. Merleau-Ponty's *Le visible et l'invisible* (1964a) [The Visible and the Invisible, 1968], which he was working on just before he died, takes up the task of modifying *Phenomenology of Perception's* near-dualistic terms.[2] In *Le visible et l'invisible*, Merleau-Ponty grounds his earlier ontology more solidly in bodiliness, in what the text refers to as corporeal reflexivity, especially in its pivotal concept of reversibility in the chapter, *L'Entrelacs-Le Chiasme* ("The Intertwining-The Chiasm"). In this text, Merleau-Ponty again compels us to reassess our ordinary understanding of the relationship between body and self. However, this time he does so by assigning to bodiliness the role of *sensible exemplaire*. In this way, he radicalizes the earlier notion of body-subject (*corps-sujet*) in *Phenomenology* (Merleau-Ponty, 1964a, pp. 179ff).[3]

In "*L'entrelacs-le chiasme*," Merleau-Ponty (1964a, p. 204) proposes "reversibility" as an "ultimate truth" (*vérité ultime*) and a central feature in his genetic phenomenology. Reversibility radicalizes the ontology of the body that was examined earlier in his *Phenomenology of Perception* (1962). What does he mean by "reversibility"? We can better understand "reversibility" within the context of the imagery of *l'entrelacs* (intertwining) and *le chiasme* (chiasm). These metaphors demonstrate an ongoing interweaving that takes place within a coalescing environment. They express the natural participation of all things in all things without being identified with the things participated in. Such natural participation and intertwining is a primal condition for any later thematic expression in thought and language. Reversibility is thus the natural interconnectedness permeating all of existence, and its fundamental feature lies in interpenetration.

Merleau-Ponty (1964a, pp. 180ff) provides us with examples such as the tactile encounter between touching and being touched. The hand that touches acts as subject upon the object of its touch. Yet, at the same time, the touching hand is also construed as object from the point of view of that which is being touched. The touching hand is both subject and object, and that which is touched is likewise. This becomes more evident when one hand touches another, demonstrating a natural ambiguity in the experience of touching in that there is both difference (between touching and being touched) and coincidence (no strict separation between touching and being touched). In another example, the visual experience of seeing and being seen is such that, in my seeing another person, I am at first acting as subject, yet I am also the object for the person I see. This visual rendezvous between seeing and being seen is both immersion and distancing, and the two allow for perception. Thus perceiving and being perceived collaborate in an inter-definitional relationship. More importantly, the fundamental basis for all of this rests upon the fact of bodiliness. And, as we have stressed earlier, the body is naturally the physical medium in medicine.

Keep in mind that there is also a natural asymmetry in that there is a distinction between the sensible and the sentient, the two components comprising what Merleau-Ponty calls "flesh." In other words, with respect to another person, we can never perceive ourselves as that other person perceives us. This is a critical point, especially in medicine. We are incapable of incorporating ourselves into another person's carnal being. My body is my own, and this unique corporeality carries with it a "dehiscent" character.[4]

In this fashion, reversibility denotes a sameness/difference, congruence/disparity rapport. Here, the crucial idea is that through reversibility Merleau-Ponty strives to escape the epistemological trap in which subject and object are dichotomized so that experience is reducible to either empiricist or idealist claims. His tactile and visual examples show that corporeal entities, what he calls "flesh", engage in an interpenetrating way. There is no encounter of an immaterial consciousness or "self" with some separate material object. Furthermore, this reversibility occurs at the unreflective level of perceptual experience, in the lived world. Reversibility essentially reaffirms, as well as radicalizes, the carnal basis of reflection, upstaging any primacy assigned to a so-called disembodied consciousness.

Reversibility thus discloses to us the corporeal nature of consciousness. The chapter *L'entrelacs-le chiasme* traces the process whereby conception is grounded upon perception, perception upon phenomena, and consciousness upon bodiliness. Merleau-Ponty's analysis of the ontogenetic development of consciousness shows that a type of corporeal reflexivity, known previously in *Phenomenology of Perception* as body-subject, is the foundation upon which self-reflection and personal identity rests. Self-reflection results from a sense of self derived from a particular body image. In other words, *my sense of self cannot be divorced from my corporeal basis.* In the most fundamental sense, I am my body. My sense of self is thereby affected by my bodily changes so that, just as my world changes as my body changes, so also does my sense of self. Nevertheless, dualism, and medical dualism in particular, is perpetuated through ignoring the implications of this bodiliness. And at no other time is my bodiliness more apparent than when my bodily reality is threatened by dysfunction or illness. In this way, Merleau-Ponty's ontology becomes a particularly appropriate basis for a phenomenology of illness.

IV. APPLYING MERLEAU-PONTY

Let us now apply these ideas to medical ethics. As we have seen, the concept of reversibility reaffirms and supports the diagnostic strategy of existential analysis introduced earlier in *Phenomenology of Perception* (1962) and radicalizes this strategy by stressing an even more obvious corporeal basis. This means that coming to terms with the patient's own personal situatedness requires all the more an understanding of the patient as a unique being-in-the-world rather than as some isolated, separate entity. Reversibility reinforces the idea that the patient and the world act together in circular causality. Reversibility also reminds us that our bodiliness is essentially project-oriented, so that the patient's ability to project meaning into the world is anchored in the patient's fundamental bodily intentionality. Illness disrupts this intentionality. It is an existential intrusion into the patient's world, a disruption of the patient's bodily intentional arc.

There are at least two principal consequences for medical ethics. First, clinical judgement now involves an approach to illness that strives to avoid a strictly mechanistic view of the patient's condition. This traditional, and still prevailing, mechanistic approach found in much of Western medicine presumes the ontological dualism that Merleau-Ponty seeks to dispel. By divorcing the patient's condition from his or her own experience of that condition, this dualism entails viewing the patient's condition in a way that short-circuits any authentic understanding of the illness. Merleau-Ponty's schema emphasizes that there is no such entity called illness separate from the experience of the illness. Therefore, the multiple conditions that require medical feeding, in and of themselves, constitute no real meaning apart from the patient's experience and perspective of these conditions. The same can also be said for the treatment measures. One needs to consider the treatment in terms of how the patient experiences that treatment and how that treatment affects the patient's world, how it primordially affects the patient's intentional arc.

Merleau-Ponty highlights corporeality in a way that requires us to respect the patient's bodily integrity.[5] This means that, since the body possesses its own subjectivity, the patient's body should not be reified, considered as some separate thing-like object. Therefore, clinical judgement not only involves the skillful application of scientific expertise to the human body, but carries with it a sufficient degree of insight and compassionate recognition of the patient as a whole. Merleau-Ponty's ontology of the body underscores the intimate relationality between body and self, and *Le visible et l'invisible* radicalizes this relationship. Clinical judgement must necessarily consider the vital link between body-image and self-awareness. In contrast, the purely technical approach assumes a misplaced ontological premise. For example, medical feeding through a nasogastric tube may well be the technically correct treatment given the variety of medical complications. However, body-self reciprocity and body-world mutuality anchors the merit of such a procedure upon how all of this impacts upon the world of that patient.

This has consequences for a second general area in medical ethics – the relationship between healthcare professional and patient, a relationship that is directly mediated through the patient's bodiliness. Here, certain ideas in *Le visible et l'invisible* provide a substantive foundation for a more genuine relationship between healthcare professional and patient. These ideas require that we broaden the patient's context in ways that incorporate the encounter with the healthcare professional. Since the patient's body offers a corporeal link to the patient's world that is real, tangible, and reversible, the patient's world IS accessible to the healthcare professional, even though the healthcare professional can never experience it herself. This is why *like-me-but-not-me* accompanies the notion of the reversibility of flesh. There can be a *connection* (but not an identity) to the other through this corporeal transfer. However, even though one can never experience another's suffering, our corporeal commonality engenders a natural accessibility along with respect for ultimate uniqueness.

Merleau-Ponty's ontology of the body also provides a corporeal basis for reinforcing the notion of personal autonomy. His body-subject compels us to recognize the intersubjective dimension of autonomy. Thus, it is imperative to reinforce the moral agency of the patient, since the realm of medicine not only addresses questions concerning the patient's condition and what can be done about it, but, most importantly, what *should* be done about it. In other words, by revealing and emphasizing the determinative agency and intentionality of the body, Merleau-Ponty's notion of body-subject expresses a bodily autonomy as the foundation and physical anchor for the more extended notion of personal autonomy.

When we apply all of this to the case of Claire Conroy, we learn some crucial lessons. To begin with, we must definitely steer away from a mechanistic approach that views Claire Conroy solely in terms of her body as a thing-like, objective entity. And even though medicine has yielded to this mechanistic approach by becoming more scientific and technical, phenomenology opposes this reification. This is clearly evident in how the notions of ordinary and extraordinary eventually developed. Although the origins of the terms considered the *entire context* of the patient with respect to possible treatment measures, medical reification altered the original meanings so that the terms themselves became attached to varying treatments, apart from the patient's overall context and condition. On the contrary, our analysis requires that, with respect to Claire Conroy (as well as for any patient undergoing medical feeding), we need to ask questions that pertain more to her experience of her condition and treatment in order to reasonably incorporate all of this into a more realistic assessment of consequences in terms of benefits and burdens. We need to ask: How invasive or noninvasive does she perceive her treatment to be? How does her treatment impact upon her world, her intentional arc? How does it influence her view of self? How does her condition affect her world? Her view of self? What are her perceived benefits and burdens? Weighed against a medical assessment of these, what is the proportion and disproportion of these benefits and burdens? What is the likelihood of success?

All of these questions are critical in how we morally assess Ms. Conroy's situation in terms of both her condition and treatment. If we tend to view her condition in mechanistic terms, we then attach undue moral weight to a strictly medical-scientific understanding of her condition. In this way, we may wrongfully assign moral meaning to what is "medical." This medical-moral confusion is evident in the use of descriptors such as ordinary and extraordinary. Confusing what is believed to be medically ordinary, such as medical feeding, with what is morally obligatory is dangerous if the focus is solely upon the treatment itself. What is morally obligatory rests upon a phenomenological understanding, one that considers the treatment in terms of the entire context of Ms. Conroy. This will constitute a more suitable groundwork for a reasonable assessment of her medical treatment in view of the proportionality of benefits and burdens.[6] Net benefits must indeed outweigh burdens. In other words, technical aspects of the treatment – whether the treatment is simple or complex, standard or unusual, natural or artificial, inexpensive or costly – are in and of themselves irrelevant considerations. The issue is not whether the treatment fits into any one of these labels. Considering the issue in this manner places Claire Conroy into a medical-technical straitjacket. Instead, the question is whether or not medical feeding stands a reasonable chance of providing more benefits than burdens for Ms. Conroy *in view of her world*. Thus, phenomenology balances the two key medical ethical principles of patient beneficence and patient autonomy. It offers an approach that seeks to consider, on the one hand, what lies in the patient's best interests in terms of that patient's world, and, on the other hand, respecting that patient's self-determination, now more fully anchored in the patient's unique bodily autonomy. In the case of Claire Conroy, the key question is not *what kind* of treatment is provided. The issue is whether the treatment is reasonable in view of Ms. Conroy's condition, interests, and preferences. If her medical feeding is medically futile and without substantive benefit in her world, then there is no morally sound justification to continue to provide such treatment. All this needs to be assessed within the entire context, bodily and personal, of the patient.

Phenomenology reminds us that illness is not simply a physiological event. It is an existential impairment, an assault upon the patient's world and meaning-giving-activity. As an existential event, illness can seriously compromise the patient's moral agency. In his *Medical Nemesis*, Ivan Ilich describes this assault in terms of a weakening of the triad of "fragility, individuality, and relatedness." Husserl and Merleau-Ponty offer us a phenomenological approach to illness that compels us to restore and strengthen this triad by enhancing the patient's best interests and moral agency through recognizing the patient's fundamental bodily basis.

Center for the Study of Ethics
La Roche College
Pittsburgh, Pennsylvania
U.S.A.

NOTES

[1] I raised this issue in a lecture to Japanese colleagues at the University of Tokyo during a fellowship in 1995. My lecture was to faculty and graduate students at the School of International Health. During this Invitational Fellowship, my research centered around Japanese perspectives on death, self, and bodiliness and issues in medical ethics. Prior to this I had done some extensive work on the controversies in Japan surrounding both brain death and heart transplants. In my opinion, Japanese views on the relationship between body and self played a prominent role in the opposition to both heart transplants as well as a brain death standard.
[2] Here, I rely primarily on the original French version of the text rather than on its English translation.
[3] Merleau-Ponty's intent to radicalize concepts from the earlier work becomes clear when we examine his working notes which accompany and constitute around 40% of the text.
[4] In Merleau-Ponty's 1950-51 Sorbonne lectures, he describes how this natural dehiscence exists despite a primordial "syncretic sociability" in the early stages of infancy (Merleau-Ponty, 1964b). This accounts for the early evolution of a perceived self-other distinction, and has implications for considerations of intersubjectivity.
[5] One reviewer pointed out that the importance attached in medicine to the patient's bodily integrity is not exclusive to Merleau-Ponty. This is certainly the case, particularly in some Asian traditions. For instance, consider the analysis in Yasuo Yuasa's *The Body: Toward an Eastern Mind-Body Theory* (Albany: SUNY Press, 1987). Nevertheless, Merleau-Ponty's phenomenological study underscores this bodily basis in a fashion that is remarkably unique among Western philosophical traditions.
[6] In her comments, Kay Toombs insightfully views the value of Merleau-Ponty's analysis in terms of

cultivating an understanding of the "selfhood of incompetent patients." At the same time, she points out the challenge in attempting to measure any perspectives that incompetent patients may have regarding their medical treatment.

BIBLIOGRAPHY

Brannigan, M.C.: 2001, 'Re-Assessing the ordinary/extraordinary distinction in withholding/withdrawing nutrition and hydration', in M.C. Brannigan and J.A. Boss (eds.), *Healthcare Ethics in a Diverse Society*, Mayfield Publishing Company, Mountain View, California, pp. 516-22.
Brannigan, M.C.: 1992, 'Reversibility as a radical ground for an ontology of the body in medicine', *Personalist Forum* **8**, 1, Supplement.
Brannigan, M.C.: 1990, 'Approaching an ontology of the body', *American Philosophical Association Newsletter on Philosophy and Medicine*, **89**, 3, 114-17.
Brannigan, M.C.: 1985, 'A phenomenological orientation to illness and ethical implications', *Contemporary Philosophy*, **10**, 8, 6-8.
In re Conroy, App. Div. 1983, 190 N.J. Super, 453, 464, A.2d 303, 304 n.1.
Stryker, J.: 1986, 'In re Conroy: History and setting of the case', in J. Lynne (ed.), *By No Extraordinary Means*, Indiana University Press, Bloomington, Indiana.
Merleau-Ponty, M.: 1968, *The Visible and the Invisible*, C. Lefort (ed.), Alphonso Lingis (trans.), Northwestern University Press, Evanston, Illinois.
Merleau-Ponty, M.: 1964a, *Le visible et l'invisible*, C. Lefort (ed.), Gallimard, Paris.
Merleau-Ponty, M.: 1964b, 'The child's relations with others', in J. M. Edie (ed.), W. Cobb (trans.), *The Primacy of Perception*, Northwestern University Press, Evanston, Illinois.
Merleau-Ponty, M.: 1962, *Phenomenology of Perception (1962)*, C. Smith (trans.), Routledge and Kegan Paul, London.
Pellegrino, E. and Thomasma, D.: 1981, *A Philosophical Basis of Medical Practice*, Oxford University Press, New York.
Zaner, R.M.: 1981, *The Context of Self: A Phenomenological Inquiry Using Medicine as a Clue*, Ohio University Press, Athens, Ohio.

SECTION SIX

RESEARCH

MAX VAN MANEN

PROFESSIONAL PRACTICE AND 'DOING PHENOMENOLOGY'

In 1975 Herbert Spiegelberg seized a title for a text on phenomenology that still speaks to the pragmatic sensibilities of many of my colleagues and graduate students. A visitor glances at my bookshelves and notices Spiegelberg's title *Doing Phenomenology;* the book is pulled and perused. Some of the section headings make attractive promises: "A new way into phenomenology: the workshop approach," "Existential uses of phenomenology," "Toward a phenomenology of experience," etc. But after a bit more browsing the book is returned to the shelf, without comment. Never has anyone asked to borrow it. And yet, *Doing Phenomenology* seems to be a text with commendable ambitions. In it Spiegelberg (1975, p. 25) decries "the relative sterility in phenomenological philosophy ... especially in comparison with what happened in such countries as France and the Netherlands."[1] Presenting essays both *on* and *in* phenomenology, he suggests that what is needed is "a revival of the spirit of doing phenomenology directly on the phenomena." And he asks: "What can be done to reawaken [this spirit] in a very different setting?" (1975, p. 25). Spiegelberg (1975, p. 26) sketches an example of the workshop approach consisting of "a small number of graduate students who would select limited, 'bitesize' topics for phenomenological exploration." It is difficult not to feel the hope that speaks in the pages of this book.

Spiegelberg's *Doing Phenomenology* did not turn into a popular phenomenological primer. His philosophical essays *on* and *in* phenomenology failed to exemplify doing it, if "doing phenomenology," as Spiegelberg suggests, means engaging phenomenological method to explore "bitesize topics" from the lifeworld. Perhaps Spiegelberg's book still signals a rarely mentioned issue in phenomenological circles: how to make phenomenology accessible and do-able by researchers who are not themselves professional philosophers or who do not possess an extensive background in the phenomenological tradition.

In recent years, there have been several developments that may not be exactly what Spiegelberg had in mind but that might have pleased him nevertheless.[2] Interest in phenomenology is indeed awakening in a very different setting – the domain of public and professional practice. This domain is characterized by priorities that arise from everyday concerns and practices, not necessarily from scholarly questions that are inherent in the traditions of phenomenological theory. In the professional fields, context sensitive research seems to have become especially relevant. It requires approaches and methodologies that are adaptive of changing social contexts and human predicaments. And that is perhaps how we may explain new non-philosophical trends in phenomenological human science methods.

For example, health science practitioners are interested in topics such as "living with obsessive compulsive disorder," "experiencing medical diagnosis," "children's experience of feeling left alone," "caring for a child with asthma," "living with schizophrenia," or "enduring chronic pain" (Madjar and Walton, 1999; van Manen, 2001). These investigations are done primarily to contribute to a more thoughtful practice of health science professionals, and secondarily to contribute to an academic knowledge base, although the latter is certainly a positive consequence of this type of inquiry. Phenomenology is, of course, essentially a philosophical discipline. Even the adjunct disciplines of phenomenological sociology, psychology, and anthropology are indisputably based in philosophy. Phenomenological research in nursing, geriatric care, clinical psychology, preventative health care, counseling, pedagogy, and human ecology is increasingly pursued by a breed of scholars who tend to hold strong backgrounds in their own disciplines, who possess less grounding in philosophical thought, yet who are "doing phenomenology." In these contexts, the disciplinary domains provide a fertile soil for the emergence of methods that diverge from conventional research practices. New forms of research that find their methodological origins in disciplinary and interdisciplinary context now transcend these boundaries and thus take on transdisciplinary perspectives oriented to the realities of everyday life and professional practices (Gibbons, et.al., 1999).

For example, treatment of Alzheimer's and other mental disorders is traditionally based on gnostic (diagnostic and prognostic) research models of medical science. In a recent report from the Canadian Medical Association Journal a list of recommendations are provided for the "Management of Dementing Disorders" (Patterson, et. al., 1999). These recommendations are formulated on an evidence based model of "Diagnosis and natural history of dementia." But what the report inadequately acknowledges is that people, as individuals, fail to match preconceived gnostic distinctions. A particular patient constantly refuses to fit diagnostic judgement and prognostic projections. The clinical path of an illness for any particular person is often different from medical assessment. The experience of disease is never experienced in exactly the same manner from person to person. Patients may continue to live when they were supposed to die or die when they were expected to recover.

This constant "defying difference" between diagnosis or prognosis on the one hand and contingency and concreteness on the other, is what makes each person, each patient, uniquely who he or she is – which is never the same as the diagnostic portrait that the medical expert constructs. The individual human being always falls to a certain extent "outside" of the dossier, the diagnosis, the description, and the prognosis. To do research that can address and reflect on the experience of illness we need to get beyond the objectifying effects of naming the things of our world with labels that distance us from them. Nurse practitioners and community service agencies tend to recognize these complexities and they try to develop support systems and advisory services that are informed by modes of knowledge and understanding gained on the basis of daily interactions with patients. As well, there is a growing literature that aims to offer insights into the lived experience of various illnesses. In fact, patients, themselves, produce some of this literature, in the form of journals or biographical accounts (van Manen, 1990/97).

So, the idea of "phenomenology of practice" as a context sensitive, eclectic, and transdisciplinary form of inquiry, employed inside and outside of traditional disciplinary boundaries, may provide a useful model to approach issues and concerns that arise in the lifeworld of professional practitioners. Of course, any methodology and its associated research strategies and techniques is only defensible to the extent that it can show ways to new knowledge, more sensitive understandings, and improved practices. How then should the methodologies, research strategies, and techniques of a phenomenological research perspective be described?

I. METHOD AS TAKING UP AN ATTITUDE OR STYLE OF THINKING

Qualitative methodology is often difficult since it requires sensitive interpretive skills and creative talents from the researcher. Phenomenological methodology, in particular, is challenging since it can be argued that its method of inquiry constantly has to be invented anew and cannot be reduced to a general set of strategies or research techniques. Methodologically speaking, every notion has to be examined in terms of its assumptions, even the idea of method itself. Heidegger (1982, p. 328) stated: "When a method is genuine and provides access to the objects, it is precisely then that the progress made by following it ... will cause the very method that was used to become necessarily obsolete." Moreover, it is difficult to describe phenomenological research methods since even within the tradition of philosophy itself "there is no such thing as one phenomenology, and if there could be such a thing it would never become anything like a philosophical technique" (Heidegger, 1982, p. 328).

Heidegger warns us against a reliance on method, yet he and others describe phenomenology in terms of method. Phenomenology "is accessible only through a phenomenological method ... each person trying to appropriate phenomenology for themselves," says Merleau-Ponty (1962, p. viii). How do we reconcile these claims? It appears that Heidegger is warning against reducing phenomenology to a set of philosophical strategies and techniques. But Merleau-Ponty (1962, p. viii) refers to method as something like an attitude: "phenomenology can be practised and identified as a manner and a style of thinking." It is best to think of the basic method of phenomenology as the taking up of a certain attitude and practicing a certain attentive awareness to the things of the world as we live them rather than as we conceptualize or theorize them. "Doing phenomenology" as a reflective method is the practice of the bracketing, brushing away or "reduction" of what prevents us from making primitive contact with the concreteness of lived reality.

The predicament of the search for method in phenomenological inquiry is somewhat like what happens in Kobo Abe's novel, *The Woman of the Dunes*. The protagonist, an entomologist, gets lost while taking a long walk along the beach and through a landscape of dunes. He happens upon a strange village with homes at the bottom of huge sand pits. It is late and he needs a place to stay. Somehow he gets trapped in one of these houses. From then on, every night he must work alongside a woman, shoveling and lifting away buckets of sand and soil that inevitably and ceaselessly keep flowing into the living space. He soon discovers that "you'll never finish, no matter how long you work at it." But that is the nature of life – a metaphor for the ceaseless task of brushing and excavating the matter that keeps drizzling and dispersing across the raw reality of our world. There is something authentic about the human task of attempting to uncover, literally by digging up and exposing the ground of our living existence – and thus to safeguard our being.

This picture of phenomenology, as constantly brushing away, in order to reach a primitive contact with the world-as-experienced, is complicated by the realization that the "matter" that constantly keeps covering the ground of our being is also the ground itself. As we lift to remove this matter, it already has taken certain shape or meaning. Lived experience is the name for that which presents itself directly – unmediated by thought or language. Yet in a fundamental sense lived experience is already mediated by thought and language, and accessed only through thought and language. Just so, the phenomenological reduction consists in the attempt not only to clear away (bracket), but simultaneously to confront the traditions, assumptions, languages, evocations, and cognitions in order to understand how the existential "facticities" of everyday lived experience are actually constituted through these assumptions and affects. Only by exploring the epistemological and ethical consequences of our perceptions, sentiments, moral imperatives, and conceptions can we strive to "free" ourselves from their perspectival effects – and yet we cannot escape them. In other words, the phenomenological reduction tries to grasp the intelligibility that lets the world "be" or come meaningfully into existence in different experiential modalities. The phenomenological reduction is indeed less a technique than a "style" of thinking and orienting, an attitude of reflective attentiveness to what it is that makes life intelligible and meaningful to us. Furthermore, language is decidedly implicit in the constitution and experience of meaning. And that is why phenomenological inquiry needs to start down-to-earth, by speaking and by letting speak the things of our everyday world, just as we experience and encounter them: see, feel, hear, touch, and sense them.

II. METHODS AND PROCEDURES

While phenomenology always involves the taking up of that attitude or style of thinking that distinguishes it from other qualitative methodologies, it does not preclude the use of certain investigative methods, techniques, and procedures.[2] It does mean, however, that whatever methods and procedures are adopted should always be "taken-up-in-the-attitude" described above.

Empirical and reflective methods and procedures can assist the practice of doing phenomenology in professional contexts.[3] Empirical methods describe the various kinds of research activities that provide the researcher with experiential material. They include personal descriptions of experiences, gathering written experiences from others, interviewing for experiential accounts, observing experiences, investigating fictional experiences, and exploring imaginal experiences from other aesthetic sources.[4] Reflective methods describe certain forms of analysis or phenomenological reflection. We may distinguish thematic reflection, guided existential reflection (corporeal, temporal, spatial, relational reflection), collaborative reflection, linguistic reflection, etymological reflection, conceptual reflection, exegetical reflection, and hermeneutic interview reflection (van Manen, 1997). In any research project the selection and usage of empirical and reflective methods and procedures depend upon the context and the nature of the study. The important point is that these methods and procedures differ from those in other forms of qualitative research.

Systematic "data" gathering through interviews, observations, descriptive accounts, etc. is rarely used in philosophy. But in professional fields experiential accounts or lived experience descriptions may provide the researcher with rich material. Written accounts that people provide of their experience may be highly recognizable. And some of the narrative accounts may be integrated in the phenomenological research text. By using an example of "memory loss in relation to name-forgetting" I will indicate how phenomenological research might deal with practical topics, and how it might speak to such concerns as understanding the experience of Alzheimer's dementia. In doing so I will outline briefly how some of the methodological features of the phenomenological reduction are engaged by methods inventively borrowed from the broader field of the social and human sciences, from the humanities, and from other practical contexts. A phenomenology of practice has to be inventive: How does one gain access to the experiential world of the elderly? How can one understand the experience of profound forgetfulness or memory loss in Alzheimer's patients? I will provide a sketch of some directions and some questions for such study.

III. AN EXAMPLE: THE EXPERIENCE OF FORGETTING SOMEONE'S NAME

Forgetting someone's name, and sometimes not recognizing his or her face, seems a common enough occurrence. On average faces seem to be remembered 70 % while only 20 % of names are remembered (Samuel, 1999, p. 63). You run into someone you recognize (or who you know you should know) and you just cannot remember the person's name. You are about to introduce guests to each other and, to your horror, the guest's name just won't come to you. Or it may happen that someone greets you by your name but you just cannot do likewise. There are two obvious aspects to this experience: the issue of forgetting and the fact that it is someone's name that is being forgotten.

As people grow older they may discover, to their dismay, that their memory for names seems to deteriorate. And yet, a failing memory (as long as it isn't a sign of some debilitating disease) does not need to threaten fundamentally one's sense of who we are: our identity or self-worth. There is something arbitrary about people's names: a name can be as easily forgotten as a phone number. We do not mind admitting to forgetfulness since we despair the irrationality of memory. There is little reason to be proud of the arbitrariness with which we remember some things and forget others. Why do we recall some seemingly trivial events from childhood and not others, more significant ones? Why do we remember one person's name but not another's? And, as far as mental faculties are concerned, one is much more inclined to admit to a poor memory than to a poor intellect. If we thought that our capacity for being thoughtful and our facility for judicious judgment were deteriorating, then we might be more alarmed. Our ability to think and feel seems to be more closely intertwined with our sense of identity and personal dignity than our ability to recall. Yet, in the case of Alzheimer's, personal dignity is very much at stake when the person can no longer remember names from the recent past or present.

Although the issue of name forgetting may seem trite, it is a common concern with professional practitioners in a variety of fields. In some situations name and face forgetting may risk loss of trust or confidence. In other contexts there are deeper issues at stake such as recognition and personal worth. It is easy to feel embarrassed about forgetting names or faces one thinks one should be able to recognize. And a person may joke slyly about Alzheimer's dementia striking. But these jokes are often uttered with a shudder of suspicion. What if one's forgetfulness is, in fact, abnormal? What if this is an early form of dementia that may strike any time during middle age? Such pathological forms of forgetfulness or dementia may call into question the very meaning of personhood, self-identity, and one's very sense of world and reality.

In order to gain a grasp of the phenomenological meaning of a phenomenon such as name-forgetting, the human science researcher finds it helpful to gather experiential accounts by using various empirical methods and procedures that are common to the social and human sciences: personal accounts, interview, and observation. As well, experiential accounts gleaned from fictional, literary, or other aesthetic sources may assist in presenting a range of experiential material that shows the variety, richness, and subtleties of lived meanings in human life. The first purpose of experiential descriptions is that they form the materials that invite phenomenological reflection and analysis. The objective is to "borrow" other people's experiences (whether factional or fictional) to explore the meaning dimensions of the phenomenon. The second purpose of gathering experiential descriptions is that they may provide narrative examples that enter the process of reflectively writing the phenomenological text (van Manen 1990/97, 1997). The following anecdotal excerpts from interview transcripts and experiential descriptions are offered here as examples of the kind of material one may want to gather. The aim is to collect experiential descriptions that are minimally interpretive, that are as close to experiences as lived through as possible, and that are somewhat like stream-of-consciousness memories. Such material may eventually reveal how particular kinds of experiences can be seen to be meaningfully constituted.

"Oh, helloooo ...! I'm fine. How are you doing?" I replied to the woman who greeted me by my name, hoping that she wouldn't notice that I didn't use her name. Outwardly, our conversation continued. Inwardly, I made desperate attempts to recall this woman's name, the context in which we must have met, anything that would end this feeling of awkwardness. My non-remembering is like a state of deafness. As she continues talking to me, I do not really hear the meaning of what she says because I am too busily engaged in chasing a clue: a circumstance, a shared happening, a common acquaintance.
She inquired, "Do you see much of her, your sister?"
"Um, every few months or so." I tried to answer in as calm a manner as possible, overcompensating with a tone of attentiveness so as not to give away my secret. I began to carefully study her face for distinguishing clues, the sound of her voice, her words. She doesn't look familiar. I don't remember telling anyone a story about my sister. Have I heard this voice before?
"I don't see much of my sister either but I guess, given our situation, that's probably all for the best...."
"Um, yeah. You're probably right."
At this point, she looked at me quizzically, "You do remember me don't you?"
"Oh, sure. I remember us talking about our sisters." The words just seemed to fall out of my mouth. I had given myself no choice but to carry on with this masquerade.
Our conversation turned to matters of children, schedules – the life of a mother. On the outside, I tried to maintain a normal flow of conversation, smiling and nodding when I thought that it was appropriate. But I wasn't all there. On the inside, I was groping desperately: where was I when I met this woman? Why was I talking about my sister? Do I know her sister? Is that where we met? Nothing. I still couldn't seem to remember her, her name, or even her face...
I experienced a strong feeling of disconnectedness, of not being able to grasp on to anything – orally or visually. I could hear the woman's voice but I wasn't making real sense of what she was saying. Putting her into any sort of context was impossible. There was no reference point, no clue that would allow me to attach myself to her. When our conversation finally ended, I stood for several minutes in the same spot hoping that her absence would suddenly trigger my memory. It didn't.[4]

Of course, every description is always already an interpretation by virtue of the fact that it is written in a certain manner, using particular words and phrases. And yet, the researcher who is engaged in exploring lived experience descriptions will soon recognize that there is a discernable difference between descriptions that are primarily experiential and descriptions that are mainly accounts containing perceptions, beliefs, opinions, interpretations, etc. The real challenge in phenomenologically oriented research is first of all to gather vivid experiential accounts. The written accounts that people provide of the experience of name or face forgetting seem highly recognizable. It appears that sometimes people engage in elaborate strategies to try to remember someone's name or to cover up for failing to remember.

PROFESSIONAL PRACTICE AND 'DOING PHENOMENOLOGY' 465

I've been in the same swimming class for two years, and I don't remember everyone's names. They just escape me. I have to listen to people calling other people by name so I can catch on. I find all kinds of ways of doing that! I've learned many ways to get around not knowing someone's name by relating to the person first and seducing them into doing more talking. If they talk about five minutes, pretty soon I have enough clues. I can usually identify more quickly with what people do, and where they come from, than their actual name.

One cannot be expected to remember every person from the past. Sometimes people get so frustrated that they simply admit their problem. But honesty can backfire as well:

"Hey, Diane! Good to see you! How have you been?" he greeted me with a smile.
Oh God, not another one! I felt I was on a trip to never-never-land. This time, I attempted to bluff my way through small talk with the young stranger. As we walked along together, he asked me how long I had worked for this firm. I hesitated, then replied I had been with them for about three years. The lad nodded approvingly, and said he was there to interview for a job as messenger or courier. Could I help him?
I threw in the towel, and smiled resignedly at him.
"Please forgive me. I know that I know you, but it is just one of those days! I simply can't bring your name to mind. I will be happy to put in a word for you, if you could write down your name and other relevant details."
"I don't get it," he muttered.
"Your name?" I did not waver.
"Diane, I'm your cousin, Rich," he said slowly.
Tears began to surface in my eyes, and I embraced my cousin, whispering, "I was just trying to keep anyone from overhearing that one of my relatives is applying. Of course I'll put in a good recommendation with the personnel department. Absolutely!"

As one explores experiential accounts through interview, writing, observations, etc. certain themes or insights into the nature of the experience of forgetting someone's name soon suggest themselves. Situations where we meet, greet and interact with people, the experience of saying someone's name, hearing one's name, forgetting someone's name, misnaming, or having no face-name recognition at all have phenomenological significance. In particular, it seems to have consequences for one's sense of self and self-identity, the recognition of others, the sense of feeling recognized or acknowledged, one's ability to be in relation, and so forth.

When the forgetfulness becomes chronic and disruptive of daily functioning then the memory failure takes on a more grimly appearance. While the above experiential accounts may seem very similar, the last two anecdotes are taken from accounts of individuals with Alzheimer's disease (Betty in Snyder, 1999, p. 114; and McGowin, 1993, p. 19). In the context of the Alzheimer's dementia these reminiscences suddenly seem much more tragic and painful. At least at first sight these accounts do not seem to differ from the many other accounts of name forgetting that I have gathered. And perhaps that is not surprising, because in the earlier stages of Alzheimer's the condition is difficult to diagnose since many of the symptoms can be interpreted as consistent of ordinary life experience. Diana was still quite young when she became unusually forgetful. But it was not until after the incident with her cousin that she was informed by a physician that she might be suffering from Alzheimer's. By that time she had realized that her name-forgetfulness was not like before. There was something unusual about it, as she describes in this instance at her work place:

As I sat down behind my polished, glass-covered desk, a brightly dressed woman approached.
"Yes?" I greeted her. "How may I help you?"
"Diane!" the woman paused in obvious surprise.
"Whatever do you mean?"
I felt a throbbing in my chest as I realized the woman thought we knew each other, and on first-name basis.
But who was she?
Staring at me intently, the woman placed a sealed payroll envelope in front of me.
"Payroll came while you were delivering the brief transcript," the woman said slowly. "I took your paycheck for you" (McGowin, 1993, p. 18).

IV. REFLECTING ON LIVED EXPERIENCE

Lived experience descriptions may seem simple, but they often contain rich material for phenomenological analysis and for phenomenological writing. To enact the reflective process one asks: how does this experiential account, this paragraph, or this line speak to the question of what it is like to forget someone's name? And indeed, upon reflection each line of a lived experience description will yield some aspect of meaning of the experience of forgetting someone's name. For every sentence one can ask: how is this line an example of the experience of meeting someone and forgetting his or her name? These meanings, insights or themes are then "used" as a resource for further phenomenological inquiry, reflection and, importantly, for writing a phenomenological text. Indeed phenomenological research is ultimately primarily a writing practice (van Manen, 1990/97, 1997, 2001).

There is not enough space here to go into the details of a full phenomenological or thematic analysis and how this analysis enters the process of phenomenological writing. But some aspects of the meaning of name forgetting readily suggest themselves: A person may not coincide with his or her name – yet the name matters. For example, when we use someone's name, then we experience a sense of connection. Naming creates relation, familiarity, and closeness. When I meet an acquaintance and the person calls me by my name, it means that my identity is being validated.

One feels recognized; one is remembered. In contrast when someone has forgotten my name, it may make me feel "forgettable." Also, to remember someone is to recognize one's past with this person. When a name is not remembered, it may strangely make that shared past blurry, while suddenly recalling the name, sometimes seems to make the past clear and merge it with the present.

"Well Hi... How are you?" I say, trying not to betray my lapse of memory. I have seen his face before, this I am sure of. But what is his name? I smile, overly cordial, as if to cover up my feeble-mindedness. On the outside I show a friendly demeanor but on the inside I feel uneasy and awkward.
As we continue talking, I purposefully adopt an intimate tone of familiarity. I nod agreeably to his comments. Yet I do not know clearly who this person is. Without a name an integral part of this person's identity seems missing. It is as if I have misplaced the key that will unlock the screen door that stands between me and my recognition. My non-remembering is like a state of deafness. As he continues talking to me, I do not really hear the meaning of what he says because I am too busily engaged in chasing a clue: a circumstance, a shared happening, a common acquaintance. While trying out several names (Bill, Mike, John, Jim, Robert) I answer in monosyllabic words. "Yes... yes... oh yes..." I say, responding only enough to maintain the conversation. I try several strategies that will hopefully help bring back to awareness the misplaced name so that I can step out of this awkward and embarrassing situation.
...Tony! Suddenly the name has cropped up. From where I do not know – Tony, that's it. What joyous occasion. What relief! I feel a genuine smile of satisfaction spread across my face. In a casual tone, acting like I knew the name all along, I say, "yes Tony," as I re-enter the conversation, but now as a fully engaged participant, ready and willing to discuss all of the other memories of our shared past that follow in the wake of remembering Tony's name.

In many accounts of forgetting someone's name, it appears that this situation is experienced as relational disorientation. While talking with the person and trying to remember his or her name, one may experience something like an inside/outside split in one's sense of self. It seems as if one's world has narrowed down to this very moment where one tries inwardly to deal with forgetfulness while one tries outwardly to maintain a normal relation and demeanor. A prevailing theme in these experiential accounts is that one's sense of self seems easily compromised in situations when one does not remember someone's name, or when the other person obviously does not remember ours. This is true for ordinary name forgetting and for Alzheimer's dementia, except that in the latter case recall strategies are less likely to work.

Indeed, sometimes the best way to study a pathological phenomenon is not to approach it from the point of view of its abnormality but as part of "normal," recognizable reality (Buytendijk, 1970, p. 569). Perhaps, an exploration of the ordinary experience of name- or face-forgetting may provide insight into the pathology of Alzheimer's memory failure and other forms of dementia that otherwise seem so unrecognizable to us. A phenomenological study of name-forgetting would have to include a reflective inquiry into the phenomenon of naming, as such, and how selfhood is constantly constituted in the lived presence through the appropriation of past and future. In this way, memory can be studied, not as an isolated object of study of behavioral science, but as an aspect of our human sense of self and selfhood.

Phenomenological research in professional domains can benefit from other scientific studies, both for what they contain and for what they do not contain. It is clear that the concern with memory, in connection with name forgetting and face naming, has a long theoretical history in the psychological literature. This research is largely experimental, developmental, neuropsychological, computational and oriented to developing cognitive maps and models. Neuroscience and particularly cognitive psychology has made extensive studies of the physiological and psychological mechanisms for analyzing, storing, and retrieving memory, but much less attention has been paid to the ways that people actually experience memory and forgetfulness. The difference between experimental cognitive research and qualitative phenomenological inquiry is that the former looks for mechanisms and behavioral patterns while the latter focuses on lived experiences and meanings.

V. THE SIGNIFICANCE OF CONTEXT OF PROFESSIONAL PRACTICE

In professional fields, questions that lend themselves to phenomenological investigation tend to find their source in situational contexts where practical action is necessary or where action poses issues of appropriateness or ethical consideration. For example, some of the literature on Alzheimer's dementia aims to offer insights and practical guidance to caregivers based on an understanding of the behaviors and inner lives of the Alzheimer's sufferers. *Living in the Labyrinth* by Diana McGowin (1993) contains a first hand experiential account during the early phases of the illness. Her story is compelling and moving. On the cover page the publisher recommends the book as: "An excellent guide for professionals, friends, and family." But in the appendix, the physician, Michael Mullin, warns the reader not to set a course on McGowin's chronicle: "Most commonly in the early stages of dementia (especially Alzheimer's), insight is mercifully lost and the victim drifts without self-awareness into the depths of the illness" (McGowin, 1993, p. 147). And he continues, "it is worth noting that Diane's piercing insight into her mental state is uncharacteristic of AD – the faculty of self-awareness often being sacrificed early" (McGowin, p. 152). I am quoting this physician since he seems to be saying that the Alzheimer's experience of most patients is quite different from McGowin's account. The pressing question is, of course: How is Alzheimer's experienced differently? What understanding does a physician have of a patient's state of being and awareness? What does the physician do with this understanding?

There still seems to be a dearth of literature dealing with the experiential or pathic dimensions of illness. After writing the simple little book of her personal account of Alzheimer's, McGowin became a much sought-after speaker. She was able to give voice to the silent sufferings and fears that people experience who come in contact with an illness that slowly seems to rob a person of his or her selfhood. In the above paragraphs I have tried to show how descriptions of personal experiences may help to construct phenomenological insights into the meaning and significance of human phenomena. Of course, the task of phenomenology as such is not to describe the psychological inner states, subjective experiences, motivations, feelings, understandings, plans, intentions, or purposes of particular individuals. Phenomenology cannot do that. But, nevertheless, experiential descriptions may serve as data for phenomenological reflection and as sources of meaning for understanding the structures of meanings that shape the inner lives of others.

The project of phenomenology not only rests on the method of the reduction and the various empirical and reflective methods mentioned above, it also rests on the method of the vocative in phenomenological texts. The vocative has to do with the recognition that a text can "speak" to us, such that we may experience a responsiveness, that we may know ourselves addressed (van Manen, 1997). There exists a relation between the writing structure of a text and the addressive effects that it may have on the reader and even the writer. So another feature of researching lived experience is the recognition that phenomenological inquiry is importantly a writing practice. Some of the insights achieved in this inquiry occur in the reflective practice of writing itself (van Manen, 1997, 2001). Through vocative techniques the phenomenologist attempts to bring dimensions of meaning into nearness while maintaining a reflective relation with them. The more vocative a text, the more strongly the meaning is embedded within it, and the more likely that it communicates complex qualitative understandings that contribute to the practical and ethical "knowledge" of the professional practitioner.

It is in the face of everyday living with an illness such as Alzheimer's that the painful difficulty of making life-sustaining treatment and support decisions are posed. Implicitly or explicitly professional practices are shaped on the basis of how one conceives the inner life and the effects of Alzheimer's on the nature of selfhood of the patients suffering from it. For example, the literature on bio-ethics discusses the withdrawal of life-sustaining treatment and the validity of advance directives in terms of the applicability and definitions of personhood to particular clinical cases (see for example, Kuhse, 1999). In these cases, the concepts of "personhood" and "selfhood" are employed as critical terms in an ethical discourse that aims to advice medical and nursing practice. Whether a severely demented person is still a "person," and therefore retains a right-to-life, is weighed against the judgment of whether the person experiences a sense of a "continuing self" with an interest in future life, even if the person clearly is "capable of experiencing states of consciousness" (Kuhse, 1999, p. 360). From a phenomenological point of view one would be extremely cautious with such reasoning. The problem is that a conceptual approach to interpreting profoundly ambiguous phenomena such as "personhood" or the experience of a "self," may promise a false sense of objectivity and clarity.

What does happen to the sense of self or selfhood of the person with dementia? "Many of my buddies have died in the last few years," an aging man says. And he continues: "Every time one of them dies it is as if a part of my self disappears. No longer do I have this person with whom to talk about the things we share and have experienced over the years." This experience of loss of self can be related phenomenologically to the Hegelian sense that the self is always constituted through others and through what is other to us. Without others there would probably not be an experience of self or selfhood. But not only do we lose others because they die or move away from us, we may lose others because we no longer recognize or know them. This is especially true for people suffering from various forms of dementia.

The ordinary experience of forgetting someone's name shows how memory is inextricably tied to selfhood. When I forget someone's name, my very sense of presence, of being in this situation, is disturbed. Not only is the other person's selfhood at stake in my forgetting of who this person is, it is as if my sense of my own selfhood is also disturbed. Forgetfulness is a temporal phenomenon. I have difficulty placing this person in my life history, and thereby my own presencing of the past and future is disturbed. The self is always a state of consciousness whereby past and future are constantly actualized in the projects and involvements wherein we are engaged. Memory is dependent on the projects we have from moment to moment. Only when "I am doing something" like preparing a meal can I have the experience of forgetfulness because now I need to recall from memory where I placed the can-opener. Perhaps memory is, in some sense, identical with temporal consciousness that structures the lifeworld.

Another feature of phenomenological inquiry, even in practical contexts such as this, is that it tends to evoke a sense of wonder and questioning attentiveness about taken-for-granted aspects of life. What would a human being "be" without memories? If selfhood is experienced in the things we do and in the projects in which we are always engaged, then what would selfhood "be" if one no longer has a sense of projects? More concretely, how (or to what extent) does a person suffering from pathological memory loss experience a sense of self? Is the "self" of the Alzheimer person who no longer recognizes friends and family members a dissolved self? Is the Alzheimers' sense of self narrowed to the extent that projects still exist? For people who are responsible for caring for a parent or spouse with Alzheimer's, it is puzzling to see how something so essential as memory seems to erode:

When my mother gets hungry she gets up and starts wandering through the house. So yesterday when she appeared hungry I said to her: "Go to the kitchen and get something to eat from the fridge." "Yes, I will do that," she pronounced in a determined voice. But when fifteen minutes later I got to the kitchen to see how she was doing she was just standing there. She was standing there, silent, as if sunken in thought. But was she? She did not seem to know why she was there. And yet she seemed to know that something was wrong, that she failed at something, that she was supposed to do something.[5]

An Alzheimer's person may have forgotten the project but the sense of failure may linger on. As our memory fails, our projects falter also, and our sense of self fails its realization in our personhood. Larry Rose (1996) wrote his own journey through Alzheimer's. But along the way his memory became so impaired that his wife Stella had to write the last part and chapter of the book. And she describes how – though Larry was still physically doing well – his mind was ebbing away. "I am still mourning the loss of Larry's intellect. I miss so much how he used to be. His personality has become more withdrawn now, and the traits that make him who he is, and what he is about, are slowly slipping away" (Rose, 1996, p. 133).

McGowin is so aware of her forgetfulness she feels indeed her-"self" diminished each day. And yet, ironically, her entire account gives evidence of an acute awareness of self. On the one hand, the self seems to erode, diminish, and on the other hand, the experience of self seems highly intensified. For example, the Alzheimer's person is acutely aware of trying to say something, but the right name or word will not come. An observer may not notice this and only see the silent, non-participatory behavior. Phenomenological research may help us to become more aware of these lived meanings of illness. Perhaps this problem is to be explored in terms of Ricoeur's (1992) distinction with reference to the meaning of self-identity. One aspect of self provides us with the experience of continuity and self-sameness, while the other aspect of self-identity is always changing, tied into our world and others around us. Thus, we can have the sensation that this person (father, mother) who we know so well is no longer the same, and yet the essence of this person seems to remain:

> My father and mother had just traveled for six hours in a bus to visit our family. And when Dad walked through the door he opened his arms and hugged me. I like it when my father gives me a big hug. I feel that I am still his little girl and that he is protective of me, even though his Alzheimer's makes him more withdrawn.
> I looked at this man who is my father and who had now become so unusually quiet. He almost seemed like a stranger. "Hi Dad," I said, to push off that thought, "Good to see you!" He nodded but did not speak. How fragile he seemed! With a shudder it struck me that I no longer really understood my father as I always thought I did. Yet I felt closer to him than ever. It was as if his forgetfulness of me became my forgetting of who he was.

Our sense of self and of the other's self may become profoundly ambiguous. But this is also true for naming. There is a deep irony in the fact that "forgetting" someone's name, in an unexpected manner, may enrich our sense of self and other. It is important to be able to remember a person's name since that is how we recognize each other and stand in reciprocal relation to each other. The uniqueness of proper names reflects the uniqueness of each relation and the unsubstitutability of each person in this relation. Forgetting people's names seems to say that there is no reason for the other person to be worth remembering. Even the social practice of snubbing is, in some sense, an acknowledgement of the Hegelian notion of recognition – the person who snubs the other says: "I see you but I have no reason to acknowledge you. You are not worth recognizing." And, indeed, sometimes it may appear as if forgetting someone's name lies unfortunately close to a snub. Professional snubbing is a common occurrence in medical practice when the "who" (name and uniqueness) of a patient is ignored in favor of the person's "whatness" (label and the diagnosis).

Without the name there would be no particular other, but even a proper name is something to which the other can never be reduced. And yet it is a paradox that precisely when we do not have to remember the other person's name, because the person (our spouse, child, close friend) belongs to the inner sphere of our being, that we may experience both the greatest distance and the greatest closeness. The uniqueness of the other person is ultimately unnameable. And herein lies our realization of the other person's essential dignity. Professional practice needs to be, not only pragmatic and effective, but also philosophic and reflective so that one can act and interact with competence and tact, and with human understanding.

PROFESSIONAL PRACTICE AND 'DOING PHENOMENOLOGY' 473

University of Alberta
Edmonton, Alberta
Canada

NOTES

[1] With respect to the Netherlands, Spiegelberg probably meant the work of an assortment of phenomenologically oriented psychologists, educators, pedagogues, pediatricians, criminologists, jurists, psychiatrists, and other medical doctors, who formed a more or less close association of like-minded academic practitioners known as the Utrecht School. This movement is unique since it predates the newly emerging interest in what I would call a "phenomenology of practice." Their practical lifeworld-oriented program required that knowledge of human existence be gathered by observations of everyday life situations and events. Some of the most important and programmatic writings of the Utrecht School are to be found in collections. For example, *Person and World: Contributions to a Phenomenological Psychology* (1953) was edited by van den Berg and Linschoten. It features classic articles such as Langeveld's "Verborgen Plaats in the Life of the Child" (The Hidden Place in the Life of the Child), "De Hotelkamer" (The Hotel Room) by van Lennep, "Over de Afkeer van de Eigen Neus" (On Being Repulsed by One's Own Nose) by Rümke, "Gelaat en Karakter" (Face and Character) by Kouwer, "Aspecten van de Sexuele Incarnatie" (Aspects of Sexual Incarnation) by Linschoten, "De Psychologie van het Chaufferen" (The Psychology of Driving a Car) by van Lennep, and "Het Gesprek" (The Conversation) by van den Berg. Another volume worth mentioning is *Rencontre, Encounter, Begegnung: A Humanistic Psychology* (1957). This book was edited by Langeveld and dedicated to professor F.J.J. Buytendijk on the occcasion of his retirement. It consists of 37 articles, more than half of which were in German or French. A selection of the above mentioned articles appear in Kockelmans, *Phenomenological Psychology: The Dutch School* (1987).

[2] For example, journals such as *Phenomenology and Pedagogy* and *Human Studies* have created opportunities for publishing phenomenological studies that have special relevance for professional fields. I also mention *Writing in the Dark: Phenomenological Studies in Interpretive Inquiry* (Van Manen, 2001), consisting of an edited collection of "bitesize" phenomenological topics by graduate students in pedagogy, clinical psychology, health sciences, adult education, etc.

[3] For an overview of a variety of phenomenological inquiry dimensions, see <http://www.phenomenologyonline.com/>

[4] Unless otherwise indicated, the transcript materials quoted in this paper have been collected as part of my SSHRC (Social Sciences and Humanities Research Council of Canada) funded Pedagogy of Recognition Project.

[5] The following quotes are taken from interviews with family members of persons afflicted with Alzheimer's.

BIBLIOGRAPHY

Abe, K.: 1991, *The Woman in the Dunes*, E.D. Saunders (trans.), Vintage Books.
Buytendijk, F.J.J.: 1970, 'Naar een existetiële verklaring van de doorleefde dwang,' *Tijdscrift voor Filosofie*, **32**, 567-608.

Gibbons, M., Limoges, C., Nowotny, H., et al.: 1999, *The New Production of Knowledge: The Dynamics of Science and Research in Contemporary Societies*, Sage Publications, London.
Heidegger, H.: 1982, *The Basic Problems of Phenomenology*. Indiana University Press, Bloomington.
Kockelmans, J.J. (ed.): 1987, *Phenomenological Psychology: The Dutch School*. Kluwer, Dordrecht, The Netherlands.
Kuhse, H.: 1999, 'Some reflections on the problem of advance directives, personhood, and personal identity,' *Kennedy Institute of Ethics Journal* **9**, 347-64.
Langeveld, M.J. (ed.): 1958, *Rencontre, Encounter, Begegnung. Bijdragen aan een Humane Psychologie Opgedragen aan Professor F.J.J. Buytendijk*. Het Spectrum, Utrecht & Antwerpen.
Madjar, I. and J. Walton (eds.): 1999, *Nursing and the Experience of Illness: Phenomenology in Practice*, Routledge, London.
McGowin, D.F.: 1993, *Living in the Labyrinth: A Personal Journey Through the Maze of Alzheimer's*, Dell Publishing, New York.
Merleau-Ponty, M.: 1962, *Phenomenology of Perception*, C. Smith (trans.), Routledge & Kegan Paul Ltd., London.
Patterson, C.J.S., Gauthier, S., Bergman, H., et al.: 1999, 'Management of dementing disorders: conclusions from the Canadian Consensus Conference on Dementia', *Canadian Medical Association Journal*, supplement 1999, 160 .
Ricoeur, P.: 1992, *Oneself as Another*, The University of Chicago Press, Chicago.
Rose, L.: 1996, *Show Me the Way to Go Home*, Elder Books, Forest Knolls, California.
Samuel, D.: 1999, *Memory*, New York University Press, New York.
Snyder, L.: 1999, *Speaking Our Minds: Personal Reflections from Individuals with Alzheimer's*, Freeman and Co, New York.
Spiegelberg, H.: 1975, *Doing Phenomenology: Essays On and In Phenomenology*, Martinus Nijhoff, The Hague.
Van den Berg, J.H. and J. Linschoten (eds.): 1953, *Persoon en Wereld. Bijdragen tot de Phaenomenologische Psychologie*. Bijleveld, Utrecht.
Van Manen, M.: 1990/97, *Researching Lived Experience: Human Science for an Action Sensitive Pedagogy*. Althouse Press, London, Ontario.
Van Manen, M. (ed.): 2001, *Writing in the Dark: Phenomenological Studies in Interpretive Inquiry*, Althouse Press, London, Ontario.
Van Manen, M.: 1999, 'The pathic nature of inquiry and nursing,' in I. Madjar and J. Walton (eds.), *Nursing and the Experience of Illness: Phenomenology in Practice*, Routledge, London, pp. 17-35.
Van Manen, M.: 1997, 'From meaning to method,' *Qualitative Health Research: An International, Interdisciplinary Journal*, Sage Periodicals Press, **7**, 345-69.

CHRISTINA PAPADIMITRIOU

FROM DIS-ABILITY TO DIFFERENCE: CONCEPTUAL AND METHODOLOGICAL ISSUES IN THE STUDY OF PHYSICAL DISABILITY

I. INTRODUCTION

This paper discusses conceptual and methodological concerns in the study of physical disability and the human body from the perspective of a phenomenologically informed sociology.[1] The study of physical disability and the human body has claimed the attention of philosophers, social scientists, psychologists, physicians, public health administrators, insurers, and so forth, each of whom brings a unique disciplinary perspective as well as distinct research interests and goals. In this paper I identify theoretical and conceptual biases in the study of disability that: (1) tend to narrow its understanding to a unitary phenomenon, i.e., as a dysfunction (either pathophysiological or psychological) affecting only the individual; (2) hamper its conceptualization as a form of difference; and (3) restrict the ability of persons with disabilities to live independent and respectful lives. Further, I demonstrate how socio-political conceptions of disability raise many theoretical and practical questions regarding research, as well as fostering the uncritical use of notions of normality and difference.[2] Contemporary research on disability and the human body is faced with a challenge: to describe and analyze the world of disability within its social, political and human contexts without perpetuating biased assumptions, ignoring bodily differences, and marginalizing the experience of disability.[3] This paper suggests that a phenomenologically-informed sociology can help researchers meet this challenge. Instead of seeing able-bodied and dis-abled persons as separate and opposed, disabled embodiment may be conceived as a form of human diversity, thus moving towards an understanding of *dis*-ability as difference. This paper discusses how researchers can reach this understanding through the phenomenological technique of bracketing and by listening to the criticisms offered by disability advocates and writers.

II. THEORETICAL AND CONCEPTUAL ISSUES

Exposing major problematic assumptions in the study of disability, such as able-bodied bias or ableism (Davis, 1995), clarifies and refines our understanding of the phenomenon of disability in its social context. These biases are most visible and pernicious in social-psychological studies of disability. In contrast, socio-political conceptualizations of disability present preferred ways of understanding disability.

The first ableist biasing assumption in the socio-psychological literature is that disability is a catastrophic and tragic event in a person's life. Persons with disabilities are assumed to be victims of unfortunate circumstances, struggling to find positive meaning in what happened to them, or blaming themselves for their misfortune (Bulman and Wortman, 1977). From this point of view, "victims" are seen as either trying to camouflage their pain and suffering, or trying to find some good in their tragedy. Assumptions about how persons with disabilities cope with chronic conditions have been drawn primarily from developmental models of "adjustment." That is, coping and adjustment theories anticipate that persons who experience physical disability will pass through a series of stages of adjustment, such as mourning, shock, denial, anger and depression, similar to those of a bereavement process (Weller and Miller, 1977). The epistemological problem with this assumption is twofold: First, such analytic formulations and explanations of behavior are deterministic and prescriptive; the theory of stages of adjustment is taken to be valid in and of itself and is superimposed on the experience of the person with a disability. Thus, this analytic model determines the phenomenon without first examining, describing, and observing it in its own terms. Second, adjustment theories result in methodological inconsistencies since researchers are already judging and categorizing persons' experiences according to their own expectations and categories. This does not allow for adequate description of the variety of responses that might occur nor does it allow us to see how persons with disabilities actually negotiate their social world.

A second ableist assumption that pervades the socio-psychological literature is that disability is a biological phenomenon, and, as such, is a personal/ individual problem. From this point of view, disability is solely a product of biological or physiological dysfunction occurring in an individual organism, rather than a phenomenon that is the result of discriminatory social and architectural barriers in society (Fine and Asch, 1988; Hahn, 1988; Oliver, 1990, 1996). To understand disability as a product of biology rather than as a social phenomenon means to concentrate on the individual and locate his/her limitations in the body. Alternatively, conceiving of disability as a social and political phenomenon exposes the environmental and social factors that may create, sustain and reinforce a dis-abling social environment. Focusing solely on the biological aspect of disability disempowers persons with disabilities and perpetuates the social injustice they experience.

A third ableist assumption is that the person with a disability is necessarily dependent, needing help and psychological support. This assumption is sustained by researchers' systematic omission of how persons with disabilities can be providers of support (Fine and Asch, 1988, p.12). There are at least three problems with this assumption: First, since the person with a disability is seen as helpless and passive, it can be deduced that s/he will never really fully adapt. Consequently, the notion of a successful disabled person who is content, independent and competent is paradoxical or impossible. At best, such a person is seen as "making the best" of a tragic event. Second, the need for assistance is generalized: it is assumed that since the person with a disability requires help in *some* spheres of life, s/he therefore needs help in *all* spheres of life. Third, it is assumed that the help transfers unidirectionally from the able-bodied to the disabled. This perpetuates the dichotomous thinking that able-bodied persons are independent and disabled persons dependent.

Fine and Asch (1988, p.16), echoing the arguments and activism of the Disability Rights Movement (DRM), suggest that:

> By thinking of the disabled persons as dependent in a given situation, and the one without disabilities as independent and autonomous, one can avoid considering how extensively people without disabilities too are dependent and 'out of control'.... Rather than the world being divided into givers and receivers of help, we are actually interdependent. Attributing neediness and lack of control to people with disabilities permits those who are not disabled to view themselves as having more control and more strength in their lives than may be the case.

III. SOCIO-POLITICAL CONCEPTIONS OF DISABILITY

Criticism of ableist assumptions and biases in social science, medicine, and everyday life has fostered grass-roots movements, self-help organizations, activism and political (legal) change. Disability groups have formed in major cities throughout North America and Europe, and Independent Living Centers have been gaining growing support and acceptance from the general public. Though disability organizations have typically formed around a specific disability (such as deafness or blindness), a growing concern with the rights of persons with disabilities has united many different groups to support and foster laws and policies such as the American with Disabilities

Act (ADA). These efforts have resulted in a variety of journals, books, theories, manifestos, workshops and proposals about the status of persons with disabilities in contemporary society and the need for change.[4] Persons with disabilities have been conceptualized as a minority group akin to other racial or ethnic populations who are oppressed by social, political and environmental barriers (Russell, 1994, pp. 11-12).

The conceptualization of disability as a minority group is in contrast to the functional limitations paradigm of the medical model and to the personal tragedy theory models of psychology (Hahn, 1988; Oliver, 1984, 1996; Charlton, 1998; Linton, 1988; Gill, 1994). Its socio-political approach has "indicated the need for strengthened laws to combat discrimination against persons with disability" (Hahn, 1988, p. 40). This point of view recognizes the restrictions of disability to be located in social and environmental surroundings rather than solely "in" the (individual) disabled person. Consequently, it focuses attention on the social, economic, and political structures that discriminate against, and marginalize, disabled citizens. Several authors (Livneh, 1982; Hahn, 1988; Oliver, 1990; Charlton, 1998) have offered explanations as to why persons with disabilities have been oppressed and can thus be seen as a minority group.

For instance, persons with disabilities violate cultural norms of appearance, beauty, and autonomy and this sets them apart from the remainder of the population (Hahn, 1988, p. 41). Issues of personal appearance are related to dominant cultural notions of appearance and attractiveness. Deviation from such standards may arouse strong feelings in nondisabled observers. Such feelings may come in the form of fear or disgust. Livneh (1982) and Hahn (1988) discuss this fear or disgust as an "aesthetic anxiety." Throughout history, visible/perceptible racial, ethnic, and gender characteristics have formed the bases of prejudice and discrimination. Similarly, for persons with disabilities, their aesthetic difference may lead to subordinate roles and treatment. Studies have shown that unattractive physical appearance affects social relations, employment, personal and sexual relationships (Shakespeare, 1996; Hahn, 1988; Safilios-Rothschild, 1970).

Hahn (1998, p. 42) suggests that unfavorable perceptions toward disabled persons may also stem from an "existential anxiety," i.e., the ableist assumption or belief that "a disability could interfere with functional capacities necessary" to pursue a satisfactory life – a threat which may trigger negative reactions and social stigma. This belief:

[R]elegates disabled individuals to the role of helpless or dependent nonparticipants in community life, and ... exacerbates nondisabled persons' worries about the potential loss of physical or behavioral capabilities that could result from disability. In a society that appears to prize liberty more than equality, and that tends to equate freedom with personal autonomy rather than with the opportunity to exercise meaningful choice, the apprehensions aroused by functional restrictions resulting from a disability often seem overwhelming (Hahn, 1988, p. 43).

For researchers of disability, the Disability Rights Movement and its critics have affected the way we theorize about disability, since they have brought to light ableist biases, socio-political problems regarding inequality, competency, and the right to a respectful life. Conceptually speaking, therefore, a new vocabulary and perspective have been suggested wherein voice is finally given to the lived experience of disability and its social and political context is appreciated. These are important advantages not only for the quality of life of persons with disabilities themselves, but also for researchers and political activists since such conceptualizations assist in the development of research and policies that may be more sensitive, non-discriminatory, accurate, and reliable.

All researchers must produce reliable, valid, and generalizeable data. For fieldworkers in the area of disability (especially able-bodied ones like myself), it is a formidable task to access and comprehend the world of persons with disabilities, while also taking into consideration ableist biases (and their criticisms) regarding normality and disability. The challenge is to avoid marginalizing, alienating, or ignoring bodily differences and the lived experience of disability. I turn now to a presentation of how I have used the phenomenological technique of bracketing to address this challenge in my own research.

IV. ACHIEVING INTERSUBJECTIVITY AND UNDERSTANDING THE OTHER

Qualitative methods of collecting and analyzing data depend heavily on the relationship between researcher and research subject. In order to demonstrate that the data are reliable, valid, and trustworthy, fieldworkers describe how they became involved with their subjects and how they managed to build trust in order to produce and legitimize a sound scholarly report. In this sense, questions of validity, reliability, and generalizability of data depend heavily upon whether the researcher-subject relationship is a trustworthy one. Trust is related to the commitment of the fieldworker to those studied and the degree of intersubjective understanding developed. In other words, trust and intersubjectivity are achieved practices during fieldwork and may at times involve conscious effort (Goode, 1994). One of the most challenging aspects of my research practices has been to access the world of physical disability as a nondisabled person and to establish intersubjective understanding and trust as an able-bodied researcher with disabled informants. In order to achieve intersubjective understanding, I have utilized the phenomenological technique of bracketing (taken from the philosophy of Edmund Husserl and elaborated by Alfred Schutz) combined with the practice of empathy (taken from the philosophy of Edith Stein and the anthropology of Gelya Frank). In my experience phenomenological bracketing required two movements, one reflexive and the other empathic. In the reflexive mode I looked within in order to "clear myself out of the way," and in the empathic mode I looked outward to the world of disability in order to listen to the Other's experience.

To "bracket" one's beliefs as a fieldworker means to suspend one's taken-for-granted assumptions about the everyday world in order to focus on the research subject's experience (see Waksler, this volume, for a detailed discussion of bracketing). In this act one necessarily becomes aware of one's own feelings, thoughts, concerns, and biases – all of which influence one's understanding of the other person's situation. As phenomenologists note, our social encounters involve a multitude of interpretations, most of which occur at a taken-for-granted level (i.e., they are not intentional). These taken-for-granted interpretations, or pre-understandings, may filter a researchers' ability to describe and analyze the other's point of view. For instance, studies on attitudes toward persons with disabilities (severe or otherwise) show that visible deformities and disability evoke powerful feelings of distress, anxiety, and fear in the nondisabled (Goffman, 1963; Fabrega and Manning, 1972; Ainlay et al, 1986; for a different perspective see Cahill and Eggleston, 1995).

As an able-bodied researcher I often experienced strong and powerful emotions about my informants. My effort to "bracket" my self, in order to understand the other from the other's point of reference, made me aware of my own pre-understandings and prejudices. This reflexive orientation toward my subject matter further developed into a mode of consciousness (to borrow from Schutz), a way of being a fieldworker, which I have come to define as the *least-able bodied* attitude.[5] "Bracketing" my self did not mean that I was protecting myself from feelings of sadness or pain. The practice of bracketing is not a more hermeneutic way of saying that I was detaching myself from my subject matter, nor is it a claim that I was being "objective" in a positivist way. In bracketing, the experience endures; it is the stance toward the experience that changes. Bracketing brought me face to face with my own emotions of loss, sadness, and fear (to mention just a few of the existential categories that became relevant to me). This allowed me to appreciate the phenomenon at hand without pre-judging it or imposing *a priori* categories. Achieving an understanding (not explanation in Ricoeur's sense) of the Other's point of view does not mean "forgetting" one's self. Rather, bracketing involves an opening of one's life-world to include the Other's perspective. Achieving an intersubjective ground of communication between myself and my informants involved common effort and interest.

Able-bodied anthropologist, Gelya Frank (1988, 1984) describes this process in her studies of severely disabled and deformed persons.[6] In her work with Diane, a quadrilateral amputee who did not use prosthetic devices, Frank (1985, p. 204) notes: "Once I could clarify my own position, I could recognize those areas of Diane's life that were different from, as well as similar to, my own. But, first of all, I had to know where I stand." Sociologist David Goode (1994, p. 24), who studied deaf and mute children, commented that "a regular part of my work with Chris was thus work on and about myself. I sought by a series of exercises to 'clear myself out of the way'."

Adopting the least-able bodied attitude allowed me to see the similarities and differences between myself and my informants, in this way expanding my own sensibilities and customary experiences. In this reflexive mode I had to clarify my own feelings and assumptions regarding loss, death, independence, competence and hope. The next challenge was to look toward the Other, the world of physical disability, and recognize the primacy of the Other's perspective.

This process was facilitated by another phenomenological practice – the practice of empathy. As a method empathy involves "the critical understanding of one's initial interests, attitudes, and expectations about the other" (Frank, 1985, p. 199). In this way, "the decisive use of empathy, leads to potentially transformative encounters ...[E]mpathy is a manner of acknowledging, in someone other than oneself, the human freedom to be, to unfold" (Frank, 1985, p. 198).

[T]he capacity to see others as human, like oneself, possessing the same essential or intrinsic qualities and needs, is proposed here as a characteristic of human consciousness. Yet there is cultural, individual, and circumstantial variation in the degree to which people develop such reflections. As a method, empathy ... must be practiced – tried and tried again – to be comprehended (Frank, 1985, pp. 189-9).

Creating empathic understanding implies a deep commitment and involvement on the part of the fieldworker. R. H. Lowie (1960, p. 159) postulates that empathy is of core importance to ethnologists. I consider his argument to be relevant for social scientists in general. He argues that they have to maintain neutrality when confronted with alien cultural practices. "Seeing from within," means rising above the partisan level in order to discover what is distinct about the other.

Edith Stein (1917/1964) makes an important point about empathic understanding that sheds light on its epistemic and methodological significance. According to Stein, empathy with another's feelings is not mediated by symbols (symbolic representations) but rather is a direct apprehension of the other's state of being. For example, the smile on someone's face expresses joy – an emotion that the perceiver immediately feels. Deceptions are possible, as in the case of the blind person trying to imagine the experiences of a sighted person, and vice versa (Stein, 1964, p. 58; Frank, 1985, p. 196). Nonetheless, empathic understanding allows one to learn about the world of another because it opens up the range of experience. Indeed, Stein (1964, p. 59) argues that empathy is the basis for intersubjective experience:

Were I imprisoned within the boundaries of my individuality, I could not get beyond 'the world as it appears to me.' ...Thus empathy as the basis of intersubjective experience becomes the condition of possible knowledge of the existing outer world ....

In order to develop empathic understanding with regard to the experience of disability, I developed close relationships – friendships even[7] – with my informants, as well as attempting to imagine the world from their point of view.[8] Rather than a strict, professional, researcher-researched orientation, my relationship can better be described as a student-teacher or expert-novice relationship, where I was the student or novice.[9] For instance, I became the "student/novice" when my informants taught me how to use a wheelchair, and I learned something of their experience by living in/being "confined" to one for a week. It may seem like "hubris" for an able-bodied person to attempt to imagine the world of wheelchair users, and I am not making any claims that I have achieved a "feeling into" the Other's experience. However intuitive, sensitive, and imaginative one may become, one cannot *fully* appreciate the Other's loss of mobility or embodiment. Toombs (2001) explains that "[a]n important component of empathy involves the effort to move into the Other's experience, to grasp it in its particularity, 'as if' one were in the Other's place. In this way I clarify for myself the content of his experience." Thus, the experience provided me with important insights into a different way of being-in-the-world. While I tried to emulate life as a wheelchair user, my informants (who actually had suggested, and then insisted, on this exercise) enjoyed showing me the ropes of maneuvering successfully. They pointed to the nuances of wheeling in the rain, on uneven pavement, forgetting one's gloves, waiting for the bus, being stared at, and so forth – all of which were themes and topics in interviews. I reported back to them and talked about my experience as a wheelchair user. Though we all recognized that this was just an exercise or experiment, a greater sense of trust and closeness developed between us. Interviews became more relaxed and open-ended, and informants became more comfortable in raising topics (such as issues of intimacy or impotency) that might otherwise have been difficult to mention. For many of my informants, my using a wheelchair demonstrated my commitment to understand their world and opened up the possibility of a less threatening context/setting for communication.[10]

However, I was not always successful in achieving the least able-bodied attitude, a failure that demonstrates the impossibility of a complete bracketing. An example taken from my field notes will make this issue clear. The following took place during an informal (i.e., not tape-recorded) coffee meeting with one of my informants, Katerina:

Katerina is a 35-year-old woman, physically disabled as a result of Lupus disease. She has been a wheelchair user for at least 3 years. During our meeting, Katerina explained to me in some detail her meditation technique. This technique is a vital source of her stability, and so she spoke passionately about it. She uses this meditation technique in order "to change my own body chemistry," as she said. Specifically, she is consistently concentrating on changing her inflexible tendons which, according to Katerina and her PT, constrict her range of motion and movement. After saying this, she paused and looked at me intensely. Perhaps she was expecting a more enthusiastic or accepting response than what my full of confusion-- and quandary -- eyes and face were able to give off. She interrupted the pause, and added: "When you're sick, you have a better body awareness and can check your body better than when you're healthy." She turned her eyes away from me, and sipped her iced-coffee.

My great sense of disbelief was obvious to Katerina. She was hurt by my inability to comprehend and appreciate her intense efforts to "change" her "body chemistry." Perhaps she thought that I would appreciate this personal "touch" in my data collection, but I was unable to treat this as "data", that is, in a non-judgmental fashion. I challenged her first through a puzzled facial expression and then I asked her, "How are you going to change your body chemistry *all by yourself* through this meditation technique?" My voice must have been full of irony, especially when I uttered the words "all by yourself." It was clear that I was not thinking in accepting ways. Finally she said, what was indeed a great lesson to me, "When you're sick you have a better body awareness... I don't expect you to understand."

I understand this as an instance of my inability to transcend my able-bodied presuppositions and perspective. I was faced with a question regarding my integrity as a researcher: How could I claim to be bracketing my self and yet still be an insensitive able-bodied person? Was there anything I could do to change?[11] I was reminded that, "One's very *inability* to think, feel, and perceive like a native can be a source of insight and, eventually, interpretation of the foreign" (Frank, 1985, pp. 197-98). Furthermore, it should never be forgotten that researchers always confront someone else's reality.[12]

A few days later I apologized to Katerina, explaining to her that I had a hard time understanding certain aspects of her life, and that I might not be ready to understand and appreciate that part of her life for a while. She accepted my apology and reassured me that I was already very different from most people she knew. I was reassured by her kindness, but was still afraid that I had betrayed her. Since Katerina and I have become good friends, we have been able to treat this incident as part of our difference and not as a point of opposition or threat. Put differently, we formed a shared world, an intersubjective ground, wherein we could successfully communicate and appreciate each other while respecting each other's difference. This common ground developed progressively and with common effort. Katerina often pointed out my ableist presuppositions. I owe her great gratitude for allowing me into her world. For my part, I offered her a social and emotional space, in which she could talk and explore her thoughts and frustrations about her disability – something she said she couldn't do elsewhere.[13]

In developing the basis for intersubjective communication and dialogue through bracketing and empathic acts, the researcher is also able to appreciate the extent to which his or her own beliefs and understandings (in my case, as an able bodied person) are shaped by the dominant culture. This (hermeneutic) self-understanding may reveal that those studied are distinct (but not deviant) actors in a common social world. As Lawrence C. Watson (1979, p. 104) states with respect to research and ethnographic analysis:

The ultimate and desired end-product in the hermeneutic sense is a synthesis in which the investigator, while never totally abandoning his [her] own preunderstandings (or, indeed, the context of [her] his own thought), now understands the life-history [or phenomenon] in a qualitatively different way, incorporating something of the text's context of reference and merging it with [her] his own.[14]

I suggest that phenomenologically informed research in the world of disability takes place in a jointly created ground developed in and through the methodological tools of reflexivity and empathy. Continuing to draw from my own research experience and the insights of phenomenologists and disability activists, I will now address the theoretical and practical consequences and provocations that the world of physical disability may advance.

V. FROM DIS-ABILITY TO DIFFERENCE: THE CHALLENGE OF PHYSICAL DISABILITY

Following Merleau-Ponty's carnal phenomenology, and the testimonies and lived experiences of persons with disabilities – while also being attentive to the criticisms offered by disability advocates and writers (Frank, 1985, 1988; Wendell, 1996; Oliver, 1990, 1996; Charlton, 1998) – I suggest that disabled embodiment, as a *sui generis* form of being, doing, and inhabiting can (potentially) upset, or even transform, normative ways of existing. Disabled embodiment may be understood as a form of human diversity that (potentially) expands the dominant or ableist notions of "normal" bodily expression. This is so to the extent that disability exposes the inadequacies of ableist norms and assumptions that assume binary distinctions between able-bodied and disabled persons.[15] For example, dominant notions of health, physical ability, beauty, competence, and independence exclude the disabled body. Shildrick and Price (1996, p.98) argue that they exclude the female and sick body as well. In this binary, the able-body is 'healthy' and there is an implicit understanding "that to be able-bodied is not to be disabled" (Shildrick and Price, 1996, p. 106). The study of embodiment (and of disabled embodiment in particular) raises questions with respect to difference, deviance, and tolerance at both theoretical and political levels.[16] With respect to disability politics and the politics of Other-ness, a common distinction puts disabled and non-disabled populations on opposite sides, holding opposing world-views. The theoretical question is: Is there an existential/perspectival difference between able-bodied and disabled persons?

Instead of seeing the "normal" and "pathological" as separate, I propose that they are varieties situated along a continuum of modes of being-in-the-world. Just as normative upright posture is a mode of being-in-the-world, so is physically disabled embodiment (Wendell, 1996).

To suggest a movement away from *dis*-ability toward difference implies the following: Focusing solely on the *"dis"* part of *dis*-ability with its implications of "I cannot," or "I am unable to," fails to capture the universal experience of persons with disabilities, some of whom do not experience disability in this way. Furthermore, focusing on *dis*-ability continues the ableist bias that construes persons with disabilities as dependent, thus alienating the person from body, family, and society since s/he is pigeonholed in the category of 'dependent,' 'deviant,' and 'incompetent.' While stigmatization and social oppression are present in the lives of persons with disabilities, concentrating on difference rather than inability/*dis*-ability fosters an understanding of disability as a form of human diversity.

"Us versus them" arguments, well known to politics, are vested in polarities that assume that able-bodied persons are the oppressors and persons with disabilities the oppressed.[17] On the one hand, both the medical model and socio-political structures pervasively treat disability as a unitary or universal phenomenon and persons with disabilities as subordinate. Treating the two populations as distinct, therefore, may serve political purposes, such as increasing awareness about disability oppression, creating solidarity among those who are being oppressed, fostering social and legal change (with the ADA being the most successful example), and so on. On the other hand, liberal tolerance and political correctness (politics of difference demagogues or those whose discourses claim to "celebrate difference") fall into the trap of extreme relativism, where it seems as if any difference (pedophiles, rapists or extreme racists alike) should or could be celebrated. The most common expression of this approach is the claim that somehow we are all disabled. Distinctions are trivialized while political agendas may remain reactionary.

I would argue that both approaches are problematic. If one argues that able-bodied and disabled persons constitute distinct ontological positions (the perspective of standpoint epistemology), then there is a problem of communication. How can the two positions interact, co-exist or converse? Translation or understanding of one perspective from the point of view of the other becomes difficult if not impossible. If, on the other hand, one focuses on similarities between the two, then how can individuals claim difference and uniqueness? The various differences among persons with disabilities may be compromised and dominant stereotypes about disability may be perpetuated.

I suggest that diversity and normality be construed on a continuum rather than in polar terms. Conceptualizing disability as difference means that though disability can offer a unique perspective on life, it nevertheless occurs in the context of an intersubjective fabric where similarities with other life circumstances and experiences exist (e.g., loss of a loved one, divorce, or other stigmatized social positions founded upon gender, class, national or racial characteristics). Persons with disabilities and able-bodied individuals do not live in different worlds, but rather live differently in the same world (Fay, 1996). As such, disability is a part of the human condition, merely a different way of being-in-the-world.

The aesthetic difference that persons with disabilities or wheelchair users may offer, for example, could expand social or cultural assumptions regarding beauty, autonomy, and competence. But it is important to see that disability's difference is neither positive nor negative (though it can be experienced as either). As a way of being-in-the-world, disability may be experienced in some cases as a second chance and in others as a disruption of one's familiar life-world; it may bring pain or atrophy, or inconvenience. There is great variation in the experiences of disability, not simply among different types of disability (most obviously between non-institutionalized SCI adults and institutionalized persons with cognitive disabilities or chronic illnesses) but also within such groups. The variety of responses has troubled able-bodied and disabled activists and writers. Charlton (1998, p. 167) explains this important point in the following long but insightful quote:

By minimizing, patronizing (hero worship), and often eradicating the essential neutrality of disability, the dominant culture trivializes the intrinsic complexity of disability (just like it does everything else).... In the real world, some people with disability have a generally good life and others a generally bad life. The condition of disability is a fork in the road of life with all its new and often difficult choices and challenges – not unlike the junctures that everyone else encounters. Some of the people with disabilities living a good life do so in spite of their disabilities; others may be living a good life because of their disabilities. This is the positive aspect of those that preach a politics of difference: everyone is different.

The differences are real; the categories and preconceptions are false. Although there are numerous differences among those with disabilities, it is on the basis of these false categories and preconceptions that the common experience of disability oppression is experienced. This paradox or contradiction—out of difference comes unity—is the basis of the disability rights movement. A contradiction that any vibrant and counter hegemonic movement must contend with and resolve in the course of its practical work.

In this fashion, I understand disability as a form of difference in the double sense of the word: as a unique and contentious phenomenon. It is unique in the sense that it is a particular, specific, and idiomorphic aspect of one's life-world, and contentious in the sense that, in an ableist society, it is unequally stratified resulting in unfair discrimination. While able-bodied and persons with disabilities have different ways of being-in-the-world that are particular and aesthetically distinct, the general features of embodiment (i.e., living in and through the body's capacities, habits, and schema) are shared by both. Nonetheless, in the social and political realm, persons with disabilities are stratified in inferior and disvalued positions, their political struggles coming into conflict with dominant/ hegemonic positions.

VI. CONCLUSION

The challenge for contemporary research on disability is to study disability, as it presents itself; that is, to stay genuine to the phenomenon. I have attempted an understanding of disability as a form of human diversity, as a form of difference, and have concentrated on the possible similarities between the conceptual categories 'able-bodied' and '*dis*-abled.' Though able-bodied and disabled persons may have different points of view, these views are not incommensurable. The DRM and its critics (as well as the reflexive and critical analyses of philosophers and social scientists) teach us that persons with disabilities collectively have accumulated various standpoints distinct from those without disabilities. But it is important to realize that these distinct standpoints do not necessarily provide epistemic advantages in general. Accessing and explicating disabled embodiment from a phenomenological perspective involves compassion or the feeling of being "with" another person (Frank, 1985, p. 204; Leder, 1990, p. 161). Being "with" involves sharing another's life and, in doing so, allowing oneself to open to the other's world. Through assuming the least able-bodied attitude via "bracketing, " I have attempted to develop a reflexive orientation and opening to the world of disability. With this orientation, intersubjective understanding and feelings of "being with" the other are possible.

American College of Thessaloniki
Greece

NOTES

[1]This jargon-laden term, "phenomenologically informed sociology," refers to theoretical and methodological influences that belong neither to philosophical nor to sociological phenomenology. It is sociological in the sense that it is interested in understanding actual social agents' practices and lived experiences in specific social contexts (see Wendell, 1996; Frank, 1985, 1988; Oliver, 1984, 1990, 1996; Goode, 1994). It is phenomenological in the sense that it draws conceptually and epistemologically from the writings of Maurice Merleau-Ponty (1962, 1968), Richard Zaner (1971), S. Kay Toombs (1993, 1995).
[2]Conceptual biases should not be seen as merely pedantic concerns. Though they do affect theory, concepts, and methodology, they also influence everyday life and political decisions. The power of

academic research, especially medical and psychological research, is clearly visible by its regular and uncritical use by the media as the "true" knowledge of "experts".

[3]This is by no means a new interest. Within the past decade many authors (social scientists, philosophers, disability activists and advocates) have criticized discourses on the body, normality and disability. This current work tries to continue this critical tradition and expand its methodological discussion, in order to inform and assist researchers in the field. This paper draws from the testimonies of people with disabilities (see Toombs, Wendell, disability journals, etc.), and the insights of Disability Rights Movement activists, in order to concentrate on the epistemic challenges that the phenomenon of disability poses for qualitative researchers.

[4]For examples see journals such as *The Disabilty Rag, The Edge, New Mobility*; and organizations such as Independent Living Centers.

[5] This is a paraphrase inspired by Nancy Mandell's (1991) article "The Least-Adult Role in Studying Children."

[6]Frank's methodology of introspection was inspired by Gaston Bachelard's psychoanalytic tools, and by being a patient in psychotherapy with a humanistic oriented psychologist (1985).

[7]Forming friendships with some of my participants was a 'natural' consequence of my research process and style. This meant exchanging home phone numbers, running errands for them, celebrating birthdays, helping each other in times of trouble, as well as answering any personal questions that my informants wanted to raise. In this way many of my relationships with informants developed beyond the limits of the interview setting. However, as Acker, Barry and Esseveld (1991, p. 141) mention, "The researcher's goal is always to gather information; thus the danger always exists of manipulating friendships to that end. Given that the power differences between researcher and researched cannot be completely eliminated, attempting to create a more equal relationship can paradoxically become exploitation and use."

[8]Though Scheler argued that comparison between self and other and thinking by analogy are not 'genuine' empathic instances, Stein seems to suggest (and Frank concurs) that "acts of empathic intention" such as the above "are nevertheless a prominent feature of attempts to understand the more complex and enduring states of being in other people" (Frank, 1985, p. 196).

[9]Other ways that may enhance/ facilitate one's ability to grasp the Other's bodily experience are "empathic listening" and "the exercise of imagination" (Toombs, 2001).

[10]In disability studies, therefore, intersubjectivity between the two worlds (ethnographer's and native's, or able-bodied and disabled) cannot, and should not, be taken for granted; rather intersubjectivity ought to become an issue of empirical and practical concern (see Waksler, 1995, and Goode, 1994, for similar points). My own involvement, spending a minimum of six hours daily at the rehabilitation center, meeting informants (formally and informally) during the week, and developing close friendships, progressively created a common communication ground in which the two worlds were not radically opposed to each other. Rather, an intersubjective world was forming. Like Goode (1994, pp. 41-42), I had the goal of making the two worlds one (as much as that was realistically/ empirically/praxiologically possible).

[11]"Clearing myself out of the way" was a difficult, but necessary, process of transformation for me. After about 6 months of constant exposure to in-patients in the rehabilitation center (paraplegics, quadriplegics, amputees, persons with developmental disabilities, etc.), I started to have intense, frightening images and dreams. It seemed important to address and investigate these images and emotions. My ability to 'bracket' my able-bodiedness was not succeeding, so (like Frank) I became a patient in psychotherapy (Rogerian therapist) where I was able to clarify my feelings of loss, existential anxiety, death and my own taken-for-granted assumptions about those with whom I was interacting during my data collection.

[12]Bittner (1973) makes a brilliant remark about objectivity and fieldwork that followed me throughout my research experience. He says: "the fieldworker ... even when [s]he dwells on the fact that this reality is to 'them' incontrovertibly real in just the same way 'they' perceive it, [s]he knows that to some 'others' it may be seen altogether different, in fact, the most impressive feature of 'the' social world is its colorful plurality."

[13] My relationship with Katerina seems to have had a therapeutic character for her. I listened and nodded a lot, and she talked. This is not an experience unique to my relationship with her. Many of my informants expressed gratitude for giving them the opportunity to talk about issues that they indicated most people do not, or cannot, hear about. Counseling services and treatments were not always available at a price that would be comfortable for them, and some informants expressed negative feelings about therapy – mostly due to unfortunate interactions with therapists during rehabilitation (see Seymour, 1998, for a similar point).
[14] During and after my encounter with the world of physical disability I have traveled well beyond my customary and traditional neighborhoods of meaning, making it almost impossible for me to truly "go home" again. This fieldwork experience has changed my horizons and perspective (see Huspek, 1994 for a similar point).
[15] Ableist norms have no room for the deaf person who can hear you perfectly as long as you don't turn your back; the wheelchair user who is an aerobics instructor; the chronically ill woman who needs twenty-four hour assistance and who is the president of a lucrative company; or the blind man who can recognize you in the evening but not in the day. Such examples expose the inadequacies of binary ableist assumptions about ability, normality and the body.
[16] The philosophical problem of the relationship between intersubjectivity and difference cannot be resolved in this present work.
[17] In order to define "oppression" I use Marilyn Frye (1996, p. 164). "The experience of oppressed people is that the living of one's life is confined and shaped by forces and barriers which are not accidental or occasional and hence avoidable, but are systematically related to each other in such a way as to catch one between and among them and restrict or penalize motion in any direction. It is the experience of being caged in: all avenues, in every direction, are blocked or booby trapped." According to this definition, it is correct to say that persons with disabilities are typically oppressed in ableist or able-bodied prejudiced societies.

BIBLIOGRAPHY

Ainlay, S.C., Becker, G., and L.M. Coleman: 1986, *The Dilemma of Difference: A Multidisciplinary View of Stigma*, Plenum Press, New York and London.
Acker, J., Barry, K., and J. Esseveld: 1991, 'Objectivity and truth: Problems in doing feminist research', in M.M. Fonow and J.A. Cook (eds.), *Beyond Methodology: Feminist Scholarship as Lived Research*, Indiana University Press, Bloomington, pp.133-53.
Bittner, E.: 1973, 'Objectivity and realism in sociology', in G. Psathas (ed.), *Phenomenological Sociology: Issues and Applications*, John Wiley and Sons, Inc., New York, pp. 109-25.
Bulman, R.J. and C.B. Wortman: 1977, 'Attributions of blame and coping in the "Real World": Severe accident victims react to their lot', *Journal of Personality and Social Psychology* 35, 351-63.
Cahill, E.S. and R. Eggleston: 1995, 'Reconsidering the stigma of disability: Wheelchair use and public kindness', *The Sociological Quarterly* 36, 681-98.
Charlton, I.J.: 1998, *Nothing About Us Without Us: Disability Oppression and Empowerment*, University of California Press, CA.
Davis, J.L.: 1995, *Enforcing Normalcy: Disability, Deafness, and the Body*, Verso, New York.
Fabrega, H., Jr. and P.K. Manning: 1972, 'Disease, illness, and deviant careers', in Scott and Douglas (eds.), *Theoretical Perspectives on Deviance*, Basic Books, Inc. Publishers, New York and London.
Fay, B.: 1996, *Contemporary Philosophy of Social Science: A Multicultural Approach*, Blackwell Publishers, Oxford, England.
Fine, M. and A. Asch (eds.): 1988, *Women with Disabilities: Essays in Psychology, Culture, and

Politics, Temple University Press, Philadelphia.
Frank, G.: 1988, 'On embodiment: A case study of congenital limb deficiency in American culture', in M. Fine and A. Asch (eds.), *Women with Disabilities: Essays in Psychology, Culture, and Politics*, Temple University Press, Philadelphia.
Frank, G.: 1985, '"Becoming the Other": Empathy and biographical interpretation', *Biography* **8**, 189-210.
Frye, M.: 1996, 'Oppression', in K. Rosenblum and T-M. Travis (eds.), *The Meaning of Difference: American Constructions of Race, Sex and Gender, Social Class, and Sexual Orientation*, McGraw-Hill Co., pp. 163-67.
Gill, C. J.: 1994, *The Disability Rag* **15**, 3-7.
Goffman, E.: 1963, *Stigma: Notes on the Management of Spoiled Identity*, Prentice Hall, Englewood Cliffs, New Jersey.
Goode, D.: 1994, *A World Without Words: The Social Construction of Children Born Deaf and Blind*, Temple University Press, Philadelphia.
Hahn, H.: 1988, 'The politics of physical differences: Disability and discrimination', *Journal of Social Issues* **44**, 39-47.
Huspek, M.: 1994, 'Critical ethnography and subjective experience', *Human Studies* **17**, 45-64.
Leder, D.: 1990, *The Absent Body*, The University of Chicago Press, Chicago and London.
Livneh, H.: 1982, 'On the origins of negative attitudes toward people with disabilities', *Rehabilitation Literature*, **43**, 338-47.
Linton, S.: 1998, *Claiming Disability: Knowledge and Identity*, New York University Press, New York.
Lowie, R.H.: 1960, 'Empathy, or "Seeing from within"', in S. Diamond (ed.), *Culture in History: Essays in Honor of Paul Radin*, Columbia University Press, New York, pp. 145-59.
Mandell, N.: 1991, 'The least-adult role in studying children', in F. Waksler (ed.), *Studying the Social Worlds of Children: Sociological Readings*, The Falmer Press, London, pp. 38-59.
Merleau-Ponty, M.: 1968, *The Visible and the Invisible*, A. Linguis (trans.), Northwestern University Press, Evanston, Illinois.
Merleau-Ponty, M.: 1962, *Phenomenology of Perception*, C. Smith (trans.), Routledge and Kegan Paul, London.
Miner, M.: 1997, '"Making up stories as we go along": Men, Women, and Narratives of Disability', in D.T. Mitchell and S.L. Snyder (eds.), *The Body and Physical Disability: Discourses of Disability*, The University of Michigan Press, Ann Arbor, pp. 41-71.
Mitchell, D.T. and S.L. Snyder (eds.): 1997, *The Body and Physical Disability: Discourses of Disability*, The University of Michigan Press, Ann Arbor.
Oliver, M.: 1996, *Understanding Disability: From Theory to Practice*, St. Martin's Press, New York.
Oliver, M.: 1990, *The Politics of Disablement: A Sociological Approach*, St. Martin's Press, New York.
Oliver, M.: 1984, 'The politics of disability', *Critical Social Theory* **4**, 21-32.
Russell, M.: 1994, 'Malcolm teaches us, too', *Disability Rag*, 11-12.
Safilios-Rothschild, C.: 1970, *The Sociology and Social Psychology of Disability and Rehabilitation*, Random House, New York.
Schutz, A.: 1962, *The Problem of Social Reality*, Vol. 1, *Collected Papers*, Martinus Nijhoff, The Hague.
Seymour, W.: 1998, *Remaking the Body: Rehabilitation and Change*, Routledge.
Shakespeare, T.: 1996, 'Power and prejudice: Issues of gender, sexuality and disability', in L. Barton (ed.), *Disability and Society: Emerging Issues and Insights*, Longman Ltd., London and New York, pp. 191-214.
Shildrick, M. and J. Price: 1996, 'Breaking the boundaries of the broken body', *Body and Society* **2**, 93-113.

Stein, E.: 1964, *On the Problem of Empathy*, E. W. Straus (trans.), Martinus Nijhoff, The Hague.
Toombs, S.K.: 2001, 'The role of empathy in clinical practice', *Journal of Consciousness Studies* **8**, 247-58.
Toombs, S.K.: 1995, 'The lived experience of disability', *Human Studies* **18**, 9-23.
Toombs, S.K.: 1993, *The Meaning of Illness: A Phenomenological Account of the Different Perspectives of Physician and Patient*, Kluwer Academic Publishers, Dordrecht, The Netherlands.
Watson, L.C.: 1976, 'Understanding a life history as a subjective document: Hermeneutical and phenomenological perspectives', *Ethos* **4**, 95-131.
Waksler, F.C.: 2001, 'The phenomenological method and medicine', in this volume.
Waksler, F.C.: 1995, 'Introductory essay: Intersubjectivity as a practical matter and a problematic achievement', *Human Studies* **18**, 1-7.
Weller, D.J. and P.M. Miller: 1977, 'Emotional reactions of patient, family, and staff in acute-care period of spinal cord injury', *Social Work in Health Care* **3**, 7-12.
Zaner, R.M.: 1971, *The Problem of Embodiment: Some Contributions to a Phenomenology of the Body*, Second Edition, Martinus Nijhoff, The Hague.

NOTES ON CONTRIBUTORS

Patricia Benner, Department of Social and Behavioral Sciences, University of California, San Francisco, California, U.S.A.

Mark J. Bliton, Center for Clinical and Research Ethics, Vanderbilt University Medical Center, Nashville, Tennessee, U.S.A.

Michael C. Brannigan, Center for the Study of Ethics and Department of Philosophy, La Roche College, Pittsburgh, Pennsylvania, U.S.A.

John B. Brough, Department of Philosophy, Georgetown University, Washington, D.C., U.S.A.

Eric J. Cassell resides in New York city. He is an attending physician at The New York Presbyterian Hospital and Clinical Professor of Public Health at Weill Medical College of Cornell University, New York, U.S.A.

Maureen Connolly, Department of Physical Education and the Center for Women's Studies, Brock University, St. Catherine's, Ontario, Canada.

Arthur W. Frank, Department of Sociology, University of Calgary, Calgary, Alberta, Canada.

Shaun Gallagher, Department of Philosophy, Canisius College, Buffalo, New York, U.S.A.

Patrick A. Heelan, Department of Philosophy, Georgetown University, Washington, D.C., U.S.A

Paul Komesaroff, Baker Medical Research Institute, Melbourne, Australia.

Irena Madjar, Faculty of Nursing, University of Newcastle, Callaghan, New South Wales, Australia.

NOTES ON CONTRIBUTORS

Catriona Mackenzie, Philosophy Department, Macquarie University, Sydney, Australia.

Glen A. Mazis, Department of Philosophy and Humanities, Soka University, Aliso Viejo, California, U.S.A. (on leave from Pennsylvania State University, Harrisburg).

Ian R. McWhinney, Centre for Studies in Family Medicine, University of Western Ontario, London, Ontario, Canada.

Christina Papadimitriou, Department of Philosophy and Social Science, American College of Thessaloniki, Thessaloniki, Greece.

Carl Edvard Rudebeck is a family physician residing in Västervik, Sweden.

Per Sundström, Medical Faculty, University of Oslo, Norway.

Fredrik Svenaeus, Department of Health and Society, University of Linköping, Sweden.

S. Kay Toombs, Baylor University, Waco, Texas, U.S.A.

Max van Manen, Faculty of Education, University of Alberta, Edmonton, Alberta, Canada.

Frances Chaput Waksler, Department of Sociology, Wheelock College, Boston, Massachusetts, U.S.A.

Jo Ann Walton, Faculty of Health Studies, Auckland University of Technology, Auckland, New Zealand

Bruce Wilshire, Philosophy Department, Rutgers University, New Brunswick, New Jersey, U.S.A.

Richard M. Zaner, Center for Clinical and Research Ethics, Vanderbilt University Medical Center, Nashville, Tennessee, U.S.A.

INDEX

Abe, K. 460
Abrams, D. 6
Agamben, G. 324-25
Alcoff, L. 183
Allen, J. 279-80
Alsop, S. 234, 236-37
Arendt, H. 323, 357
Asch, A. 477
Astbury, J. 282
Auden, W.H. 117
Bacon, F. 336
Bakan, D. 273, 387
Bakhtin, M.M. 318, 319, 320, 321, 322-323, 324
Balint, E. 314
Balint, M. 338, 347, 348
Banja, J. 268
Bartky, S.L. 184, 185
Basch, M.F. 314
Bazermann, C. 60
Beck, C.T. 289-90
Beckman, H.B. 346
Behnke, E.A. 5
Benner, P. 15, 184, 351-69
Benson, P. 431
Berger, P.L. 70-71
Berlin, I. 418
Bernard, C. 135
Bhavani, K-K. 181, 188
Black Elk 217-19, 222
Bliton, M. 16, 393-415
Bordo, S. 183
Bordieu, P. 183
Brannigan, M. 17, 441-454
Brisson, L. 48
Brough, J. 3, 29-46
Broyard, A. 229, 235, 241
Bugbee, H. 215
Bunch, W.H. 397
Burch, R. 282
Burnett, M. 128, 136
Burney, F. 264-65
Burtt, E.A. 332
Butler, J. 183
Buytendijk, F.J.J. 87, 151, 467
Cairns, D. 394
Callahan, D. 242
Camus, A. 142
Carr, D. 73
Casey, E.S. 356, 358, 359, 361
Casper, M.J. 402, 404, 405
Cassell, E.J. 15-16, 311, 352, 371-90
Caton, D. 267
Charlton, I.J. 487
Christman, J. 430, 431
Clendinnen, I. 269, 270
Connolly, M. 11, 177-96
Craig, T. 190
Crombie, A.C. 56, 59
Crookshank, F.G. 331
Darwin, C. 136
Davidson, L. 284
De Beauvoir, S. 184
Derrida, J. 130, 325
Descartes, R. 331, 351, 364-66
Doubt, K. 79
Drane, J.F. 335, 348
Dreyfus, H.L. 355, 356, 358
Du Bois, P. 183, 192
Duden, B. 50, 51, 59, 61-62, 335, 402, 411
Dworkin, G. 430
Eccles, J. 128, 136, 142
Emerson, R.W. 216, 224
Engel, G. 311, 313, 339
Engelhardt, H.T. 5, 79-80
Entralgo, P. L. 332, 334, 335, 339, 371
Fabrega, H. 339
Fine, M. 477
Fisher, W.F. 279
Fleck, L. 52-53, 55, 59, 331
Foss, L. 339
Foucault, M. 8, 79, 179, 183, 325, 351, 355, 364
Frank, A. 12, 190, 229-45, 340, 353, 363
Frank, G. 480, 481, 483, 488
Frankel, R.M. 346
Frankfurt, H. 428

Franklin, S. 402
Frede, D. 285-86
Freud, S. 373
Friedman, M. 429
Gadamer, H.G. 49, 52, 88-89, 94-95, 264, 351, 364
Galen 334
Gallagher, S. 5, 10-11, 147-75,
Gilbert, W. 128, 137, 142
Gilman, S.L. 280
Goode, D. 479-80, 481
Groddeck, G. 374
Grosz, E. 180
Hahn, H. 478-79
Halliday, M.A.K. 50, 60
Halpern, J. 351
Hawkins, A.H. 340
Head, H. 148
Heelan, P.A. 3, 47-66, 51
Heidegger, M. 48, 49, 50, 52, 54, 55, 56-57, 90-94, 96-99, 100, 287, 288, 300, 301, 310, 355, 356, 459-60
Held, R. 165-66
Hippocrates 332-33
Hoffmaster, B. 242
Husserl, E. 1, 5, 29-37, 49, 67, 68, 69, 70, 71, 73, 74, 75-76, 77, 78, 82-83, 88, 89, 137, 161, 248, 299, 393, 394, 399, 443-45, 480
Irigaray, L. 183
James, W. 155, 220
Jamison, K. 281
Jeffreys, T. 190
Johnson, J. 360
Johnson, M. 50, 305-306, 376-77
Jonas, H. 5, 115, 135, 136, 409
Karp, D.A. 282
Kass, L. 140
Katz, J. 135
Kenyon, J. 44-45
Kern, S. 41-42
Kirk, S.A. 77, 78
Kleinman, A. 190, 339, 351
Koch, E. 337
Kohák, E. 73, 79, 84
Komesaroff, P. 14, 317-30
Kristeva, J. 180, 184, 193
Kuhn, T.S. 56
Kuhse, H. 470
Kutchins, H. 77, 78

Laennec, R. 336-37
Lakoff, G. 50,
Lane, H.C. 80
Lannigan, R. 179, 185
Leder, D. 5, 8, 50, 51, 181, 205-206, 234, 298, 300, 301, 308, 488
Levenstein, J.H. 342
Levin, D.M. 351, 358
Levinas, E. 14, 241, 320, 322, 324, 326, 327, 328, 356, 361-62
Lévi-Strauss, C. 130
Liebrich, J. 281, 283
Lingis, A. 323, 328
Livneh, H. 478
Locke, J. 156
Longden, D. 191
Lorde, A. 185, 237-38, 240, 243
Lowie, R.H. 482
Luckmann, T. 70-71
MacIntyre, A. 354, 355, 358, 367
Mackenzie, C. 16-17, 417-39
Madjar, I. 13, 263-77, 458
Mairs, N. 185, 190, 234
Malone, R. 366-67
Martin, E. 184
Martin, J. 50, 60
Mawer, S. 131-33, 140, 141, 142
May, W. 242
Mazis, G. 11, 197-214
McGowin, D.F. 465, 466, 468, 471
McLellan, M.F. 281
McWhinney, I.R. 15, 331-50
Mellor, C.S. 165
Meltzoff, A. 157, 159
Melzack, R. 191
Merleau-Ponty, M. 2, 49, 50, 51, 99, 147, 148, 155, 156, 164, 183, 193, 201, 202-204, 220, 234, 239, 248, 254, 255, 256, 268, 298, 300, 303, 306, 308, 320, 328, 356, 366, 445-50, 460
Meyers, D. 429
Meyerstein, F.W. 48
Miller, J. 49, 50
Mitchell, D. 81,
Mogel, J.S. 272
Moore, M.K. 157
Morgan, L. 404
Morse, J. 184
Mullin, M. 468
Murphy, R. 234

Murray, T.J. 38, 39, 40, 41
Neihardt, J. 217, 218
Nielsen, C. 215
Nussbaum, M. 334
Olson, E. 147
Orgass, B. 148, 149
Paget, M.A. 71, 72, 74, 77, 183, 190, 191
Paillard, J. 153
Papadimitriou, C. 18, 475-92
Parfit, D. 421, 422-25
Pasteur, L. 337
Patterson, C.J.S. 458
Payne, L. 273-74
Peirce, C.S. 113
Pellegrino, E. 9, 446
Peplau, I.A. 363
Perls, F. 302
Piaget, J. 156, 305
Pinker, S. 224
Poeck, K. 148, 149
Polyani, M. 347
Porkert, M. 75, 76
Porter, R. 265, 280
Priestley, J.B. 43
Reiser, S. 115
Rheinberger, H.J. 128, 129-30, 136, 137
Ricoeur, P. 421, 422-25, 471
Rilke, R.M. 317, 321, 329
Robillard, A. 234, 235
Rogers, G. 229, 235
Rose, L. 471
Rubin, Z. 363
Rudebeck, C.E. 13-14, 297-316
Sacks, H. 72
Sacks, O. 81, 82, 199, 207, 208, 210, 213, 340, 349, 353
Santopinto, M.D.A. 290-91
Sartre, J.P. 5, 248, 254, 256, 257, 298
Sass, L.A. 283-84
Sauvages, F. 336
Scannell, K. 38, 39, 40-41
Scarry, E. 186-87, 191, 268, 269, 401
Schutz, A. 138-39, 229, 231-32, 234, 235-236, 239, 240, 241, 248, 398, 399, 403, 480
Shaaban, A.F. 402
Sheets-Johnstone, M. 50, 51, 183, 400
Sherwin, S. 398
Shoemaker, S. 147

Simmel, M. 154
Snyder, L. 465
Spicker, S. 5,
Spiegelberg, H. 140, 141, 399, 457
Spiro, H. 380-81
Stein, E. 482
Steiner, R. 12, 220-21, 222
Stetten, D. 340
Stewart, M. 342, 346
Straus, E. 5, 7,
Stryker, J. 442
Styron, W. 282
Sundström, P. 4, 109-126,
Svenaeus, F. 4, 87-108, 300, 301, 309
Sydenham, T. 336
Tarski, A. 54
Taylor, C. 354, 355, 428
Tiemersma, D. 148
Todes, S. 360
Tomkins, S.S. 304, 305
Toombs, S.K. 1-26, 2, 5, 8, 12-13, 37, 39, 42, 43, 119, 190, 209, 210, 211, 234, 247-61, 298, 301, 305, 311, 340, 482
Torgerson, W.S. 191
Toulmin, S. 338
Tymieniecka, A.T. 68
Van den Berg, J.H. 257
Van Manen, M. 17-18, 272, 275, 457-74
Varela, F. 50, 301
Von Linné, C. 336
Waksler, F. 4, 67-86
Wall, P. 191
Walton, J.A. 13, 279-93, 458
Watson, G. 428
Watson, L. 484
Weiss, G. 180
Weissmann, G. 137
Wendell, S. 180, 181, 182, 185, 187, 189, 191
Whitehead, A.N. 335-36
Wilshire, B. 11-12, 215-225
Wittgenstein, L. 114
Wolpert, L. 282
Young, I. 183, 190
Young, K. 50, 51
Zaner, R.M. 2, 4, 5, 6, 7, 50, 51, 127-44, 100, 201, 209, 230, 241, 242, 393, 394, 397, 398, 399, 405-406, 446

Philosophy and Medicine

1. H. Tristram Engelhardt, Jr. and S.F. Spicker (eds.): *Evaluation and Explanation in the Biomedical Sciences.* 1975 ISBN 90-277-0553-4
2. S.F. Spicker and H. Tristram Engelhardt, Jr. (eds.): *Philosophical Dimensions of the Neuro-Medical Sciences.* 1976 ISBN 90-277-0672-7
3. S.F. Spicker and H. Tristram Engelhardt, Jr. (eds.): *Philosophical Medical Ethics. Its Nature and Significance.* 1977 ISBN 90-277-0772-3
4. H. Tristram Engelhardt, Jr. and S.F. Spicker (eds.): *Mental Health. Philosophical Perspectives.* 1978 ISBN 90-277-0828-2
5. B.A. Brody and H. Tristram Engelhardt, Jr. (eds.): *Mental Illness. Law and Public Policy.* 1980 ISBN 90-277-1057-0
6. H. Tristram Engelhardt, Jr., S.F. Spicker and B. Towers (eds.): *Clinical Judgment. A Critical Appraisal.* 1979 ISBN 90-277-0952-1
7. S.F. Spicker (ed.): *Organism, Medicine, and Metaphysics.* Essays in Honor of Hans Jonas on His 75th Birthday. 1978 ISBN 90-277-0823-1
8. E.E. Shelp (ed.): *Justice and Health Care.* 1981
   ISBN 90-277-1207-7; Pb 90-277-1251-4
9. S.F. Spicker, J.M. Healey, Jr. and H. Tristram Engelhardt, Jr. (eds.): *The Law-Medicine Relation.* A Philosophical Exploration. 1981 ISBN 90-277-1217-4
10. W.B. Bondeson, H. Tristram Engelhardt, Jr., S.F. Spicker and J.M. White, Jr. (eds.): *New Knowledge in the Biomedical Sciences.* Some Moral Implications of Its Acquisition, Possession, and Use. 1982 ISBN 90-277-1319-7
11. E.E. Shelp (ed.): *Beneficence and Health Care.* 1982 ISBN 90-277-1377-4
12. G.J. Agich (ed.): *Responsibility in Health Care.* 1982 ISBN 90-277-1417-7
13. W.B. Bondeson, H. Tristram Engelhardt, Jr., S.F. Spicker and D.H. Winship: *Abortion and the Status of the Fetus.* 2nd printing, 1984 ISBN 90-277-1493-2
14. E.E. Shelp (ed.): *The Clinical Encounter.* The Moral Fabric of the Patient-Physician Relationship. 1983 ISBN 90-277-1593-9
15. L. Kopelman and J.C. Moskop (eds.): *Ethics and Mental Retardation.* 1984
    ISBN 90-277-1630-7
16. L. Nordenfelt and B.I.B. Lindahl (eds.): *Health, Disease, and Causal Explanations in Medicine.* 1984 ISBN 90-277-1660-9
17. E.E. Shelp (ed.): *Virtue and Medicine.* Explorations in the Character of Medicine. 1985 ISBN 90-277-1808-3
18. P. Carrick: *Medical Ethics in Antiquity.* Philosophical Perspectives on Abortion and Euthanasia. 1985 ISBN 90-277-1825-3; Pb 90-277-1915-2
19. J.C. Moskop and L. Kopelman (eds.): *Ethics and Critical Care Medicine.* 1985
    ISBN 90-277-1820-2
20. E.E. Shelp (ed.): *Theology and Bioethics.* Exploring the Foundations and Frontiers. 1985 ISBN 90-277-1857-1

Philosophy and Medicine

21. G.J. Agich and C.E. Begley (eds.): *The Price of Health.* 1986
    ISBN 90-277-2285-4
22. E.E. Shelp (ed.): *Sexuality and Medicine.* Vol. I: Conceptual Roots. 1987
    ISBN 90-277-2290-0; Pb 90-277-2386-9
23. E.E. Shelp (ed.): *Sexuality and Medicine.* Vol. II: Ethical Viewpoints in Transition. 1987 ISBN 1-55608-013-1; Pb 1-55608-016-6
24. R.C. McMillan, H. Tristram Engelhardt, Jr., and S.F. Spicker (eds.): *Euthanasia and the Newborn.* Conflicts Regarding Saving Lives. 1987
    ISBN 90-277-2299-4; Pb 1-55608-039-5
25. S.F. Spicker, S.R. Ingman and I.R. Lawson (eds.): *Ethical Dimensions of Geriatric Care.* Value Conflicts for the 21th Century. 1987   ISBN 1-55608-027-1
26. L. Nordenfelt: *On the Nature of Health.* An Action-Theoretic Approach. 2nd, rev. ed. 1995   SBN 0-7923-3369-1; Pb 0-7923-3470-1
27. S.F. Spicker, W.B. Bondeson and H. Tristram Engelhardt, Jr. (eds.): *The Contraceptive Ethos.* Reproductive Rights and Responsibilities. 1987
    ISBN 1-55608-035-2
28. S.F. Spicker, I. Alon, A. de Vries and H. Tristram Engelhardt, Jr. (eds.): *The Use of Human Beings in Research.* With Special Reference to Clinical Trials. 1988
    ISBN 1-55608-043-3
29. N.M.P. King, L.R. Churchill and A.W. Cross (eds.): *The Physician as Captain of the Ship.* A Critical Reappraisal. 1988   ISBN 1-55608-044-1
30. H.-M. Sass and R.U. Massey (eds.): *Health Care Systems.* Moral Conflicts in European and American Public Policy. 1988   ISBN 1-55608-045-X
31. R.M. Zaner (ed.): *Death: Beyond Whole-Brain Criteria.* 1988
    ISBN 1-55608-053-0
32. B.A. Brody (ed.): *Moral Theory and Moral Judgments in Medical Ethics.* 1988
    ISBN 1-55608-060-3
33. L.M. Kopelman and J.C. Moskop (eds.): *Children and Health Care.* Moral and Social Issues. 1989   ISBN 1-55608-078-6
34. E.D. Pellegrino, J.P. Langan and J. Collins Harvey (eds.): *Catholic Perspectives on Medical Morals.* Foundational Issues. 1989   ISBN 1-55608-083-2
35. B.A. Brody (ed.): *Suicide and Euthanasia.* Historical and Contemporary Themes. 1989   ISBN 0-7923-0106-4
36. H.A.M.J. ten Have, G.K. Kimsma and S.F. Spicker (eds.): *The Growth of Medical Knowledge.* 1990   ISBN 0-7923-0736-4
37. I. Löwy (ed.): *The Polish School of Philosophy of Medicine.* From Tytus Chałubiński (1820–1889) to Ludwik Fleck (1896–1961). 1990
    ISBN 0-7923-0958-8
38. T.J. Bole III and W.B. Bondeson: *Rights to Health Care.* 1991
    ISBN 0-7923-1137-X

Philosophy and Medicine

39. M.A.G. Cutter and E.E. Shelp (eds.): *Competency. A Study of Informal Competency Determinations in Primary Care.* 1991 ISBN 0-7923-1304-6
40. J.L. Peset and D. Gracia (eds.): *The Ethics of Diagnosis.* 1992
ISBN 0-7923-1544-8
41. K.W. Wildes, S.J., F. Abel, S.J. and J.C. Harvey (eds.): *Birth, Suffering, and Death.* Catholic Perspectives at the Edges of Life. 1992 [CSiB-1]
ISBN 0-7923-1547-2; Pb 0-7923-2545-1
42. S.K. Toombs: *The Meaning of Illness.* A Phenomenological Account of the Different Perspectives of Physician and Patient. 1992
ISBN 0-7923-1570-7; Pb 0-7923-2443-9
43. D. Leder (ed.): *The Body in Medical Thought and Practice.* 1992
ISBN 0-7923-1657-6
44. C. Delkeskamp-Hayes and M.A.G. Cutter (eds.): *Science, Technology, and the Art of Medicine.* European-American Dialogues. 1993 ISBN 0-7923-1869-2
45. R. Baker, D. Porter and R. Porter (eds.): *The Codification of Medical Morality.* Historical and Philosophical Studies of the Formalization of Western Medical Morality in the 18th and 19th Centuries, Volume One: Medical Ethics and Etiquette in the 18th Century. 1993 ISBN 0-7923-1921-4
46. K. Bayertz (ed.): *The Concept of Moral Consensus.* The Case of Technological Interventions in Human Reproduction. 1994 ISBN 0-7923-2615-6
47. L. Nordenfelt (ed.): *Concepts and Measurement of Quality of Life in Health Care.* 1994 [ESiP-1] ISBN 0-7923-2824-8
48. R. Baker and M.A. Strosberg (eds.) with the assistance of J. Bynum: *Legislating Medical Ethics.* A Study of the New York State Do-Not-Resuscitate Law. 1995
ISBN 0-7923-2995-3
49. R. Baker (ed.): *The Codification of Medical Morality.* Historical and Philosophical Studies of the Formalization of Western Morality in the 18th and 19th Centuries, Volume Two: Anglo-American Medical Ethics and Medical Jurisprudence in the 19th Century. 1995 ISBN 0-7923-3528-7; Pb 0-7923-3529-5
50. R.A. Carson and C.R. Burns (eds.): *Philosophy of Medicine and Bioethics.* A Twenty-Year Retrospective and Critical Appraisal. 1997 ISBN 0-7923-3545-7
51. K.W. Wildes, S.J. (ed.): *Critical Choices and Critical Care.* Catholic Perspectives on Allocating Resources in Intensive Care Medicine. 1995 [CSiB-2]
ISBN 0-7923-3382-9
52. K. Bayertz (ed.): *Sanctity of Life and Human Dignity.* 1996
ISBN 0-7923-3739-5
53. Kevin Wm. Wildes, S.J. (ed.): *Infertility: A Crossroad of Faith, Medicine, and Technology.* 1996 ISBN 0-7923-4061-2
54. Kazumasa Hoshino (ed.): *Japanese and Western Bioethics.* Studies in Moral Diversity. 1996 ISBN 0-7923-4112-0

Philosophy and Medicine

55. E. Agius and S. Busuttil (eds.): *Germ-Line Intervention and our Responsibilities to Future Generations.* 1998　ISBN 0-7923-4828-1
56. L.B. McCullough: *John Gregory and the Invention of Professional Medical Ethics and the Professional Medical Ethics and the Profession of Medicine.* 1998
　ISBN 0-7923-4917-2
57. L.B. McCullough: *John Gregory's Writing on Medical Ethics and Philosophy of Medicine.* 1998 [CiME-1]　ISBN 0-7923-5000-6
58. H.A.M.J. ten Have and H.-M. Sass (eds.): *Consensus Formation in Healthcare Ethics.* 1998 [ESiP-2]　ISBN 0-7923-4944-X
59. H.A.M.J. ten Have and J.V.M. Welie (eds.): *Ownership of the Human Body. Philosophical Considerations on the Use of the Human Body and its Parts in Healthcare.* 1998 [ESiP-3]　ISBN 0-7923-5150-9
60. M.J. Cherry (ed.): *Persons and Their Bodies.* Rights, Responsibilities, Relationships. 1999　ISBN 0-7923-5701-9
61. R. Fan (ed.): *Confucian Bioethics.* 1999 [APSiB-1]　ISBN 0-7923-5853-8
62. L.M. Kopelman (ed.): *Building Bioethics.* Conversations with Clouser and Friends on Medical Ethics. 1999　ISBN 0-7923-5853-8
63. W.E. Stempsey: *Disease and Diagnosis.* 2000　PB ISBN 0-7923-6322-1
64. H.T. Engelhardt (ed.): *The Philosophy of Medicine.* Framing the Field. 2000
　ISBN 0-7923-6223-3
65. S. Wear, J.J. Bono, G. Logue and A. McEvoy (eds.): *Ethical Issues in Health Care on the Frontiers of the Twenty-First Century.* 2000　ISBN 0-7923-6277-2
66. M. Potts, P.A. Byrne and R.G. Nilges (eds.): *Beyond Brain Death.* The Case Against Brain Based Criteria for Human Death. 2000　ISBN 0-7923-6578-X
67. L.M. Kopelman and K.A. De Ville (eds.): *Physician-Assisted Suicide.* What are the Issues? 2001　ISBN -7923-7142-9
68. S.K. Toombs (ed.): *Handbook of Phenomenology and Medicine.* 2001
　ISBN 1-4020-0151-7; Pb 1-4020-0200-9